Heat Transfer

Heat Transfer: A Systematic Learning Approach presents valuable tools for understanding heat transfer mechanisms and provides a clear understanding of complex Heat transfer mechanisms. It gives a comprehensive introduction to topics of heat transfer, including conduction, convection, thermal radiation, and nanofluids.

Covering both traditional analytical models for canonical flows and modern turbulence modeling approaches for heat transfer, the book discusses complex impinging jet flow, phase change flows, nanofluids, and convective mass transfer flow. The text includes numerous end-of-chapter problems to enhance student understanding and different solving approaches. It offers the basic flow and energy analysis along with useful MAPLE code to facilitate the learning process.

The book is intended for senior undergraduate mechanical, aerospace, and chemical engineering students taking courses in heat transfer.

Instructors will be able to utilize a Solutions Manual, Jupyter Notebook programmes, and Figure Slides for their courses.

Heat Transfer
A Systematic Learning Approach

Naseem Uddin

CRC Press
Taylor & Francis Group
Boca Raton London New York

CRC Press is an imprint of the
Taylor & Francis Group, an **informa** business

Designed cover image: Pexels

Maple is a trademark of Waterloo Maple Inc.

First edition published 2024
by CRC Press
2385 NW Executive Center Drive, Suite 320, Boca Raton FL 33431

and by CRC Press

4 Park Square, Milton Park, Abingdon, Oxon, OX14 4RN

CRC Press is an imprint of Taylor & Francis Group, LLC

Library of Congress Cataloging-in-Publication Data

Names: Uddin, Naseem, author.
Title: Heat transfer : a systematic learning approach / Naseem Uddin.
Description: Boca Raton : CRC Press, 2024. | Includes bibliographical references and index.
Identifiers: LCCN 2023027014 | ISBN 9781032549828 (hardback) | ISBN 9781032549835 (paperback) | ISBN 9781003428404 (ebook) | ISBN 9781032549934 (ebook other)
Subjects: LCSH: Heat--Transmission--Textbooks.
Classification: LCC QC320 .U44 2024 | DDC 536/.2--dc23/eng/20231012
LC record available at https://lccn.loc.gov/2023027014

ISBN: 978-1-032-54982-8 (hbk)
ISBN: 978-1-032-54983-5 (pbk)
ISBN: 978-1-003-42840-4 (ebk)
ISBN: 978-1-032-54993-4 (eBook+)

DOI: 10.1201/9781003428404

Typeset in Nimbus font
by KnowledgeWorks Global Ltd.

Publisher's note: This book has been prepared from camera-ready copy provided by the authors.

Access the Instructor and Student Resources: routledge.com/9781032549828

<h1 align="center">Units and Conversion Factors</h1>

Mass	**Length**	**Time**
$1\ lb_m = 453.6$ g	1 in. = 2.54 cm	1 minute = 60 s
1 ton (short) = $2000\ lb_m$	1 angstrom = 10^{-8} cm	1 hour = 60 min
1 ton (long) = $2400\ lb_m$	1 micron = 10^{-4} cm	1 day = 24 hr
1 kg = 1000 g	1 ft = 0.3048 m	
1 slug = $32.17\ lb_m$	1 mile = 1.609 km	
	1 m = 10^2 cm	
	1 yard = 0.9144 m	

Force	**Area**	**Volume**
1 N = 10^5 dyne	$1\ in^2 = 6.452\ cm^2$	$1\ in^3 = 16.39\ cm^3$
1 dyne = $2.248 \times 10^{-6}\ lb_f$	$1\ ft^2 = 9.29 \times 10^{-2}\ m^2$	$1\ ft^3 = 2.83 \times 10^{-2}\ m^3$
1 poundal = $3.108 \times 10^{-2}\ lb_f$	1 acre = $4.35 \times 10^4\ ft^2$	1 gal (US) = $231\ in^3$
$1\ lb_f = 4.448$ N	1 square mile = $2.59\ km^2$	1 quart (liq) = 0.25 gal (US)
		1 barrel = 31.5 gal (US)
		1 gal (Imperial) = 1 gal (US)

Pressure	**Energy**	**Power**
1 atm = $14.7\ lb_f/in^2$	1 cal = 4.187 J	1 horsepower = 745.7 W
$1\ lb_f/in^2 = 6.89 \times 10^3\ N/m^2$	1 Btu = 252 cal	1 horsepower = 42.6 Btu/min
1 torr = $133.3\ N/m^2$	1 erg = 10^{-7} J	1 W = 9.51×10^{-4} Btu/s
1 atm = $1.013 \times 10^6\ dyne/cm^2$	1 Btu = 1055 W·s	$1\ ft\ lb_f/s = 1.356$ W
	$1\ ft\ lb_f = 1.356$ J	

Temperature	**Viscosity**	**Thermal Conductivity**
$1°C = 1.8\ °F$	1 poise = 1 g/cm·s	1 Btu/hr ft°F = 1.731 W/m °C
$K = °C + 273.16$	1 centipoise = 10^{-3} kg/m·s	1 (kcal/h)/m·K = 1.163 W/m·K
$°R = °F + 459.69$	1 poise = $6.72 \times 10^{-2}\ lb_m/ft·s$	1 cal/cm·s °C = 242 Btu/hr ft °F
$°F = (1.8)°C + 32$	1 centipoise = $2.09 \times 10^{-5}\ lb_f·s/ft^2$	
	1 poise = 0.1 N·s/m²	

Thermal Diffusivity and Kinematic Viscosity	**Heat Capacity**	**Heat Transfer Coefficient**
$1\ m^2/s = 3.875 \times 10^4\ ft^2/hr$	1 J/kg °C = 2.39×10^{-4} Btu/lb_m °F	$1\ W/m^2\ °C = 0.1761$ Btu/hr ft^2 °F
$1\ cm^2/s = 10^{-4}\ m^2/s$	$1\ Btu/lb_m\ °F = 1$ cal/g °C	$1\ cal/cm^2·s\ °C = 7373$ Btu/hr ft^2 °F
$1\ stokes = 1\ cm^2/s$	$1\ Chu/lb_m°C = 1\ Btu/lb_m°F$	$1\ J/s\ cm^2°C = 1761$ Btu/hr ft^2 °F

Dedication

This book is dedicated to my parents

Contents

Nomenclature

English Symbols

Symbol	Unit	Meaning
A	m^2	Area
$A = H/L$	–	Aspect ratio
a	m	Side length, radius of coiled tube
b	$W.s^{1/2}/m^2.K$	Heat intrusion coefficient
C	$(J/m^3 \cdot K)$	Volumetric heat capacity, ρc_p
C_f	–	Fanning friction factor
c_o	m/s	Speed of light
c_d	–	Drag coefficient
c_p	J/kgK	Specific heat at constant pressure
c_v	J/kgK	Specific heat at constant volume
d, D	m	Diameter
E	J	Total energy
E_i	J	Energy
E_b	$W/m^2 \cdot m$	Blackbody emissive power
e	J/kg	Specific energy
F	N	Force
F_{ij} or F_{i-j}	–	View factor from surface i to surface j
$f''(0)$	–	Dimensionless wall shear stress
f	–	Darcy friction coefficient
$f(\eta)$	–	Similarity function
g	m/s^2	Gravitational acceleration
H	J	Enthalpy = $I+P/\rho$
H	m	Height
h	J/kg	Spec. enthalpy
h	W/m^2K	Convective heat transfer coefficient
I	kJ/kg	Internal energy
I	A	Electric current
J	W	Radiosity
k	W/mK	Thermal conductivity
k_B	J/K	Boltzmann constant
L	m	Length
L_o	$W \cdot \Omega K^{-2}$	Lorenz number = 2.44×10^8
L_{hyd}	m	Hydraulic entrance length
L_{th}	m	Thermal entrance length
m	kg	Mass
M	kg	Molecular mass
$\dot{m}$	kg/s	Mass flow rate
m	–	Exponent
n	–	Exponent
P	m	Perimeter

Symbol	Units	Description
P	Pa	Pressure
p'	Pa	Pressure fluctuations
Δp	Pa	Pressure difference or pressure drop
Q	J	Heat interaction in energy equation
$Q, \dot{Q}, \dot{\mathsf{Q}}$	J	Heat rate
q	W/m^2	Heat-flux
q^R	W/m^2	Reynolds heat flux
$\tilde{q}$	W/m^3	Internal energy conversion
R	$J/kg \cdot K$	Gas constant
R	K/W	Thermal resistance
R_c	$m^2 \cdot K/W$	Thermal contact resistance
R	m	Radius, radius of curvature of a coil
R_f	$m^2 \cdot K/W$	Fouling resistance
R_u	$J/kg \cdot K$	Universal Gas constant
r	m	Radial coordinate or radius
S	m	Conduction shape factor
T	K	Temperature
T'	K	Temperature fluctuations
Tu	—	Turbulence intensity
t	s	Time
t	K	Temperature on shell side
t	m	Thickness
U	$W/m \cdot K$	Overall heat transfer coefficient
$U(x)$	m/s	Velocity
U_∞	m/s	Free stream velocity
u	m/s	Velocity (x-component)
u'	m/s	Velocity fluctuations in x-direction
u_τ	m/s	Frictional velocity
$\vec{V}$		Velocity vector
v	m/s	Velocity (y-component)
v'	m/s	Velocity fluctuations in y-direction
W	J	Work
$\dot{W}, \mathsf{P}$	W	Rate of Work
w or W	m/s	Velocity in z-direction
w'	m/s	Velocity fluctuations in z-direction
W	J	Work interaction in energy equation
x	m	Cartesian coordinate
y	m	Cartesian coordinate
z	m	Cartesian coordinate, depth

Greek Symbols

Symbol	Unit	Meaning
α	W/m^2K	Thermal diffusivity
α	-	Absorptivity
β	$1/K$	Isobaric thermal expansion coefficient
	-	Fin parameter
β	$^\circ$	Angle
γ	$-$	Specific heat ratio, parameter in transitional flow correlation
δ	m	Thickness of momentum boundary layer
δ_t	m	Thickness of thermal boundary layer
ε	-	Emissivity
ε	-	Effectiveness
μ	Pa s	Dynamic viscosity
$\eta(x,y)$	$-$	Similarity variable
η	$-$	Efficiency
θ	$-$	Dimensionless temperature
λ	m	Mean free path, wavelength, separation constant
v	m^2/s	Kinematic viscosity or momentum diffusivity
Φ	s^{-2}	Viscous dissipation
ρ	kg/m^3	Density
ρ	-	Reflectivity
σ	W/m^2K^4	Stefan-Boltzmann constant
σ	Pa	Normal stress
σ	N/m	Surface tension
θ	-	Dimensionless temperature
τ	Pa	Shear stress
τ	s	Time
τ	-	Transmissivity
υ	m^3/kg	Specific volume
φ	-	Volumetric concentration, dimensionless temperature

Special Symbols and Expressions

Symbol	Unit	Meaning
$\forall$	m^3	Volume
$\vee$	$m+$	Voltage
$\dot{\forall}$	m^3/s	Volume flow rate
ℓ	m	Liquid, Condensate, Length
$\rho\overline{u_i'u_j'}$	Pa	Reynolds stress tensor
$\rho c_p\overline{u'T}$	W/m^2	Reynolds heat vector or scalar-fluxes

Dimensionless Parameters

$Bi = h \cdot L / k_s$	Biot number
$j = St \cdot Pr^{2/3}$	Colburn j-factor
$De = Re\sqrt{(R/a)}$	Dean number
$Fo = \alpha \cdot t / L^2$	Fourier number
$Gr = gL^3 \beta \Delta T / v^2$	Grashof number
$Grz = (Re \cdot Pr) / (L/D)^2$	Graetz number
$Ja = c_{p,f}(T_w - T_{sat})/h_{fg}$	Jakob or Jacob number
$Nu_D = h \cdot D / k_f$	Nusselt number
$Nu_L = h \cdot L / k_f$	Nusselt number
$Pe = Re \cdot Pr$	Péclet number
$Pr = v / \alpha$	Prandtl number
P	Temperature difference ratio
R	Temperature difference ratio
$Ra = Gr \cdot Pr$	Rayleigh number
$Re_D = u \cdot D / v = \rho u \cdot D / \mu$	Reynolds number
$Re_L = u \cdot L / v$	Reynolds number
S_T	Dimensionless transverse pitch
S_L	Dimensionless longitudinal pitch
$St = Nu / Re \cdot Pr$	Stanton number
$Str = U \cdot f / D$	Strouhal number

Subscript or Superscripts

$(.)_{atm}$	Atmosphere
$(.)_b$	Blackbody
$(.)_c$	Corrected
$(.)_{clean}$	New heat exchanger
$(.)_{crit}$	Critical
$(.)_D, (.)_d$	Diameter
$(.)_f$	Fluid
$(.)_{fg}$	Phase change
$(.)_{fouled}$	Resistance due to fouling
$(.)_{hyd}$	Hydrodynamic entrance
$(.)_i$	Inner, initial
$(.)_{lam}$	Laminar
$(.)_L$	Length
$(.)_m$	Mean
$(.)_{m,T}$	Mean value at constant temperature
$(.)_{m,H}$	Mean value at constant heat-flux
$(.)_{max}$	Maximum
$(.)_{min}$	Minimum

$(.)_{mol}$	Molecular
$(.)_{nf}$	Nanofluids
$(.)_o$	Outside
$(.)_{th}$	Thermal entrance
$(.)_{turb}$	Turbulence related quantity
$(.)_w$	Wall condition
$(.)_\infty$	Free stream condition
$(.)_\sigma$	Electrical
$(.)_\lambda$	Based on wavelength

Acronym

LMTD	Log mean temperature difference
MWCNT	Multiwalled carbon nanotubes
NS	Navier-Stokes equation
NTU	Number of transfer units
PBI	Polybenzimidazole
RANS	Reynolds Averaged Navier-Stokes equations
SWCNT	Single-walled carbon nanotubes
TIM	Thermal interface material
WP	Wetted-perimeter

Author

Dr.-Ing. Naseem Uddin, CEng MIMechE, MIEAust CPENG, earned his PhD in aerospace engineering from Universitaet Stuttgart, Germany in 2008. He is a senior assistant professor at the Mechanical Engineering Programme, Universiti Teknologi Brunei (UTB), Brunei Darussalam. Previously, he worked as a full professor at NED University of Engineering and Technology, Pakistan from 2010–2019. Dr. Naseem is a registered chartered engineer with Engineers Australia and the Institution of Mechanical Engineers, UK. He is also listed in the National Engineering Register (NER) of Australia and recognised as a professional engineer by the Board of Professional Engineers of Queensland, Australia and by the Pakistan Engineering Council. He is a member of the American Society of Mechanical Engineers (ASME), USA. Dr. Uddin authored *Fluid Mechanics: A Problem Solving Approach*, CRC Press, 2022. He teaches fluid mechanics and heat transfer to both graduate and undergraduate students at UTB, and his research interests are in the areas of heat transfer and computational turbulence.

Prologue

The cooling of devices is now one of the most urgent requirements in many industrial and engineering applications. Imagine that your computer processor heats up and it stops performing as it should due to a lack of a proper cooling mechanism. Imagine that a car with a perfectly working engine cannot be driven just because of radiator issues. Imagine that your car batteries have flared up just because of excessive heat. Modern gas turbines function because of the cooling of the blades and combustor walls, which are constantly being exposed to high temperatures. The performance of a gas turbine engine is directly connected to efficient heat transfer. Thus, the need for effective heat transfer is obvious.

This book is meant for engineering students in mechanical, chemical, biomedical, and aerospace engineering disciplines. The book neither claims to be an encyclopedic tome on heat transfer nor can it address all heat transfer problems encountered in engineering applications. This textbook not only offers content for a one-semester course on heat transfer at the undergraduate level, but also can be used for teaching intermediate/advanced level heat transfer courses. The purpose of the book is to make the learning process enjoyable and assist the reader in becoming a problem solver. Thus, it is hoped that it will inculcate in students critical thinking, training them to use mathematically derived results and empirical correlations for the solution of engineering and industrial applications.

The author would like to express deep and sincere gratitude to previous researchers in the field of heat transfer and mathematics whose work has been mentioned or cited. There is ongoing progress in the field of heat transfer and the judicious selection of content presented in this book is only based on the need to provide the necessary and basic knowledge of this field to engineering students.

There are a total of **15** chapters in this book. The presentation of the topics is in a classical manner emphasising the basic principles of heat transfer. Wherever need arises, the results are corroborated with experimental data. Six chapters are devoted to conductive heat transfer. Both analytical and numerical techniques are discussed for the solution of the conduction heat transfer problems. Five chapters are devoted to discussion on single-phase forced, free, and mixed convection. A list of necessary and useful correlations is provided at the end of the convective heat transfer-related chapters. A chapter is provided with a focus on phase-change heat transfer. Two chapters are devoted to discussions related to the radiative heat transfer, and finally, the last chapter we have focused on the designing of heat exchangers.

We have used the MAPLE software for solution of some problems, and the snippets of MAPLE codes are provided in the text wherever the need arises.

Naseem Uddin
2023

1 Heat Transfer

Heat transfer is a branch of thermodynamics that deals with the calculation of the amount and rate of heat transfer that has interacted with the surrounding. Thermodynamically, heat interaction means the gain and loss of the heat from the system in consideration. In engineering analysis, we are often more interested in the rates of heat transfer rather than the amount of heat transfer. In this chapter, we will review the historical developments related to temperature measurement and the theories of heat transfer. After finishing this chapter, one should be able to:

- Understand the heat conduction process and write an expression for Fourier Law of heat conduction.
- Understand the convection and express mathematically Newton's law of cooling.
- Understand the radiative heat transfer and express the basic law of radiative heat exchange.

In the modern era, we cannot imagine the design of certain devices unless we calculate the amount or rate of heat transfer. Some examples of devices that need this important heat transfer analysis are gas turbines, rocket nozzles, space shuttle surfaces, electronic cooling, furnaces, etc. We know that amount of heat transfer can be calculated for a system using thermodynamic laws and principles. However, the heat as an interaction between the system and surroundings involves a lot of variables, which must look into. In many applications, we are interested in the rate of heat transfer rather than the amount of heat transfer. Over the last 200 years, heat transfer has been developed as a distinct subject in engineering.

We look into the design of certain devices which demand careful consideration of heat transfer analysis. As an example, we mention first the case of modern gas turbines. Advanced gas turbine engines operate at high temperature to improve power output and thermal efficiency. The combustion firing temperature increases with the advances in the material used for the gas turbines. As the firing temperature for combustion increases, the heat transferred to the turbine components also increases. The operating temperatures are far above the permissible metal temperatures. For the safe operations, turbine component cooling is mandatory. To accomplish this, the complex passages inside the turbine vanes/blades are created. In the complex passages inside the high pressure nozzle guide vane (NGV), the blades are cooled both internally and externally by the air extracted from compressor. Typical high pressure bleed flows are at $600\,^oC$ and used to cool the turbine immersed in gas having total temperature of $1400\,^oC$ (Han et al., 2020) (Han et al. 2000). The gas turbines component cooling by impinging jets is not possible at every location, as the construction of an impinging jet system weakens the structural strength. They are used only at the

DOI: 10.1201/9781003428404-1

Figure 1.1   CPU-cooler with surface enlargement elements. An array of impinging jets is used for cooling. (Courtesy of Institute for Aerospace Thermodynamics, Universitaet Stuttgart, Germany.)

locations where the thermal loads are excessively high like turbine guide vanes, rotor blades, rotor disks, combuster case and combuster walls etc.

As a second example on importance of heat transfer, we can mention the case cooling by using impinging jets. In a single phase flows impinging jets are a promising way to achieve high heat transfer from a surface. Compact High Intensity Coolers are the kind of heat-exchange devices, where an array of impinging jets strikes the heated surface with orifices. The meandering flow through plates creates further impingement possibilities, and this in turn improves the heat transfer. They are used in applications where the primary criterion is the weight of exchanger, for example, air conditioning units in airplanes.

As a third example, we would like to mention the case of the cooling of electronic components by the impingement of cold fluid is a promising approach. The increase in computational resources has brought the advances in the electronic components cooling approaches. The jet impingement is a well tested approach for electronic cooling. A CPU-cooler with surface enlargement and jet impingement is shown in Figure 1.1.

1.1   HEAT TRANSFER – A HISTORICAL PERSPECTIVE

There were substances that ancient philosophers and medieval scientists had created. One was Ether, which was supposed to fill up the space. Ether was the fifth element that filled up the entire universe. The other four elements were earth, water, air, and fire. The concepts of hotness of a body were discussed by Aristotle (384-322 B.C.). However, it was a crude in its format.

The the sense of warm and cold is present in human skin. This understanding of hot and cold surfaces is apparently has been known since antiquity.

Galileo Galilei (1564-1642), Evangelista Torricelli (1608-1647), Otto von Guericke (1602-1686), and others tried to construct apparatus or devices which measure the degree of hotness and coldness. In 1592, Galileo made the first air thermoscope.

Heat is not a substance like ether, but rather a mode of motion. This was the conceptualization push forward by Pierre Gassendi (1592-1655), Rene Descartes (1596-1650), and Robert Boyle (1627-1691). Isaac Newton (1642-1727), Christiaan Huygens (1629-1695), and Robert Hooke (1635-1702). In 1714, Gabriel Daniel Fahrenheit (1686-1736) developed the modern thermometers using alcohol and mercury. Fahrenheit in his temperature scale used $0°$ as fixed points which was obtained for a mixture of ice, water, and cooking salt; and he took $212°$ for temperature level corresponded to the boiling point of water, with $32°$ given by a mixture of water and ice. Swedish astronomer, Anders Celsius (1701-1744), in 1742, introduced his centigrade scale also called the Celsius scale.

Another mysterious substance was called Caloric, which was hypothesised to describe the heat transfer phenomenon. Both Ether and Caloric vanished as science progressed, and it has been found that there are more than five elements and rarefied gases fill up space. Up till 2021, there are 118 elements in periodic table. The existence of caloric was denied around 1840 when the British scientist James Joule showed that heat is indeed energy rather than a material substance that can flow. This discovery or explanation assisted the scientific community in developing a different and deeper insight which eventually helped in shaping up the phenomenon of heat transfer.

> **Pinoneers of Heat Transfer**
>
> **James Prescott Joule** (December 24, 1818, Salford, England - 11 October 1889, Sale) was a British physicist. His discovery of the relationship between heat and work interaction led him to the theory of conservation of energy (the first law of thermodynamics). He also found the law of Joule. He worked with Lord Kelvin to develop the absolute temperature scale.
>
> **James Watt** (January 19, 1736 in Greenock, Scotland - 1819) was a renoned professor who in 1764, added a condenser and a water pump to a steam engine in order to increase the power output. He also worked on vaporization of water and, in 1769, he patented his improvements for steam engine.

Knowledge of heat transmission principles is essential for designing many devices, like boilers, gas turbines, air conditioners, car engines, and heat exchangers, etc. Many phenomena in physics occur when the driving potential is present, like diffusion of heat requires temperature difference. The flow of current requires the voltage potential difference. Mass diffusion requires a high and low concentration of species in gaseous forms.

In this chapter, we introduce the three modes of heat transfer conduction, convection, and radiation and their fundamental equations.

Year	Developments in theories about temperature and heat
1593	Galileo, thermoscope (air thermometer)
1643	Torrieelli conducted experiment on vacuum
1665	Boyle's qualitative demonstration of the principle later known as Newton's law of cooling

1687	Newton published *Principia*
1700	Savery's engine, steam in contact with the water pumped
1701	Newton published 'A Scale of the Degrees of Heat', a paper describing Newton's law of cooling and quantity of heat
1702	Amontons, air thermometer, the concept of absolute zero degrees
1704	Newton released *Opticks* discussing the nature and properties of light
1720	Newcomen's beam engine with piston
1724	Fahrenheit, temperature scale
1740	Martine explained the limitation of Newton's law of cooling
1742	Celsius, temperature scale
1760 - 62	Black discussed specific and latent heats
1782	Watt's double-acting expansion expansion engine
1803	Black gave lectures on the elements of chemistry and discussed Newton's law of cooling
1804	Biot proposed a proportionality between heat transfer rate and temperature gradient
1807	Fourier proposed the law of heat conduction
1820	Fourier gave derivation of the energy equation for moving fluids
1822	Fourier, *Analytical Theory of Heat*
1822	Navier developed his viscous flow equations
1824	Carnot's motive power of heat
1830	Steam engine
1845	Stoke's proposed Constitutive relation, Navier-Stokes equations.
1848	Kelvin, concept of absolute temperature
1874	Reynolds analogy for turbulent heat transfer
1875	Grashof, equation for heat exchanger
1879	(1884) Stefan-Boltzmann law for black-body radiation
1881	Lorenz, analytical theory for natural convection
1883	(1885) Graetz problem
1894	Reynolds conducted experiments on laminar and turbulent flow
1897	Stanton number
1900	Multiple expansion engine
1900	Planck, laws of black-body radiation
1901-05	Boussinesq gave the first analytical solution for cooling of a heated body
1904	Prandtl, boundary-layer theory
1910	Prandtl analogy for turbulent heat transfer; Nusselt discussed thermal entrance region problem in tube (also known as Graetz problem)
1915	Nusselt, basic law of heat transfer (Nusselt number)
1916	Rayleigh, application of similitude to forced convection
1916	Taylor, analogy for turbulent heat transfer
1920	Steam turbine power plant (large)
1921	Pohlhausen, laminar forced convection on a flat plate (Blasius flow)
1925	Steam turbine power plant (large)
1939	von Karman analogy for turbulent heat transfer

The list above outlines some of the significant developments in theories of heat and temperature.

Atomic and molecular level movement is called Brownian motion, and the temperature of an object is directly related to this phenomenon. It means that the higher the temperature of the object, the higher will be the Brownian motion.

Temperature is a degree of hotness or coldness of a substance.

It has been found that certain physical quantities can help us in understanding the state of the system. The important and directly measurable quantities are temperature pressure and volume. We can further divide the state variables into extensive or intensive quantities. Mass, volume, entropy, and amount of heat are considered extensive quantities, whereas pressure, temperature and chemical potential are considered intensive quantities.

1.2   ENERGIES IN THERMODYNAMICS

There are several forms of energies, like sound energy, wave energy, solar energy, wind energy, nuclear energy, etc. The energy E is a conserved quantity and will remain same. Energy can exist in several forms such as chemical, mechanical, kinetic, potential, electric, nuclear and magnetic, etc. The total energy of a system (E) on a unit mass basis is denoted by e.

Thermodynamically a system can have intrinsic and extrinsic energies:

i *Intrinsic energies* which depend on the nature of the media inside system. All molecular and atomic level energies are collectively called as *internal energy.* Intrinsic energies cannot be measured directly. Internal energy is the total energy of a system due to the motion and interactions of its particles, and it is considered as a thermodynamic property of a system. The molecular level kinetic energy is a function of the velocity of the molecules or atoms present inside the system, while potential energy depends on the atomic level or intermolecular interactions. Internal energy in thermodynamics is often symbolised as (U) however, in this book to avoid confusion, it is symbolised as I.

ii *Extrinsic energies* are independent of the nature of media present within the system boundaries. Examples of extrinsic energies include gravitational potential energy, electric potential energy, and magnetic potential energy, and they are classified as *work.*

In thermodynamics, heat energy is not same as internal energy. Heat can be defined as:

Heat cannot be contained by the system and heat energy is significant only at the boundaries of the system. Heat is an interaction of system with its surrounding due to temperature difference.

The energy present inside a system is its internal energy not the heat. In human and animals, the conversion of food's chemical energy inside the system (part of internal energy) into thermal energy is known as metabolism. Eating food can give us the feeling of *warmth* but we cannot call it as heat inside the body.

Heat in thermodynamics is indicated as (Q). In engineering applications, the rate of heat transfer ($\dot{Q}$) can have more importance than the amount of heat transfer (Q).

1.3   FIRST LAW OF THERMODYNAMICS

> The **law of conservation of energy** is also known as **First Law of Thermodynamics** says that Energy cannot be created or destroyed, but it can be transferred from one object to another and may change forms.

According to first law of thermodynamics, heat cannot be generation (created) inside a system. We now consider an open system through which energy and mass are moving. Let say there is a thermodynamic quantity of interest Φ, which got transported, convected or diffused due to sources or sinks. Mathematically one can write the balance of quantity Φ as:

$$
\frac{D\Phi_{system}}{Dt} = \underbrace{\sum_i (Convection)_i}_{at\ System\ boundaries} + \underbrace{\sum_j (Diffusion)_j}_{at\ System\ boundaries}
$$

$$
+ \underbrace{\sum_k (Source)_k + \sum_\ell (Sink)_\ell}_{in\ system}
$$

$$
+ \underbrace{\sum_m (Field)_m}_{acting\ on\ System\ volume}
\tag{1.1}
$$

Figure 1.2 shows the schematic representation of energy transport processes.

Convection represents the macroscopic transport of the thermodynamic quantities with flow or mass. In a close system, there would be no convection.

Diffusion is a phenomenon which happens due to gradient of the potential. It happens with the molecular movements. Any aspect of diffusion inside system will be an aspect of internal energy and covered in it. In case of mass transfer, the potential is the concentration difference. In case of heat transfer, the potential is temperature difference.

Source and sink terms represents the transport due to effects internal to a system. For example, when the chemicals energy present in the explosives converts the point source of thermal energy is released. The internal energy conversion in the nuclear fuel rods is also an example of same category.

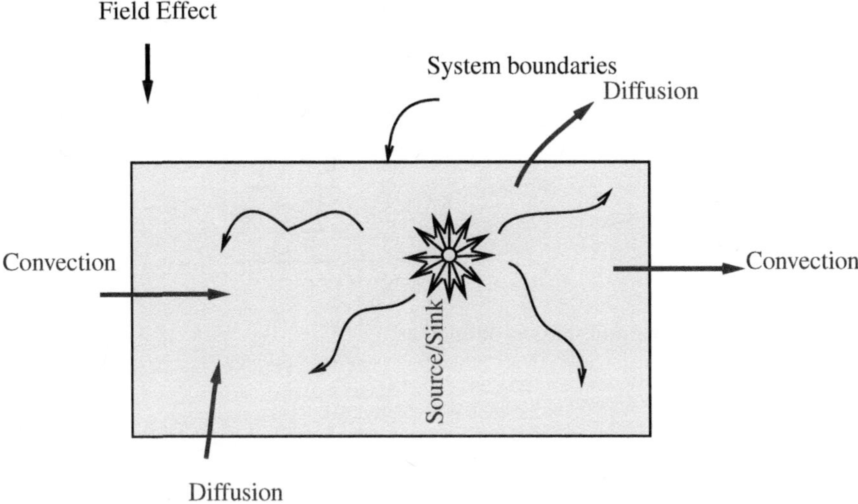

Figure 1.2 The energy transfer through control volume.

Field effect represents the transport which arises due to fields affecting directly the volume of the system. Transports that are induced due to gravity or magnetic fields come under this category.

Note that in heat transfer literature and even in textbook, we see the erroneous terminology of heat generation and heat storage. It is better to avoid this jargon as it defies the basic principles of thermodynamics. The so-called heat generation is actually the conversion of molecular chemical energy into internal energy. Similarly, the so-called heat storage is internal energy increase. Any energy present inside the control volume and system is its internal energy and not heat. Energy is called heat when it is energy in transit at the boundaries of a system and interacts with surrounding due to temperature change.

$$\left(\frac{DE}{Dt}\right)_{System} \underbrace{- \sum (e\dot{m})_{in} + \sum (e\dot{m})_{out}}_{at\ System\ boundaries}$$

$$= \sum \dot{Q} + \sum \dot{W}$$

$$\underbrace{- \sum P\frac{d\forall}{dt}}_{at\ System\ boundaries} \tag{1.2}$$

In a process the system volume may increase or decrease in time ($dV < 0$ or $dV > 0$). Since $\dot{V} = \dot{m}\upsilon$, the term $P(dV/dt)$ is called the flow work that can be written as

$\dot{W}_{flow} = P\dot{m}\upsilon$ and equation can be written as:

$$\left(\frac{DE}{Dt}\right)_{System} \underbrace{- \sum(e\dot{m})_{in} + \sum(e\dot{m})_{out}}_{at\ System\ boundaries}$$
$$= \sum\dot{Q} + \sum\dot{W}$$
$$+ \underbrace{\sum_{in}(P\upsilon\dot{m}) - \sum_{out}(P\upsilon\dot{m})}_{at\ System\ boundaries} \qquad (1.3)$$

The Energy E per unit mass is defined as:

$$e = \left(\mathbf{u} + gz + \frac{\tilde{u}^2}{2}\right) \qquad (1.4)$$

We can define a property enthalpy(h) as u+Pυ.

$$\left(\frac{DE}{Dt}\right)_{System} \underbrace{- \sum((e+P\upsilon)\dot{m})_{in} + \sum((e+P\upsilon)\dot{m})_{out}}_{at\ System\ boundaries}$$
$$= \sum\dot{Q} + \sum\dot{W} \qquad (1.5)$$

$$\left(\frac{DE}{Dt}\right)_{open} - \sum_{in}\dot{m}\left(h + gz + \frac{\tilde{u}^2}{2}\right) + \sum_{out}\dot{m}\left(h + gz + \frac{\tilde{u}^2}{2}\right)$$
$$= \sum\dot{Q} + \sum\dot{W} \qquad (1.6)$$

This is First Law of Thermodynamics for Open Systems where energy and mass can flow through system. However, we are not interested in distribution of energy inside system, rather we are interested in overall bulk energy balance. We write $\vec{V}$ in the equation:

$$\left(\frac{DE}{Dt}\right)_{sys} - \sum_{in}\dot{m}\left(h + gz + \frac{\vec{V}^2}{2}\right) + \sum_{out}\dot{m}\left(h + gz + \frac{\vec{V}^2}{2}\right) = \sum\dot{Q} + \sum\dot{W}$$

The LHS term in the above equation represent the energy change of the close system or control. In heat transfer literature, it is sometimes called energy stored. The bracketed terms on RHS are simply total energy coming in and going out of the system. Therefore, we can express the balance of energy in more simplified manner as

$$\left(\frac{DE}{Dt}\right)_{stored} = \sum\dot{Q} + \sum\dot{W} + \dot{E}_{total,in} - \dot{E}_{total,out}$$

In most of the heat transfer processes, work interactions are not present and we can further simply it as

$$\left(\frac{DE}{Dt}\right)_{stored} = \sum \dot{Q} + \dot{E}_{total,in} - \dot{E}_{total,out}$$

In case there is no energy stored with time or we assume a steady state conditions, then the balance of energy will be further reduced to form

$$\sum \dot{Q} = \dot{E}_{total,out} - \dot{E}_{total,in}$$

$\dot{Q}$ we recall is the heat interaction of system with surroundings or simply the rate of heat transfer. The aim of heat transfer investigations is to find this rate of heat transfer. In thermodynamics, we express the amount of heat transfer as Q, however, we label $\dot{Q}$ as Q in the following section and chapters, as in heat transfer literature, the rate of heat transfer is ususally expressed as Q. Note that the units of Q are W (watts) in SI system, and Btu/h or Btu/hr in USC units.

1.4 TYPES OF HEAT

Matter exist in nature in distinct three forms, solid, liquid and gas. If the energy transferred to material causes the change in temperature, we call it sensible energy. If the heat provided to a system is used up in increasing the temperature of the system, we call it Sensible heat. Referring back to first law of thermodynamics, if the system has no energy storage and work interaction, we can write

$$\sum \dot{Q} = -\sum_{in} \dot{m}\left(h + gz + \frac{\vec{V}^2}{2}\right) + \sum_{out} \dot{m}\left(h + gz + \frac{\vec{V}^2}{2}\right)$$

Also, assuming no appreciable macroscopic kinetic and potential energy changes are occurring, we have

$$\sum \dot{Q} = -\sum_{in} \dot{m}\left(h\right) + \sum_{out} \dot{m}\left(h\right)$$

Substituting $h = c \cdot T$, in above equation, where c is the specific heat of the matter, the rate of heat transfer can be computed as

$$\boxed{Q_{sensible} = \sum \dot{Q} = \dot{m} \cdot c \cdot \Delta T} \tag{1.7}$$

where $\dot{m}$ is the mass flow rate through the system, c is the specific heat, and ΔT is the change in temperature. Specific heat is a thermophysical property that plays an important role in thermal energy absorption. The tri and di-atomic gases can store more thermal energy than the inert gases.

Compared to sensible heating, the latent heat transfer is based on the energy liberation or addition associated with phase change. There is an internal energy associated with various binding forces between the atoms within a molecule and also between the inter-molecular forces. The system in a gas phase is at a higher internal energy level than it is in the solid or the liquid phase. If the heat supplied to the system is absorbed as energy and causes the change in phase of a system, we call this latent energy. Using similar assumptions as taken for the sensible energy balance, we can write

$$\sum \dot{Q} = -\sum_{in} \dot{m}\,(h) + \sum_{out} \dot{m}\,(h)$$

Since in and out for latent energy are phases, and since $H = m \cdot h$, we can write

$$\sum \dot{Q} = -\sum_{phase,in} \dot{H} + \sum_{phase,out} \dot{H} = \dot{m} \cdot \Delta h_{pc}$$

The Δh_{pc} is the phase change enthalpy in (J/kg) and we can write the first law of thermodynamics for phase change problem as

$$Q_{latent} = \dot{m} \cdot \Delta h_{pc} \tag{1.8}$$

where Δh_{pc} is the enthalpy or heat of vaporization, sublimation, fusion, vaporization, etc. For boiling conditions, we can write the above equation as

$$\boxed{Q_{latent} = \dot{m} \cdot (h_g - h_f) = \dot{m} \cdot h_{fg}} \tag{1.9}$$

where f represents the liquid state (nomenclature probably adopted from German word for liquid *flüssigkeit*), and g represents the gas state. The above equation is also valid for the condensation process.

1.5 VOLUMETRIC HEAT CAPACITY

Owing to the fact that material temperature will not vary while the phase change process is going on, this characteristic is exploited in design of energy storage systems where the heat can be added or extracted without affecting the material's temperature. Gels, Salts, polymers, paraffin based waxes, and metal alloys are some of the examples of such Phase-Change Materials (PCM). PCMs can be classified as organic, inorganic, and eutectic materials. Organic PCMs have high storage capacity and thermal conductivities.

Volumetric heat capacity of a material is an important quantity in applications, like energy storage, temperature regulation, heat dissipation, and thermal comfort,

TABLE 1.1

Volumetric Heat Capacities of Some Substances

Material	c_p (J/kg·K)	ρ (kg/m^3)	$(\rho \cdot c_p)$ 10^6 (J/m^3·K)
Water	4182	988	4.17
Magnetite	752	5177	3.69
Steel	465	7840	3.68
Aluminium	896	2710	2.43
Glass	837	2710	2.27
Concrete	880	2000	1.76
Wood	2390	700	1.67
Sandstone	712	2200	1.507
Brick	833	1810	1.51
Clay	879	1458	1.28

etc. Materials or substances such as concrete or water have high volumetric heat capacity and can help to maintain a stable temperature in a building by absorbing and releasing thermal energy as needed. Materials with high volumetric heat capacity can also help to dissipate heat in industrial processes. Materials with high volumetric heat capacity can also be used in clothing and personal protective equipment to improve thermal comfort. Volumetric heat capacities of some substances are tabulated in Table 1.1.

This bring us to the discussion on the greenhouse effect. The greenhouse effect is a natural phenomenon that occurs when certain gases in the Earth's atmosphere, like carbon dioxide, water vapor, methane, and nitrous oxide, absorb thermal energy from the sun and prevent it from escaping back into space. These gases are known as greenhouse gases, and this process of absorbing heat helps to regulate the planet's temperature and make it suitable for life as we know it.

The heat capacities of greenhouse gases vary depending on their molecular structure and the conditions under which they are measured. The heat capacities of some of the most important greenhouse gases are listed in Table 1.2.

The rise in the amount of greenhouse gases present in atmosphere due to rapid industrialization is posing a challenge to us today. As can be seen in Table 1.2, the sulfur dioxide ranked highest in greenhouse gases in terms of volumetric Heat Capacity. Refrigerant also contribute in global warming as they are leaked into atmosphere. The refrigerant global warming index (RGWI) is a metric that is established to assess the global warming potential (GWP) of refrigerants, which are used in various cooling and refrigeration applications. The refrigerant R134a (1,1,1,2-tetrafluoroethane) has a global warming potential (GWP) of 1,430.

TABLE 1.2

Volumetric Heat Capacities of Some Greenhouse Gases at 25 °C and 1 atm Pressure (kJ/m^3·K)

Gases	(ρc_p)
Sulfur dioxide (SO_2)	2.51
Nitrogen dioxide (NO_2)	1.63
Methane (CH_4)	1.58
Carbon dioxide (CO_2)	1.56
Water vapor (H_2O)	1.50
Nitrogen monoxide (NO)	1.40

1.6   MODES OF HEAT TRANSFER

So far we have learned how to do the balance of energy and arrive at simple equations for calculation of rate of heat transfer. We will now learn about the modes of heat transfer and the mathematical laws related to these modes in this section.

1.6.1   CONDUCTION

Conduction

Conduction is a heat transfer process in which thermal diffusion occurs through solid body, static liquid, and gases trapped in between narrow gaps.

In 1811, Joseph Fourier (1768-1830) submitted a manuscript on *Théorie analytique de la chaleur* in Institut de France for a prize problem of the French Academy of Science on heat diffusion in 1812. By heat conduction, we mean an energy transfer phenomenon that happens at atomic and molecular levels due to interaction under the influence of temperature difference.

Biot (1804, 1816) and Fourier (1822) hypothesised the following form of heat conduction equation, which now called as Fourier's law of heat conduction:

$$\mathbf{q} = \frac{Q}{A} = -k \, \mathbf{grad} \, T$$

Here, Q is the amount of heat transfer (scalar), $\mathbf{q}$ is the heat flux (vector), and T is the temperature (scalar). The term **grad** T is the vector quantity as gradient operator always increases the order of a tensor. The quantity k is called thermal conductivity, and it is defined in SI units as W/m.K or W/m.^{o}C and in US units it is defined as $Btu/h \cdot ft \cdot ^{\circ}F$ or $Btu/h \cdot ft \cdot ^{\circ}R$.

The negative sign helps in bringing the correct sign for amount of heat transfer depending on how the gradient is measured. Consider the slab shown in Figure 1.3 (a) shows the case in which the temperature is decreasing in increasing x direction, and thus dT/dx is negative. Figure 1.3 (b) shows the case in which the temperature is increasing in increasing x direction, and thus dT/dx is positive. According to the convention, if the heat transfer direction is in the positive direction of a coordinate axis, then the transfer is considered positive, whereas if the direction of the transfer and the coordinate system is opposite to each other, then it is taken as negative. In simple words, a positive value for heat transfer indicates heat transfer is happening in the positive coordinate direction and a negative value indicates heat transfer is happening in the negative direction.

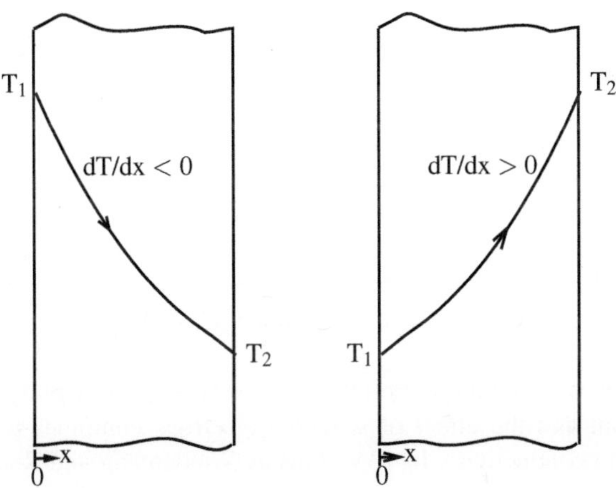

Figure 1.3   The case of temperature variation inside a slab.

Thermal diffusion through solids, gases between slots and stationary liquid is called heat conduction. Heat conduction can happen in all forms of matter.

Joseph Fourier

Born in 1768, Joseph Fourier studied mathematics and proposed his Theory of heat. In 1789, he came to Paris, to learn algebraic equations. Joseph Fourier then went to teach first in Auxerre and then at the Polytechnic School of Paris. He also participated in Napolean's expedition to Egypt and on his return to France, he was appointed to write the historical preface to the work, which brought together all observations made by savants during this expedition In 1826, he entered the Academie Francaise and, despite his illness, he worked tirelessly until the end of his life, on 17 May 1830.

Fourier's law of heat conduction (1807), $q = -k$ **grad** T is very similar to Newton's law of viscosity (1687), $\tau = \mu \, du/dy$ in terms of modelling of diffusion phenomenon. Note that in 1802, the law of the conservation of energy was not yet established.

1.6.1.1 Thermal Conductivity

We will now discuss the heat conduction through solids, liquids and gases.

Solid: The heat conduction through solids is because of the lattice vibration waves which are induced by the vibrational motion of the molecules, and the energy transferred through the movement of free electrons in the solid. The total thermal conductivity of a solid is the sum of the lattice and electron-related conductivities. Metals have high thermal conductivity because of the movement of free electrons. The lattice-related thermal conductivity depends on the structure of the molecules. For example, a diamond is a highly ordered crystalline structure, and therefore it has the highest number conductivity. The lattice vibration is interpreted as sound waves movement which acts like weakly interacting quasi-particles called *phonons*. Heat conduction through solids lattice crystalline structure is interpreted as collective excitation in a elastic arrangement of atoms or molecules – a phenomenon known as phonon. According to quantum mechanics, phonons have dual interpretation as wave and a particle. The scientific concept of a phonon was introduced first by Soviet physicist and Nobel Prize winner in Physics, Igor Tamm, in 1932. At higher temperatures, electron scattering becomes dominant. When phonons are dealt as particles, they are called bosons, which is named after Jagadish Chandra Bose and Einstein theory.

At elevated temperatures, the electron scattering by phonons becomes even more dominant and the effect of scattered electrons continues to make a major contribution to conductivity. In 1853, Gustav Wiedemann and Rudolph Franz discovered that ratio of the electronic contribution of the thermal conductivity (or simply thermal conductivity) to the electrical conductivity of a metal (k_σ) has approximately the same value at the same temperature irrespective of the metals. This is known as **Wiedemann–Franz law.**

$$\frac{k}{k_\sigma} \propto \cdot T$$

$$\frac{k}{k_\sigma} = L_0 \cdot T$$

In 1872, Ludvig Lorenz discovered the value of constant $L_0 = 2.44 \times 10^{-8} W \cdot \Omega K^{-2}$.

In some cases, the thermal conductivity of an alloy can be higher than that of its base metal due to the unique combination of properties that the alloy exhibits. For example, some alloys can have a more ordered microstructure, with fewer defects and impurities than their base metals, which can improve thermal conductivity. Additionally, some alloys can have a more uniform distribution of atoms, which can lead to improved heat transfer.

Quenching can sometimes reduce the thermal conductivity of a material, particularly if it results in the formation of defects, such as voids, cracks, or dislocations. These defects can interfere with the flow of heat through the material, reducing its thermal conductivity. However, in some cases, quenching can indirectly improve the thermal conductivity of a material by improving its microstructure. For example, if the material has a high concentration of impurities or defects, quenching can help to remove these defects and promote the formation of a more uniform and ordered microstructure. This can lead to an improvement in the thermal conductivity of the material, as there are fewer obstacles to the flow of heat.

Insulation Materials: The thermal conductivity of insulation materials is very low (less than 0.03 W/m·K). Practically, this can be achieved by creating an insulation material based on either fibrous materials or material based on cellular or granular structure. Cellular structure-based installations are as extended flexible or rigid boards. They have a low-density, low heat capacity, and good compressive strength. Polyurethane and expanded polystyrene foam are examples of cellular structure-based insulation. The granular insulations comprise of inorganic material, and they are in the form of powder. Diatomaceous silica and vermiculite are examples of granular-type insulation. The most common kind of insulation is fibrous materials of high porosity. Wool and fiberglass are examples of fibrous insulations.

In anisotropic continua like crystals, laminates, and oriented fiber composites, the direction of the heat flux vector may not necessarily be normal to a surface. For heterogeneous anisotropic continua, we can express Fourier law in vector form as

$$\mathbf{q} = -k \cdot \nabla T$$

where k is the thermal conductivity tensor and the components of thermal conductivity tensor are called the conductivity coefficients:

$$q_x = -\left[k_{11}\frac{\partial T}{\partial x} + k_{12}\frac{\partial T}{\partial y} + k_{13}\frac{\partial T}{\partial z} \right]$$

$$q_y = -\left[k_{21}\frac{\partial T}{\partial x} + k_{22}\frac{\partial T}{\partial y} + k_{23}\frac{\partial T}{\partial z} \right]$$

$$q_z = -\left[k_{31}\frac{\partial T}{\partial x} + k_{32}\frac{\partial T}{\partial y} + k_{33}\frac{\partial T}{\partial z} \right]$$

Gases: From the experiments, it is confirmed that the thermal conductivity of the gases is directly related to the square root of the absolute temperature and inversely proportional to the square root of the molecular mass. This has already been established by the kinetic molecular theory of gases, and it shows that the thermal conductivity of the gas will increase with the increase in temperature and decrease with the molecular weight. That is why the thermal conductivity of Helium is much higher than the other gases.

Using ideal gas, Eucken proposed to relate viscosity and thermal conductivity as:

$$k = \left(\frac{15\mu}{4} \right)\left(\frac{R_u}{M_{mol}} \right)\left[\frac{4}{15}\left(\frac{c_p M_{mol}}{R_u} \right) + \frac{1}{3} \right]$$

where

R_u = Universal gas constant = 8.314 kJ/kmol· K

A_v = Avogadros number = 6.022×10^{23} molecules/mole,

k_B is the Boltzmann constant (1.380649×10^{-23} J/K in SI units),

T is the absolute temperature,

d_{mol} is the particle hard-shell diameter,

γ is ratio of c_p/c_v,

M_{mol} is molecular mass, and

c_v is the of the gas at constant volume.

Chapman-Enskog proposed the following model for the estimation of thermal conductivity:

$$k_{trans} = \frac{15}{4} \cdot \left(\frac{R_u}{M} \right) \cdot \mu \tag{1.10}$$

$$k = k_{trans} + 1.32 \cdot \left(c_p - \frac{5}{2} \cdot \frac{R_u}{M} \right) \cdot \mu \tag{1.11}$$

Eucken proposed the relation for the computation of thermal thermal conductivity of gases:

$$k = \frac{c_p \cdot \mu}{4} \left[9 - \frac{5}{\gamma} \right]$$

Annaratone (2010) proposed the relation for the estimation of thermal conductivity of air in the range 0 to 300°C:

$$k_{air} = 0.02326 + 0.006588 \left(\frac{T(°C)}{1000} \right)$$

Thermal conductivity and thermal diffusivity values for some substances and liquids is provided in Table 1.3. The variation of thermal conductivity with temperature is shown in Figure 1.4.

Example 1.1

Estimate the thermal conductivity for air at 300 K and 1 atm pressure. Estimate viscosity and thermal conductivity using the power law relation:

TABLE 1.3

Thermal Conductivity k and Thermal Diffusivity

	k (W/m.K)	α (10^{-6} m^2/s)
Diamond	2200	-
Metals	5 to 400	3 to 100
Inorganic solids	0.5 to 10	0.5 to 1
Rocks	1.6 to 2.9	1 to 1.4
Organic solids	0.1 to 1	0.1
Liquids	0.1 to 1	0.1
Gases	0.01 to 0.2	3 to 100

$$\mu = B \cdot T^n$$

where B=0.4093$\times 10^{-6}$ and n=2/3.

$$k = \left(\frac{15\mu}{4}\right)\left(\frac{R_u}{M_{mol}}\right)\left[\frac{4}{15}\left(\frac{c_p M_{mol}}{R_u}\right) + \frac{1}{3}\right]$$

Take c_p=1005 J/kg·K.

Solution We compute the air's viscosity using power law equation:

$$\mu = 0.4093 \times 10^{-6} \cdot (300)^{2/3} = 18.342 \times 10^{-6} Pa \cdot s$$

$$k = \frac{15 \cdot \mu}{4} \cdot \frac{R_u}{M_{mol}} \cdot \left(\frac{4}{15} \cdot \left(\frac{c_p \cdot M_{mol}}{R_u}\right) + \frac{1}{3}\right)$$

The molecular weight of air is 28.97 kg/kmol, and γ=1.4, this gives

$$k = \frac{15 \cdot (18.342 \times 10^{-6})}{4} \times \frac{8314}{28.97} \times \left(\frac{4}{15} \cdot \left(\frac{1005 \times 28.97}{8314}\right) + \frac{1}{3}\right)$$

$$= 0.02501 W/m \cdot K$$

Using Eucken proposal, we can compute the thermal thermal conductivity of air:

$$k = \frac{c_p \cdot \mu}{4}\left[9 - \frac{5}{\gamma}\right]$$

$$k = 0.02501 W/m \cdot K$$

Comparing with the value of thermal conductivity at 300 K in Table 25 (appendix) we have k=0.02624 $W/m \cdot K$.

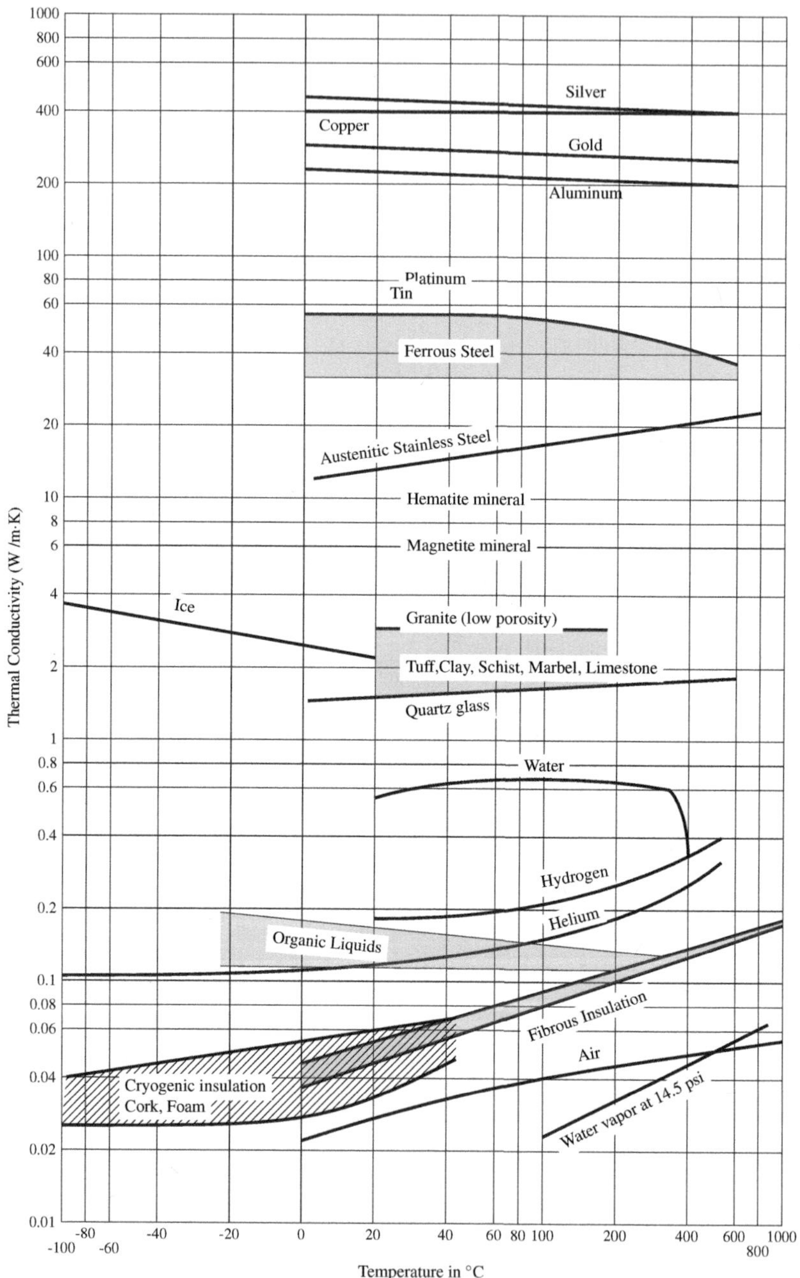

Figure 1.4　Thermal conductivity variation with temperature (SI units).

Liquids: The mechanism of heat conduction in liquids is a bit complicated as the molecules neither follow the kinetic molecular theory of the gases nor act like a solid. In liquids, the intermolecular forces are still high and the thermal conductivity values lie between those we know for solids and gases. The thermal conductivity of the liquids is inversely proportional to the temperature except in the case of water. Liquid metals have high thermal conductivities and act as good heat transfer fluids, especially for high-temperature applications like those encountered in nuclear power plants. Smith (1936) proposed the relation for the computation of thermal thermal conductivity for water, paraffin alcohols, hydrocarbons, petroleum fractions:

$$k_\ell = 4.602 \times 10^{-3} + 0.03676[c_{p,\ell}(kJ/kg \cdot K) - 1.88]^3$$
$$+ 0.519 \left(\frac{SG}{M_{mol}}\right)^{1/3} + 0.09\left(\frac{\mu_\ell}{SG}\right)^{1/9} \tag{1.12}$$

where SG is the specific gravity of liquid, and subscript ℓ indicates the properties for the liquid in consideration. For USC units, Smith proposed relation:

$$k_\ell = 0.00266 + 1.56(c_{p.\ell} - 0.45)^3 + 0.3\left(\frac{SG}{M_{mol}}\right)^{1/3} + 0.0242\left(\frac{\mu_\ell}{SG}\right)^{1/9} \tag{1.13}$$

where k_ℓ unit is Btu/hr·ft°F. The $c_{p,\ell}$ is the specific heat in units of Btu/lb-°F, SG is the specific gravity, M_{mol} is the average molecular weight; μ_ℓ is the viscosity in centipoise.

The thermal conductivity of water is basically independent from pressure and Annaratone (2010) proposed the relation for the estimation of thermal conductivity and specific heat as

$$k_{water} = 0.5755 + 0.1638\left(\frac{T(°C)}{100}\right) - 0.05767\left(\frac{T(°C)}{100}\right)^2$$

$$c_{p,water} = 4219.58 - 187.25\left(\frac{T(°C)}{100}\right) + 172.17\left(\frac{T(°C)}{100}\right)^2$$

Example 1.2

Estimate the thermal conductivity of water using the Smith relation at $T=80°C$.

Solution At $T=80°C$, we have the water viscosity $\mu = 0.000354536$ Pa·s, and SG=0.974. The molecular weight of water is 18. We estimate the specific heat as

$$c_{p,\ell} = 4219.58 - \frac{187.25 \cdot T}{100} + 126.17 \cdot \left(\frac{T}{100}\right)^2$$

$$c_{p,\ell} = 4219.58 - \frac{187.25 \cdot 80}{100} + 126.17 \cdot \left(\frac{80}{100}\right)^2 = 4157 \, J/kg \cdot K$$

$$k_\ell = 4.602 \times 10^{-3} + 0.03676[c_{p,\ell}(kJ/kg \cdot K) - 1.88]^3 + 0.519 \left(\frac{SG}{M_{mol}}\right)^{1/3}$$

$$+ 0.09 \left(\frac{\mu_\ell}{SG}\right)^{1/9}$$

$$k = 4.602 \times 10^{-3} + 0.03676 \cdot (4.157 - 1.88)^3 + 0.519 \cdot \left(\frac{0.974}{18}\right)^{1/3}$$

$$+ 0.09 \cdot \left(\frac{0.000354536}{0.974}\right)^{\frac{1}{9}}$$

This gives the thermal conductivity of water as

$$k = 0.668 W/m \cdot K$$

From Table A14: Properties of Saturated Liquids (Water) we see that water has thermal conductivity value k = 0.668 W/m·K. From Annaratone proposal, we have

$$k_{water} = 0.5755 + 0.1638 \left(\frac{T(°C)}{100}\right) - 0.05767 \left(\frac{T(°C)}{100}\right)^2$$

$$k_{water} = 0.669 W/m \cdot K$$

Both Smith and Annaratone correlation are good enough for the thermal conductivity estimation.

The range of thermal conductivities for different states of matter

$$0.015 \, W/m \cdot K \leq k_{gases} \leq 0.15 \, W/m \cdot K$$
$$0.1 \, W/m \cdot K \leq k_{liquids} \leq 0.65 \, W/m \cdot K$$
$$1 \, W/m \cdot K \leq k_{solid} \leq 450 \, W/m \cdot K$$

Note that the definition of thermal conductivity is based on temperature difference, and therefore we do not require the unit conversion from centigrade to kelvin.

$$1\frac{W}{m \cdot K} = 1\frac{W}{m \cdot °C}$$

Similarly,

$$1\frac{Btu}{h \cdot ft \cdot °F} = 1\frac{Btu}{h \cdot ft \cdot °R}$$

The conversion constant for k between the SI and English unit is

$$1\frac{W}{m \cdot K} = 0.578\frac{Btu}{h \cdot ft \cdot °F}$$

1.6.1.2 Some parameters used in industry

The construction, clothing, building industries are often interested in heat resistant materials. They have devised some parameters to asses the potential of the material as being *heat-resistant*. For an interested reader, here we mention a few of them.

Heat Intrusion Coefficient is defined as:

$$b = \sqrt{k.\rho.c_p}$$

This parameter represents the capacity of a material to withstand applied heat-flux or temperature, and it is used often in German heat transfer literature where it is called *Wärmeeindringkoeffizient*. The parameter b helps in assessing on how quickly the heat might penetrate into the building material. Table 1.4 lists the values of heat penetration coefficient for different substances and materials.

TABLE 1.4

Heat Intrusion Coefficient, b (W.s$^{1/2}$/m^2.K)

Material	b
Copper	36000
Iron	15000
Concrete	1600
Water	1400
Sandy soil	1200
Wood	400
Foam	40
Gases	6

Heat intrusion coefficient is also called thermal inertia in English heat transfer literature.

The thermal inertia of a material is often used to characterise its thermal response to sudden changes in temperature. Materials with high thermal inertia require more energy to heat up or cool down and therefore have a slower response time to changes in temperature. Thermal inertia is important for the design of protective clothing for fire-fighters, which is made with materials (aramid or para-aramid fibers) with higher thermal inertia and thus helping in reducing the rate of heat transfer to the skin. In some applications where rapid heating or cooling is required, a material having lower thermal inertia is more appropriate. The thermal inertia of human skin can range from about 500 to 1000 $J/m^2 \cdot K \cdot s^{1/2}$, depending on the specific properties of the skin.

An important parameter for the insulation selection is the *Heat Deflection Temperature* (HDT), or Heat Distortion Temperature. This is the measure of a polymer's potential to resistance to alteration under a given heat load at an elevated temperature. It is also called the *Deflection temperature under load* (DTUL) or *heat deflection temperature under load* (HDTUL). It is also defined as the temperature at which a polymer or plastic product deforms under a specified load.

Some approximate HDT values for common materials:

Polyethylene (PE): 50-80 °C

Polypropylene (PP): 80-130 °C

Polyvinyl chloride (PVC): 70-100 °C

Polycarbonate (PC): 135-155 °C

Nylon (PA): 80-160 °C

1.6.2 CONVECTION

Convective heat transfer occurs due to the bulk fluid movement over a surface. Convection, therefore, takes place in fluids (liquids and gases), and the most important region is the near-wall region as heat will be added or removed from the surface. The phenomena of convection depend on the fluid properties, flow conditions, as well as geometry of the surface or duct.

> **Convection**
>
> Convection is a heat transfer process in which thermal diffusion occurs between solid surface and fluid layers due to temperature difference.

There are two kinds of convection: forced convection and free convection. Forced convection is a phenomenon in which the fluid is set into motion by an external agency like the blowing of the wind, the fan, the pump, or a blower. Note that the blowing of the wind is a natural phenomenon but the heat transfer that happens due to it is classified as forced convection, as the fluid movement is generated externally. On the other hand, if the fluid movement is generated due to a change in density (or temperature) of convecting fluid in a gravitational potential field, then the heat transfer associated with such a phenomenon is classified as free or natural convection. Figure 1.5 shows the thermal boundary layer formation over a hot and cold surfaces.

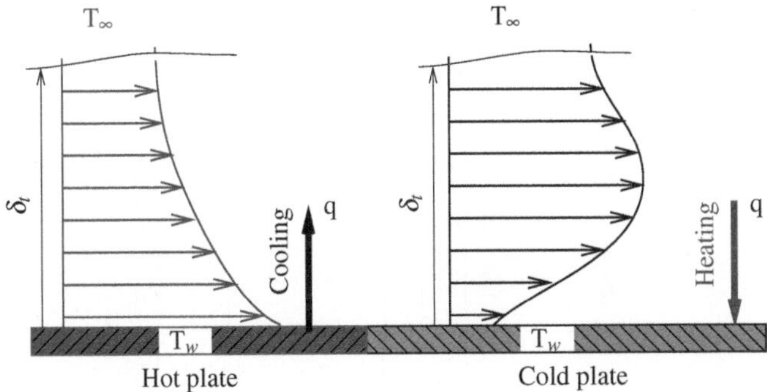

Figure 1.5   The thermal boundary layer formation over hot and cold plates along with temperature distribution.

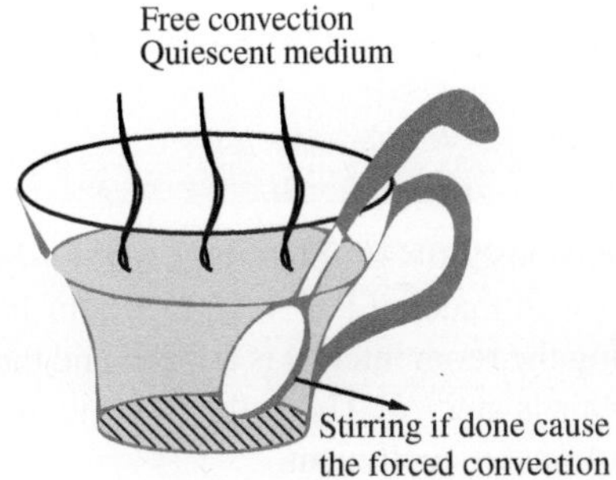

Figure 1.6   The tea in the cup placed in a quiescent or dormant medium where free convection happening.

Figure 1.6 shows the hot tea cup placed in a quiescent air medium and major heat transfer to surrounding is happening due to either radiation or free convection. There is conduction of heat going on in the cup and spoon and if the spoon is stirred there will be turbulent mixing inside tea which will increasing forced convection from tea to the cup walls.

Newton in 1701, proposed a law to describe the thermal convection phenomena:

$$\mathbf{q} = \frac{Q}{A} = h\left(T_w - T_\infty\right)$$

where h is called the convective heat transfer and it has the units of $W/m^2.K$, or $W/m^2.{}^oC$. This relationship is known as Newton's law of Cooling. Figure 1.7 shows the free convection boundary layer formation over hot and cold surfaces.

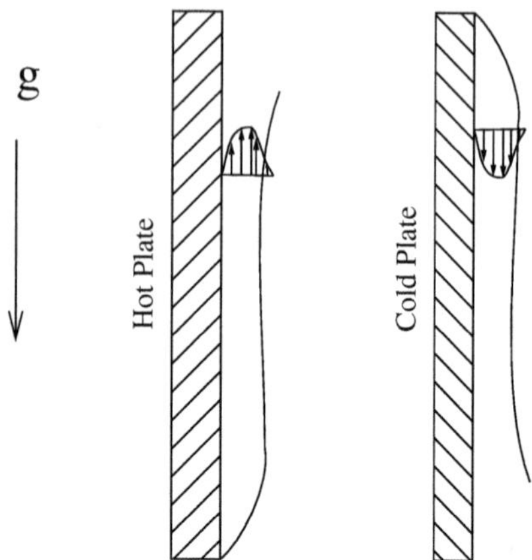

Figure 1.7   Natural convection boundary layer formation over hot and cold plates.

Example 1.3

A wall having dimensions 9 ft×15 ft has thickness L=7 inch. The mean thermal conductivity of wall material is k = 0.012 Btu/hr·ft·°F. The temperature of wall surface facing the room interior is at 72°F and the temperature of wall surface exposed to air is at 27°F. The ambient air temperature is 20°F. Find the convective heat transfer coefficient.

Solution

The area of the wall is, A = 9 ft×15 ft = 135 ft^2.

$$T_1 = 72°F, \ T_2 = 27°F, \ and \ T_\infty = 20°F$$

The heat conducted through wall is

$$Q = -k \cdot A \cdot \frac{dT}{dx} \equiv -k \cdot A \cdot \frac{T_2 - T_1}{L}$$

$$Q = -k \cdot A \cdot \frac{T_2 - T_1}{L} \equiv -0.012 \frac{Btu}{hr \cdot °F \cdot ft} \cdot 135 ft^2 \cdot \left(\frac{27 - 72}{7/12}\right) \frac{°F}{ft}$$

$$= +124.971 \frac{Btu}{hr}$$

where L is the thickness of the wall, and A is the area perpendicular to the heat-flux vector. The amount of heat transfer convected to air according to Newton's law of cooling is:

$$Q = h \cdot A \cdot (T_w - T_\infty)$$

where T_w is the wall temperature exposed to convective environment. In this problem, $T_w=T_2$, hence

$$Q = h \cdot A \cdot (T_2 - T_\infty)$$

$$124.971\frac{Btu}{hr} = h \cdot 135\,ft^2 \cdot (27 - 20)^\circ F$$

$$h = 0.1322\frac{Btu}{hr \cdot ft^2 \cdot {}^\circ F}$$

Note that in heat transfer literature (in English language), h is used for hours in USCU or English units, whereas h is also used to label convective heat transfer coefficient. To avoid confusion in this book, hours are labeled as hr, whenever convective transfer coefficient is computed in US customary units.

1.6.2.1 Convection with Nano-fluids

The discovery of Nanofluids is attributed to Argonne National Laboratory, USA. The first-ever report on nanofluids was published by Choi (1995). Nanofluids contain suspended nanoparticles that have high thermal conductivity. They are colloidal suspensions engineered by including the nano-sized particles in some base fluid. They are classified into distinct categories depending on dispersed particles: metal, metal oxide, metal hybrid, and carbon-based. The most common choice of host fluid or base fluid is water, ethylene glycol, oils, and methanol. Suspensions based on hybrid nanofluids are also investigated, which used a combination of different nanoparticles. Nanofluid quickly gained popularity in the heat exchanger industry, as they give better thermal performance than conventional fluids. Because of the growing importance of nanofluids in this book, the properties' correlations are collected. The flow and thermal characteristics of nanofluid are strongly influenced by the properties of the nanofluid. A great amount of research has been done on the properties of nanofluids (see Eastman et al. (1999), Ding et al. (2006), Eastman et al. (1999), Jang and Choi (2006), Koo and Kleinstreuer (2004)). The Al_2O_3-water nanofluid, with nanoparticles concentration of less than 10%, behaves as newtonian fluid for the temperature range of 0–90 °C. The relationship proposed by Brinkman (1952) can be used to calculate the effective viscosity of the nanofluid.

$$\mu_{nf} = \frac{\mu_f}{(1 - \varphi)^{2.5}} \tag{1.14}$$

where φ is the nano-particles concentration and subscript nf and f refer to the nanofluid and the base fluid properties, respectively. Equation is valid as long as the volumetric concentrations $\varphi < 4\%$.

The effective density and specific heat capacity of the nanofluid can be modelled as per proposal of Xuan (2003):

$$(c_p)_{nf} = (1 - \varphi)(c_p)_f + \varphi(c_p)_p \tag{1.15}$$

TABLE 1.5

Thermophysical Properties of Base Fluids and Nanoparticles, a Calculation Based on Pak and Cho (1998) and Brinkman (1952) Models

	ρ (kg/m^3)	c_p (J/kg.K)	k (W/m.K)	μ (Pa·s)
Water (base fluid)	997	4179	0.613	0.0008206
Ethylene glycol (base fluid)	1115	2428	0.253	0.021404
Copper (Cu)	8933	385	401	
Silver (Ag)	10,500	235	429	
Alumina (Al$_2$O$_3$)	3970	765	40	
Titanium oxide (TiO$_2$)	4250	686.2	8.9538	

$$\rho_{nf} = (1 - \varphi)\rho_f + \varphi\rho_p \tag{1.16}$$

The thermal conductivity of nanofluid can be modeled by use of Maxwell theory which was proposed in 1873. For the two-component mixture of the spherical-particle suspension, the Maxwell proposed the following model:

$$\frac{k_{nf}}{k_f} = \frac{\left(k_p + 2k_f\right) - 2\varphi(k_f - k_p)}{\left(k_p + 2k_f\right) + \varphi(k_f - k_p)}$$

However, this model only account for the static dilute suspension and conductivity due to Brownian motion is not accounted for in this model. A simplified model for thermal conductivity of nanofluid is

$$k_{nf} = (1 - \varphi)k_f + \varphi k_p \tag{1.17}$$

The thermal and momentum diffusivities of nanofluid can be calculated as:

$$\frac{\alpha_{nf}}{\alpha_f} = \frac{\left(k_{nf}/k_f\right)}{(1 - \varphi) + \varphi\left[(\rho c_p)_p/(\rho c_p)_f\right]} \tag{1.18}$$

$$\frac{v_{nf}}{v_f} = \frac{1}{(1 - \varphi)^{2.5}\left[(1 - \varphi) + \varphi(\rho_p/\rho_f)\right]} \tag{1.19}$$

Table 1.5 lists the properties of two base fluids and some nano-particles. Table 1.6 lists the properties of nano-fluid formed by adding Al$_2$O$_3$ particles in water. The concentration of nano-particles will charge the properties of the mixture.

The density of nanofluid depends upon nanoparticle concentration in the fluid, and therefore, the more particles we add into the base fluid denser will be the nanofluid and added pressure loss should be expected. The mechanism of enhancement of thermal conductivity is attributed to the transfer of energy due to the collision of higher-temperature particles with lower-temperature ones. The improvement

TABLE 1.6

Thermophysical Properties of the Water-Al$_2$O$_3$ Nanofluid for Different Volumetric Concentrations

φ %	ρ_{nf} kg/m^3	k_{nf} W/m.K	$\mu_{nf} \times 10^4$ Pa.s	$c_{p,nf}$ J/kg.K
0	996.24	0.616	8.21	4179
0.2	1003	0.671	8.247	4172.4
0.7	1018.02	0.8187	8.351	4155.90
1.5	1042.04	1.0538	8.52	4129.51
3	1087.09	1.4946	8.85	4080

in thermal conductivity helps in augmenting the heat transfer rates for nanofluid. The ethylene-glycol-based nanofluids have higher heat transfer rates than the water-based nanofluids. The effect of the Brownian motion is to reduce the rates of heat transfer in the case of nanofluids. In the near-wall region of the wall, the viscous effects are very strong and hence the Brownian motion is not playing a big role here in terms of heat transfer. In some cases, the Brownian motion may have an indirect role in producing particle clustering, which in turn could improve thermal conductivity. Note that it is possible that after certain particle concentration levels, the heat transfer can be deteriorated.

Figure 1.8 shows the varation of thermal conductivity of nanofluid with the volumetric concentration.

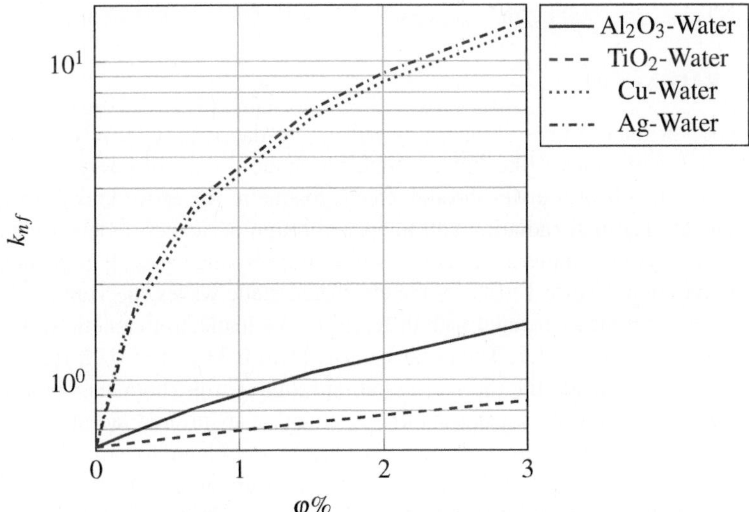

Figure 1.8   Thermal conductivity of nanofluids at different volumetric concentrations.

There is a growing interest in carbon nanotubes. Carbon nanotubes (CNTs) have cylindrical carbon molecules origin, with very high thermal conductivities. The diameter of CNTs ranges from $\mathscr{O}(1\text{ nm})$ to $\mathscr{O}(100\text{ nm})$ and has lengths in micrometer. The thermal conductivity of single-wall carbon nanotubes can be as high as 6,600 W/m·K and for multi-wall carbon nanotubes is it can be as high as 3,000 W/m·K. Xue (2005) proposed a model based on Maxwell's theory accounting the rotational elliptical nanotubes with very large axial ratios and compensating for the effects of the space distribution on carbon nanotubes:

$$\frac{k_{\text{nf}}}{k_{\text{f}}} = \frac{1 - \varphi + 2\varphi\left(\frac{k_{\text{CNT}}}{k_{\text{CNT}} - k_{\text{f}}}\right)\ln\left(\frac{k_{\text{CNT}} + k_{\text{f}}}{2k_{\text{f}}}\right)}{1 - \varphi + 2\varphi\left(\frac{k_{\text{f}}}{k_{\text{CNT}} - k_{\text{f}}}\right)\ln\left(\frac{k_{\text{CNT}} + k_{\text{f}}}{2k_{\text{f}}}\right)} \tag{1.20}$$

The rest of the properties can be estimated as per proposal of Hone (2004), and Ruoff and Lorentz (1995):

$$\begin{aligned}
\frac{v_{nf}}{v_f} &= \frac{1}{(1-\phi)^{2.5}[(1-\phi) + \phi(\rho_s/\rho_f)]} \\
\rho_{nf} &= (1-\phi)\rho_f + \phi\rho_{CNT} \\
(\rho c_p)_{nf} &= (1-\phi)(\rho\, c_p)_f + \phi(\rho\, c_p)_{CNT} \\
\frac{\alpha_{nf}}{\alpha_f} &= \frac{k_{nf}/k_f}{(1-\phi) + \phi((\rho c_p)_s/(\rho c_p)_f)}
\end{aligned} \tag{1.21}$$

Carbon nanotubes are classified as the single long wrapped graphene sheet also called single-walled carbon nanotubes (SWCNT) and multiwalled carbon nanotubes (MWCNT). The nominal values usually taken in the numerical analysis for SWCNT and MWCNT are

SWCNT: $\rho_{CNT} = 2600$ kg/m^3, $c_{p,CNT} = 425$ J/kg·K, $k_{CNT} = 6600$ W/m·K
MWCNT: $\rho_{CNT} = 1600$ kg/m^3, $c_{p,CNT} = 796$ J/kg·K, $k_{CNT} = 3000$ W/m·K.

1.6.3 RADIATION

We encounter many forms of energy in our day-to-day life; some of these energies we perceive with senses like light, sound, but some modes of energy transfer are more subtle, like heat transfer through electromagnetic waves we know as radiative heat transfer. Thermal radiation can happen through a vacuum as electromagnetic waves do not require matter to transmit. Solids, liquids, and gases all emit and absorb thermal radiation. Figure 1.9 shows the electromagnetic waves spectrum.

The wavelengths associated with different waves in electromagnetic waves spectrum are listed in Table 1.7. The color band is from 0.38 μm to 0.76 μm in wavelength, and it lies inside the thermal radiation range. Unlike humans, some insects, such as bees and butterflies, can see ultraviolet and near-infrared radiation as part of their vision, which helps them locate food sources and identify each other. Humans can use devices called infrared cameras or thermal imaging cameras to detect and visualise infrared radiation. Ultraviolet (UV) cameras are used in forensic science, mineralogy, detection of skin damage, sterilisation, and water purification.

Infrared heating is often used in industrial and commercial applications, such as in manufacturing processes, drying applications, and space/residential heating applications.

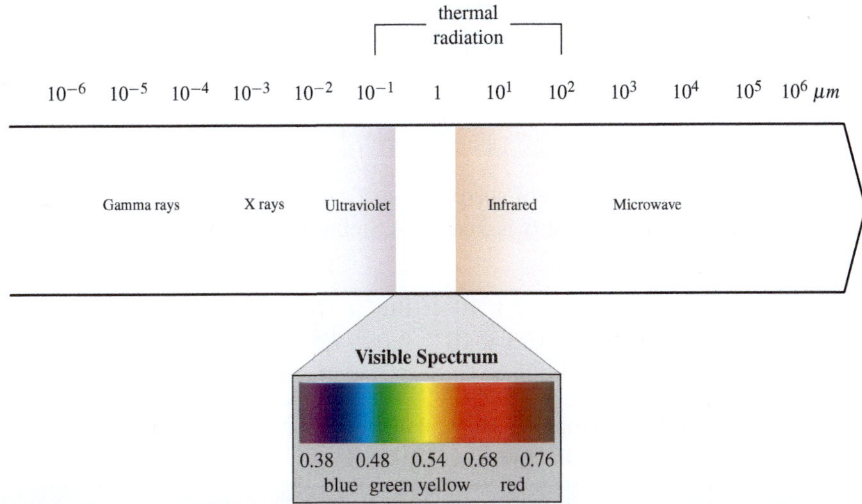

Figure 1.9 The electromagnetic waves spectrum.

TABLE 1.7

The Range of Wavelengths

Electromagnetic Waves	Wavelengths
Cosmic rays	$\lambda < 10^{-7}\ \mu m$
Gamma rays	10^{-6} to $10^{-5}\ \mu m$
X rays	10^{-5} to $10^{-2}\ \mu m$
Ultraviolet	$10^{-2}\ \mu m$ to $0.38\ \mu m$
Thermal Radiation	$0.1\ \mu m$ to $100\ \mu m$
Infrared	$0.76\ \mu m$ to $1000\ \mu m$
Microwave	$1000\ \mu m$ to 1 m

Radiative heat transfer is the mode of heat transfer that occurs due to temperature difference and happens due to electromagnetic waves. Note that a radiative exchange between the surfaces happens all the time, at all temperatures, irrespective of the presence of the medium. It is a net exchange of energy that determines the direction of the heat transfer, and it is always from high to low temperature.

Radiative heat transfer is unique among the other modes of heat transfer as it is related to the absolute temperature of the power four. In radiative heat transfer, we define an idealised body called *blackbody* whose emission and absorption of thermal radiation is maximum at a given temperature. Stefan-Boltzmann proposed a law for emissions from a blackbody:

$$q_{blackbody} = \frac{Q}{A} = \sigma(T_w^4 - T_{surr}^4) \tag{1.22}$$

where T_{surr} is the temperature of surrounding, T_w is the surface temperature of radiating object, and the constant σ is known as Stefan-Boltzmann constant, which has the values:

$$\begin{aligned}
\sigma &= 1712 \times 10^{-12} \text{ Btu/hr} \cdot ft^2 \cdot (^\circ R)^4 \\
&= 5.673 \times 10^{-8} \text{ W}/m^2 \cdot K^4 \\
&= 1.355 \times 10^{-12} \text{ cal/s} \cdot cm^2 \cdot K^4
\end{aligned}$$

Note that the temperature in above equation is always substituted as absolute temperature (either in Kelvin or degree Rankine)

<hr>

Example 1.4

A new type of telescope is launched in space with surface area of m². It is planned that the telescope will dissipate 4,000 W/m² during operating time. Find the temperature of the telescope. Assume that the telescope is acting like a black body.

Solution
According to Stefan-Boltzmann Law, the emission from

$$Q_{b,rad} = \sigma \cdot A \cdot (T_w^4 - T_{space}^4)$$

T_{space} is the space temperature and since in space the density of gases is extremely low, the space is taken at 0 K temperature in heat transfer analysis. We have

$$\frac{Q_{b,rad}}{A} = \sigma \cdot T_w^4$$

The heat flux dissipated by telescope is

$$q_w = \frac{Q}{A} = 4000 \ W/m^2$$

Comparing the heat-fluxes, we have

$$\sigma \cdot T_w^4 = 4000 \ W/m^2$$

$$5.67 \times 10^{-8} W/m^2 \cdot K^4 \times \cdot T_w^4 = 4000 \ W/m^2$$

This gives temperature of the surface of the telescope as T_w=515.37 K = 242.37°C.

TABLE 1.8

Heat Flux in Different Cases

Bearable solar radiation on the skin	40 W/m^2
Heat released by the human body	50 W/m^2
Electric heating	70 - 350 W/m^2
Hot water heating	500 W/m^2
Solar constant	1366 W/m^2
Maximum solar heat flux or *insolation*	500-800 W/m^2
Human skin pain threshold for thermal radiation	20-30 kW/m^2
Microwave	2-10 kW/m^2
Cooling of rocket nozzle	45 kW/m^2
Heating systems for household appliances	10 - 80 kW/m^2
Boiler Tubes	0.5 MW/m^2
Fuel rods in the nuclear reactor/Gas turbine blade	1 MW/m^2
Central solar receiver	1-10 MW/m^2

With real surfaces, electromagnetic radiation is not maximum like that in a blackbody. To accommodate this lack of radiative emission, compared with the blackbody is mentioned as a *gray* nature. This low radiative emission nature of gray bodies or surfaces is accommodated by introducing a dimensionless factor, called the Emissivity, and donated as ε. Its values for various surfaces are provided in the Appendix. We must then amend the equation 1.22 to account for gray-body behaviour.

$$\mathbf{q}_{gray} = \varepsilon\sigma(T_w^4 - T_{surr}^4) \tag{1.23}$$

Note that $\varepsilon=1$ for blackboday and for gray body it is always less than one.

The amount of heat transfer is a scalar quantity, but the amount of heat-flux (heat per unit area) is a vector quantity. It is helpful to have some ball-park figures for the heat flux associated with different heat transfer situations (see Table 1.8).

The thermophysical properties of air, water and engine oil are listed in Table 1.9.

TABLE 1.9

Commonly Used Properties of Air and Water

PROPERTY	SI	USCS & ENGLISH
Air at $20°C$ $(68°F)$ **and 1 atm**		
Specific gas constant	$R_{air} = 0.2870$ kJ/kg $\cdot$ K $R_{air} = 287.0$ m^2/s^2 $\cdot$ K	$R_{air} = 0.06855$ Btu/lb$_m$ $°$R $R_{air} = 53.34$ ft $\cdot$ lb$_f$/lb$_m$ $°$R $R_{air} = 1717$ ft $\cdot$ lb$_f$/slug $°$ R

Specific heats	$c_p = 1.007 \text{ kJ/kg} \cdot \text{K}$	$c_p = 0.2404 \text{ Btu/lb}_\text{m}°\text{R}$
	$c_p = 1007 \text{ m}^2/\text{s}^2 \cdot \text{K}$	$c_p = 187.1 \quad \text{ft} \cdot \text{lb}_f/\text{lb}_m \, °\text{R}$
	$c_v = 0.7200 \text{ kJ/kg} \cdot \text{K}$	$c_v = 6019 \text{ ft}^2/\text{s}^2°\text{R}$
	$c_v = 720.0 \text{ m}^2/\text{s}^2 \cdot \text{K}$	$c_v = 0.1719 \text{ Btu/lb}_\text{m}°\text{R}$
		$c_v = 133.8 \text{ ft} \cdot \text{lb}_\text{f}/\text{lb}_\text{m}°\text{R}$
		$c_v = 4304 \text{ ft}^2/\text{s}^2 \cdot °\text{R}$
Thermal conductivity	$k = 0.025 W/m \cdot K$	$k = 0.01495 Btu/h \cdot ft \cdot° F$
Density	$\rho = 1.204 \text{ kg/m}^3$	$\rho = 0.07518 \text{ lbm/ft}^3$
		$\rho = 0.00237 \text{ slug/ft}^3$
Abs. Viscosity	$\mu = 1.825 \times 10^{-5} \text{ kg/m} \cdot \text{s}$	$\mu = 1.227 \times 10^{-5} \text{ lb}_\text{m}/\text{ft} \cdot \text{s}$
Kinematic Viscosity	$v = 1.516 \times 10^{-5} \text{ m}^2/\text{s}$	$v = 1.632 \times 10^{-4} \text{ ft}^2/\text{s}$

Water at 20°C (68°F) and 1 atm

Specific heat	$c = 4.182 \text{ kJ/kg} \cdot \text{K}$	$c = 0.9989 \text{ Btu/lb}_\text{m} \cdot °\text{R}$
	$c = 4182 \text{ m}^2/\text{s}^2 \cdot \text{K}$	$c = 777.3 \text{ ft} \cdot \text{lb}_\text{f}//\text{lb}_\text{m} \cdot °\text{R}$
		$c = 25,009 \text{ ft}^2/\text{s}^2 \cdot °\text{R}$
Thermal conductivity	$k = 0.597 W/m \cdot K$	$k = 0.3455 Btu/h \cdot ft \cdot° F$
Density	$\rho = 998.0 \text{ kg/m}^3$	$\rho = 62.30 \text{ lb}_\text{m}/\text{ft}^3$
		$\rho = 1.936 \text{ slug/ft}^3$
Abs. Viscosity	$\mu = 1.002 \times 10^{-3} \text{ kg/m} \cdot \text{s}$	$\mu = 6.733 \times 10^{-4} \text{lb}_\text{m}/\text{ft} \cdot \text{s}$
Kinematic Viscosity	$v = 1.004 \times 10^{-6} \text{ m}^2/\text{s}$	$v = 1.081 \times 10^{-5} \text{ft}^2/\text{s}$

Engine oil at 20°C (68°F)

Specific heat	$c = 1880 \text{ kJ/kg} \cdot \text{K}$	$c = 0.449 \text{ Btu/lb}_\text{m} \cdot °\text{R}$
Thermal conductivity	$k = 0.145 W/m \cdot K$	$k = 0.084 Btu/h \cdot ft \cdot° F$
Density	$\rho = 888 \text{ kg/m}^3$	$\rho = 55.436 \text{ lb}_\text{m}/\text{ft}^3$
		$\rho = 1.723 \text{ slug/ft}^3$
Kinematic Viscosity	$v = 0.00090 \text{ m}^2/\text{s}$	$v = 0.0097 \text{ft}^2/\text{s}$

PROBLEMS

1P-1 Figure shows a 100-W incandescent lamp, with 4.3 cm long filament and has a diameter of 0.52 mm. The glass bulb can be approximated as an sphere with the diameter of the glass bulb of the lamp around 8 cm. Find the heat flux (W/m^2), (a) on the surface of the filament and (b) on the surface of the glass bulb.

1P-2 Estimate the thermal conductivity of air at T=850K using Chapman-Enskog model. Take absolute viscosity as 3.76×10^{-5} Pa.s.

1P-3 It is planned to construct a stone wall either of granite or limestone with thickness of L=20 cm to prevent heat transfer from outside to room. The room wall is supposed to be maintained at 20°C, and the wall side exposed to air is maintained at 40°C. What type of stone wall you would like to recommend. Granite properties: k= 3.1 W m^{-1} K^{-1}, α= 1.6 × 10^{-6} m^2 s^{-1}, $(\rho \cdot c)$ =1.9 × 10^6 J m^{-3} K^{-1} and c= 720.7 J kg^{-1} K^{-1}.
Limestone properties: k=1.3 W/(m·K), ρ=2750 kg/m^3, c=840 J/g·K.

1P-4 In your opinion what modes of heat transfer are important during the launch sequence of the space shuttle, and later when it reaches in the orbit. What role the ablative materials play in the design of a space shuttle, when it *reenter* the planet's atmosphere and , the speeding shuttle surface temperature reaches in range 2000°C to 2500°C, due to shock formation?

1P-5 Fish are cold-blooded animals with body temperature of 37°C. Total surface area of fish is 0.1 m^2. Find the heat transfer from fish to the sea water if it is flowing at a depth of sea where temperature is 10°C.

1P-6 A satellite is orbiting around planet and due to a small nuclear plant inside it the satellite has temperature of 300°C. Assume satellite as a sphere and a blackbody find the radiative heat flux from satellite to the space.

1P-7 It is planned to replace the heat transfer fluid moving over a flat surface from water to a nanofluid comprised of water-Al$_2$O$_3$. The thermophysical properties of nanofluid are ρ_{nf}=1017.30 kg/m^3, k_{nf}=0.822 W/m·K, μ_{nf}=0.000835 Pa·s, cp_{nf}=4158.9 J/kg·K, and the thermophysical properties of pure water are k_f=0.597, ρ_f=998 kg/m^3, μ_f=1×10^{-3} Pa·s, c_p=4182 J/kg·K. The freestream temperature is both cases is 300 K and surface is maintained at wall temperature of 500 K. If the convective heat transfer coefficient for pure water flow is h_f=10.5 W/m^2·K, and with use of nanofluid 10% increase in heat transfer is obtained compared with the pure fluid, find the convective heat transfer coefficient (h_{nf}) for nanofluid flow. Take area of flat surface as 4 m^2.

[Ans: 46.2 W/m^2·K]

1P-8 Prandtl number is an important dimensionless number in convective heat transfer analysis. It is defined as Pr=$c_p\mu$/k. Using the data presented in Table 1.5, estimate the Prnadtl number for Ethylene glycol-Al$_2$O$_3$ nanofluid for different volumetric concentrations of φ =0.35, 0.46, 0.78 %. What trend have you noticed for Prandtl number of nanofluids?

1P-9 Estimate the thermophysical properties of SWCNT-water solution at volumetric concentration of φ=5%.

[Ans: $\alpha_{nf} = 2.908 \times 10^{-7}$ m^2/s, $k_{nf} = 1.1673$ W/m·K]

1P-10 Estimate the thermophysical properties of MWCNT-water solution at volumetric concentration of $\varphi=4\%$.

[Ans: $\alpha_{nf}=2.497\times 10^{-7}$ m^2/s, $k_{nf}=1.0115$ W/m·K]

1P-11 A roof approximated as a flat surface is losing heat to the open sky and atmosphere via both convection and radiation. The emissivity of the surface (ε) is 0.95 and the convective heat transfer coefficient is 12 W/m²·K. The sky temperature (T_{sky}) is $-12\,°C$ and the blowing cold air temperature (T_∞) is $4\,°C$. If the net heat-flux loss from roof is 30 W/m², find the roof's surface temperature.

1P-12 A mercury thermometer with glass covering ($\varepsilon=0.9$) hangs in a duct and indicates the temperature value of 18 °C. The air is flowing through the duct with convective heat transfer coefficient of 10 W/m²·K. The temperature indicated by the thermometer is not the true temperature but an equilibrium temperature arrived at after radiation and convection balance out each other. If the duct walls are at 6 °C, find the error in air temperature in degree centigrade using this mercury thermometer.

[Ans: $Error = 5.6727\,°C$]

1P-13 What do you know about Heat Deflection Temperature? Can it be an important consideration in an injection molding process?

1P-14 Using Smith relation (Eq. 1.13) find the thermal conductivity of water in unit of Btu/hr·ft°F at 68 °F.

[Ans: k=0.399 Btu/hr·ft°F]

1P-15 Compare and contrast the heat capacities for water, gold, iron, asbestos, corck, and ebonite rubber at 20 °C. Water at 20 °C: $\rho =1000$ kg/m³, $c_p= 4181$ J/k·K
Gold at 20 °C: $\rho = 19300$ kg/m³, $c_p= 129$ J/k·K
Iron at 20 °C: $\rho = 7870$ kg/m³, $c_p= 452$ J/k·K
Asbestos at 20 °C: $\rho = 383$ kg/m³, $c_p= 816$ J/k·K
Corck at 20 °C: $\rho = 150$ kg/m³, $c_p= 1880$ J/k·K
Rubber (Ebonite) at 20 °C: $\rho =1150$ kg/m³, $c_p= 2009$ J/k·K

[Ans: Water at 20 °C: $\rho \cdot c_p = 4.18$ MJ/m³·K]

REFERENCES

J. C. Han, S. Dutta, and S. Ekkad, Gas turbine heat transfer and cooling technology, Taylor & Francis, USA, 2000.

B. Lakshminarayana, Fluid dynamics and heat transfer of turbomachinery, John Wiely & Sons, USA 1996.

M. P. Boyce, Gas Turbine Engineering Handbook, Gulf Professional Publishing, 2002.

L. Prandtl, *Über* Flüssigkeitsbewegungen bei sehr kleiner Reibung, Internationaler Mathematischer Kongress, Heidelberg, 1904.

W. Nusselt, Das Grundgesetz des Wärmeübergangs. GesundheitsIngenieur. 38, 477–482, 490–496, 1915.

M. Kaviany, Heat Transfer Physics, Cambridge, UK, 2008.

W. H. McAdams, Heat Transmission, Third Edition, McGraw-Hill Book Co., Inc., New York, USA, 1954.

C. L. Tien and J. H. Lienhard, Statistical Thermodynamics, Hemisphere, New York, USA, 1985.

C. Truesdell, Essays in the History of Mechanics, Springer-Verlag New York Inc., Germany, 1968.

J. B. J. Fourier, Theorie Analytique de la chaleur, Gauthier-Villars, 1822, English translation by Alexander Freeman, Dover Publications, Inc., New York, USA, 1955.

N. V. Tsederberg, Thermal Conductivity of Gases and Liquids, The M.I.T. Press, Cambridge, Mass., USA, 1965.

H. Gröber, S. Erk, and U. Grigull, Fundamentals of Heat Transfer. New York: McGraw-Hill, USA, 1961.

F. P. Incropera and D. P. DeWitt, Introduction to Heat Transfer. 4th ed. New York: John Wiley & Sons, USA, 2002.

V. S. Arpaci, Conduction Heat Transfer, Addison-Wesley, Reading, MA, USA, 1966.

W. Sutherland, The viscosity of gases and molecular force, Philosophical Magazine, S. 5, 36, pp. 507–531, 1893.

S. U. S. Choi and J. A. Eastman, Enhancing Thermal Conductivity Of Fluids With Nanoparticles, SME International Mechanical Engineering Congress and Exposition, November 12–17, San Francisco, CA, 1995.

H. C. Brinkman, The Viscosity of Concentrated Suspensions and Solutions, J. Chem. Phys., vol. 20, no. 4, pp. 571, 1952.

B. C. Pak and Y. I. Cho, Hydrodynamic and Heat Transfer Study of Dispersed Fluids with Submicron Metallic Oxide Particles, Exp. Heat Transfer, vol. 11, no. 2, pp. 151–170. 1998.

Y. Xuan and Q. Li, Investigation on convective heat transfer and flow features of nano-fluids, ASME J. Heat Transfer, vol. 125, no. 1, pp. 151–155, 2003.

Y. Ding, H. Alias, D. Wen, and R. A. Williams, Heat transfer of aqueous suspensions of carbon nanotubes (CNT nanofluids), Int. J. Heat Mass Transf., vol. 49, no. 1–2, pp. 240–250, 2006.

J. A. Eastman, U. S. Choi, S. Li, G. Soyez, L. J. Thompson, and R. J. Melfi, Novel thermal properties of nanostructured materials, J. Metastable Nanocryst. Mater., vol. 2–6, pp. 629–634, 1999.

S. P. Jang and S. U. S. Choi, Cooling performance of a micro channel heat sink with nanofluids, Appl. Therm. Eng., vol. 26, no. 17–18, pp. 2457–2463, 2006.

J. Koo and C. Kleinstreuer, A new thermal conductivity model for nanofluids, J. Nanopart. Res., vol. 6, no. 6, pp. 577–588, 2004.

J. C. Maxwell, A Treatise on Electricity and Magnetism. Oxford: Clarendon Press, 1873.

S. Chapman and T. G. Cowling, The Mathematical Theory of Non-Uniform Gases (3rd ed.), Cambridge University Press, 1970.

Q. Xue, Model for thermal conductivity of carbon nanotube-based composites. Phys B Condens Matter 368:302–307, 2005.

J. Hone, Carbon nanotubes: thermal properties, in Dekker Encyclopedia of Nanoscience and Nanotechnology, pp. 603–610, 2004.

R. S. Ruoff and D. C. Lorentz, Mechanical and thermal properties of carbon nanotubes, Carbon, vol. 33, no. 7, pp. 925–930, 1995.

D. Annaratone, Dimensional Analysis. In: Engineering Heat Transfer. Springer, Germany, 2010.

J. F. D. Smith, The Thermal Conductivity of Liquids,Trans. ASME, 58, 719, 1936.

2 Heat Conduction Equation

This chapter deals with the theoretical and mathematical aspects of thermal diffusion. We will develop a generalised three-dimensional heat conduction equation. We will also discuss some important boundary conditions used in heat conduction analysis. After finishing this chapter one should be able to:

- Drive the three-dimensional heat conduction equation in Cartesian coordinates.
- Understand different boundary conditions used in mathematical analysis of heat conduction problems.

We can classify the heat transfer problems as either one-dimensional, two-dimensional, or three-dimensional depending on the boundary conditions and the size of the object in consideration. The three-dimensional heat conduction problem is that the temperature is three major coordinate directions. The temperature distribution at any location in this general can be specified through the Cartesian coordinate system; the cylindrical coordinate system; and the spherical coordinate system, depending on the geometry of the object. In the previous chapter, we learned that heat conduction can be described as a transfer of thermal energy at atomic, molecular, and lattice structural levels. This transfer of energy is driven by the thermal potential which is described by the temperature difference. Unlike temperature, heat-flux is a directional quantity and also has the magnitude.

2.1 THREE-DIMENSIONAL HEAT CONDUCTION EQUATION

We can develop the heat conduction equation in Cartesian spherical and cylindrical coordinates. However, a first course in heat transfer may resort only to the development of partial differential equations for three-dimensional heat conduction problems in Cartesian coordinates. Heat flux is a vector quantity and in Cartesian coordinates:

$$\mathbf{q} = q_x \hat{i} + q_y \hat{j} + q_z \hat{k}$$

where $\hat{i}, \hat{j}, \hat{k}$ are unit vectors in x, y, and z-directions, respectively.

Figure 2.1 shows the three dimensional system used for the derivation of heat conduction equation.

DOI: 10.1201/9781003428404-2

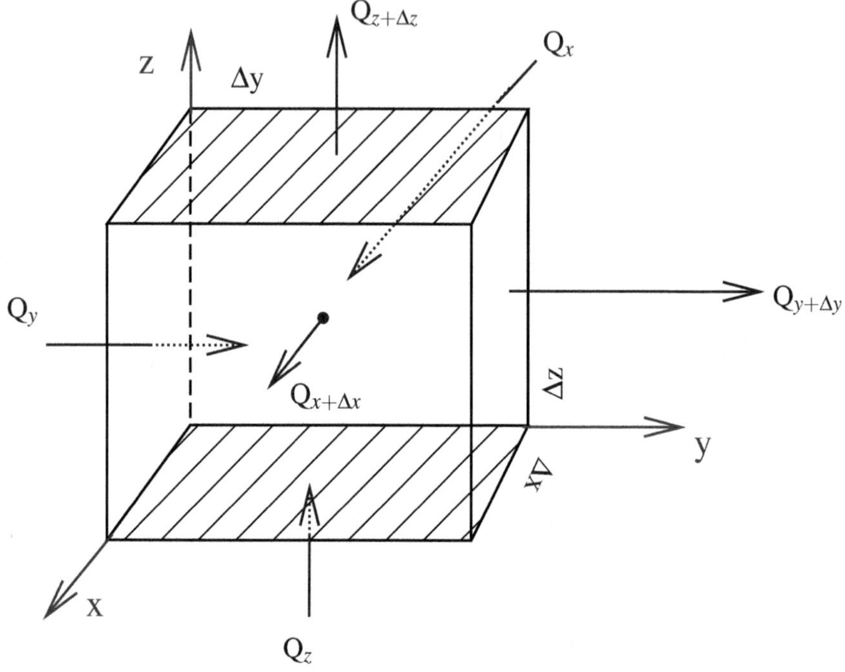

Figure 2.1   The system used for heat conduction equation derivation.

We use the First law of thermodynamics

$$\dot{E}_{in} - \dot{E}_{out} + \dot{E}_g = \dot{E}_{stored}$$

Internal energy conversion into heat can be defined as

$$\dot{E}_g = \tilde{q}\forall = \tilde{q}(\Delta x \Delta y \Delta z)$$

and the energy stored is

$$\dot{E}_{stored} = \rho c \forall \frac{\partial T}{\partial t}$$

where $\forall$ we have volume of system as $\forall = (\Delta x \Delta y \Delta z)$. We use Brook Taylor's series (1715) to estimate the heat going out of the system

$$Q_{x+\Delta x} = Q_x + \frac{\partial Q_x}{\partial x}\frac{(\Delta x)}{1!} + \frac{\partial^2 Q_x}{\partial x^2}\frac{(\Delta x)^2}{2!} \cdots$$

Ignoring higher order terms we have

$$Q_{x+\Delta x} = Q_x + \frac{\partial Q_x}{\partial x}\Delta x$$

Similarly, we can write for $Q_{y+\Delta y}$, and $Q_{z+\Delta z}$:

$$Q_{y+\Delta y} = Q_y + \frac{\partial Q_y}{\partial y}\Delta y$$

and

$$Q_{z+\Delta z} = Q_z + \frac{\partial Q_z}{\partial z}\Delta z$$

Inserting above terms into the energy equation, we have

$$\left[\cancel{Q_x}+\cancel{Q_y}+\cancel{Q_z}\right]_{in} - \left[\cancel{Q_x}+\frac{\partial Q_x}{\partial x}\Delta x+\cancel{Q_y}+\frac{\partial Q_y}{\partial x}\Delta y+\cancel{Q_z}+\frac{\partial Q_z}{\partial z}\Delta z\right]_{out} + \tilde{q}\forall = \rho c \forall \frac{\partial T}{\partial t}$$

$$-\frac{\partial Q_x}{\partial x}\Delta x - \frac{\partial Q_y}{\partial x}\Delta y - \frac{\partial Q_z}{\partial z}\Delta z + \tilde{q}\forall = \rho c \forall \frac{\partial T}{\partial t}$$

Introducing the Fourier law of heat conduction in energy balance, we have

$$Q_x = -k\frac{\partial T}{\partial x}(\Delta y \Delta z)$$

$$Q_y = -k\frac{\partial T}{\partial y}(\Delta x \Delta z)$$

$$Q_z = -k\frac{\partial T}{\partial z}(\Delta x \Delta y)$$

This leads to

$$-\frac{\partial}{\partial x}\left[-k\frac{\partial T}{\partial x}\right]\Delta y \Delta z \Delta x - \frac{\partial}{\partial y}\left[-k\frac{\partial T}{\partial y}\right]\Delta x \Delta z \Delta y - \frac{\partial}{\partial z}\left[-k\frac{\partial T}{\partial z}\right]\Delta x \Delta y \Delta z + \tilde{q}\forall = \rho c \forall \frac{\partial T}{\partial t}$$

$$\frac{\partial}{\partial x}\left[k\frac{\partial T}{\partial x}\right]\forall + \frac{\partial}{\partial y}\left[k\frac{\partial T}{\partial y}\right]\forall + \frac{\partial}{\partial z}\left[k\frac{\partial T}{\partial z}\right]\forall + \tilde{q}\forall = \rho c \forall \frac{\partial T}{\partial t}$$

$$\frac{\partial}{\partial x}\left(k\frac{\partial T}{\partial x}\right) + \frac{\partial}{\partial y}\left(k\frac{\partial T}{\partial y}\right) + \frac{\partial}{\partial z}\left(k\frac{\partial T}{\partial z}\right) + \tilde{q} = \rho c \frac{\partial T}{\partial t}$$

$$\frac{\partial}{\partial x}\left(\frac{\partial T}{\partial x}\right) + \frac{\partial}{\partial y}\left(\frac{\partial T}{\partial y}\right) + \frac{\partial}{\partial z}\left(\frac{\partial T}{\partial z}\right) + \frac{\tilde{q}}{k} = \frac{\rho c}{k}\frac{\partial T}{\partial t}$$

$$\frac{\partial}{\partial x}\left(\frac{\partial T}{\partial x}\right) + \frac{\partial}{\partial y}\left(\frac{\partial T}{\partial y}\right) + \frac{\partial}{\partial z}\left(\frac{\partial T}{\partial z}\right) + \frac{\tilde{q}}{k} = \frac{1}{\alpha}\frac{\partial T}{\partial t} \tag{2.1}$$

The $\tilde{q}$ represents the volumetric Internet energy conversion. This term is often called the heat generation term, but this classification is not correct as per the first law of thermodynamics. The $\tilde{q}$ can be understood as rise in internal energy. For example, an enormous amount of internal energy is liberated from the fuel elements of the nuclear rods due to the nuclear fission process. This large amount of energy is transferred to the water or steam that acts as a heat source for the nuclear power plants. Another example of internal energy conversion is the metabolism process in which the food chemical energy is converted into the internal energy which is absorbed by the human or animal body. In heat transfer analysis, we consider the internal energy conversion process as a volumetric phenomenon that is it happens throughout the body of the medium or an object. the rate of heat generation in a medium is specified as per unit volume and is reported in units of W/m^3 or $Btu/h{\cdot}ft^3$.

The quantity ρc_p is called the heat capacity, and it represents the thermal energy storage per unit volume for the material.

Heat Conduction Equation in Cartesian, Cylindrical and Spherical Coordinates

Heat conduction equation in Cartesian coordinates (x,y,z):

$$\frac{\partial}{\partial x}\left(\frac{\partial T}{\partial x}\right) + \frac{\partial}{\partial y}\left(\frac{\partial T}{\partial y}\right) + \frac{\partial}{\partial z}\left(\frac{\partial T}{\partial z}\right) + \frac{\tilde{q}}{k} = \frac{1}{\alpha}\frac{\partial T}{\partial t}$$

Heat conduction equation in cylindrical coordinates (r,θ,z):

$$\frac{1}{r}\left[\frac{\partial}{\partial r}\left(kr\frac{\partial T}{\partial r}\right)\right] + \frac{1}{r^2}\left[\frac{\partial}{\partial\theta}\left(k\frac{\partial T}{\partial\theta}\right)\right] + \frac{\partial}{\partial z}\left(k\frac{\partial T}{\partial z}\right) + \tilde{q} = \rho c\frac{\partial T}{\partial t}$$

In case of constant thermal conductivity, the equation will take form:

$$\frac{\partial^2 T}{\partial r^2} + \frac{1}{r}\frac{\partial T}{\partial r} + \frac{1}{r^2}\frac{\partial^2 T}{\partial\theta^2} + \frac{\partial^2 T}{\partial z^2} + \frac{\tilde{q}}{k} = \frac{1}{\alpha}\frac{\partial T}{\partial t}$$

Heat conduction equation in spherical coordinates (r,θ,φ):

$$\frac{1}{r^2}\left[\frac{\partial}{\partial r}\left(kr^2\frac{\partial T}{\partial r}\right)\right] + \frac{1}{r^2\sin\theta}\left[\frac{\partial}{\partial\theta}\left(k\sin\theta\frac{\partial T}{\partial\theta}\right)\right] + \frac{1}{r^2\sin^2\theta}\left[\frac{\partial}{\partial\varphi}\left(k\frac{\partial T}{\partial\varphi}\right)\right]$$

$$+\tilde{q} = \rho c\frac{\partial T}{\partial t}$$

In case of constant thermal conductivity, the equation will take form:

$$\frac{1}{r}\frac{\partial^2(rT)}{\partial r^2} + \frac{1}{r^2\sin\theta}\frac{\partial}{\partial\theta}\left(\sin\theta\frac{\partial T}{\partial\theta}\right) + \frac{1}{r^2\sin^2\theta}\frac{\partial^2 T}{\partial\theta^2} + \frac{\tilde{q}}{k} = \frac{1}{\alpha}\frac{\partial T}{\partial t}$$

Definitions

Steady-state or quasistatic heat transfer: Heat conduction in a medium is considered as steady when the temperature does not change with time.

Unsteady or transient heat transfer: If the medium is experiencing sudden or drastic temperature changes in time then such a thermal diffusion process cannot be modeled as a steady-state phenomenon.

On-dimensional (1D) heat conduction: Thermal diffusion in a medium can be modeled as a one-dimensional phenomenon if diffusion of heat is significant in one dimension only.

Two-dimensional (2D) heat conduction: If the object is thin in one direction like that in the case of a thin strip or tape, then thermal diffusion in a medium can be assumed to be limited to a two-dimensional process only.

Three-dimensional (3D) heat conduction: should be considered if an object has considerable dimensions and diffusion can happen in any direction.

2.2   SOME COMMON BOUNDARY CONDITIONS

The common types of heat transfer boundary conditions are either specified temperature or heat flux or the combination of both. Mathematically we called these boundary conditions as Dirichlet, Neumann, and Robin boundary conditions.

2.2.1   DIRICHLET BOUNDARY CONDITION

This is most widely used boundary condition and easiest to work with. Practically, temperature of the surface is first state variable to measure for heat transfer problems. A Dirichlet boundary condition at a wall (x = 0) can be represented by T(0, t) = Tw where Tw is the known or specified wall temperature.

2.2.2   NEUMANN BOUNDARY CONDITION

Let say the solid boundary starts at (x = 0) where a constant specified wall heat flux ($\mathbf{q}_w$) is applied. Applying Fouriers law, the Neumann condition is

$$\left.\frac{\partial T}{\partial x}\right|_{x=0} = \frac{-\mathbf{q}_w}{k}$$

For case of insulation, the heat flux is zero and the temperature gradient is also zero.

2.2.3   ROBIN BOUNDARY CONDITION

Robin boundary condition is a combination of Dirichlet boundary condition and Neumann boundary condition. Sometimes the convective heat transfer coefficient

is known, and it is desired to know the heat transfer in the wall. This is often the case
if conjugate heat transfer needs to be estimated.

$$-k\frac{\partial T}{\partial x}\bigg|_{x=0} = h\left[T(0,t) - T_\infty\right]$$

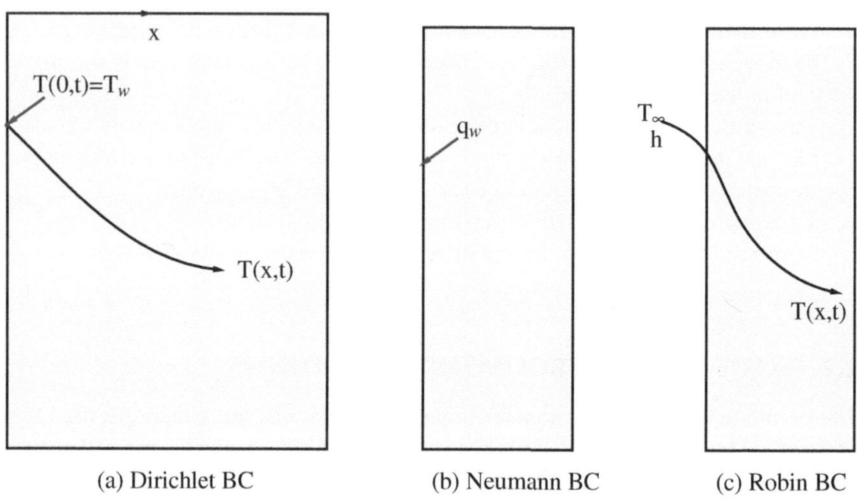

Figure 2.2   Three types of boundary conditions often used in heat transfer analysis.

Figure 2.2 shows schematically the three types of boundary conditions.

2.3   SCALE ANALYSIS FOR HEAT TRANSFER PROBLEMS

We can represent the flow and heat transfer quantities, properties, and dimensions
into basic or fundamental dimensions like Mass (M), Length (L), Time (τ), Temperature (θ), Force (F), Enthalpy, or heat (H). The length or displacement related
quantities, the flow and medium related quantities, and the heat and mass diffusion
related quantities are tabulated in Table 2.1.

In analysing heat transfer problems, we can arrive at useful dimensionless numbers using a balance of basic forces of heat fluxes, performing scale analysis on
differential equations, and also by dimensional analysis. Here we show the benefit of
such methods as they give speedy and convenient tools to reach meaningful results.
If we balance the convective and conductive heat fluxes at the solid wall exposed to
convective environment, we have

$$q_{conv} \sim q_{cond}$$

$$h \cdot \Delta T \sim k_b \cdot \frac{\Delta T}{\ell}$$

TABLE 2.1

Some Quantities and Their Dimensions

Length-related quantities

Symbol	Quantity	Dimensions
A, S	Area of surface, cross section	L^2
b	Breadth, perimeter	L
D, r	Diameter, radius	L
L	Length	L
x_m	Mean free path of molecule	L

Flow and material related quantities

Symbol	Quantity	$M-L-\tau$ $T-F-H$	$M-L-\tau-\theta$
E	Elastic modulus	F/L^2	$M/L\tau^2$
F	Force	F	ML/τ^2
g, a	Acceleration	L/τ^2	L/τ^2
G	Mass velocity, $\dot{m}/A$	$M/\tau L^2$	$M/\tau L^2$
p	Pressure	F/L^2	$M/L\tau^2$
V	Velocity	L/τ	L/τ
$\forall$	Volume	L^{-3}	L^{-3}
$\dot{m}$	Mass flow rate	M/τ	M/τ
μ	Viscosity	$M/L\tau$	$M/L\tau$
ρ	Density (mass per unit volume)	M/L^3	M/L^3
σ	Surface tension	F/L	M/τ^2
τ_{xy}	Shear force per unit area	F/L^2	$M/L\tau^2$

Heat and mass diffusion-related quantities

Symbol	Quantity	$M-L-\tau$ $T-F-H$	$M-L-\tau-\theta$
c	Specific heat, on mass basis	$H/M\theta$	$L^2/\tau^2\theta$
D_v	Diffusivity, volumetric	L^2/τ	L^2/τ
dT/dx	Temperature gradient	θ/L	θ/L
h, U	Coefficients of heat transfer	$H/\tau L^2\theta$	$M/\tau^3\theta$
k	Thermal conductivity	$H/\tau L\theta$	$ML/\tau^3\theta$
$k/\rho c$	Thermal diffusivity	L^2/τ	L^2/τ
Q	Quantity of heat	H	ML^2/τ^2
Q	Steady rate of heat flow	H/τ	MLL^2/τ^3
R	Thermal resistance	$\theta\tau/H$	$\theta\tau^3/ML^2$
T	Temperature	θ	θ
β	Coefficient of expansion	$1/\theta$	$1/\theta$

where h is convective heat transfer coefficient, ℓ is the characteristic length scale, ΔT is the temperature difference, and k_b is the thermal conductivity of body.

$$\frac{h \cdot \ell}{k_b} \sim 1$$

This shows that lefthand side term is dimensionless number. In heat transfer literature, we identify it as *Biot number* if thermal conductivity of solid material used, on the other hand if thermal conductivity of fluid is used the same parameter is called the Nusselt number.

$$Bi = \frac{h \cdot \ell}{k_b}$$

In engineering applications, we can also arrive at dimensionless parameters by analysing the differential equations and doing a representative scale analysis. For example, in a thermal energy storage application, one may consider only the 1D heat conduction equation of the form without internal energy conversion term:

$$\frac{\partial}{\partial x}\left(\frac{\partial T}{\partial x}\right) = \frac{1}{\alpha}\frac{\partial T}{\partial t} \tag{2.2}$$

The representative scales for x, time, and temperature are

$$\partial x \sim \ell,\ t \sim \tau,\ \partial T \sim \Delta T$$

where τ has the dimensions of time. We now substitute these scales into the differential equation 2.2, which leads to form

$$\frac{\partial}{\partial x}\left(\frac{\partial T}{\partial x}\right) = \frac{1}{\alpha}\frac{\partial T}{\partial t}$$

$$\frac{1}{\ell}\left(\frac{\Delta T}{\ell}\right) \sim \frac{1}{\alpha}\left(\frac{\Delta T}{\tau}\right)$$

$$\left(\frac{\alpha \cdot \tau}{\ell^2}\right) \sim 1$$

The left-hand side term is dimensionless number, and in heat transfer literature we identify this parameter as *Fourier number*. This shows that scale analysis offers an elegant way to arrive at insightful parameters, which can shed light on the underlying physics of the problem.

$$Fo = \left(\frac{\alpha \cdot \tau}{\ell^2}\right)$$

PROBLEMS

2P-1 The heat transfer in cylindrical coordinates:

$$\frac{1}{r}\left[\frac{\partial}{\partial r}\left(kr\frac{\partial T}{\partial r}\right)\right] + \frac{1}{r^2}\left[\frac{\partial}{\partial \theta}\left(k\frac{\partial T}{\partial \theta}\right)\right] + \frac{\partial}{\partial z}\left(k\frac{\partial T}{\partial z}\right) + \tilde{q} = \rho c\frac{\partial T}{\partial t}$$

Answer briefly:

(i) How many directions the above equation represent?

(ii) Is the above equation applicable for steady state conduction?

(iii) What is the meaning of the term $\tilde{q}$?

(iv) Is this equation applicable for materials with variable thermal conductivity?

2P-2 The heat transfer in spherical coordinates:

$$\frac{1}{r^2}\left[\frac{\partial}{\partial r}\left(kr^2\frac{\partial T}{\partial r}\right)\right] + \frac{1}{r^2\sin\theta}\left[\frac{\partial}{\partial \theta}\left(k\sin\theta\frac{\partial T}{\partial \theta}\right)\right]$$
$$+ \frac{1}{r^2\sin^2\theta}\left[\frac{\partial}{\partial \varphi}\left(k\frac{\partial T}{\partial \varphi}\right)\right] + \tilde{q} = \rho c\frac{\partial T}{\partial t}$$

Answer briefly:

(i) How many directions the above equation represent?

(ii) Is the above equation applicable for steady state conduction?

(iii) What is the meaning of the term $\tilde{q}$?

(iv) Is this equation applicable for materials with variable thermal conductivity?

2P-3 Melting and freezing are the heat conduction problems in which the phase change is happening. The important thermophysical properties and quantities are the temperature difference, latent heat of melting/freezing, specific heat, etc. Suggest a dimensionless parameter which in your opinion might be important.

REFERENCES

B. Taylor, Methodus Incrementorum Directa et Inversa (Direct and Reverse Methods of Incrementation) (in Latin) 1715.

R. Aris, Vectors, Tensors, and the Basic Equations of Fluid Mechanics, Prentice-Hall, Inc., Englewood Cliffs, N.J., USA, 1962.

H. S. Carslaw and J. C. Jaeger, Conduction of Heat in Solids, Oxford Press, UK, 1957.

J. D. Knudsen and D. L Katz, Fluid dynamics and heat transfer. McGraw-Hill, New York, USA, 1958.

3 Steady-State One-Dimensional Heat Conduction

Analytical models are very important in heat transfer, especially they are used to validate the experimental data and to calibrate the experiments setups. We devote this chapter to steady-state models for simple one-dimensional cases. After finishing this chapter, one should be able to:

The learning outcomes:
- Calculate the heat transfer by conduction through solids, and static liquids.
- Calculate the critical thickness of insulation.
- Analyse the one-dimensional heat transfer problems with variable thermal-conductivity.

In this chapter, we consider the situations in which heat is transmitted by thermal diffusion. For the sake of simplicity, we focus in this chapter only on one-dimensional heat transfer with quasi-static or steady-state conditions. The term one-dimensional means that only one coordinate is enough to describe the heat transfer. Such an assumption can provide meaningful results for seemingly complex heat transfer situations. The system is assumed to be in quasi-static or steady-state conditions if the temperature at each point is not depending on time. Mathematically, we can write $\partial T / \partial t$.

We first develop a mathematical model for the conduction through plane walls. We will then proceed to the concept of analogy between heat transfer and current flow, which helps us in solving a variety of engineering heat transfer problems like multilayer walls and estimation of the overall heat transfer coefficient. We will also solve the hollow cylinder and sphere problems including the composite cylinder or sphere cases. Many materials have thermal conductivity varying in them and the assumption of constant thermal conductivity is no longer valid. We will solve a simple case of such kind of material as an exemplary case. Further, we will also focus on thermal contact resistance and internal heat conversion cases. Finally, heat transfer through fins or extended surfaces will be discussed.

We can simplify the three-dimensional heat conduction equation developed in the previous chapter for case of constant thermal conductivity and no internal heat conversion:

DOI: 10.1201/9781003428404-3

$$\frac{1}{\alpha}\left(\frac{\partial T}{\partial t}\right) = \nabla^2 T \tag{3.1}$$

For a steady-state condition with constant thermal conductivity, the heat conduction equation can be cast into form called **Poisson Equation**:

$$\nabla^2 T + \frac{\tilde{q}}{k} = 0 \tag{3.2}$$

In case of steady-state with constant thermal conductivity, and no internal energy conversion the heat conduction equation can be cast into form called **Laplace Equation**:

$$\nabla^2 T = 0 \tag{3.3}$$

3.1 CONDUCTION THROUGH PLANE WALL

Fourier proposed the model of one-dimensional conduction heat transfer as:

$$\mathbf{q} = \frac{Q}{A} = -k\,\mathbf{grad}\,T$$

We consider first the conduction through a plane wall, i.e. we ignore the conduction in the depth of the wall and we also assume homogeneous, isotropic thermal conductivity value.

The difference in wall temperature $T_1 > T_2$, causing the heat to flow from high temperature to low temperature.

$$\int_0^L \left(\frac{Q}{A}\right) dx = -k \int_{T_1}^{T_2} dT$$

$$\left(\frac{Q}{A}\right) L = -k(T_2 - T_1)$$

$$Q = \frac{kA(T_1 - T_2)}{L} = \frac{(T_1 - T_2)}{L/kA}$$

3.2 ANALOGY BETWEEN HEAT TRANSFER AND CURRENT FLOW

According to Fourier Law, Heat conduction is formulated as:

$$Q_{cond} = \frac{kA(T_1 - T_2)}{L} = \frac{(T_1 - T_2)}{L/kA}$$

Ohm's law states that electric current through a conducting body between two potentials is directly proportional to the voltage across the two potential points as:

$$\Delta V = I . \sum R$$

where V is the voltage potential. The current flow is therefore:

$$I = \frac{\Delta V}{\sum R}$$

Thermal conduction and electricity conduction phenomenon has similarities as both require a potential difference we can cast the heat transfer equations and formulate heat transfer equations like Ohm's law.

$$Q_{cond} = \frac{\Delta T}{L/kA}$$

where conduction resistance is

$$R_{cond} = L/kA$$

The convection according to Newton's law of cooling is

$$Q_{conv} = hA(T_w - T_\infty)$$

We can rearrange this equation as

$$Q_{conv} = \frac{(T_w - T_\infty)}{1/hA} = \frac{\Delta T}{1/hA}$$

The convective resistance is

$$R_{conv} = 1/hA$$

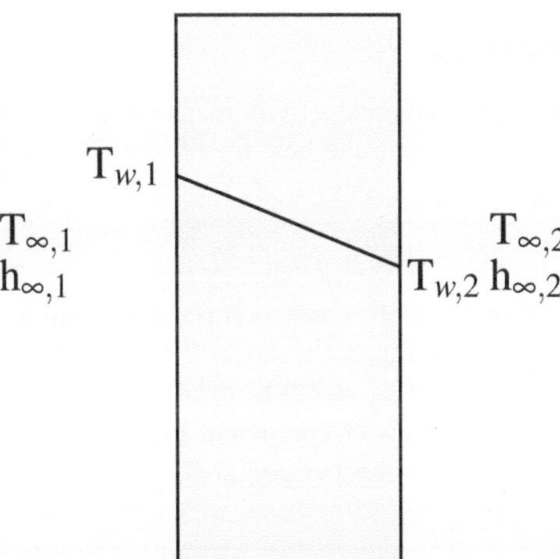

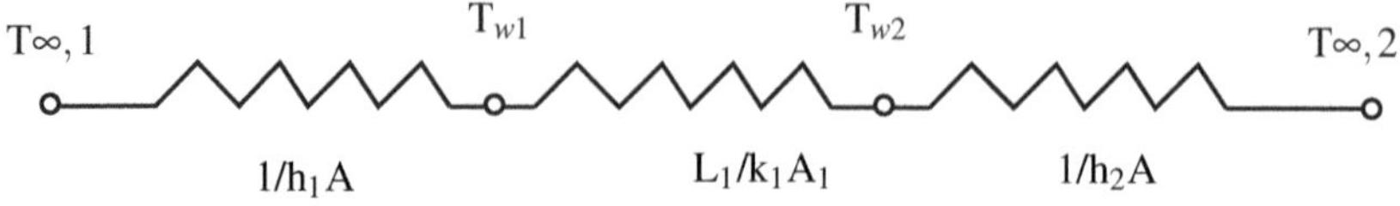

Figure 3.1   Thermal network.

Figure 3.1 shows a heat conducting wall with wall temperatures and surrounding fluid temperatures along with respective thermal network.

The radiative heat transfer from surface to surrounding is defined as

$$Q_{rad} = \varepsilon \sigma A (T_w^4 - T_{surr}^4)$$

We can expand the temperature terms algebraically as

$$Q_{rad} = \varepsilon \sigma A (T_w^4 - T_{surr}^4) \simeq \left[\varepsilon \sigma A (T_w^2 + T_{surr}^2)(T_w + T_{surr}) \right] (T_w - T_{surr})$$

Introducing radiative heat transfer coefficient h_{rad} into above equation

$$Q_{rad} = \frac{(T_w - T_{surr})}{1/\left[\varepsilon \sigma A (T_w^2 + T_{surr}^2)(T_w + T_{surr})\right]} \approx \frac{(T_w - T_{surr})}{1/h_{rad}.A}$$

$$h_{rad} = \left[\varepsilon \sigma (T_w^2 + T_{surr}^2)(T_w + T_{surr})\right]$$

and the radiative resistance is

$$R_{rad} = \frac{1}{h_{rad}.A}$$

Using these resistances, we can formulate a thermal network analogous to electrical network.

3.3   MULTILAYER WALL

The layers of different thermal conductivity materials are often encountered in heat transfer applications. To estimate the heat transfer, we consider now the following example.

Example 3.1

Consider the case of the wall which is formed by combining the layers of material 1 (k_1 =0.3 W/m.K) and material 2 (k_2 =3 W/m.K) together. The ambient temperatures are $T_{\infty,1}$ = 400 K and $T_{\infty,2}$ = 300 K, respectively. The outside convective heat transfer coefficient h_1 is 20 W/m^2 K and h_2 is 10 W/m^2 K. Find the heat transfer per unit area (W/m^2)). Ignore contact resistance.

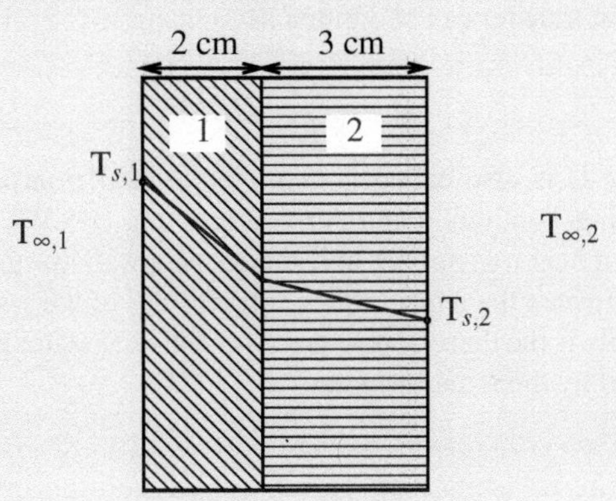

Solution Using the thickness of layers $L_1 = 0.02$ m and $L_2 = 0.03$ m, we can formulate the heat conduction resistances as:

$$R_{cond,1} = \frac{L_1}{k_1 \cdot A_1} = \frac{0.02 \, m}{0.2 \, \frac{W}{K.m} \times 1 m^2} = 0.1 \frac{K}{W}$$

$$R_{cond,2} = \frac{L_2}{k_2 \cdot A_2} = \frac{0.03 \, m}{3 \, \frac{W}{K.m} \times 1 m^2} = 0.01 \frac{K}{W}$$

$$R_{conv,1} = \frac{1}{h_1 \cdot A_1} = \frac{1}{20} \frac{K}{W}$$

$$R_{conv,2} = \frac{1}{h_2 \cdot A_2} = \frac{1}{10} \frac{K}{W}$$

$$R_T = \sum R = R_{cond,1} + R_{cond,2} + R_{conv,1} + R_{conv,2} = 0.26 \frac{K}{W}$$

$$Q = \frac{T_{\infty,1} - T_{\infty,2}}{R_T} = 384.61 \, W$$

3.4 OVERALL HEAT TRANSFER COEFFICIENT

It is convenient to describe the heat transfer via a cumulative coefficient that represents both heat conduction and convective heat transfer in many heat transfer applications. This combined heat transfer coefficient is used in designs of heat exchangers and for walls comprised of layers of several materials. A symbol U historically represents it in the heat transfer literature.

The overall heat transfer coefficient is defined as:

$$\frac{1}{UA} = R_T = \sum R = \sum_{k=1}^{n} \frac{1}{h_k A_k} + \sum_{k=1}^{n} \frac{L_k}{k_k A_k}$$

and the the heat transfer can be written as

$$Q = \frac{\Delta T}{\sum R} = UA\Delta T$$

The parameter U is also called *U-factor* or *thermal transmittance* in the heat transfer and Heating-Ventilation and Air conditioning (HVAC) literature. It can be seen that the rate of heat transfer per unit surface of a wall assembly per unit temperature difference between the temperatures at both sides of the assembly. The R-value of the wall assembly is the thermal resistance for a given assembly of that wall, which is usually provided by the manufacturers.

Clothing Comfort Level

For human clothing, the industry has introduced the unit of *clo*, defined as the amount of thermal insulation in a clothing material that keep a human comfortable in a room kept at 70 °F (21.11 °C) with normal ventilation conditions. Note that a zero *clo* value corresponds to a naked human.

$$1clo = 0.155\, m^2 \cdot {}^\circ C/W$$

$$1clo = 0.88\, ft^2 \cdot h \cdot {}^\circ F/Btu$$

Example 3.2

Figure 3.2 shows a composite wall with brick sandwiched between coating layers of 1/5 inch. The brick thickness is 6 inch. The inner temperature at one

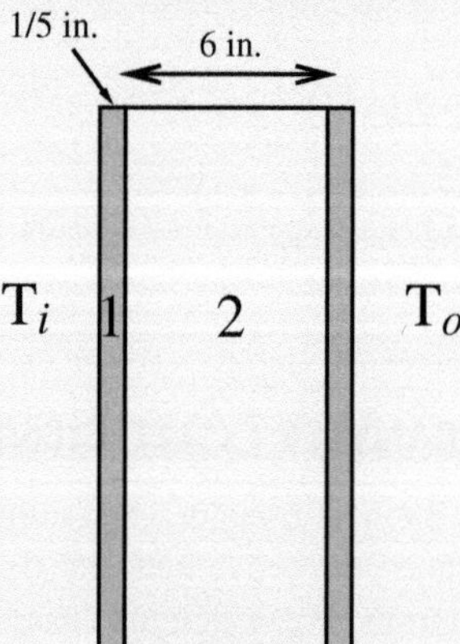

Figure 3.2   Composite wall.

side of wall is $T_i = 82°F$, and on the outer side of wall, the temperature is $T_o = 46°F$. The area of wall is 9 ft^2. The brick is labelled as 2 and coating layers are indicated as 1 in the figure. The thermal conductivities and convective heat transfer coefficients are

k_1 = 0.083 Btu/hr · ft · °F, k_2 = 0.025 Btu/hr · ft · °F, h_i = 0.7 Btu/hr · ft^2 · °F, h_o = 1.45 Btu/hr · ft^2 · °F.

Solution

Thermal resistance associated with inner convective environment:

$$R_1 = \frac{1}{h_i \cdot A} = 0.1587 \frac{°F}{Btu/hr}$$

Combined thermal resistance associated with two layers of coated material:

$$R_2 = \frac{2 \cdot L_1}{k_1 \cdot A} = 0.044622 \frac{°F}{Btu/hr}$$

Thermal resistance associated with brick:

$$R_3 = \frac{L_2}{k_2 \cdot A} = 2.222 \frac{°F}{Btu/hr}$$

Thermal resistance associated with outer convective environment:

$$R_4 = \frac{1}{h_o \cdot A} = 0.07662 \frac{°F}{Btu/hr}$$

The heat transfer is

$$Q = \frac{(T_i - T_o)}{R_1 + R_2 + R_3 + R_4} = 14.38 Btu/hr$$

In case layer of two different materials are connected in parallel and heat is passing through them, then this is the case of parallel resistances.

Figure 3.3 shows parallel conduction through composite wall made up of two different materials.

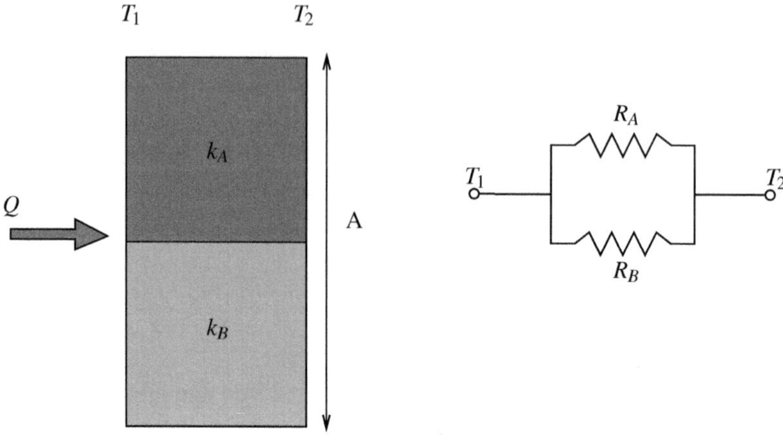

Figure 3.3   Heat diffusion through a composite wall with two materials giving heat flow two paths in parallel.

The parallel resistance can be estimated as

$$\frac{1}{R_{parallel}} = \sum_{n=1}^{m} \frac{1}{R_n}$$

Example 3.3

Figure 3.4 shows a cross-section of a long wall consist of Type 347 stainless steel cladding on one side of the hollow bricks. The indoor and the outdoor temperatures are $40°C$ and $10°C$, and the convection heat transfer coefficients on the inner and the outer sides are $h_i = 10$ W/m²·°C and $h_\infty = 25$ W/m²·°C, respectively. Find the rate of heat conduction through the wall. The thermal conductivities of steel, brick, and air are 14.3, 0.38, and 0.025 W/m·K, respectively. If the wall is comprised to 50 such pattern in vertical direction then the total heat transfer.

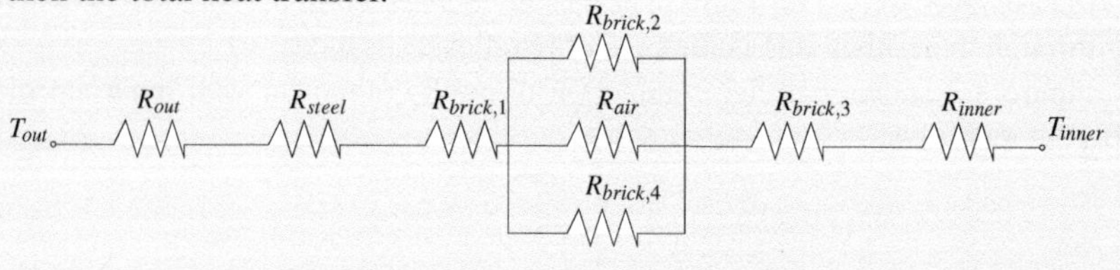

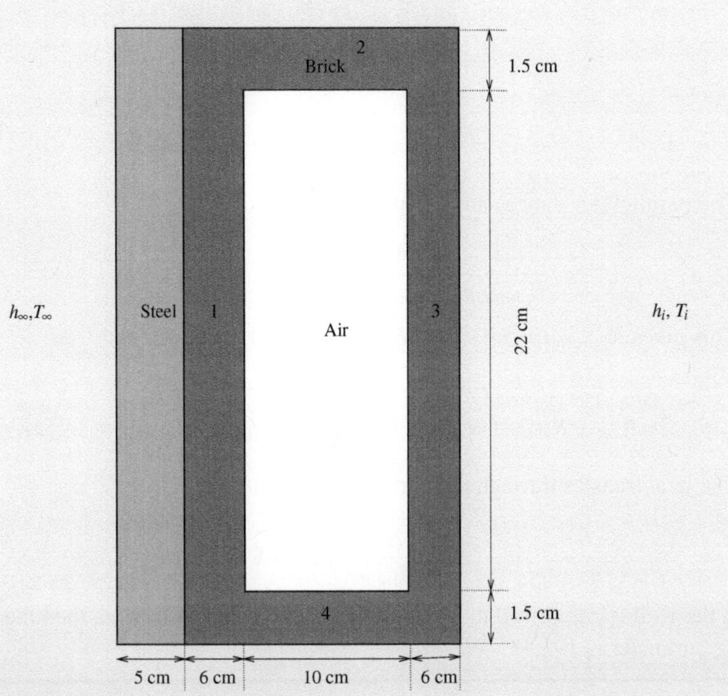

Figure 3.4   Cross section of wall.

Solution We consider the repetitive part of the wall only as shown in the schematic diagram Figure 3.4.
The individual resistances are estimated as

$$R_{out} = \frac{1}{h_\infty \cdot A} = \frac{1}{25 \times (25/100)} = 0.16 \, K/W$$

$$R_{steel} = \frac{\Delta x}{k \cdot A} = \frac{\frac{5}{100}}{14.3 \times (25/100)} = 0.01398 \, K/W$$

$$R_{brick,1} = R_{brick,3} = \frac{\frac{6}{100}}{0.38 \times (25/100)} = 0.63157 \, K/W$$

$$R_{brick,2} = R_{brick,4} = \frac{\frac{10}{100}}{\frac{0.38 \times 1.5}{100}} = 17.5438 \, K/W$$

$$R_{air} = \frac{\frac{10}{100}}{\frac{0.025 \times 22}{100}} = 18.18 \; K/W$$

$$R_{inner} = \frac{1}{h_i \cdot A} = \frac{1}{10 \times (25/100)} = 0.4 \; K/W$$

The parallel resistances are computed as

$$\frac{-1}{R_{parallel}} + \frac{1}{R_{brick,2}} + \frac{1}{R_{brick,4}} + \frac{1}{R_{air}} = 0$$

This gives $R_{parallel}$ = 5.9171 K/W, and the total resistance is

$$R_{total} = R_{out} + R_{inner} + R_{steel} + R_{brick,1} + R_{parallel} + R_{brick,3} = 7.754 K/W$$

The heat transfer through the pattern is computed as

$$Q = \frac{\Delta T}{R_{total}} = 3.8688 \; W$$

If the wall is comprised to 50 such pattern in vertical direction, then the total heat transfer is 193.44 W.

3.5 HEAT TRANSFER THROUGH A HOLLOW CYLINDER

The heat conduction through cylindrical coordinates is

$$\frac{1}{r}\left[\frac{\partial}{\partial r}\left(kr\frac{\partial T}{\partial r}\right)\right] + \frac{1}{r^2}\left[\frac{\partial}{\partial \theta}\left(k\frac{\partial T}{\partial \theta}\right)\right] + \frac{\partial}{\partial z}\left(k\frac{\partial T}{\partial z}\right) + \tilde{q} = \rho c \frac{\partial T}{\partial t}$$

Assuming steady state with internal energy conversion and one-dimensional heat conduction in radial direction only, we have

$$\frac{1}{r}\left[\frac{\partial}{\partial r}\left(kr\frac{\partial T}{\partial r}\right)\right] = 0$$

Since radial coordinate r cannot be zero, we have

$$\frac{\partial}{\partial r}\left(r\frac{\partial T}{\partial r}\right) = 0$$

Integrating once we have

$$r\frac{\partial T}{\partial r} = C_1$$

Rearranging and integrating we get

$$T(r) = C_1 \ln(r) + C_2$$

Referring to Figure 3.5 the applicable boundary conditions are

$$r = r_i, \qquad T = T_i$$
$$r = r_o, \qquad T = T_o$$

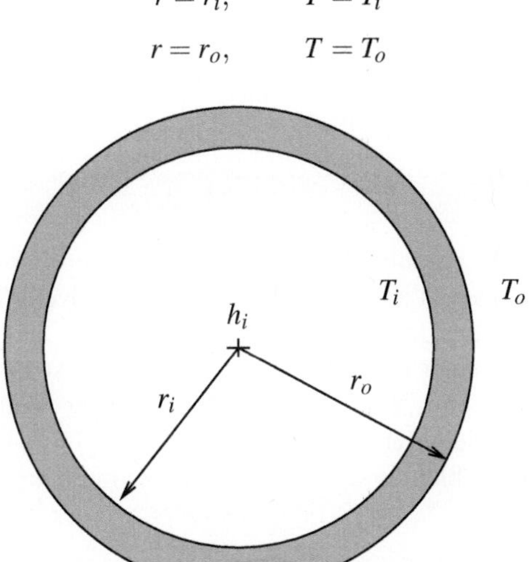

Figure 3.5 The boundary conditions for the heat conduction in the hollow cylinder case.

This give a set of equations

$$T_i = C_1 \ln(r_i) + C_2$$
$$T_o = C_1 \ln(r_o) + C_2$$

Solving above equations, we can have the constant of integrations

$$C_1 = \frac{T_o - T_i}{\ln\left(\frac{r_o}{r_i}\right)}$$

and

$$C_2 = \frac{-\ln(r_i)\left(T_o - T_i\right)}{\ln\left(\frac{r_o}{r_i}\right)} + T_i$$

Substituting them into T(r) relation gives

$$T(r) = T_i - \frac{\ln\left(r/r_i\right)}{\ln\left(r_o/r_i\right)}\left(T_i - T_o\right)$$

Using Fourier law of heat conduction with the surface area of cylinder as $A = 2\pi r L$ we get

$$Q = -kA\frac{dT}{dr} = -k(2\pi r L)\left(-\frac{(T_i - T_o)}{r\ln(r_o/r_i)}\right)$$

Rearranging

$$Q = \left(\frac{k(2\pi L)}{\ln(r_o/r_i)} \right)(T_i - T_o) = \frac{(T_i - T_o)}{R_{cyl}}$$

where R_{cyl} is thermal resistance for heat flow in wall of cylinder.

$$R_{cyl} = \frac{\ln(r_o/r_i)}{2\pi k L}$$

Example 3.4

Water is flowing in a smooth horizontal 10 m long copper pipe having 2.3 cm and 2.65 cm internal and outer diameters, respectively. The pipe is covered with a 2-cm layer of polystyrene foam insulation, and it is exposed to ambient air. If the water is flowing at 70°C and air is at 19°C, estimate the heat loss from this pipe. The convective heat transfer coefficients for pipe flow and atmospheric air are $h_i = 7895$ W/m²·K, and $h_{atm} = 3.24$ W/m²·K, respectively. The problem is schematically shown in Figure 3.6.

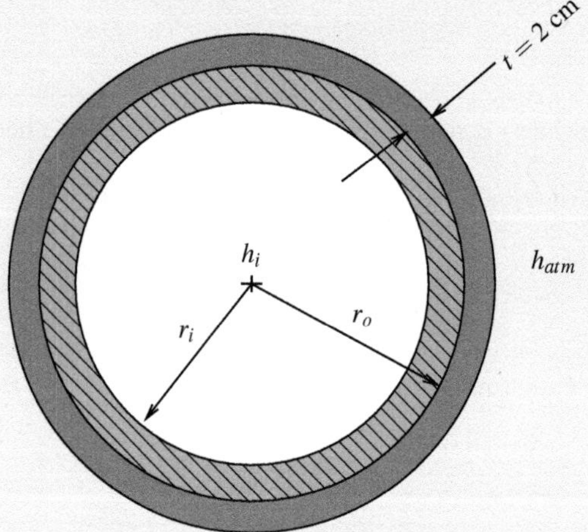

Figure 3.6   Schematic diagram of Example problem.

Solution
We can make a thermal network for heat flow as shown in circuit below. Here thermal resistances are labelled as R. Figure 3.7 shows the corresponding thermal network.

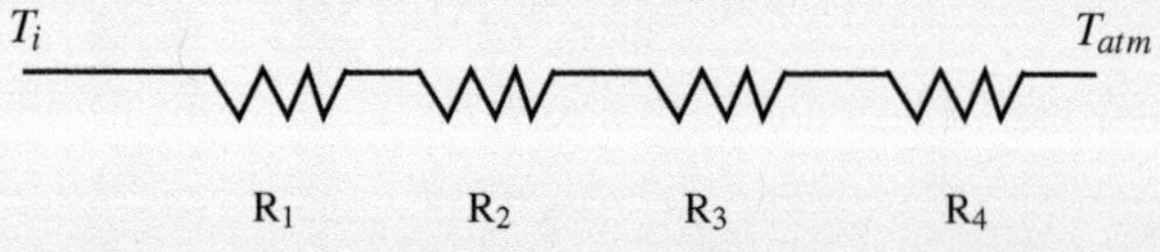

Figure 3.7   Thermal network.

Note that R_1 and Note that R_4 are convective resistances, and R_2, R_3 are conductive resistances. Thermal conductivities of copper and insulation layer are taken as $k_{copper} = 400 \ W/m \cdot K, \ k_{layer} = 0.038$.
The heat transfer can be computed as

$$Q = \frac{T_i - T_{atm}}{\frac{1}{2\pi h_i r_i L} + \frac{\ln\left(\frac{r_o}{r_i}\right)}{2 \cdot \pi \cdot k_{copper} \cdot L} + \frac{\ln\left(\frac{r_3}{r_o}\right)}{2 \cdot \pi \cdot k_{layer} \cdot L} + \frac{1}{2\pi h_{atm} r_3 L}}$$

The thermal resistances are

$$R_1 = \frac{1}{7895 \cdot 2 \cdot \pi \cdot 0.0115 \cdot 10}$$

$$R_2 = \frac{\ln\left(\frac{0.01325}{0.0115}\right)}{2 \cdot \pi \cdot 400 \cdot 10}$$

$$R_3 = \frac{\ln\left(\frac{0.03325}{0.01325}\right)}{2 \cdot \pi \cdot 0.038 \cdot 10}$$

$$R_4 = \frac{1}{3.24 \cdot 2 \cdot \pi \cdot 0.03325 \cdot 10}$$

$$Q = \frac{(70 - 19)}{\Sigma R}$$

$$Q = 84.386 \ W$$

3.6 HEAT TRANSFER THROUGH A HOLLOW SPHERE

In many engineering applications, the the fluid is stored in spherical vessel and knowing the heat transfer through such bodies is important. For such geometries, it is easy to use the heat conduction equation in spherical coordinates for engineering analysis. The heat conduction through spherical coordinates is

$$\frac{1}{r^2}\left[\frac{\partial}{\partial r}\left(kr^2\frac{\partial T}{\partial r}\right)\right] + \frac{1}{r^2 \sin\theta}\left[\frac{\partial}{\partial \theta}\left(k\sin\theta\frac{\partial T}{\partial \theta}\right)\right]$$
$$+ \frac{1}{r^2 \sin^2\theta}\left[\frac{\partial}{\partial \varphi}\left(k\frac{\partial T}{\partial \varphi}\right)\right] + \tilde{q} = \rho c\frac{\partial T}{\partial t}$$

Assuming steady state, one-dimensional heat conduction in radial direction only, we have

$$\frac{1}{r^2}\frac{d}{dr}\left(kr^2\frac{dT}{dr}\right) + \tilde{q} = 0$$

For no internal energy conversion mechanism, we have

$$\frac{1}{r^2}\frac{d}{dr}\left(kr^2\frac{dT}{dr}\right) = 0$$

For constant thermal conductivity case, we can further simplify the equation

$$\frac{d}{dr}\left(r^2\frac{dT}{dr}\right) = 0$$

Now integrating it once we have

$$r^2\frac{dT}{dr} = C_1$$

Rearranging we have

$$\frac{dT}{dr} = \frac{C_1}{r^2}$$

Now integrating it second time, we have equation of radial distribution of temperature

$$T(r) = \frac{-C_1}{r} + C_2$$

The term C_1 and C_2 appear as constants of integration. The applicable boundary conditions are

$$r = r_i, \qquad T = T_i$$

$$r = r_o, \qquad T = T_o$$

Introducing the boundary condition we can write inner and outer temperatures as

$$T_i = \frac{-C_1}{r_i} + C_2$$

$$T_o = \frac{-C_1}{r_o} + C_2$$

Solving these equations, we can write the constants of integrations in terms of known functions as

$$C_1 = \frac{(T_i - T_o)}{\left(\frac{1}{r_o} - \frac{1}{r_i}\right)}$$

$$C_2 = T_i - \frac{(T_i - T_o)}{\left[1 - \left(\frac{r_i}{r_o}\right)\right]}$$

Finally, the radial temperature distribution is

$$T(r) = T_i - \left[\frac{1 - \left(\frac{r_i}{r_o}\right)}{\left(\frac{1}{r_o} - \frac{1}{r_i}\right)}\right](T_i - T_o)$$

Using the surface area of the sphere $(A = 4\pi r^2)$, we have from Fourier's law of heat conduction

$$Q = -kA\frac{dT}{dr}$$

$$Q = -k(4\pi r^2)\frac{1}{r^2}\left[\frac{(T_i - T_o)}{\left(\frac{1}{r_o} - \frac{1}{r_i}\right)}\right]$$

Rearranging we can cast it as

$$Q = k(T_i - T_o)\left[\frac{4\pi}{\left(\frac{1}{r_i} - \frac{1}{r_o}\right)}\right] = \frac{(T_i - T_o)}{R_{sphere}}$$

where R_{sphere} is the thermal resistance of a sphere.

$$R_{sphere} = \frac{\left(\frac{1}{r_i} - \frac{1}{r_o}\right)}{4\pi k}$$

The heat equation can be rearranged as

$$\frac{Q}{4\pi r_i r_o}(r_o - r_i) = k(T_i - T_o)$$

$$\frac{Q}{A_m}(r_o - r_i) = k_m(T_i - T_o)$$

where $A_m = 4\pi r_i r_o = \sqrt{A_i A_o}$, and k_m is the mean thermal conductivity.

$$\frac{Q}{A_m} = \frac{k_m(T_i - T_o)}{(r_o - r_i)} = \frac{k_m(T_i - T_o)}{t}$$

where t is the thickness of annular gap region.

3.7 HEAT CONDUCTION THROUGH COMPOSITE PIPE OR SPHERE

We now develop the relation for heat transfer through pipe or sphere comprised of two layers of different material or metal. The thermal conductivities are k_1 and k_2. We assume that there is no gap between the layers of the material or metal, and we ignore the contact resistance as well. The fluid in the inner core of pipe has convective heat transfer coefficient h_i and fluid in ambient has convective heat transfer coefficient h_o. The Temperature are different at the surface of metal/material layers, and they are indicated in Figure 3.8. The corresponding thermal network is presented in Figure 3.9.

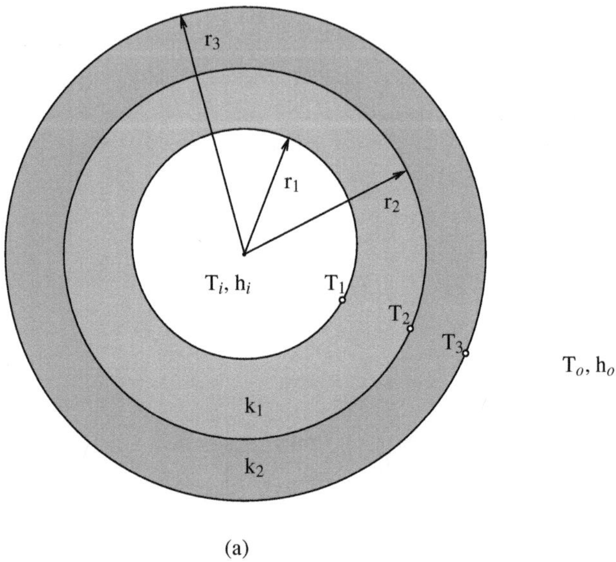

Figure 3.8   The case of composite cylinder and sphere.

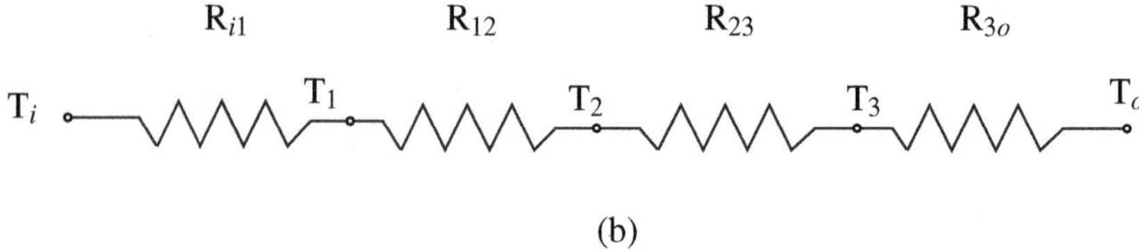

Figure 3.9   Thermal network for the case of composite cylinder and sphere.

The overall heat transfer is

$$Q = \frac{T_i - T_o}{R_{i1} + R_{12} + R_{23} + R_{3o}}$$

where R_{i1}, R_{12}, R_{23}, and R_{3o} are the thermal resistances. We first intrduce the convective heat transfer resistances

$$Q = \frac{T_i - T_o}{\frac{1}{h_i A_1} + R_{12} + R_{23} + \frac{1}{h_o A_3}}$$

For a cylinder, the area is $A = 2\pi L r$, which gives

$$Q = \frac{T_i - T_o}{\frac{1}{h_i(2\pi L r_1)} + \frac{\ln(r_2/r_1)}{2\pi k_1 L} + \frac{\ln(r_3/r_2)}{2\pi k_2 L} + \frac{1}{h_o(2\pi L r_3)}}$$

3.8 THE CONCEPT OF CRITICAL RADIUS

In many engineering applications, the insulation are added over wires, pipes and other objects to prevent heat loss or as a safety precaution. In case of electrical wires, it is desired that the wire dissipate the heat. Further, the addition of insulation over the pipe may lead to increase the heat loss from the pipe rather than reducing it. Figure 3.10 shows the schematic presentation of the insulated wire case.

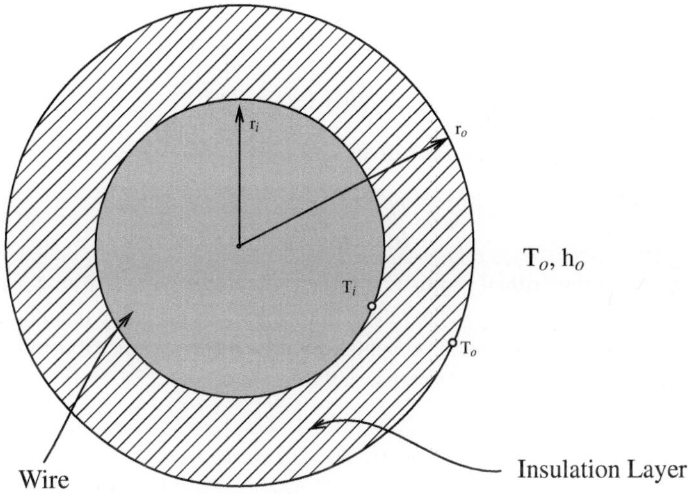

Figure 3.10   Schematic diagram of an insulated wire.

$$Q = \frac{2\pi k L(T_i - T_o)}{\ln\left(r/r_i\right) + \frac{(k/h_o \cdot r_i)}{(r/r_i)}}$$

$$Q(r) = \frac{2\pi k L(T_i - T_o)}{\ln\left(r/r_i\right) + \frac{(1/Bi)}{(r/r_i)}}$$

where Biot number is defined as

$$Bi = h_o \cdot r_i / k$$

> **Pioneers of Heat Transfer**
>
> **Jean-Baptiste Biot** works of Jean-Baptiste Biot are marked by a great diversity. Biot conducted a scientific balloon expedition, the first of its kind – in the company of Gay-Lussac, who aimed to study the variation the intensity of Earth's magnetic field altitude. Biot also wrote books on Indian and Chinese astronomy. Biot was greatly influenced by works done by Laplace. Jean-Baptiste Biot died in 1862.

For an extremum position, we set

$$\frac{\mathrm{d}}{\mathrm{d}\,r}\left(Q(r)\right) = 0$$

$$\frac{\mathrm{d}}{\mathrm{d}\,r}\left(Q(r)\right) = -\frac{2\pi kL\cdot\Delta T\left(\frac{1}{r}-\frac{r_i}{Bi\cdot r^2}\right)}{\left(\ln\left(\frac{r}{r_i}\right)+\frac{r_i}{Bi\cdot r}\right)^2} = 0$$

Rearranging the numerator for r gives

$$r = \frac{r_i}{Bi} = \frac{r_i}{h\cdot r_i/k} = \frac{k}{h}$$

This is called the critical radius and often indicated r_c in heat transfer literature, and this result is valid as long as the radius of the pipe or tube being insulated is small. Figure 3.11 shows the schematic presentation of heat transfer distribution versus the radial position in case of insulation over a wire.

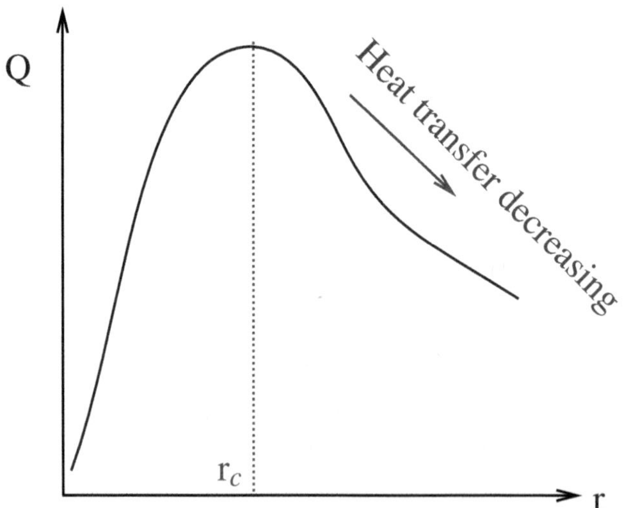

Figure 3.11   Schematic representation of heat transfer near critical thickness radius.

The important points to be noted are

If the outer radius is smaller than the critical radius value, then increasing insulation layer thickness will increase the heat loss from the wire or small cylinder.

If the outer radius is larger than the critical radius, then improving the insulation thickness will cause a reduction in heat loss from the wire or small cylinder.

Example 3.5

A wire has been insulated with material which has the thermal conductivity k=0.1 W/m·K. The outside convective heat transfer coefficient is h=25 W/m^2·K. The temperature difference between the wire surface and the outside is 300K. Find the critical thickness of insulation if wire radius is 0.001 m.

Solution The Biot number for this case is

$$Bi = \frac{h_o \cdot r_i}{k} = 2.5$$

The critical thickness radius is

$$r_c = \frac{k}{h_o} = 0.004$$

The heat transfer is

$$Q(r) = \frac{2\pi k L (T_i - T_o)}{\ln(r/r_i) + \frac{(1/Bi)}{(r/r_i)}}$$

$$Q(r) = \frac{188.49}{\ln(100r) + \frac{0.004}{r}} \tag{3.4}$$

Thus, the heat transfer loss shall be maximum when insulation thickness equal to $(r_c\text{-}r_i)$=0.003m is used.

3.9 VARIABLE THERMAL CONDUCTIVITY

In some cases, the thermal conductivity is varying inside material as an anisotropic quantity. Also, thermal conductivity can be temperature dependent. We will here analyse few simplified cases.

3.9.1 PLANE WALL

We consider the case of one-dimensional heat transfer through the plane wall without internal energy conversion. It is assumed that thermal conductivity is varying as function of temperature as $k(T) = a + b \cdot T$. From Fourier's law of heat conduction, we have

$$Q = -kA \frac{dT}{dx}$$

$$\left(\frac{Q}{A}\right) dx = -k dT$$

$$q'' dx = -k dT$$

Inserting thermal conductivity as function of temperature into above equation, we have

$$k(T) = a + b \cdot T$$

$$q'' \cdot L = -\int_{T_1}^{T_2} [a + b \cdot T] \, dT$$

$$q'' = \frac{\frac{-b(T_2^2 - T_1^2)}{2} + a(T_2 - T_1)}{L}$$

Rearranging we have

$$q'' = \frac{(bT_1 + bT_2 + 2a)(T_1 - T_2)}{2L}$$

We can treat the terms containing a and b as a new relationship for thermal conductivity

$$\tilde{k} = \frac{(bT_1 + bT_2 + 2a)}{2} = a + b\left(\frac{T_1 + T_2}{2}\right)$$

leading to form

$$q'' = \frac{\tilde{k}(T_1 - T_2)}{L}$$

$$Q = \frac{\tilde{k}A(T_1 - T_2)}{L}$$

This shows that thermal conductivity can be treated as a constant in special case of linear variation of thermal conductivity.

3.9.2 THERMAL CONDUCTIVITY VARIATION IN A CYLINDER

$$Q = -kA \frac{dT}{dr}$$

$$Q = -k(2\pi r L) \frac{dT}{dr}$$

$$Q \int_{r_i}^{r_o} \frac{dr}{r} = -2\pi L \int_{T_i}^{T_o} k\, dT$$

$$Q \ln\left(\frac{r_o}{r_i}\right) = 2\pi L \left[\int_{T_o}^{T_i} k\, dT \right]$$

$$Q = \frac{2\pi L \left[\int_{T_o}^{T_i} k\, dT \right]}{\ln\left(\frac{r_o}{r_i}\right)}$$

3.10   THERMAL CONTACT RESISTANCE

When we create the wall comprising of several material layers, the temperature is not evenly distributed across the interface between material layers. This phenomenon is called thermal contact resistance as it poses extra resistance to the heat flow. Note that we need good contact between the layers for conduction between solids. In case there is a microscope gaps due to surface unevenness, the air is filled up in these gaps, and as the air has low thermal conductivity, the heat transfer in gap region is less compared to the heat transfer in the contact region. Figure 3.12 shows a composite wall formed by bring layer of material (M1) and material (M2) together. The zoom-in view shows of the interface shows that there are microscopic ridges and materials do not have a uniform contact with each other.

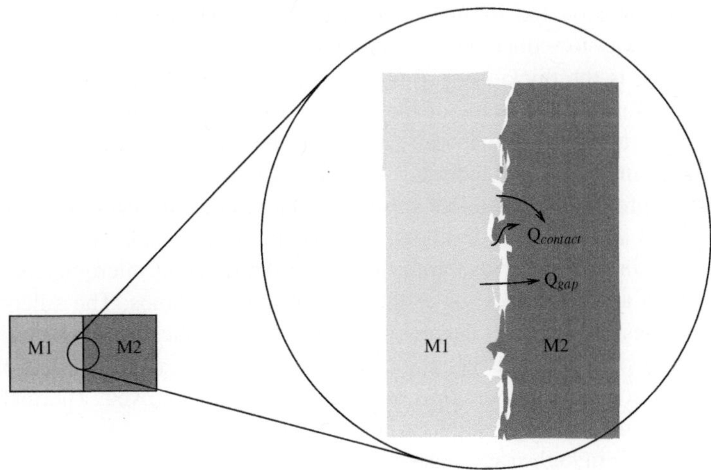

Figure 3.12   Temperature is not uniformly distributed at the interface between materials (M1) and (M2).

We can reduce the thermal contact resistance by introducing an interfacial fluid of high thermal conductivity, known as thermal grease. Silicon oil is one of the type of thermal grease, which is applied on the surfaces before they are brought closer.

TABLE 3.1

Contact Resistance for Some Material Surfaces

Material	Gap fluid	Roughness μm	$T\,°C$	P (atm)	R_c $m^2 \cdot °C/W \times 10^4$	R_c $h \cdot ft^2 \cdot °F/Btu$
Stainless steel 416, ground	air	1.14	20	40-70	5.28	0.003
Stainless steel 304, ground	air	2.54	90-200	3-25	2.64	0.0015
Aluminium, ground	air	2.54	150	12-200	1.23	0.0007
Aluminium, ground	air	2.54	150	12-25	0.88	0.0005
Copper, milled	vacuum	0.25	30	7-70	0.88	0.0005
Aluminium, ground	air	0.25	150	12-25	0.18	0.0001
Copper, milled	air	3.81	20	10-50	0.18	0.0001
Copper, ground	air	1.27	20	12-200	0.07	0.00004

Mathematically, we can express the contact resistance as

$$R_c = \left(\frac{|\Delta T|_{interface}}{Q/A} \right)$$

where $\Delta T = (T_{M1} - T_{M2})$. Table 3.1 lists the Contact conductance of some surfaces.

Thermal Interface Material (TIM) is a substance (or material) that is used to enhance the heat transfer between two surfaces. In most applications, a TIM layer is used between a heat-generating component, such as a microprocessor or power transistor, and a heat sink or cooling system. The purpose of a TIM is to fill the microscopic air gaps and irregularities that exist between the two surfaces. They offer a quick way to reduce the thermal contact resistance. An important factor in TIM selection is the thickness and uniformity of the TIM material along with its ability to conform to the shape of the two surfaces it is in contact with. TIMs can be made from a variety of materials, including thermal greases, thermal adhesives, phase change materials, and thermally conductive tapes.

Thermal grease is a paste-like substance that is typically made from a mixture of silicone oil and metal oxide fillers, such as aluminium oxide or zinc oxide. The particle sizes in thermal grease are typically in the range of micrometers or larger and thus they are not comes under the category of nanofluids. The silicone oil in thermal greases offers a high degree of wetting of the surface and its helps in filling up the microscopic gaps between the two surfaces. The metal oxide fillers offers the high thermal conductivity, and thus the heat transfer is improved. Thermal greases are now often used in applications where a high degree of thermal performance is required, such as in high-performance gaming computers.

Thermal paste or thermal adhesives is also used in industry over the thermal contact resistance issue. Thermal paste usually consists of a polymerisable liquid matrix with large volume fractions of electrically insulating, but thermally conductive filler. They can cure or harden to form a more solid and permanent bond between the two surfaces. They now used in semiconductor devices, power transistors, CPUs, GPUs, and LED lights. It is important to ensure that thermal adhesive is applied correctly at the right location.

3.11 CONDUCTION WITH INTERNAL ENERGY CONVERSION INTO INTERNAL ENERGY

A medium through which thermal energy is conducted may also involve the conversion of nuclear, chemical, and electrical, energies into internal energy. In heat conduction literature, such conversion processes are erroneously characterised as heat generation. According to the first of thermodynamics, heat cannot be generated or created as it is the form of energy. Therefore, the right terminology is to specify this energy is a conversion process as an increase in the internal energy of the system or the medium.

We consider the case of internal energy conversion into heat in a plane wall. The schematic diagram of temperature distribution is shown in Figure 3.13. The internal energy converted inside the plate is transferred from inside to the surroundings by convection. The ambient conditions are assumed to be maintained at temperature T_∞. The temperature is maximum inside the wall and the applicable boundary condition are

$$\left(\frac{\partial T}{\partial x}\right)_{x=0} = 0$$

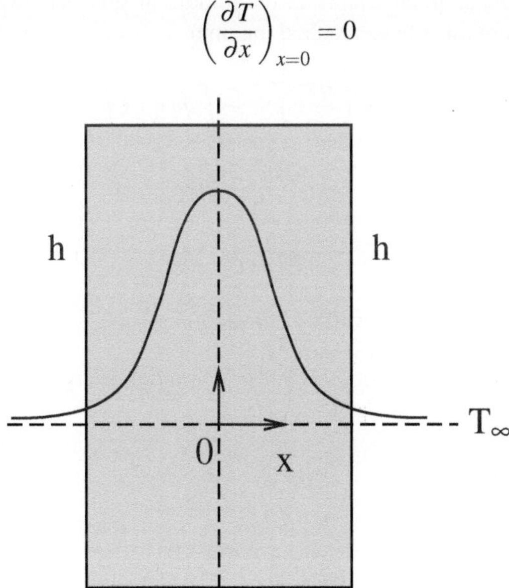

Figure 3.13   Schematic representation of temperature distribution inside a plane wall with internal energy conversion into heat.

The steady three dimensional energy equation is

$$k\nabla^2 T + \tilde{q} = 0$$

We consider the case of one-dimensional heat transfer and the equation can be reduced to form

$$k\left(\frac{\partial^2 T}{\partial x^2}\right) + \tilde{q} = 0$$

We integrate this equation twice and consider that thermal conductivity and internal energy conversion is a function of temperature

$$k\left(\frac{\partial T}{\partial x}\right) = -\int \tilde{q}\,dx + C_1$$

$$k(T)\cdot dT = -\int\left(\int \tilde{q}(x)dx\right)dx + C_1 x + C_2$$

$$T = -\frac{\tilde{q}x^2}{2k} + \frac{C_1}{k}x + \frac{C_2}{k}$$

$$T = -\frac{\tilde{q}x^2}{2k} + C_1^* x + C_2^* \tag{3.5}$$

where $\bar{C}_1^* x = C_1/k$ and $C_2^* x = C_2/k$.

i We investigate the temperature at the middle of the plane wall (x=0), where the temperature must have reached maxima.

$$k\left(\frac{\partial T}{\partial x}\right) = -\int \tilde{q}\,dx + C_1$$

$$-\tilde{q}(0) + C_1 = k(0) = 0$$

$$C_1 = C_1^* = 0$$

ii Heat loss at the boundary wall by convection at x=L

$$q''\big|_{x=L} = -k\left(\frac{\partial T}{\partial x}\right)_{x=L} = h\left(T - T_\infty\right)$$

$$k\left(\frac{\partial T}{\partial x}\right)_{x=L} = -\tilde{q}\cdot L$$

$$q''\big|_{x=L} = -k\left(\frac{\partial T}{\partial x}\right)_{x=L} = \tilde{q}\cdot L = h\left(T(L) - T_\infty\right)$$

$$\tilde{q}\cdot L = h\left(T(L) - T_\infty\right)$$

We estimate the temperature as

$$T(L) = -\frac{\tilde{q}L^2}{2k} + C_2^*$$

this gives

$$C_2^* = \frac{\tilde{q} \cdot L}{h} + \frac{\tilde{q}L^2}{2k} + T_\infty$$

We substitute both $C_1^* x$ and $C_2^* x$ into temperature distribution equation 3.5, which takes the form

$$T(x) - T_\infty = \frac{\tilde{q}(L^2 - x^2)}{2k} + \frac{\tilde{q} \cdot L}{h}$$

The temperature at x=0 is

$$T(0) - T_\infty = \frac{\tilde{q}L^2}{2k} + \frac{\tilde{q} \cdot L}{h}$$

and the temperature at x=L is

$$T(L) - T_\infty = \frac{\tilde{q} \cdot L}{h}$$

The difference between the extreme temperatures is

$$T(0) - T(L) = \frac{\tilde{q}L^2}{2k}$$

We can also rearrange the temperature distribution

$$T(x) - T_\infty = \frac{\tilde{q}L^2 \left[1 - \left(\frac{x}{L}\right)^2\right]}{2k} + \frac{\tilde{q} \cdot L}{h}$$

Introducing $\zeta = x/L$ and Biot number $Bi = hL/k$ into above equation and we arrive at dimensionless temperature

$$\frac{T(\zeta) - T_\infty}{\tilde{q}L^2/k} = \frac{(1 - \zeta^2)}{2} + \frac{1}{Bi}$$

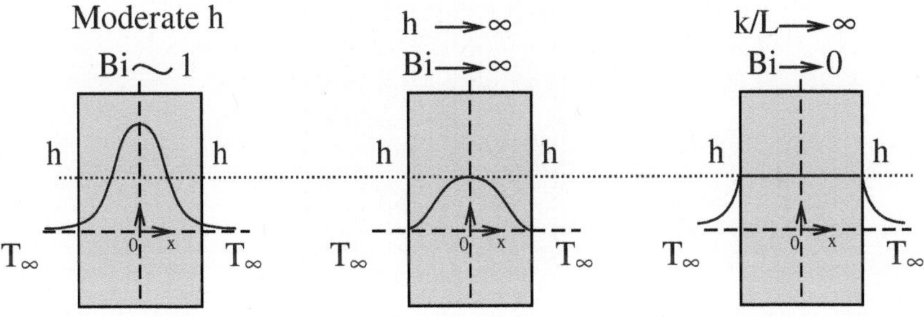

Figure 3.14   Influence of Biot number on temperature distribution inside a plane wall with internal energy conversion into heat.

Figure 3.14 shows the influence of Biot number on temperature distribution inside a plane wall as Biot number takes moderate to extreme values.

Example 3.6

A nuclear fuel rod is modelled as a plane wall of thickness 2L. The both sides of wall are covered with a metallic cladding of thickness b, and one side is well insulated. The internal heat energy conversion into thermal energy is happening, converting the nuclear energy into thermal energy at a rate $\widetilde{q}$ W/m^3. The heat is removed by a fluid on once side having convective condition h_∞ and T_∞. The thermal conductivities of fuel and metal are designated as k_f, and k_m, respectively. Do the energy balance for this system and derive an expression for the temperature distribution. Take the depth of this system as unity. Plot the expected temperature distribution under using the given data:

$$k_f = 70\,W/m \cdot K, L = 17 \times 10^{-3}m, b = 5 \times 10^{-3}m, k_m = 200\,W/m \cdot K,$$

$$h_\infty = 1 \times 10^4\,W/m^2.K; T_\infty = 300°C,$$

$$\tilde{q} = 3 \times 10^7\,W/m^3$$

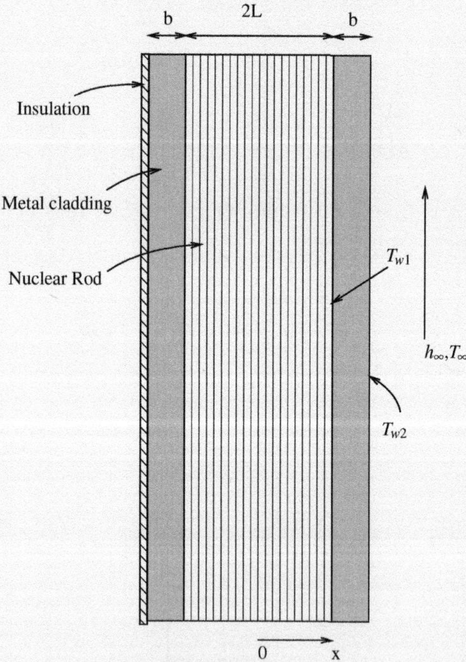

Figure 3.15   Schematic of example problem.

Figure 3.15 shows the schematic of example problem.

Solution The steady three dimensional energy equation is

$$k\nabla^2 T + \tilde{q} = 0$$

We consider the case of one-dimensional heat transfer and the equation can be reduced to form

$$k\left(\frac{\partial^2 T}{\partial x^2}\right) + \tilde{q} = 0$$

We integrate this equation twice and consider that thermal conductivity and internal energy conversion is a function of temperature

$$k\left(\frac{\partial T}{\partial x}\right) = -\int \tilde{q}\, dx + c_1$$

$$k(T)\cdot dT = -\int \left(\int \tilde{q}(x)\, dx\right) dx + c_1 x + c_2$$

$$T = -\frac{\tilde{q}x^2}{2k} + \frac{c_1}{k}x + \frac{c_2}{k}$$

As thermal conductivity is constant, we introduce $C_1 = c_1/k$, and $C_2 = c_2/k$

$$T = -\frac{\tilde{q}x^2}{2k} + C_1 x + C_2$$

The differentiation of temperature gives:

$$\frac{dT}{dx} = -\frac{\tilde{q}x}{k_f} + C_1$$

Substituting x=L we have

$$\left(\frac{dT}{dx}\right)_{x=L} = -\frac{\tilde{q}L}{k_f} + C_1$$

We get

$$C_1 = -\frac{\tilde{q}L}{k_f}$$

We update the temperature distribution:

$$T(x) = -\frac{\tilde{q}x^2}{2k_f} - \frac{\tilde{q}Lx}{k_f} + C_2$$

Boundary Conditions:

i *Energy balance*: There is a balance between wall conduction and internal energy conversion. As one side is insulated and at insulation dT/dx=0, the heat will flow towards the convective side. We can do the balance of energy for A=1 m² plane wall area:

$$Q = \tilde{q} \cdot \forall = \tilde{q} W/m^3 \cdot (2L)m \cdot 1m^2$$

where $\forall$ is the volume of the plane wall.

$$Q = k_m \frac{(T_{w1} - T_{w2})}{b} = h(T_{w2} - T_\infty)$$

We can arrive at two equations:

$$\tilde{q} \cdot 2L = \frac{k_m}{b}(T_{w1} - T_{w2}) \qquad \bigg| \qquad \tilde{q} \cdot 2L = h(T_{w2} - T_\infty)$$
$$T_{w1} = \frac{\tilde{q} \cdot 2L \cdot b}{k_m} + T_{w2} \qquad \bigg| \qquad T_{w2} = \frac{\tilde{q} \cdot 2L}{h} + T_\infty$$

This will Give

$$T_{w1} = \frac{2\tilde{q} \cdot L \cdot b}{k_m} + \frac{2\tilde{q} \cdot L}{h} + T_\infty$$

ii *Wall condition*: At x=L, T=T_{w1}:

$$T(x = L) = T_{w1} = -\frac{\tilde{q}L^2}{2k_f} - \frac{\tilde{q}L^2}{k_f} + C_2$$

Rearranging:

$$C_2 = T_{w1} + \frac{\tilde{q}L^2}{2k_f} + \frac{\tilde{q}L^2}{k_f}$$

Inserting T_{w1} we have

$$C_2 = \frac{2\tilde{q} \cdot L \cdot b}{k_m} + \frac{2\tilde{q} \cdot L}{h} + T_\infty + \frac{\tilde{q}L^2}{2k_f} + \frac{\tilde{q}L^2}{k_f}$$

We now update the temperature distribution:

$$T(x) = T_\infty - \frac{\tilde{q} \cdot x^2}{2 \cdot k_f} - \frac{\tilde{q} \cdot L \cdot x}{k_f} + \tilde{q} \cdot L \left[\frac{2 \cdot b}{k_m} + \frac{2}{h} + \frac{3 \cdot L}{2 \cdot k_f} \right]$$

Using the given data:

$$k_f = 70\,W/m \cdot K, L = 17 \times 10^{-3} m, b = 5 \times 10^{-3} m, k_m = 200\,W/m \cdot K,$$

$$h_\infty = 1 \times 10^4\,W/m^2.K; T_\infty = 300°C,$$

$$\tilde{q} = 3 \times 10^7\,W/m^3$$

we plot the temperature distribution in Figure 3.16.

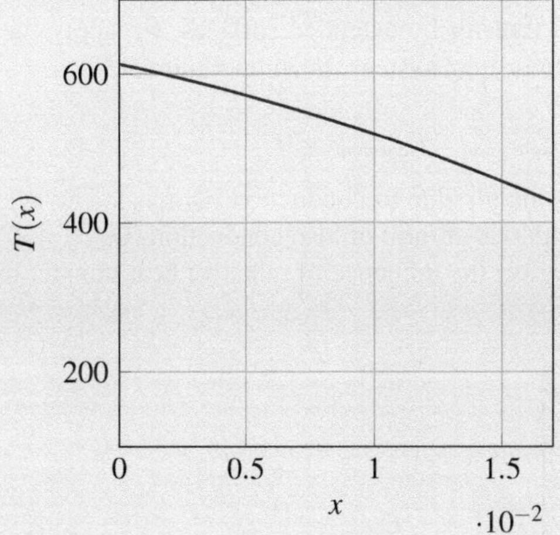

Figure 3.16 Temperature distribution inside nuclear fuel element (plane wall model).

3.12   HEAT TRANSFER BY EXTENDED SURFACES

The heat transfer from the surface is a function of temperature difference, thermal conductivity, area, and convective heat transfer coeffcient. In most engineering applications, it is rarely possible to control these dependencies except the area. Therefore, in order to increase heat transfer from the surface, the most simple option is to increase the surface area. This is often done in case of boilers, air-cooled engines, heating or cooling radiators, resistance heaters, liquid-to-gas, or gas-to-gas heat exchangers.

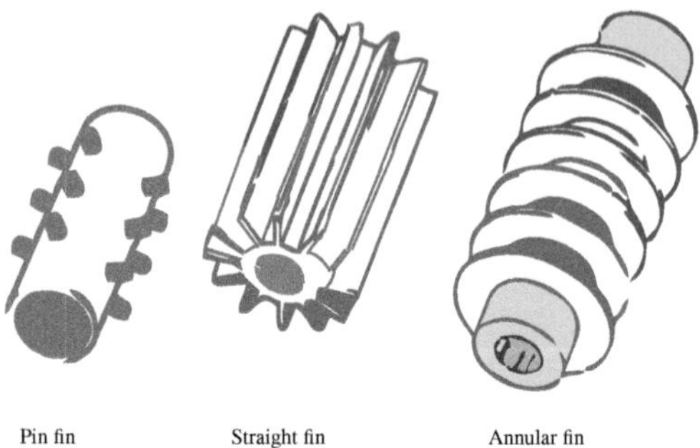

Figure 3.17   Types of extended surfaces.

The most common types of fins are straight fins, annular fins, and pin fins (see Figure 3.17). To analyse the heat transfer by extended surfaces, we apply the first law of thermodynamics on an infinitesimal system shown in Figure 5.3.

$$E_{in} - E_{out,cond} - E_{out,conv} = 0$$

The heat energy coming in is mainly due to conduction ($Q_{cond,x}$). The heat energy going out of the system boundaries in form of heat conduction ($Q_{cond,x+\Delta x}$) and surface convection (Q_{conv}). We neglect the influence of radiative heat transfer in this analysis.

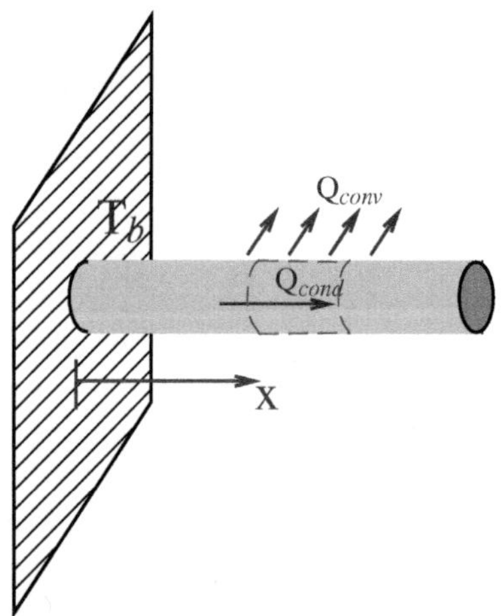

Figure 3.18   Schematic representation for heat flow from a fin.

Heat transfer inside fins is conduction and according to Fourier Law of heat conduction we have (see Figure 3.18)

$$Q_{x,cond} = -kA_c \frac{dT}{dx}$$

This is the heat energy entering the system. For the energy leaving the system, we use Taylor's series

$$Q_{x+\Delta x} = Q_x + \frac{d(Q_x)}{dx}\frac{(\Delta x)^1}{1!} + \frac{d^2(Q_x)}{dx^2}\frac{(\Delta x)^2}{2!} + \cdots$$

Ignoring the higher order terms, we have

$$E_{out,cond} = Q_{x+\Delta x} = Q_x + \frac{d(Q_x)}{dx}\Delta x$$

Heat transfer from the surface of the fins is mainly by convection.

$$Q_{conv} = hA_s(T - T_\infty)$$

The balance of conduction and convection is given as

$$E_{in} - E_{out,cond} - E_{out,conv} = 0$$

$$Q_x - \left[Q_x + \frac{d(Q_x)}{dx}\frac{(\Delta x)^1}{1!} \right] - hA_s(T - T_\infty) = 0$$

$$-\frac{d(Q_x)}{dx}\Delta x - hA_s(T - T_\infty) = 0$$

Introducing Fourier's Law of heat conduction:

$$-\frac{d}{dx}\left[-kA_c\frac{dT}{dx} \right]\Delta x = hA_s(T - T_\infty)$$

where A_c is cross-sectional area and A_s is the surface area which can be related to the perimeter of the fins.

$$A_s = P.\Delta x$$

$$-\frac{d}{dx}\left[-kA_c\frac{dT}{dx} \right]\Delta x = hP.\Delta x(T - T_\infty)$$

We can combine convective heat transfer coefficient, thermal conductivity, perimeter, and area as a constant parameter β^2:

$$\beta^2 = \left(\frac{hP}{kA_c} \right)$$

$$\frac{d^2T}{dx^2} = \left(\frac{hP}{kA_c} \right)(T - T_\infty)$$

$$\frac{d^2T}{dx^2} - \beta^2(T - T_\infty) = 0$$

Since T_∞ is constant, we can introduce it into differential equation:

$$\frac{d^2(T - T_\infty)}{dx^2} - \beta^2(T - T_\infty) = 0$$

This is a second-order ordinary differential equation, whose solution is

$$T(x) - T_\infty = C_1 \exp(\beta \cdot x) + C_2 \exp(-\beta \cdot x)$$

The temperature at the base of the fin is T_b and boundary condition is

$$x = 0, \quad T = T_b$$

At the end of the fin, the area available for conduction and convection is same; hence the second boundary condition is

$$x = L, \quad -kA_c\frac{dT}{dx} = hA_s(T - T_\infty)$$

$$\frac{d(T - T_\infty)}{dx} = -\frac{h}{k}(T - T_\infty)$$

The solution of equation, after boundary conditions substitutions is

$$T(x) - T_\infty = (T_b - T_\infty)\left[\frac{\cosh[\beta(L - x)] + \left(\frac{h}{\beta k}\right)\sinh[\beta(L - x)]}{\cosh(\beta L) + \left(\frac{h}{\beta k}\right)\sinh(\beta L)}\right]$$

and corresponding heat transfer is

$$Q = (T_b - T_\infty)\sqrt{hPkA}\left[\frac{\sinh(\beta L) + \left(\frac{h}{\beta k}\right)\cosh(\beta L)}{\cosh(\beta L) + \left(\frac{h}{\beta k}\right)\sinh(\beta L)}\right]$$

Some values of $\beta.L$ along with corresponding exponential and hyperbolic functions are listed in Table 3.2.

Example 3.7

An iron fin is experiencing the convective environment where convective heat transfer coefficient (h) is 28W/m$^2\cdot$K and T_∞=10 °C. The fin's diameter is 10 mm and length is 200 mm. The temperature at base of fin is 300 °C. Find the heat transfer from the fin.
The thermal conductivity of iron k = 80 W/m$\cdot$K.

Solution We first calculate the β parameter $\beta = \sqrt{\frac{h \cdot P}{k \cdot Ac}} = 11.832$
It is given that length and temperatures are
$T_\infty = 10\,°C, T_b = 300\,°C, L = 200mm$
so the heat transfer from the fin is

TABLE 3.2

Useful Exponential and Hyperbolic Functions

$\beta \cdot L$	$e^{\beta L}$	$e^{-\beta L}$	$\sinh(\beta L)$	$\cosh(\beta L)$	$\tanh(\beta L)$
0	1	1	0	1	0
0.1	1.105	0.905	0.1	1.005	0.0997
0.2	1.221	0.819	0.2	1.02	0.1973
0.3	1.35	0.741	0.305	1.045	0.2913
0.4	1.492	0.67	0.411	1.0311	0.38
0.5	1.649	0.607	0.521	1.128	0.462
0.6	1.822	0.549	0.637	1.186	0.537
0.7	2.014	0.497	0.759	1.255	0.6044
0.8	2.226	0.449	0.888	1.337	0.664
0.9	2.46	0.407	1.027	1.433	0.7163
1	2.718	0.368	1.175	1.543	0.7616
1.1	3.004	0.333	1.336	1.669	0.8005
1.2	3.32	0.301	1.509	1.811	0.8337
1.3	3.67	0.272	1.698	1.971	0.862
1.4	4.055	0.247	1.904	2.151	0.8854
1.5	4.482	0.223	2.129	2.352	0.905
1.6	4.953	0.202	2.376	2.577	0.922
1.7	5.474	0.1827	2.646	2.828	0.9354
1.8	6.05	0.1653	2.942	3.107	0.947
1.9	6.686	0.15	3.268	3.418	0.956
2	7.389	0.1353	3.627	3.762	0.964
2.1	8.166	0.1224	4.022	4.144	0.9705
2.2	9.025	0.111	4.457	4.568	0.976
2.3	9.974	0.1	4.937	5.037	0.98
2.4	11.02	0.0907	5.466	5.557	0.984
2.5	12.18	0.0821	6.05	6.132	0.987
2.6	13.46	0.074	6.695	6.77	0.989
2.7	14.88	0.067	7.406	7.473	0.991
2.8	16.445	0.061	8.192	8.253	0.9926
2.9	18.174	0.055	9.06	9.115	0.994
3	20.09	0.05	10.018	10.068	0.995
4	54.6	0.0183	27.29	27.31	0.9993
5	148.4	0.00673	74.2	74.21	0.9999

$$Q = (T_b - T_\infty)\sqrt{hPkA}\left[\frac{\sinh(\beta L) + \left(\frac{h}{\beta k}\right)\cosh(\beta L)}{\cosh(\beta L) + \left(\frac{h}{\beta k}\right)\sinh(\beta L)}\right]$$

$$Q = (T_b - T_\infty)\cdot(0.0743)\cdot\left(\frac{5.282 + 0.02958 \times 5.376}{5.376 + 0.02958 \times 5.282}\right) = 21.1941W$$

3.12.1 INFINITELY LONG FIN

The temperature at the end of the fin will approach the temperature of the surrounding fluid.

$$x \to \infty, T \to T_\infty$$

The temperature distribution for the infinitely long fin is

$$T(x) - T_\infty = (T_b - T_\infty)\exp(-\beta x)$$

The heat transfer to the fluid is

$$Q = (T_b - T_\infty)\sqrt{h \cdot P \cdot k \cdot A_c}$$

Example 3.8

A fin made up of an unknown material has diameter 5.5 cm. The temperatures in fin at two locations 15.5 cm apart are 145°C and 80°C. The surrounding temperature T_∞ is 30 °C. Assuming convective heat transfer coefficient is 30 W/m$^2 \cdot$ K, find the thermal conductivity of material.

Solution

The temperature distribution for the infinitely long fin is

$$T(x) - T_\infty = (T_b - T_\infty)\exp(-\beta x)$$

Using the data at two locations, we can formulate the equations

$$145 - 30 = (T_b - T_\infty)\exp(-\beta x_1)$$

$$80 - 30 = (T_b - T_\infty)\exp(-\beta x_2)$$

Taking the ratio of these equations, we have

$$\left(\frac{145 - 30}{80 - 30}\right) - \exp[\beta \cdot (x_2 - x_1)] = 0$$

$$\left(\frac{145 - 30}{80 - 30}\right) - \exp[\beta \cdot (0.155)] = 0$$

Using Maple *fsolve* function, we can solve the equation and find the unknown quantity β:

```
T1 := 145;   T2 := 80;   Tinfi := 30;   L := 15.5/100;
beta := fsolve((T1 - Tinfi)/(T2 - Tinfi) - exp(beta*L))
```

This gives $\beta = 5.373$.

$$\beta = \sqrt{\frac{h \cdot P}{k \cdot Ac}} = 5.373$$

where $P = \pi D = 0.1727$ m, $A_c = 0.0023746$ m^2. Solving β equation, we have k = 75.55 W/m $\cdot$ K.

3.12.2   FIN INSULATED AT END

If the tip of the fin is insulated, then the boundary condition at the end of the fin is

$$x = L \qquad Q = -k \cdot A_c \frac{dT}{dx} = 0$$

The solution for the temperature distribution is

$$T(x) - T_\infty = (T_b - T_\infty)\left(\frac{\cosh\left[\beta(L-x)\right]}{\cosh(\beta L)}\right) \tag{3.6}$$

and heat transfer is

$$Q = \tanh(\beta L) \cdot (T_b - T_\infty) \cdot \sqrt{h \cdot P \cdot k \cdot A_c}$$

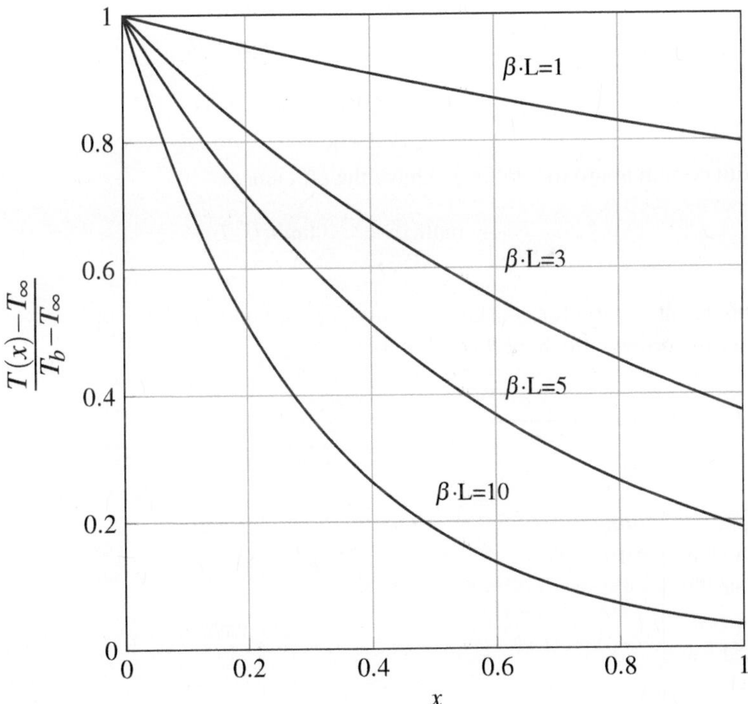

Figure 3.19   Fin temperature distribution as per Equation 3.6.

Figure 3.19 shows the plot of equaion 3.6 at $\beta \cdot L = 1,5,10$. For large β values, temperature will approach the fluid temperature at the tip of the fin.

3.13   EFFICIENCY AND EFFECTIVENESS OF FIN

$$\eta_{fin} = \frac{Heat\ flux\ from\ surface\ with\ fins}{Heat\ flux\ from\ surface\ without\ fins}$$

$$\eta_{fin} = \frac{Q_{fin}}{Q_{T=T_b}}$$

$Q_{T=T_b}$ is the rate at which heat would be dissipated if the entire surface is at the base temperature.

$$Q_{fin} = \tanh(\beta \cdot L) \cdot \theta_b \cdot \sqrt{h \cdot P \cdot k \cdot A_c}$$

Fin efficiency is defined as the ratio of the heat transfer rate from surface with fins to the heat transfer rate from surface with no fins at surface at base temperature:

$$\eta_{fin} = \frac{Q_{fin}}{h \cdot A_w \cdot (T_b - T_\infty)} = \frac{Q_{fin}}{h \cdot A_w \cdot \theta_b}$$

$$\eta_{fin} = \frac{\tanh(\beta L) \cdot (\theta_b) \cdot \sqrt{h \cdot P \cdot k \cdot A_c}}{h \cdot A_w \cdot \theta_b}$$

With certain approximations, we have the efficiency:

$$\eta_{fin} \equiv \frac{\tanh(\beta \cdot L)}{\sqrt{\beta^2 \cdot L^2}} = \frac{\tanh(\beta \cdot L)}{\beta \cdot L}$$

This result is valid for constant base temperature and insulated tip. We can compute the hyperbolic functions:

$$\sinh x = \frac{e^x - e^{-x}}{2}, \quad \cosh x = \frac{e^x + e^{-x}}{2}, \quad \tanh x = \frac{e^x - e^{-x}}{e^x + e^{-x}}$$

Case	θ/θ_b	$Q/(B\theta_b)$
Infinitely Long fin	$\exp(-\beta x)$	1
Fin Insulated at End	$\left(\dfrac{\cosh[\beta(L-x)]}{\cosh(\beta L)} \right)$	$\tanh(\beta L)$
Convective Boundary	$\dfrac{\cosh[\beta(L-x)] + \left(\frac{h}{\beta k}\right)\sinh[\beta(L-x)]}{\cosh(\beta L) + \left(\frac{h}{\beta k}\right)\sinh(\beta L)}$	$\left[\dfrac{\sinh(\beta L) + \left(\frac{h}{\beta k}\right)\cosh(\beta L)}{\cosh(\beta L) + \left(\frac{h}{\beta k}\right)\sinh(\beta L)} \right]$

where, $\theta = T(x) - T_\infty$, $\theta_b = T_b - T_\infty$, $B = \sqrt{h \cdot P \cdot k \cdot A_c}$, $\beta^2 = h \cdot P / k \cdot A_c$.

Example 3.9

Consider an infinite fin (L=20 cm) with sectionalism area A_c=0.00785 m^2 is subjected to very high convective environment (h=5000 W/$m^2 \cdot$ K). The thermal conductivity of the fin material is $k = 16W/m \cdot K$. Find the fin efficiency. What will be the fin efficiency if the h values is just 5 W/$m^2 \cdot$ K?

Solution We compute the parameter β:

$$\beta = \sqrt{\frac{h \cdot P}{k \cdot A_c}} = 111.8317494$$

The efficiency of the fin is

$$\eta_{fin} = \frac{\tanh(\beta \cdot L)}{\beta \cdot L} = 0.0447$$

This shows that efficiency is low and putting fins on base surface is not very useful in this highly convective environment. On the other hand, if h=5 W/$m^2 \cdot$ K, η_{fin}=0.86.

The fins are helpful in heat transfer applications where convection heat transfer coefficient has low values like those associated with free convection. In a highly convective forced convection environment fin might work in opposite way, i.e. instead of dissipating heat they would increase the heat transfer to the surface. Another useful quantity to decide about the fin's efficacy is the fin's effectiveness, defined as

$$\varepsilon_{fin} = effectiveness = \frac{Rate\ of\ heat\ transfer\ from\ the\ fin's\ base\ surface}{Rate\ of\ heat\ transfer\ from\ the\ surface\ without\ fins}$$

$$\varepsilon_{fin} = \left(\frac{A_{fin}}{A_b}\right) \eta_{fin}$$

Note that the use of fins can only be justified if ε_{fin} is greater than 1.

3.14 CORRECTED FIN LENGTH

Corrected fin length L_c is an increased length plus original length needed to convert the case of heat transfer from the actual fin of length L with convection at the fin's tip to the case of heat transfer from a insulated-tip fin of length L_c. Figure 3.20 shows the concept of corrected fin length with convective and insulated boundary conditions at the tip of the fin.

$$L_c = L + \left(\frac{A_c}{P}\right)$$

$$L_{c,cylindrical} = L + \left(\frac{r}{2}\right)$$

$$L_{c,rect} = L + \left(\frac{t}{2}\right)$$

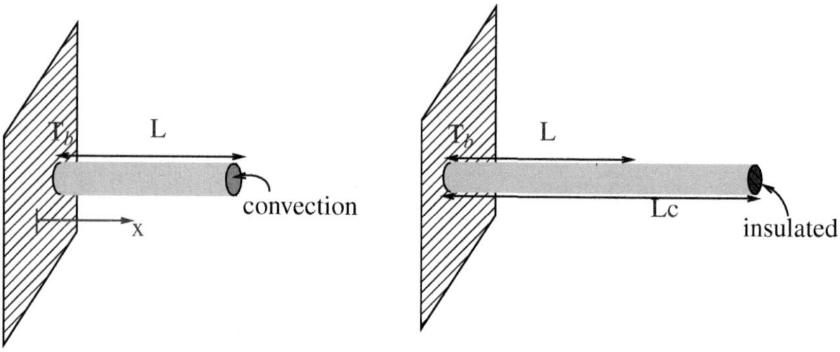

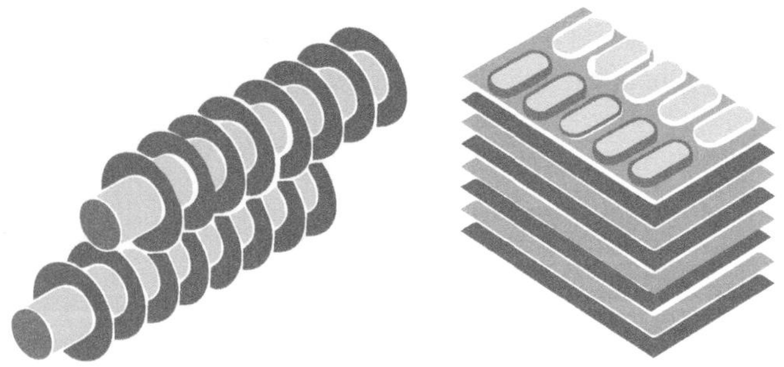

Figure 3.20   The concept of corrected fin length.

where r is the radius of the cylindrical fins. Typical annular and rectangular fins are shown in Figure 3.21.

Figure 3.21   Schematic representation of annular and rectangular fins.

Referring to Figure 3.22, for annular Fin, we use the relations:

$$L_{c,annular} = L + \left(\frac{t}{2}\right)$$

where t is the thickness of the rectangular fin.

$$r_{2c} = L + r_1, \; and \; A_m = t \cdot L_c$$

Once we the efficiency from the graph, the heat loss can be estimated as

$$Q_{fin} = h \cdot A_w \cdot (T_w - T_\infty) \cdot \eta_{fin}$$

The efficiency of circumferential and annular fins are plotted in Figure 3.23.

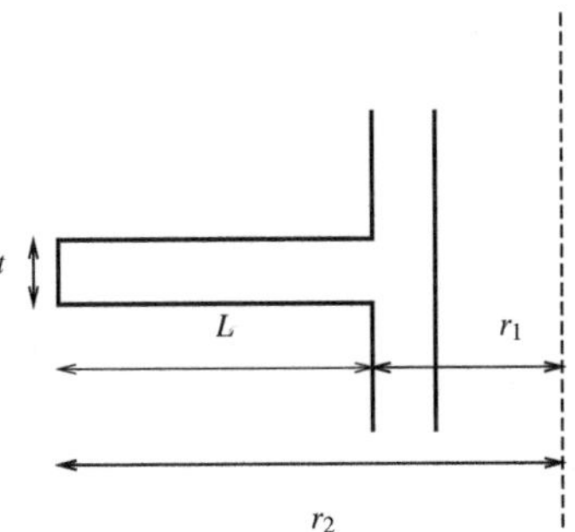

Figure 3.22   Annular fin nomenclature.

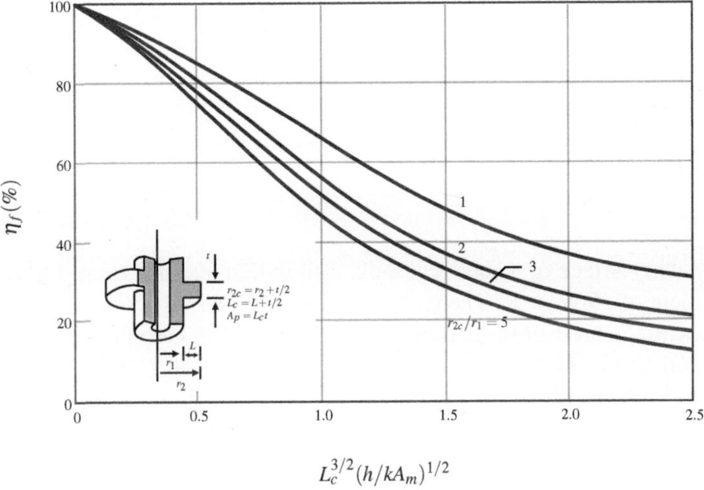

$$L_c^{3/2}(h/kA_m)^{1/2}$$

Figure 3.23   Efficiency of circumferential or annular fins.

PROBLEMS

3P-1 The thermal conductivity measurements are being conducted in an instrument that consists of two spheres of diameter d_i=50mm, and d_o=25 mm, as shown in figure. The test material on unknown thermal conductivity is filled between the gap region of concentric spheres. In the core of the instrument, an electric heating element is installed which can generate heat flux of 80 W/m^2. The temperatures at inner and outer spheres walls is uniform and around 500K and 320K, respectively. Assuming steady-state operation find the the thermal conductivity of the filled material. Ignore the spheres wall thickness effects. Figure 3.24 shows the schematic of problem 3P-1.

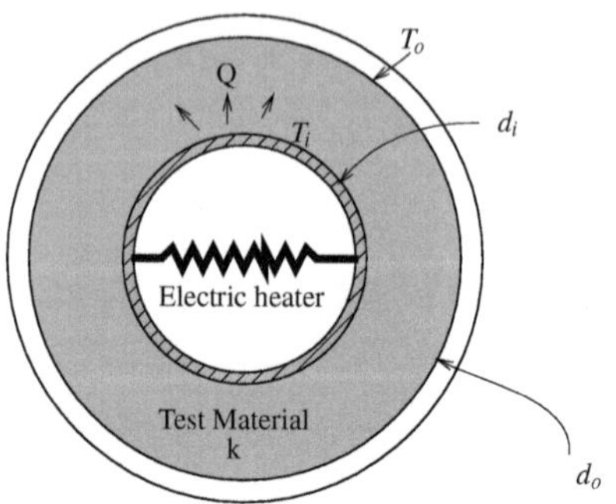

Figure 3.24 Thermal conductivity measurement apparatus.

3P-2 A steel pipe having internal and external diameters as 2.375-in and 2.067-in is covered with a fiber glass insulation of 2-in. The pipe is carrying liquid having 400°F and the inner and outer convective heat transfer coefficients are 40 and 5 Btu/h.ft^2 ·°F. The ambient temperature is 70°F. The thermal conductivity of steel and fiber glass are k=28 Btu/h·ft·°F, and k=0.025 Btu/h·ft·°F. Find the heat loss if pipe is 8 ft long.

[Ans: 603.72 Btu/hr·ft]

3P-3 The surface of a spherical tank, 2 m in diameter is exposed to convective environment of 25 W/m^{2}K. The tank surface temperature is 150°C and it decided to cover it with the layer of insulation (k=0.018 W/m·K) to keep the insulation surface temperature as low as 50°C. The ambient temperature is 20°C. Find the thickness of the insulation layer, also, find the heat transfer from tank to surrounding if tank is not insulated and if the tank is insulated.

[Ans: thickness= 2.39 mm, $Q_{no\ insulation}$=40.84 kW, $Q_{with\ insulation}$= 9.469 kW]

3P-4 The walls of a fridge is constructed in such a way that a layer of insulation has been sandwiched between two metal sheets. The insulation has thermal conductivity value of k_{ins}=0.046 W/m·K, and the insulation thickness is t_{ins}=7 cm. The thermal conductivity of the steel is k_s=66 W/m·K, and thickness of steel sheet is 5 mm. The compartment of fridge has temperature of 3°C and the ambient air has 35°C. The inner and outer convective heat transfer coefficient are 5 and 15 W/m^2·K. Find the heat transfer from outside to the fridge compartment.

[Ans: Q=17.891W/m^2]

3P-5 An oven is being used with inner temperature around 220 °C, and the outside kitchen temperature is 20 °C. The wall of oven is comprised of 1-in stainless steel plate, 3-in insulation and 5-in wood layer. The thermal conductivities of wood, insulation and steel are 0.19, 0.038, and 14.3 W/m·K, respectively. The convective heat transfer coefficient outside oven is 5 W/m^2·K. It is suggested to install a fan inside oven, which will increase the air movement inside oven and hence, the inner convective heat transfer coefficient can be controlled in range from 10 to 120 W/m^2·K.

Explore the influence of this *convection-mode* of oven on heat transfer, whether it will bring any benefit? Ignore radiative heat transfer.

3P-6 An assembly is created by pressing two 1.5 cm thick stainless Steel plates, having thermal conductivity k= 16.6 W/m·K at a pressure of 10 atm. If the temperature drop across assembly is 20 °C, find the heat transfer through plates, and temperature drop at the contact plane between the plates.

3P-7 The gas turbine blades are internally cooled with *cooler* air from the compressor. A thermal barrier coating of zirconia (k=1.3 W/m·K) is applied on the gas turbine blade which are made up of inconel (k=25 W/m·K). The air is moving inside blade at T_{in}=127 °C and blade is exposed to convective environment of T_∞=1527 °C, h_∞=2342 W/m²·K. The blade inner passage convective heat transfer coefficient is h=700 W/m²·K. If the contact resistance between zirconia coating and inconel is 5×10^4 m².K/W, find the heat transfer from blade to air in internal passage. Take the zirconia coating and blade surface thicknesses as 0.3 and 0.9 inches.

3P-8 A clothing company is designing a jacket to protect from cold environment of $T_\infty = -3$ °C. The surrounding convective coefficient is h_o=15 W/m²·K. The proposed jacket area is 1.3 m^2, and it is comprised of three layers of insulation with 0.2 mm thickness (k=0.123 W/m·K). The air between the insulation layers is estimated to be around 0.5 mm. The thermal conductivity of air is 0.025 W/m·K. Assuming that the body feels the thermal comfort condition (T= 25 °C), find the heat transfer associated with this design. In second round of design it is decided to replace the three layers of insulation with a single layer of another new insulation material having k=0.038 W/m·K. Estimate the thickness of new insulation layer if same level of thermal comfort is achieved as promised by initial design.

3P-9 A composite wall is comprised of 6 mm mild steel and 3 mm of polyvinylchloride (PVC) layers. The thermal conductivity of mild steel and for polyvinylchloride (PVC) are 64.5 W/m K, and 0.095 W/m K, respectively. The outside temperature is 45 °C and inside temperature is 20 °C. The inner and outer convective heat transfer coefficients are 3 and 30 W/m²·K. Find the U-factor for this wall.

3P-10 A plane wall is exposed to heat flux q_w. The material thermal conductivity is varying according to relation

$$k = k_o(1 + \beta(T - T_o))$$

Find the amount of heat transfer and investigate the influence of β on thermal conductivity.

3P-11 It is desired to enhance the heat loss from a surface by installing circumferential or annular fins. The fin thickness is 15 mm with radius r_1=25 mm. The length of fin is 90 mm and the fin material has $k = 300$ W/m·K. The wall temperature is 600 °C and surrounding air is at 20 °C with h = 50 W/m²·K. Find heat loss from the fins.

3P-12 Radioactive material is stored in a thin-walled, long cylindrical vessel of radius r_w which is submerged in a liquid which provides a uniform convection environment of h_o and T_o. The material is going into decay and converting nuclear energy into thermal energy non-uniformly across the vessel as per the following relation:

$$\frac{\widetilde{q}}{\widetilde{q}_o} = \left[1 - \left(\frac{r}{r_w}\right)^2\right]$$

where, $\widetilde{q}_o$ is the constant local rate of energy conversion per unit volume. Assume steady-state conditions and find the temperature distribution in the vessel as function of radius.

3P-13 Consider a 1.3-m-high and 4-m-wide single and double-pane windows with the glass plates having the thickness of 3 mm. The air space in double pane window is 14 mm. The room temperature is maintained at 24°C, and the outside temperature is 35°C. The convection heat transfer coefficients in room and outside as h_1=10 W/m^2.K and h_2=25 W/m^2.K. Take thermal conductivity of air and glass are k =0.026 W/m·K and k=0.75 W/m·K, respectively. Ignoring radiation, find the steady-state heat transfer through the single and double-pane windows. What window you would recommend to be installed. Figure 3.25 shows the schematic of problem 3P-13.

[Ans: Q$_{single}$ = 397.222 W, Q$_{double}$ = 83.325 W]

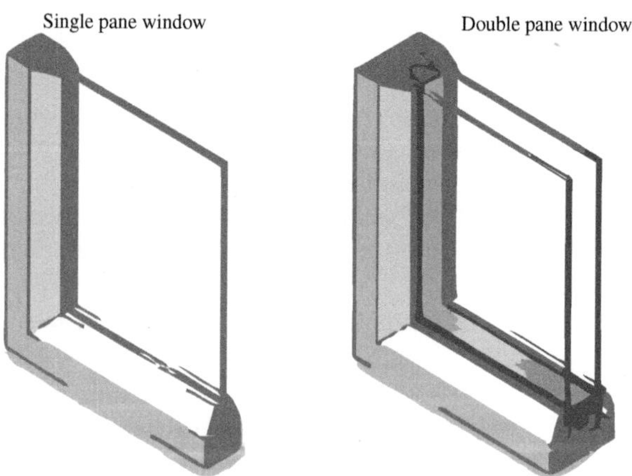

Figure 3.25   Single and double pane window.

3P-14 Internal energy is converted uniformly into thermal energy at rate of 6 MW/m^3 in a cylindrical nuclear fuel rod at a steady rate. The radius of the rod is 34.6 mm, and the temperature distribution inside rod is

$$T(r) = 900 - 3.13 \times 10^5 r^2$$

where T is reported here in °C. The uranium thermophyscial properties are c_p=113 J/kg K, k=27 W/m·K, ρ = 19000 W/m·K. (i) Find rate of heat transfer per unit rod length, and (ii) the temperature change rate if volumetric energy conversion is suddenly reduced to 4 MW/m^3.

[Ans: $(i)Q/L = 31.78kW/m$, $(ii)dT/dr = -13.881$ °C/m]

3P-15 Calculate the increase in heat transfer from a steel cylinder surface on which 9 longitudinal steel fins are installed. The diameter of cylinder is 120 mm and its surface is at 200 °C. The ambient air conditions are 40 °C, and 20 W/m^2·K. The fins are 1.5 mm thick and 9 mm in length.

3P-16 Pin fins which are insulated at the end are installed on a plate of 2 m × 2 m. The diameter of fins are 5 mm and length of fin are 5 cm. The temperature on which the fins are installed is at 50°C, and the surrounding temperature is 20°C. The thermal conductivity of pin fins is k=237 W/m·K. The convective heat transfer coefficient 33 W/m^2 · °C.

3P-17 A 200 mm long and 10 mm in diameter cylindrical aluminium strut is installed between two wall. One wall is maintained at 30 °C and another one is maintained at 10 °C. The hot air at 120 °C is blowing over the strut with convective heat transfer coefficient 130 W/m^2 · °C.

3P-18 A stack of aluminium fins is 232 mm wide and 115 mm deep contains 50 fins, each having length L=10 mm. The fins are installed between the plates in order to dissipate heat from compact heat exchanger cores. The air is moving over the fins with h = 134 W/m^2·K and T_∞=300 K. If the maximum allowable plate temperatures is T_o = 500 K, and T_L=325 K, how much shall be the heat loss? Figure 3.26 shows the schematic of problem 3P-18.

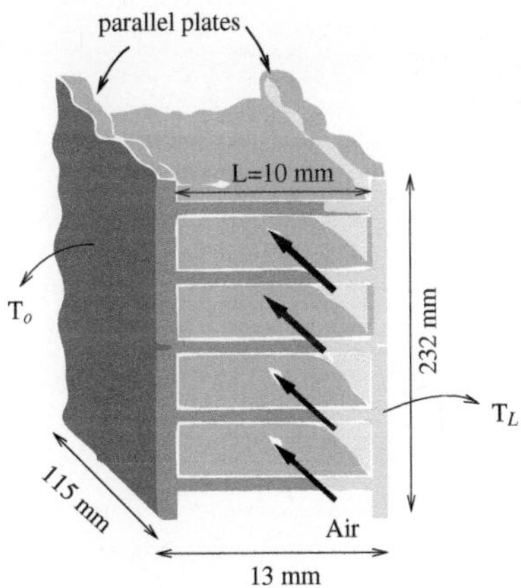

Figure 3.26   Schematic for problem 3P18.

3P-19 Consider a 30 mm long aluminium fin (k=210 W/m·°C) with a rectangular cross section 2 mm × 40 mm. The fin base temperature is 90°C, and the ambient temperature is 288 K. Assuming the combined convection and radiation heat transfer coefficient is 40 W/m^2°, find the temperature distribution in this fin and heat transfer rate from it (a) if the fin is infinitely long, (b) if the fin tip is adiabatic, (c) if the temperature at the fin tip is 50 °C.

REFERENCES

H. D. Baehr and K. Stephan, Heat and mass transfer. Springer, Berlin Heidelberg, Germany, 2011.

A. Bejan, Heat transfer. John Wiley & Sons, New York, USA, 1993.

A. Bejan, Convective heat transfer. John Wiley & Sons, New York, USA, 1995.

P. J. Schneider, Conduction Heat Transfer, Addison-Wesley, Reading, MA, USA, 1955.

D. Q. Kern and A. D. Kraus, Extended Surface Heat Transfer, McGraw-Hill, New York, USA, 1972.

4 Steady-State Two-Dimensional Heat Conduction

In this chapter, we will develop the analytical models for some simple two-dimensional conduction heat transfer problems. We will learn about the separation of variables procedure which is used for solution of partial differential equations. Also, we will learn about the conduction shape factors, which are used for fast calculations for complex situations. After finishing this chapter, one should be able to:

The learning outcomes:
- Understand the differences between 1D and 2D conduction problems.
- Drive 2D conduction problems with homogeneous boundary conditions using separation of variables procedure.
- Calculate the heat transfer using conduction shape factors.

In many engineering applications, we can no longer assume one-dimensional heat transfer. Consider the high-temperature furnaces which are often used for metallurgy experiments in the laboratory (see Figure 4.1). Although the walls of the furnace are highly insulated but one cannot eliminate the heat loss completely. In such scenarios, we cannot consider the heat conduction as one-dimensional heat transfer but we must analyse them as two and three dimensional problems. Now since we want to retain the dependence of temperature on two space coordinates, the heat conduction equation must be partial differential equations and we can no longer use the ordinary differential equation based models. Classical analytical techniques for solving the partial differential equations of heat transfer known since the times of Fourier and Biot. In the past, many books have been published on analytical solutions, with classical analytical techniques to solve the partial differential equations for heat transfer problems. Some of the noteworthy contributions are by Schlichting (1955), Carslaw and Jaeger (1959), Arpaci (1966), and Myers (1971).

Analytical models or solutions offer ready-made solutions for heat conduction problems with various kinds of boundary conditions. They are also used to check the correctness of numerical schemes. Another useful approach to tackle the complex heat conduction problems is to use the conduction shape factors (CSF), which are devised on the basis of complex conformal mapping theory or experiments. Conduction shape factor is defined as the ratio of a heat-transfer-area to a heat-transfer-length. The major benefit of using conduction shape factors is that it is function of

DOI: 10.1201/9781003428404-4

geometrical characteristics of the domain.

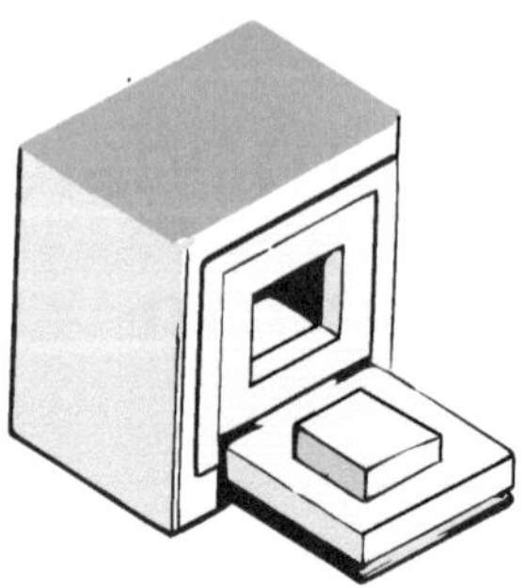

Figure 4.1 The multidimensional heat transfer in a laboratory furnace.

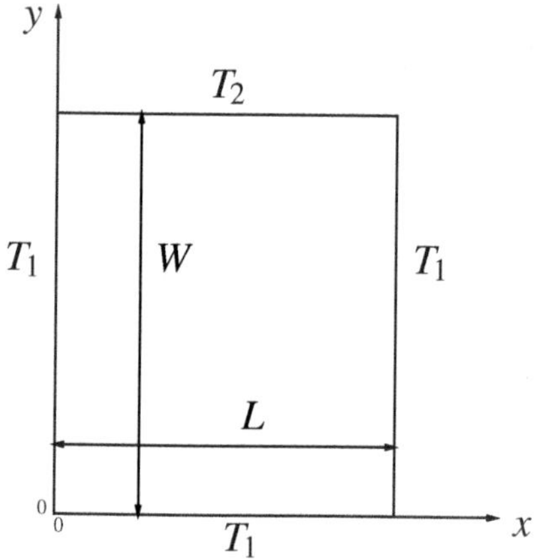

Figure 4.2 The two-dimensional heat conduction problem.

Consider the rectangular plate in Figure 4.2. The plate dimensions are L and W. For a constant thermal conductivity and no internal heat generation, the two-dimensional steady temperature distribution of the plate is governed by the second-order partial differential equation

$$\frac{\partial^2 T}{\partial x^2} + \frac{\partial^2 T}{\partial y^2} = 0 \tag{4.1}$$

The three sides of the plate are maintained at a constant temperature, T_1, while the fourth side is maintained at a temperature, $T_1 \neq T_2$

To simplify the solution, we introduce the transformation

$$\theta = \frac{T - T_1}{T_2 - T_1} \tag{4.2}$$

$$\frac{\partial^2 \theta}{\partial x^2} + \frac{\partial^2 \theta}{\partial y^2} = 0 \tag{4.3}$$

The four boundary conditions in terms of the variable, θ, are

$$\theta(0,y) = 0; \theta(x,0) = 0 \tag{4.4}$$

$$\theta(L,y) = 0; \theta(x,W) = 1 \tag{4.5}$$

Next a separation of variables solution is assumed with X being a function of x alone and Y being a function of y alone.

$$\theta(x,y) = X(x)Y(y) \tag{4.6}$$

We introduce this equation into Eq. 4.31

$$Y(y)\frac{\partial^2 X(x)}{\partial x^2} + X(x)\frac{\partial^2 Y(y)}{\partial y^2} = 0 \tag{4.7}$$

In next step, we divide this equation by θ

$$\frac{Y(y)\frac{\partial^2 X(x)}{\partial x^2} + X(x)\frac{\partial^2 Y(y)}{\partial y^2}}{X(x)Y(y)} = 0 \tag{4.8}$$

$$\frac{X''(x)}{X(x)} + \frac{Y''(y)}{Y(y)} = 0 \tag{4.9}$$

We now can separate the two parts of Eq. 4.34. Here we introduce separation constant λ^2 as:

$$\frac{X''(x)}{X(x)} = -\lambda^2 \tag{4.10}$$

and leading to a new equation as:

$$\boxed{X'' + \lambda^2 X(x) = 0} \tag{4.11}$$

Also we do for the Y:

$$\frac{Y''(y)}{Y(y)} = +\lambda^2 \tag{4.12}$$

and leading to a new equation as:

$$\boxed{Y'' - \lambda^2 Y(y) = 0} \tag{4.13}$$

The Equation is Ordinary differential equation, whose solution is known:

$$X(x) = C_1 sin(\lambda x) + C_2 cos(\lambda x) \tag{4.14}$$

The Equation is Ordinary differential equation, whose solution is known:

$$Y(y) = C_3 e^{(\lambda y)} + C_4 e^{(-\lambda y)} \tag{4.15}$$

We can write

$$\theta = X(x).Y(y) \tag{4.16}$$

$$\boxed{\theta = [C_1 sin(\lambda x) + C_2 cos(\lambda x)] . \left[C_3 e^{(\lambda y)} + C_4 e^{(-\lambda y)}\right]} \tag{4.17}$$

We apply boundary conditions one by one. Apply $\theta(0, y) = 0$:

$$0 = [C_1 sin(\lambda 0) + C_2 cos(\lambda 0)] . \left[C_3 e^{\lambda y} + C_4 e^{-\lambda y}\right]$$

$$0 = [C_1 sin(0) + C_2 cos(0)] . \left[C_3 e^{\lambda y} + C_4 e^{(-\lambda y)}\right]$$

$$0 = [C_2] . \underbrace{\left[C_3 e^{\lambda y} + C_4 e^{(-\lambda y)}\right]}_{Y(y)}$$

As Y(y) function cannot be zero the only possibility is that C_2 is zero. The θ solution takes new form:

$$\theta = [C_1 sin(\lambda x)] . \left[C_3 e^{(\lambda y)} + C_4 e^{(-\lambda y)}\right] \tag{4.18}$$

We apply boundary condition $\theta(x, 0) = 0$:

$$0 = [C_1 sin(\lambda x)] . \left[C_3 e^{(0)} + C_4 e^{(-0)}\right] \tag{4.19}$$

As $e^0 = 1$, we have

$$0 = [C_1 sin(\lambda x)] . [C_3 + C_4] \tag{4.20}$$

We already know that this possible only when $C_4 = -C_3$ or vice versa
      We can write:

$$\theta = [C_1 sin(\lambda x)] . \left[C_3 e^{(\lambda y)} - C_3 e^{(-\lambda y)}\right]$$

$$\theta = C_3 [C_1 sin(\lambda x)] . \left[e^{(\lambda y)} - e^{(-\lambda y)}\right] \tag{4.21}$$

We apply now BC $\theta(L, y) = 0$:

$$0 = C_3 [C_1 sin(\lambda L)] . \left[e^{(\lambda y)} - e^{(-\lambda y)}\right] \tag{4.22}$$

At this stage neither C_1 nor C_3 can be zero so only possibility is that $sin(\lambda L)$ is zero!
      As $sin(\lambda L) = 0$, and sinusoidal wave is a repetitive wave where it can be zero infinite times so we introduce

$$\lambda_n = n\pi/L$$

where n = 1, 2, 3,...∞. The integer n = 0 is excluded as it results in a trivial solution.

Also from Trigonometry we know

$$sinh(x) = (e^x - e^{-x})/2$$

This leads to form:

$$\theta = 2.C_3 C_1 sin(\frac{n\pi}{L}x)sinh(\frac{n\pi}{L}y) \tag{4.23}$$

We can merge $C_3.C_1$ as C_n and write:

$$\theta = 2.C_n sin(\frac{n\pi}{L}x)sinh(\frac{n\pi}{L}y) \tag{4.24}$$

This being repetivite for n times we introduce Σ in equation.

$$\boxed{\theta = 2.\sum_{n=1}^{\infty} C_n sin(\frac{n\pi}{L}x)sinh(\frac{n\pi}{L}y)} \tag{4.25}$$

We now apply the last boundary condition $\theta(x,W) = 1$:

$$1 = 2.\sum_{n=1}^{\infty} C_n sin(\frac{n\pi}{L}x)sinh(\frac{n\pi}{L}W) \tag{4.26}$$

Because $sin(n\pi\ x/L)$ is orthogonal in the interval $x = 0$ to L, we multiply both sides of the forgoing result by $sin\ (m\pi\ x/L)$ and integrate from $x = 0$ to L. We use property of orthogonal functions to estimate the solution. Multiply both sides of equation by $sin\ (m\pi\ x/L)$ and integrate from 0 to L.

$$\int_0^L sin(\frac{m\pi}{L}x)dx = 2.\int_0^L \sum_{n=1}^{\infty} C_n sin(\frac{m\pi}{L}x)sin(\frac{n\pi}{L}x)sinh(\frac{n\pi}{L}W)dx \tag{4.27}$$

Noting that the integral under the summation vanishes except when m = n, the constants C_n can be evaluated as follows.

$$C_n = \frac{2[cos(n\pi) - 1]}{sinh(\frac{n\pi}{L}W)(-cos(n\pi)sin(n\pi) + n\pi)} \tag{4.28}$$

so final form of solution is

$$\boxed{\theta = 2.\sum_{n=1}^{\infty} sin(\frac{n\pi}{L}x)sinh(\frac{n\pi}{L}y)\left[\frac{2[cos(n\pi) - 1]}{sinh(\frac{n\pi}{L}W)(-cos(n\pi)sin(n\pi) + n\pi)}\right]} \tag{4.29}$$

To obtain numerical results, we put $L = W = 1$ and sum up the series to 200 terms

MAPLE CODE

```
restart;

# Define the partial differential equation
Eq1 := diff(theta(x, y), x, x) + diff(theta(x, y), y, y) = 0;

# Define the separable variables
theta(x, y) := X(x) * Y(y);

# Divide Eq1 by theta(x, y)
Eq2 := Eq1 / theta(x, y);

# Expand Eq2
Eq3 := expand(Eq2);

# Expand the left-hand side of Eq3
Eq4 := expand(X(x) * (op(1, lhs(Eq3)) + lambda^2));

# Solve Eq4 for X(x)
sol1 := dsolve(Eq4, X(x));

# Expand the left-hand side of Eq3
Eq5 := expand(Y(y) * (op(2, lhs(Eq3)) - lambda^2));

# Solve Eq5 for Y(y)
sol2 := dsolve(Eq5, Y(y));

# Substitute constants
sol3 := subs(_C1 = _C3, _C2 = _C4, sol2);

# Calculate the product of solutions
sol4 := rhs(sol1) * rhs(sol3);

# Evaluate the result at x=0
eval(subs(x = 0, sol4)) = 0;

# Set _C2 to 0
_C2 := 0;

# Copy the result to sol5
sol5 := sol4;

# Evaluate the result at y=0
```

```
eval(subs(y = 0, sol5)) = 0;

# Set _C4 to -_C3
_C4 := -_C3;

# Collect the result with respect to _C3
sol6 := collect(sol5, _C3);

# Substitute x=L
subs(x = L, sol6) = 0;

# Define lambda as n*Pi/L
lambda := n * Pi / L;

# Convert the solution to trigonometric form
sol7 := convert(sol6, trig);

# Define the theta function as a series
theta(x, y) := sum(C[n] * sin(n * Pi * x / L) *
 sinh(n * Pi * y / L), n = 1 .. infinity);

# Set y=W
1 = subs(y = W, theta(x, y));

# Calculate the integral equation
sum(Int(C[n] * sin(n * Pi * x / L) * sin(m * Pi * x / L) *
 sinh(n * Pi * y / L), x = 0 .. L), n = 1 .. infinity) =
 Int(sin(m * Pi * x / L), x = 0 .. L);

# Define C[n] in terms of integrals
C[n] := int(sin(n * Pi * x / L), x = 0 .. L) / int(sin(n *
 Pi * x / L)^2 * sinh(n * Pi * W / L), x = 0 .. L);

theta(x,y):=subs(C[n]=C[n],theta(x,y));
theta2(x,y):=subs(W=1,L=1,infinity=200,theta(x,y));
 theta3:=evalf(theta2(x,y)):
 plot3d(theta3,x=0..1,y=0..1,axes=box);
 with(plots):
 contourplot(theta3,x=0..1,y=0..1,contours=
[seq(0.1*i,i=1..9)], axes=box,color=black);
```

Figures 4.3 and 4.4 show the temperature distribution as obtained from solution of partial differential equations in the two-dimensional region.

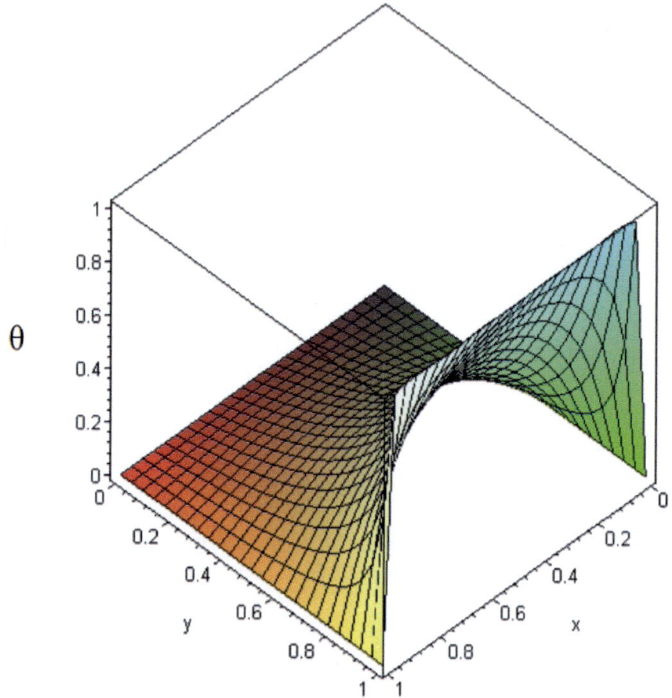

Figure 4.3   The surface view of temperature field inside the 2D region.

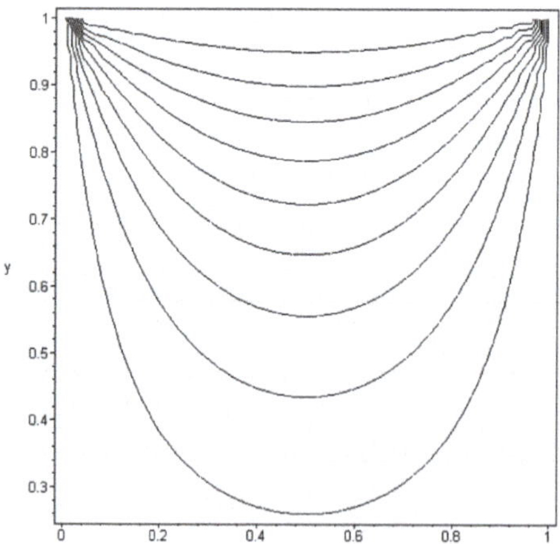

Figure 4.4   The planar view of temperature field inside the 2D region.

4.1   INFINITE STRIP CASE

$$\frac{\partial^2 T}{\partial x^2} + \frac{\partial^2 T}{\partial y^2} = 0 \tag{4.30}$$

To simplify the solution, we introduce the transformation

$$\theta = T - T_\infty$$

and rendered equations are

$$\frac{\partial^2 \theta}{\partial x^2} + \frac{\partial^2 \theta}{\partial y^2} = 0 \tag{4.31}$$

We now consider case of infinite strip in y direction with four boundary conditions in terms of the variable, θ, as:

$$\theta(0, y) = 0$$

$$\theta(b, y) = 0$$

$$\theta(x, 0) = f(x) - T_\infty = \phi(x)$$

$$\theta(x, \infty) = 0$$

Next a separation of variables solution is assumed with X being a function of x alone and Y being a function of y alone.

$$\theta(x, y) = X(x)Y(y) \tag{4.32}$$

We introduce this equation into Eq. 4.31

$$Y(y)\frac{\partial^2 X(x)}{\partial x^2} + X(x)\frac{\partial^2 Y(y)}{\partial y^2} = 0 \tag{4.33}$$

$$\frac{X''(x)}{X(x)} + \frac{Y''(y)}{Y(y)} = 0 \tag{4.34}$$

We now can separate the two parts of Eq. 4.34. Here we introduce separation constant λ^2 as:

$$\frac{X''(x)}{X(x)} = -\lambda^2 \tag{4.35}$$

and leading to a new equation as:

$$\boxed{X'' + \lambda^2 X(x) = 0} \tag{4.36}$$

Also we do for the Y:

$$\frac{Y''(y)}{Y(y)} = +\lambda^2 \tag{4.37}$$

and leading to a new equation as:

$$\boxed{Y'' - \lambda^2 Y(y) = 0} \tag{4.38}$$

The general solution of these equations gives

$$\theta = X(x).Y(y) \tag{4.39}$$

$$\boxed{\theta = [C_1 sin(\lambda x) + C_2 cos(\lambda x)] . \left[D_1 e^{(\lambda y)} + D_2 e^{(-\lambda y)}\right]} \tag{4.40}$$

We apply boundary conditions one by one. Apply $\theta(0,y) = 0$, which leads to C_2 is zero. We apply boundary condition

$$\theta(x,\infty) = 0$$

$$0 = [C_1 sin(\lambda x)] . \left[D_1 e^{(\lambda \infty)} + D_2 e^{(-\lambda \infty)}\right] \tag{4.41}$$

As

$$e^\infty = \infty$$

and

$$e^{-\infty} = 0$$

$$0 = [C_1 sin(\lambda x)] . [D_1(\infty) + D_2(0)] \tag{4.42}$$

leading to declare $D_1 = 0$

So equation takes form:

$$\theta = C_1 D_2 [sin(\lambda x)] . \left[e^{(-\lambda y)}\right] \tag{4.43}$$

to simplify we take $C_1 D_2 = C$ and equation becomes:

$$\theta = C[sin(\lambda x)] e^{(-\lambda y)} \tag{4.44}$$

We now apply BC

$$\theta(b,y) = 0$$

and this leads to:

$$0 = C[sin(\lambda b)] e^{(-\lambda y)} \tag{4.45}$$

Only possibility is that $sin(\lambda b) = 0$ which can be repetetively zero so for n times we can write $sin(\lambda_n b) = 0$ where $\lambda_n = n\pi/b$ and n=1,2,3,.... All these values satisfy equation $sin(\lambda b) = 0$ and are called eigenvalues.

$$\theta = C_n \sum_{n=1}^{\infty} \left[sin(\lambda_n x)\right] e^{(-\lambda_n y)} \tag{4.46}$$

We now apply BC

$$\theta(x,0) = f(x) - T_\infty = \phi(x)$$

and this leads to:

$$\phi(x) = C_n \sum_{n=1}^{\infty} \left[sin(\lambda_n x)e^{(-\lambda_n 0)}\right] \tag{4.47}$$

note $e^{(-\lambda_n 0)} \approx 1$

To proceed further, we invoke Orthogonality property of sine function:

$$\int_0^b sin(\frac{m\pi x}{b})sin(\frac{n\pi x}{b})dx = 0$$

if m $\neq$ n else

$$\int_0^b sin(\frac{m\pi x}{b})sin(\frac{n\pi x}{b})dx = b/2$$

if m = n.

$$\int_0^b sin(\frac{m\pi x}{b})\phi(x)dx = \sum_{n=1}^{\infty} \int_0^b C_n sin(\frac{m\pi x}{b})sin(\frac{n\pi x}{b})dx \tag{4.48}$$

All terms on RHS are zero except when m=n which gives

$$\int_0^b sin(\frac{m\pi x}{b})\phi(x)dx = C_m \frac{b}{2} \tag{4.49}$$

leading to define C_n back as:

$$\boxed{C_n = \frac{2.\int_0^b sin(\frac{n\pi x}{b})\phi(x)dx}{b}} \tag{4.50}$$

$$\theta = C_n \sum_{n=1}^{\infty} \left[sin(\lambda_n x)\right] e^{(-\lambda_n y)} \tag{4.51}$$

$$\theta = \sum_{n=1}^{\infty} \left[\frac{2.\int_0^b sin(\frac{n\pi x}{b})\phi(x)dx}{b}\right] \left[\left[sin(\lambda_n x)\right] e^{(-\lambda_n y)}\right] \tag{4.52}$$

Recalling

$$\theta = T - T_\infty$$

we get

$$T(x,y) = \sum_{n=1}^{\infty} \left[\frac{2 \cdot \int_0^b sin(\frac{n\pi x}{b})\phi(x)dx}{b} \right] \left[[sin(\lambda_n x)]\, e^{(-\lambda_n y)} \right] + T_\infty \qquad (4.53)$$

Using Fourier Law of heat conduction

$$q''(x) = -k \left(\frac{\partial T}{\partial y} \right)_{y=0} \qquad (4.54)$$

As if

$$f(y) = e^{\alpha y}$$

then

$$f' = \alpha e^{\alpha y}$$

so derivative of

$$e^{-n\pi y/b}$$

is

$$-\frac{n\pi}{b} e^{-n\pi y/b}$$

$$q''(x) = \frac{2\pi k}{b^2} \sum_{n=1}^{\infty} n \left[\int_0^b sin(\frac{n\pi x}{b})\phi(x)dx \right] [sin(\lambda_n x)] \qquad (4.55)$$

We cannot solve this until we specify function $\phi(x)$ in following equation:

$$q''(x) = \frac{2\pi k}{b^2} \sum_{n=1}^{\infty} n \left[\int_0^b sin(\frac{n\pi x}{b})\phi(x)dx \right] [sin(\lambda_n x)] \qquad (4.56)$$

Assuming $\phi(x) = T_o - T_\infty$, where T_o is a constant temperature at the base of fin. Also replacing sine function with infinite series we get:

$$T(x,y) = T_\infty + \frac{4}{\pi}(T_o - T_\infty) \sum_{n=0}^{\infty} \left(\frac{1}{2n+1} \right) sin \left[\frac{(2n+1)\pi x}{b} \right] e^{-\frac{(2n+1)\pi}{b}y} \qquad (4.57)$$

Example 4.1

The temperature maintained at 0 ^{o}C along three surfaces of a rectangular bar shown in figure, while fourth surface at y=b is held at temperature of 100 ^{o}C. If a =2b the centre line temperature under steady-state condition is:

$$\frac{T(x,y)}{T_o} = \left(\frac{2}{\pi} \right) \sum_{n=1}^{\infty} \left(\frac{1-(-1)^n}{n} \right) \left[\frac{sin(n\pi x/a)sinh(n\pi y/a)}{sin(n\pi b/a)} \right]$$

$$T(\frac{a}{2},\frac{a}{4}) = \left(\frac{2 \times 100}{\pi}\right) \sum_{n=1}^{\infty} \left(\frac{1-(-1)^n}{n}\right) \left[\frac{sin(n\pi a/2a)sinh(n\pi/4)}{sin(n\pi/2)}\right]$$

$$T(\frac{a}{2},\frac{a}{4}) = \left(\frac{2 \times 100}{\pi}\right) \sum_{n=1}^{\infty} \left(\frac{1-(-1)^n}{n}\right) \left[\frac{sin(n\pi/2)sinh(n\pi/4)}{sin(n\pi/2)}\right]$$

$$T(\frac{a}{2},\frac{a}{4}) = \left(\frac{2 \times 100}{\pi}\right) \sum_{n=1}^{\infty} \underbrace{\left(\frac{1-(-1)^n}{n}\right)}_{odd-terms} \left[\frac{\cancel{sin(n\pi/2)}sinh(n\pi/4)}{\cancel{sin(n\pi/2)}}\right]$$

$$T(\frac{a}{2},\frac{a}{4}) = \left(\frac{2 \times 100}{\pi}\right) \sum_{n=1,3,5...}^{\infty} \{[sinh(n\pi/4)]\}$$

4.2 CONDUCTION SHAPE FACTOR (CSF)

Shape factors for steady heat conduction offer a way for quick calculation of heat transfer rates within objects which are experiencing the combination of adiabatic and isothermal boundary conditions. The idea of Conduction Shape Factor was first given by Langmuir et al. in 1913. Langmuir determined the the conductance of copper sulphate solution in a container shaped like the plane wall and related it with the heat conduction phenomenon and formulated the concept of conduction shape factor. Although the concept of conduction shape factor comes from experiments, the later developments in this regards were purely mathematical. Conduction shape factors have been determined analytically for many geometrical configurations, and they are available in work of Andrews (1955), Sunderland and Johnson (1964), and Hahne and Grigull (1975) for common configurations encountered in engineering practices. The conduction shape factors are invariant when an object is conformally mapped onto another object Lienhard (2019)).

Pioneer of Heat Transfer

Irving Langmuir (31 January 1881 - 16 August, 1957) was an American scientist worked in many fields. Langmuir addressed the engineering impact on human behaviour and morality. He discovered that by twisting the filament into a tight coil the efficiency can be increased. He was one of the first scientists to work with ionized gas, and he was the first who gave it the name of *plasma* since it reminded him of plasma present in the blood. He invented an electrostatic probe, now known as Langmuir probe and it is used in plasma physics studies. He was awarded the Nobel Prize in Chemistry in 1932 for his work on chemistry of oil films.

The heat flow from an isothermal object at temperature T_1 to another body maintained at temperature T_2 can be expressed as

$$Q = \frac{(T_1 - T_2)}{R_{conduction}}$$

where conductive resistance and we know from previous discussion that conductive resistance is is inversely proportional to the thermal conductivity (k) of the medium between the two isothermal objects.

$$R_{conduction} \propto \frac{1}{k}$$

We can now define a new relationship between heat transfer and temperature difference without using area as

$$Q = kS(T_1 - T_2)$$

where S is the conduction shape factor (S) which has units of length. The heat transfer can be estimated as

$$Q = kS(T_1 - T_2) = \left(\frac{W}{m \cdot K}\right)(m)\,K = W$$

$$Q = kS(T_1 - T_2) = \left(\frac{Btu}{h \cdot ft \cdot {}^\circ F}\right) \times (ft) \times {}^\circ F = Btu/h$$

For SI system, the Q has units of watts and for USCS units Q has units of Btu/h. Note that some of the conduction shape factors need the evaluation of anti-cosine hyperbolic function which can easily be calculated as

$$\cosh^{-1}x = \ln(x \pm \sqrt{x^2 - 1})$$

$$S_{wall} = A/t, \quad S_{edge} = 0.54\ell, \quad S_{corner} = 0.15t$$

where A= wall Area, t=wall thickness, ℓ= wall length (see Figure 4.5).

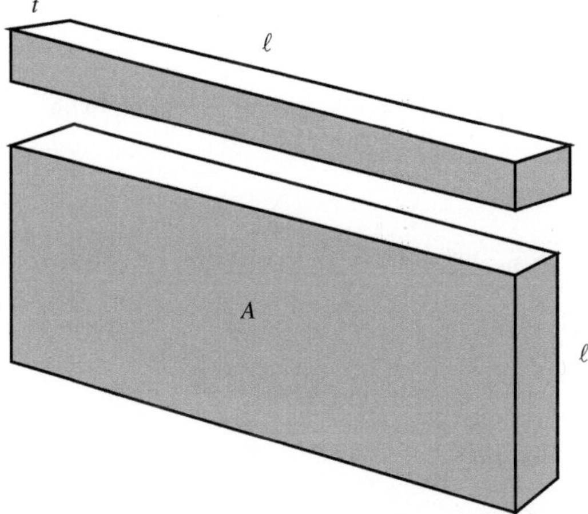

Figure 4.5　Schematic presentation for the important length to be used for the computation of shape factor for the wall, edge, and corner shape factors.

Example 4.2

A furnace of 0.6 m by 0.5 m by 0.7 m inside dimensions is constructed of a material having a thermal conductivity of 0.865 W/m·K as shown in Figure 4.6. The thickness of the furnace wall is 7 in, and the inner and outer wall temperatures are 500°C and 100°C, respectively. Estimate the heat loss through the furnace wall.

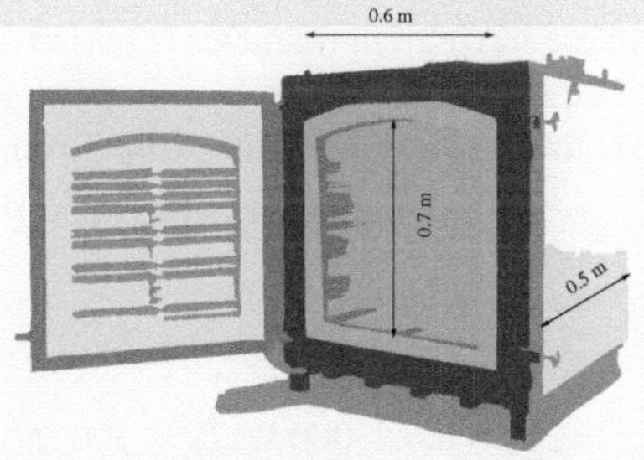

Figure 4.6　A furnace of 0.6 m by 0.5 m by 0.7 m inside dimensions is constructed of a material having a thermal conductivity of 0.865 W/m · K.

Solution　The inside dimensions of the furnace are: $L = 0.6m$, $W = 0.5m$, $H = 0.7m$, and thickness t=0.1778 m. The areas inside furnace are

$$A_1 = L \cdot W = 0.3m^2, \; A_2 = W \cdot H = 0.35m^2, \; A_3 = H \cdot L = 0.42m^2$$

There are six walls inside furnace:

$$S_{walls} = 6\left[\frac{(A_1 + A_2 + A_3)}{t}\right] = 36.10794m$$

There are four edges:

$$S_{edges} = 4 \times [0.54 \cdot (L + W + H)] = 3.888m$$

There are eight corners:

$$S_{corners} = 8 \times (0.15 \cdot t) = 0.2133m$$

The total shape factor is

$$S_{total} = S_{walls} + S_{edges} + S_{corners} = 40.20m$$

The total heat loss is

$$Q_{total} = k \cdot S_{total} \cdot (T_2 - T_1)$$

$$Q_{total} = 0.865\frac{W}{m \cdot K} \times 40.20m \times (500 - 100)^{\circ}C = 13911.96W$$

The conduction shape factors for several geometric configurations are tabulated in Table 4.1–4.3.

Example 4.3

Buried sphere A sphere having 2m diameter is buried 2 m in the ground has isothermal temperature of 300°C. The sphere is generating heat at rate of 4 kW. Find the ground thermal conductivity, if the ground temperature is 25 °C.

Solution
The Given data are

$$r = 1m, \quad z = 2m, \quad Q = 4,000\,W, T_1 = 300\,^{\circ}C, T_2 = 25\,^{\circ}C$$

We use Case 10 for solution of this problem. The thermal shape factor is

$$S = \frac{2\pi r}{1 + \left(\frac{r}{2z}\right)}$$

$$S = \frac{2\pi(1)}{1 + \left(\frac{1}{4}\right)} = 16.755m$$

$$Q = kS(T_1 - T_2) = k \times 16.755(300 - 25) = 4607.669k$$

TABLE 4.1

Conduction Shape Factors $q = S \cdot k(T_1\text{-}T_2)$

Case	Conduction Shape Factor
Case 1: Infinite square pipe 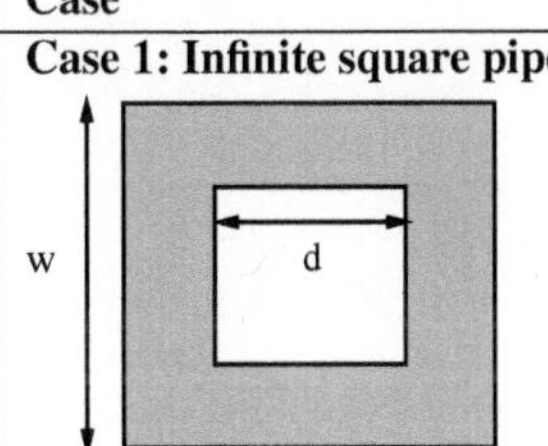	$$S = \frac{2\pi L}{0.93\ln(w/d) - 0.0502} \quad w/d > 1.4$$ $$S = \frac{2\pi L}{0.785\ln(w/d)} \quad w/d < 1.4$$
Case 2: Infinite hollow cylinder 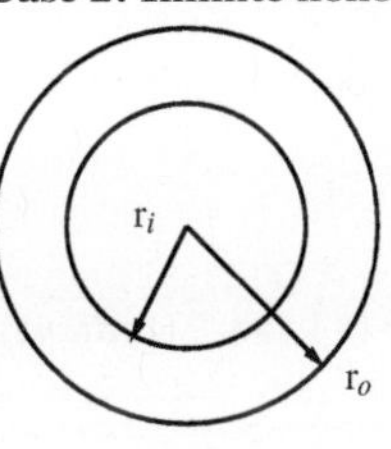	$$S = \frac{2\pi L}{\ln\left(\frac{r_o}{r_i}\right) + \frac{1}{Bi_i} + \frac{1}{Bi_o}}$$
Case 3: Hemisphere buried in semi-infinite medium 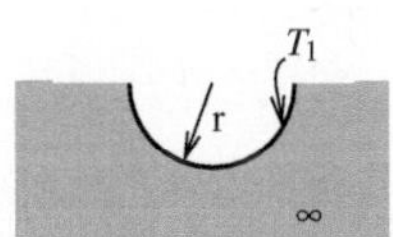	$$S = 2\pi r$$
Case 4: Buried cylinder 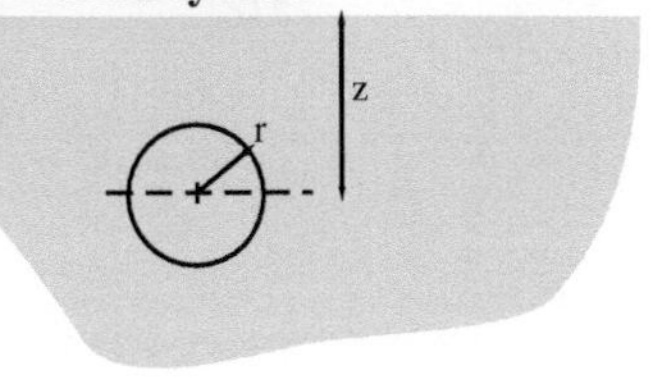	$$S = \frac{2\pi L}{\cosh^{-1}(z/r)}$$ $$S = \frac{2\pi L}{\ln(2z/r)} \quad z > 3r$$

TABLE 4.2

Conduction Shape Factors

Case	Conduction Shape Factor
Case 5: Buried rectangular parallelepiped in semi-infinite medium at depth h 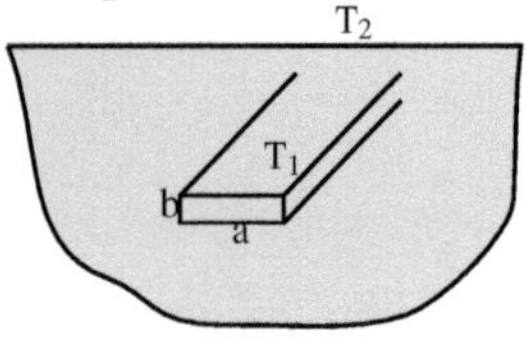	$S = 1.685L\left[\log\left(1+\dfrac{h}{a}\right)\right]^{-0.59}\left(\dfrac{h}{b}\right)^{-0.078}$
Case 6: Buried thin horizontal disk 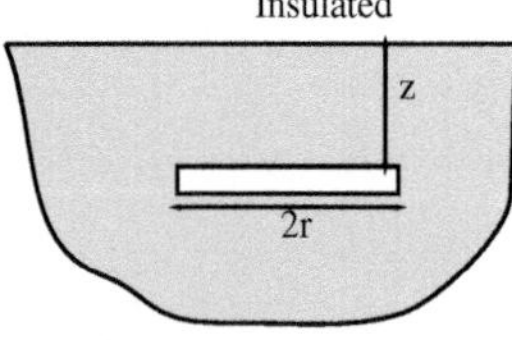	$S = \dfrac{4\pi r L}{\frac{\pi}{2}+\tan^{-1}\left(\frac{r}{2z}\right)} \quad z/2r > 1, \tan^{-1}(radians)$
Case 7: Eccentric cylinders 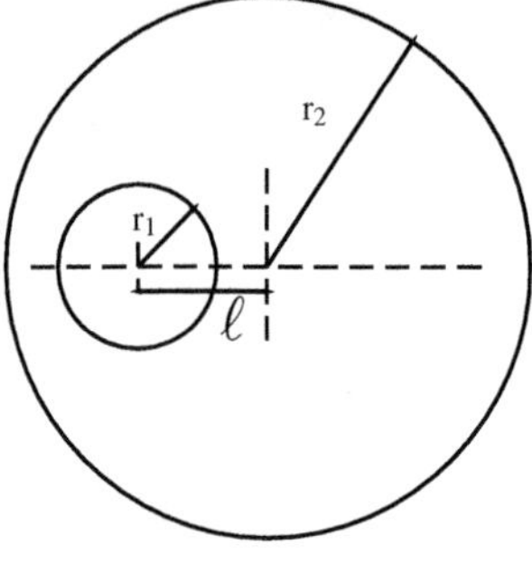	$S = \dfrac{2\pi L}{\cosh^{-1}\left(\frac{r_1^2+r_2^2-\ell^2}{2r_1 r_2}\right)}$
Case 8: Two buried cylinders 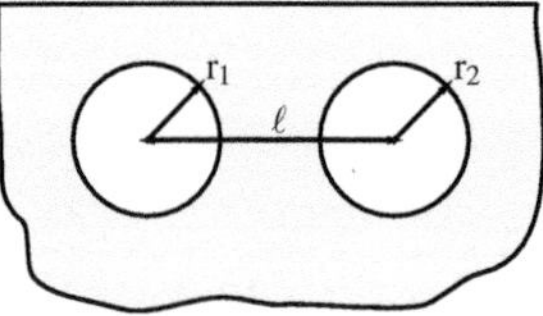	$S = \dfrac{2\pi L}{\cosh^{-1}\left(\frac{\ell^2-r_1^2-r_2^2}{2r_1 r_2}\right)}$

TABLE 4.3

Conduction Shape Factors

Case	Conduction Shape Factor
Case 9: Vertical cylinder in a semi-infinite medium 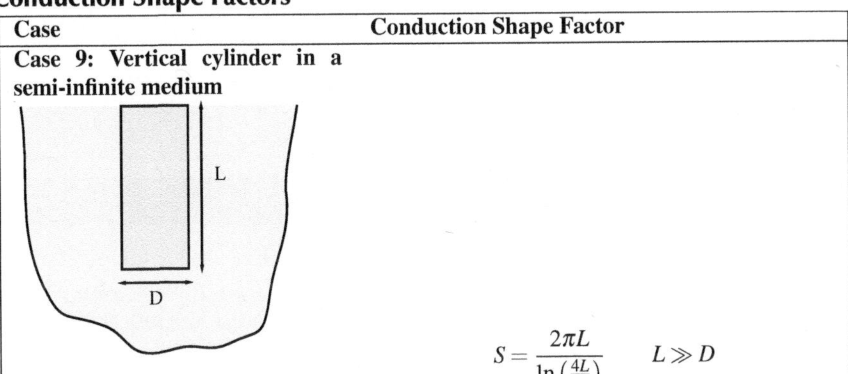	$S = \dfrac{2\pi L}{\ln\left(\frac{4L}{D}\right)} \qquad L \gg D$
Case 10: Buried isothermal sphere 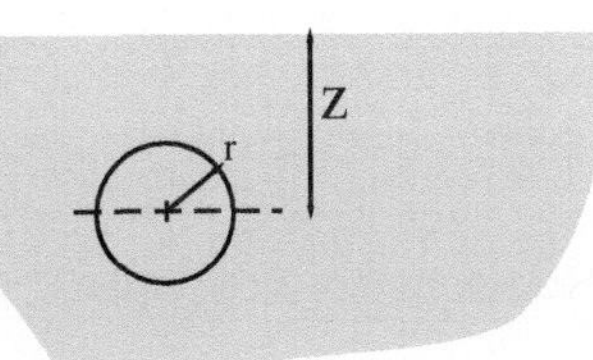	$S = \dfrac{4\pi r}{1-(r/2z)} \qquad h > r$
Case 11: Horizontal isothermal cylinder of dia D 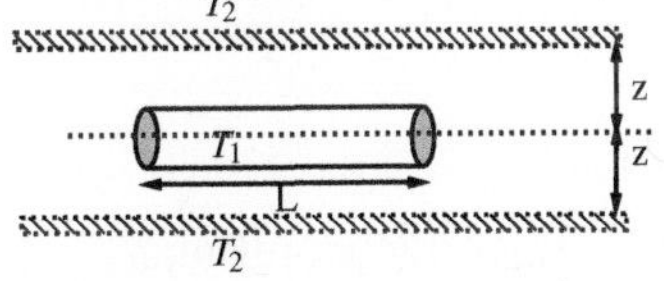	$S = \dfrac{2\pi L}{\ln(8z/\pi D)}$
Case 12: Concentric spheres 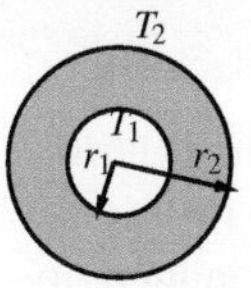	$S = \dfrac{4\pi r_1 \cdot r_2}{(r_2 - r_1)}$
Case 13: Two isothermal spheres buried in infinite medium 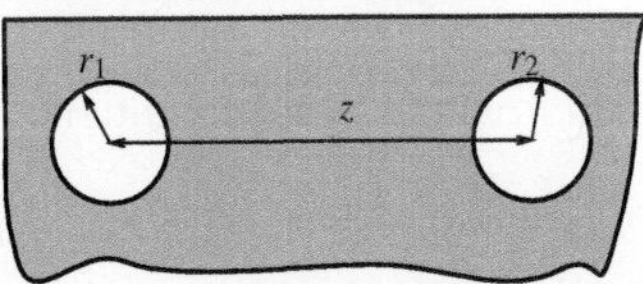	$\dfrac{4 \cdot \pi \cdot r_2}{\dfrac{r_2}{r_1}\cdot\left(1 - \dfrac{\left(\frac{r_1}{z}\right)^4}{1-\left(\frac{r_2}{z}\right)^2}\right) - \dfrac{2 \cdot r_2}{z}} \qquad z > 5 r_{max}$

where k is thermal conductivity. $Q = 4$ kW so equation will take form

$$4000. - 4607.669k = 0$$

$$k = 0.868 W/m \cdot K$$

This is the thermal conductivity of ground.

Example 4.4

Buried pipe A 90-ft-long pipe buried 3 ft in ground has surface temperature of 200 °F. The ground temperature is 37 °F. Find the heat transfer from pipe to the ground if the the ground thermal conductivity is 0.876 Btu/h.ft.°F and the pipe radius is 0.06 ft.

Solution
The data given are

$$r = 0.06 ft, \quad z = 3 ft, \quad T_1 = 200\,°F, T_2 = 37\,°F$$

We use the Case 4 for the solution of this problem. Since z>3r, the conduction shape factor is

$$S = \frac{2\pi L}{\ln\left(\frac{2z}{r}\right)}$$

$$S = \frac{2\pi(90)}{\ln\left(\frac{2\times 3}{0.06}\right)} = 122.79 ft \qquad z > 3r$$

The heat transfer is

$$Q = kS(T_1 - T_2) = 0.876 \times 122.79 \times (200 - 37) = 17533.49 Btu/h$$

Example 4.5

A bare pipe having 2.5 inch external pipe radius is 5 statue mile long is buried in ground. The pipe's surface temperature is 345 °F and pipe is positioned z=5 ft. The ground temperature is 40 °F and ground's thermal conductivity is 1.5 Btu/hr·ft·°F.

i Calculate the conduction shape factor.

ii Calculate the amount of heat lost to the ground in steady-state condition.

Solution

This is the Case 4 of buried cylinder. The pipe is uninsulated and heat flows from pipe's surface to the surrounding ground.

Since z>3r, we calculate the conduction shape factor as

$$S = \left(\frac{2 \cdot \pi \cdot L}{\text{arccosh}\left(\frac{z}{r}\right)} \right) = \left(\frac{2 \cdot \pi \cdot 26400}{\text{arccosh}\left(\frac{5}{2.5/12}\right)} \right) = 42853.54$$

The heat transfer associated with this case is

$$Q = k \cdot S \cdot (T_w - T_\infty)$$

$$Q = 1.5 \frac{Btu}{hr \cdot ft \cdot {}^\circ F} \times 42853.54 \ ft \times (345 - 40){}^\circ F$$

$$Q = 1.960 \times 10^7 \frac{Btu}{hr}$$

PROBLEMS

4P-1 A thin rectangular plate of length a and width b is subject to following boundary conditions:

At x=0: T = 400 K, and at x=a: T = 500 K

At y=0: T = 430 K, and at y=b: T = 480 K

Solve from 2D conduction heat equation and obtain the expression for temperature distribution.

4P-2 A thin rectangular plate of length a and width b is subject to following boundary conditions:

y=0: dT/dy=0; y=b: T = 100 C

x=0: T = 0 C; x=a: T = 0 C

Solve from 2D conduction heat equation and obtain the expression for temperature distribution.

4P-3 Two long pipes of 3 m are placed in a thick concrete block during construction of the wall. The pipe radii are 35 mm and 50 mm, respectively, with centre-to-centre distance of 100 mm. One of the pipe is carrying steam at $100{}^\circ C$ and the another is carrying the city water at $20{}^\circ C$. If the thermal conductivity of concrete is 0.8 W/m.K, find the rate of conductive heat transfer between the pipe through the concrete.

[Ans: $Q = -1015.35W$]

4P-4 A horizontal cylinder of 5 m length and 60 mm diameter is placed in the middle of two parallel plates each having temperature of 300 K. The space between the cylinder and plates is filled up with material having thermal conductivity $k=8$ W/m·K. Find the temperature of the cylinder wall if the heat transfer rate between the cylinder and plates is 800 W, and the distance from one plate to cylinder centre is 150 mm.

[Ans: T_{cyl}=21.108°C]

4P-5 Two off centre cylinders are arranged as one inside another. The large cylinder has radius 5 cm and the smaller one has radius 1.4 cm. The distance between cylinders' centers is 1 cm. The thermal conductivity of large cylinder is k=15 W/m·K. Find the conduction shape factor and the heat transfer between the cylinders if large cylinder is at 300 K and smaller one is maintained at 298 K.

[Ans: $S = 5.11m, Q = -30W/m$]

4P-6 Two sphere of diameters 4 cm and 6 cm are placed 15 cm apart. The temperatures of spheres are 300°C and 500°C. Find the heat transfer from the spheres.

[Ans: Q=1964.14 W]

4P-7 A rectangular slab of width 30 cm and height 10 cm in cross-section is of 1 m length. The slab is placed at the depth of 2 m. The temperature at the surface is 40°C and temperature of the slab is 5°C.

[Ans: Q=8.72044 W]

REFERENCES

P. J. Schneider, Conduction Heat Transfer, Addison-Wesley, 1955.

H. S. Carslaw and J. C. Jaeger, Conduction of Heat in Solids, Oxford University Press, 1959.

V. S. Arpaci, Conduction Heat Transfer, Addison-Wesley, 1966.

G. E. Myers, Analytical Methods in Conduction Heat Transfer, McGraw-Hill, 1971.

R. V. Andrews, Solving conductive heat transfer problems with electrical-analogue shape factors, Chemical Engineering Progress, Vol. 51, pp. 67–71, 1955.

I. Langmuir, E. Q. Adams, and G. S. Meikle, Flow of heat through furnace walls: the shape factor, Transactions American Electrochemical Society, Vol. 24, pp. 53–81, 1913.

J. E. Sunderland and K. R. Johnson, Shape factors for heat conduction through bodies with isothermal boundaries, Transactions ASHRAE, Vol. 70, pp. 237–241, 1964.

E. Hahne and U. Grigull, Formfaktor und Formwiderstand der stationären mehrdimensionalen Wärmeleitung, International Journal Heat Mass Transfer, Vol. 18, 1975, pp. 751–767, 1975.

A. F. Mills. Basic Heat and Mass Transfer. 2nd ed, Upper Saddle River, NJ: Prentice Hall, 1999.

J. H. Lienhard V, Exterior Shape Factors From Interior Shape Factors. Journal of Heat Transfer 141, 6, American Society of Mechanical Engineers, 2019, USA.

A. Aziz, Heat Conduction with Maple, R.T. Edwards, USA, 2005.

5 Transient Heat Conduction

In certain situations, the assumption of steady-state heat transfer cannot be invoked and we must consider the variation of thermal field with time. In this chapter, we will discuss the heat transfer in bodies in which the temperature varies with time. After finishing this chapter, one will be able to:

The learning outcomes:
- Analyse the heat conduction in a solid object using lumped capacity technique (if $Bi \leq 0.1$).
- Use analytical models for some commonly encountered transient heat conduction problems irrespective of Biot number.

Inherently heat transfer problems are unsteady in nature, but we may arrive at meaningful results using the steady-state or quasi-static approximation, as such approximations can help in making the analysis simple. Also, in most cases, the steady-state analysis is easier to arrive at. The steady-state analysis is good enough for engineering practices at the ground level; however, care must be taken in making assumptions and approximations as the complexity of the problem may drastically change if the temperatures vary both in space and time. For example, we can treat the heat transfer through the solid walls in a typical house as a steady-state problem but it would not be a good idea to consider the steady-state heat transfer in the oven.

During the unsteady heat transfer process, the temperature varies with time as well as in space position. In some special cases, we can assume that the object is not going through the variation of temperature with position rather temperature change is happening only with time. This crude approximation is valid as long as the temperature of the medium changes uniformly in space with time. The analysis based on such approximation is called the lumped analysis of the problem and it is a type of an integral analysis. The lumped analysis is an old technique but found to be a useful tool specially in case of unsteady heat conduction phenomenon. The technique is applicable as long as the conducting body's internal resistance to heat diffusion is small compared to the convective resistance at the conducting body's surface. The dimensionless Biot number can be used for quick calculation to check the validity of analysis as

$$Bi = \frac{hL}{k} < 0.1 \quad (\textit{Lumped Analysis is valid})$$

The above limit is a simply a rule of thumb.

DOI: 10.1201/9781003428404-5

5.1   ONE-DIMENSIONAL UNSTEADY HEAT CONDUCTION (BI<0.1)

Using Lumped analysis approach, we now analyse the transient heat conduction for a body of volume $\forall$ shown in Figure 5.1. The body initial temperature is T_i and the specific heat is c. The heat stored inside the body is increased/decreased as the body comes in contact with the fluid and convection occurs. The phenomenon is unsteady, and the body will quickly attain an equilibirium state.

$$Q_{storage} = -\rho c \forall \frac{dT}{dt}$$

According to Newton's law of cooling, we have

$$Q_{conv} = h \cdot A(T - T_\infty)$$

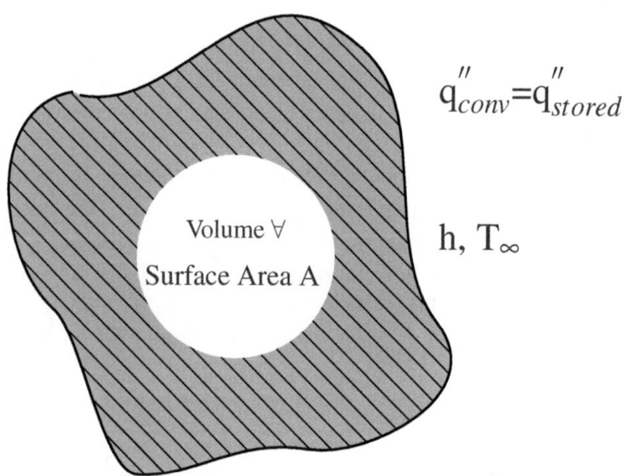

Figure 5.1   Schematic representation of lumped analysis of transient conduction phenomenon process.

There will be a balance between convection and internal energy stored:

$$Q_{conv} = Q_{storage}$$

$$hA(T - T_\infty) = -\rho c \forall \frac{dT}{dt}$$

$$-\left(\frac{hA}{\rho c \forall}\right) dt = \frac{dT}{(T - T_\infty)}$$

For sake of simplification, we set $\beta = \frac{hA}{\rho c \forall}$ and we can write

$$-\int_0^t \beta \cdot dt = \int_{T_i}^T \frac{dT}{(T - T_\infty)}$$

Integration will lead to form

$$\ln\left(\frac{T - T_\infty}{T_i - T_\infty}\right) = -\beta t$$

$$\frac{T - T_\infty}{T_i - T_\infty} = \exp(-\beta t)$$

Rearranging the temperature we get

$$T(t) = (T_i - T_\infty)\exp(-\beta t) + T_\infty$$

The total energy stored in the body is

$$Q_{total} = -\int \left(\rho c \forall \frac{dT}{dt}\right) dt = -\int_{T_i}^T (\rho c \forall) dT$$

$$Q_{total} = -\rho c \forall (T - T_i)$$

Substituting the temperature in above equation and rearranging, we have

$$Q_{total} = \rho c \forall (T_i - T_\infty)\left[1 - \exp(-\beta t)\right]$$

We can also introduce the Biot number into β parameter by introducing characteristic length L:

$$\beta = \frac{hA}{\rho c \forall} \approx \frac{hL^2}{\rho c L^3} \approx \frac{hL}{\rho c L^2} \approx \frac{\alpha hL}{kL^2} \equiv \left(\frac{hL}{k}\right)\left(\frac{\alpha}{L^2}\right)$$

Further, the product of β with time gives us

$$\beta \cdot t = \left(\frac{hL}{k}\right)\left(\frac{\alpha t}{L^2}\right) = Bi \cdot Fo$$

Here, Fo is Fourier number, which is another useful dimensionless number in transient heat diffusion problems, often defined as dimensionless time for diffusion process.

$$\frac{T(t) - T_\infty}{T_i - T_\infty} = \exp(-Bi \cdot Fo) \tag{5.1}$$

$$Q_{total} = \rho c \forall (T_i - T_\infty)\left[1 - \exp(-Bi \cdot Fo)\right] \tag{5.2}$$

$$Q_{total} = \frac{h \cdot (A_s) \cdot (T_i - T_\infty) \cdot (1 - \exp(-Bi \cdot Fo)) \cdot t}{Bi \cdot Fo} \tag{5.3}$$

In practice, the characteristic length can be estimated as L=∀/A. The characteristic lengths for sphere, cylinder, and cube are listed in Table 5.1:

TABLE 5.1

Characteristic Lengths for Sphere, Cylinder, and Cube

Sphere of Radius r	Cylinder of radius r	Cube of side a
r/3	r/2	a/6

lumped parameter analysis useful in case of measurements of temperature by a thermocouple. The response of a thermocouple is the time taken by the thermocouple to attain the source temperature. It is clear from equation that the larger the quantity β, more rapid will be the response of the temperature measuring device as faster shall be the exponential term will approach zero. The reciprocal of parameter β is actually called time constant of the thermocouple. In order to achieve the rapid response of the temperature measuring device or thermocouple the viable option is to decrease the wire diameter, and reduce the heat capacity of the thermocouple material. Hence, a very thin wire is recommended to be used in thermocouples to ensure a rapid response. The response times for most of the thermocouple wires falls in between 0.04 and 2.5 seconds.

Example 5.1

A hot steel spherical ball having diameter 22cm, at uniform surface temperature of 650°C is taken out of a furnace and immersed into an oil bath. The oil bath temperature is around 40°C, and the convective heat transfer coefficient surrounding ball is around 120 W/m^{2}K. Estimate the time the steel ball will take to reach temperature of 200°C.
The properties of steel are: ρ= 7830 kg/m^3, c=465 J/kg·K, k=46W/m·K.

Solution The radius of spherical ball is 11cm and the characteristic length is $L = R/3$. The applicable equation is

$$\frac{T(t) - T_\infty}{T_i - T_\infty} = \exp(-Bi \cdot Fo)$$

We calculate the Biot and Fourier numbers

$$Bi = \frac{hL}{k} = \frac{120\,W/m^2 \cdot K \times (11/300)m}{46W/m \cdot K} = 0.0956 < 0.1$$

As Biot number is less than 0.1, the lumped-capacity method is valid for this problem. The Fourier number is

$$Fo = \frac{\alpha t}{L^2} = \frac{0.00001263406529 \times t}{(11/300)^2} = 0.00939t$$

We calculate the temperature difference ratio:

$$\frac{T(t) - T_\infty}{T_i - T_\infty} = \frac{200° - 40°}{650° - 40°} = 0.2622$$

The equation

$$\frac{T(t) - T_\infty}{T_i - T_\infty} = \exp(-Bi \cdot Fo)$$

takes the form

$$0.2622 - \exp[0.0956 \times 0.00939t] = 0$$

The solution of this equation gives the time as

$$t = 1488.858s = 24.814min$$

5.2 SEMI-INFINITE SOLID WITH DIFFERENT BOUNDARY CONDITIONS

5.2.1 SEMI-INFINITE PLANE WALL: CONSTANT WALL TEMPERATURE

We now consider the case of one-dimensional, unsteady conduction in a semi-infinite plane wall as shown in Figure 5.2. The wall one side is maintained at temperature T_w and the initial uniform temperature inside wall is T_i. The boundary conditions are written as

$$T(0,t) = T_w \quad for\ t > 0$$

$$T(x,0) = T_i$$

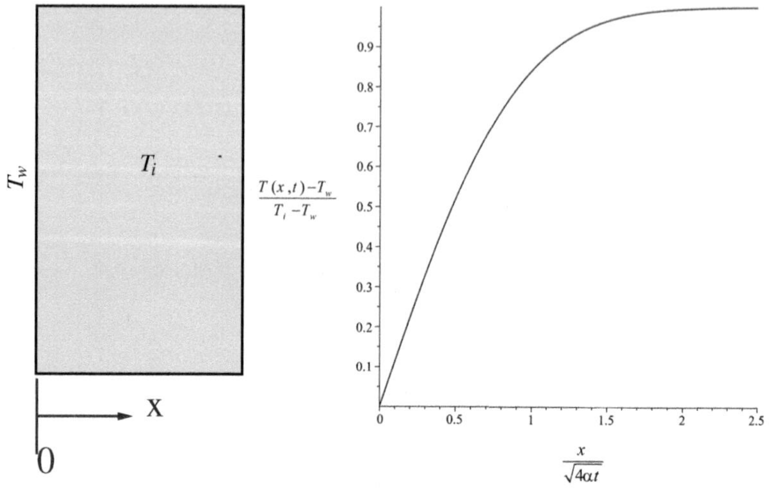

Figure 5.2   Schematic representation for transient heat flow in a semi-infinite solid.

The conduction equation is

$$\alpha \frac{\partial^2 T}{\partial x^2} = \frac{\partial T}{\partial t}$$

> The exact solution of this problem is
>
> $$\frac{T(x,t) - T_w}{T_i - T_w} = erf\left(\frac{x}{\sqrt{4\alpha t}}\right)$$

where, error function is defined as

$$erf\left(\frac{x}{\sqrt{4\alpha t}}\right) = \frac{2}{\sqrt{\pi}} \int\limits^{x/\sqrt{4\alpha t}} e^{-\eta^2} d\eta$$

The temperature derivative is

$$\frac{\partial T}{\partial x} = \frac{e^{-\frac{x^2}{4\alpha t}}(T_i - T_w)}{\sqrt{\pi}\sqrt{\alpha t}}$$

At the surface (x = 0), the heat flow is

$$q'' = -k\frac{\partial T}{\partial x} = \frac{-k(T_i - T_w)}{\sqrt{\pi}\sqrt{\alpha t}}$$

The total heat flow (Q_o) in time t can be obtained by integrating above equation as

$$Q_o = \frac{-kA}{\sqrt{\pi\alpha}}(T_i - T_w) \int\limits_0^t \frac{dt}{\sqrt{t}} = \frac{-2kA\sqrt{t}}{\sqrt{\pi\alpha}}(T_i - T_w)$$

Example 5.2

The surface temperature of large slab of copper is suddenly increased to $200°C$. The initial temperature of copper was at a uniform temperature of $30°C$. The cooper material properties are k = 388 W/m·K, $\alpha = 113 \times 10^{-6}$ m^2/s.

(i) Find temperature of the slab a depth 20.0 cm, after 30s.

(ii) Total heat transfer to the slab per unit surface area after passing of 30s.

Solution

$$\frac{T(x,t) - T_w}{T_i - T_w} = erf\left(\frac{x}{\sqrt{4\alpha t}}\right)$$

$$\frac{T(x,t) - T_w}{T_i - T_w} = \frac{T - 200}{30 - 200}$$

$$\frac{T(x,t) - T_w}{T_i - T_w} = -0.00588T + 1.176470588$$

$$erf\left(\frac{x}{\sqrt{4 \cdot \alpha \cdot t}}\right) = 0.9848565792$$

$$\frac{T(x,t) - T_w}{T_i - T_w} = erf\left(\frac{x}{\sqrt{4\alpha t}}\right)$$

This gives the equation

$$-0.00588T + 0.1916140088 = 0$$

Solving this we get

$$T = 32.574°C$$

We now calculate the total heat transfer as

$$Q_o = \frac{-2kA\sqrt{t}}{\sqrt{\pi\alpha}}(T_i - T_w)$$

$$q_o = -2 \cdot k \cdot (T_i - T_w) \cdot \sqrt{\frac{t}{\alpha\pi}} = 3.835 \times 10^7 J/m^2$$

where q_o is Q_o/A.

5.2.2   SEMI-INFINITE SOLID: HEAT FLUX BOUNDARY CONDITION

The semi-infinite solid initially at uniform temperature T_i,

$$T(x,0) = T_i$$

and suddenly the surface of solid is exposed to a constant surface heat flux Q_o. The temperature at a depth x at any time is

$$Q_o = -kA\left(\frac{\partial T}{\partial x}\right)_{x=0} \qquad for\ t > 0$$

the temperature distribution inside solid is

$$T(x,t) - T_i = \frac{Q_o}{kA}\left\{\sqrt{\frac{4\alpha t}{\pi}}\exp\left(\frac{-x^2}{4\alpha t}\right) - x\left[1 - erf\left(\frac{x}{\sqrt{4\alpha t}}\right)\right]\right\}$$

Example 5.3

A semi-infinite steel initially at a uniform temperature of 35°C is exposed to a constant surface heat flux of 5×10^5 W/m^2. Estimate the temperature at a depth of 1 cm after a time of 30 s. The properties of steel are k = 46 W/m · °C, $\alpha = 1.43 \times 10^{-5}$ m^2/s]

Solution
The given constant surface heat flux is $Q_o/A = 5 \times 10^5$ W/m^2, x = 1/100 m and t = 30 s. The applicable solution for this problem is

$$T(x,t) - T_i = \frac{Q_o}{kA}\left\{\sqrt{\frac{4\alpha t}{\pi}}\exp\left(\frac{-x^2}{4\alpha t}\right) - x\left[1 - erf\left(\frac{x}{\sqrt{4\alpha t}}\right)\right]\right\}$$

$$T(x,t) - 35 = \frac{Q_o}{kA}(0.02337 \times 0.9434 - 0.007328)$$

$$T = 195.0633349°C$$

5.2.3   SEMI-INFINITE SOLID: CONVECTION BOUNDARY CONDITION

The semi-infinite solid initially at uniform temperature Ti and wall is exposed to a convective environment with fluid at T_∞ and a convective coefficient h. The temperature at a depth x from the surface at any time t will be

$$\frac{T(x,t) - T_i}{T_\infty - T_i} = erfc\left(\frac{x}{\sqrt{4\alpha t}}\right) - \left\{\exp\left(\frac{hx}{k} + \frac{\alpha t h^2}{k^2}\right) \cdot erfc\left(\frac{x}{\sqrt{4\alpha t}} + \frac{\sqrt{\alpha t h^2}}{k}\right)\right\}$$

> ### Example 5.4
>
> It is planned to construct a seed vault in a cold region. The vault is planned to be an under ground storage space. The atmospheric temperature is -13°C considered to be remain like this for 5 weeks. How deep should the vault be to prevent the temperature to be less than -8°C? The combined convective and radiative coefficient at the ground's surface is estimated to be 30 W/m^2·K. Assume that the ground is initially at a uniform temperature of 10°C. The ground properties are k=2.2 W/m·K, ρ=1700 kg/m^3, c=2100 J/kg·K
>
> ***Solution***
> Time = t = 5 weeks $\times$ 7 $\frac{days}{week}$ $\times$ 24 $\frac{hr}{day}$ $\times$ 3600 $\frac{s}{hr}$ = 3024000 s
> The applicable equation is
>
> $$\frac{T(x,t) - T_i}{T_\infty - T_i} = erfc\left(\frac{x}{\sqrt{4\alpha t}}\right)$$
>
> $$-\left\{\exp\left(\frac{h\,x}{k} + \frac{\alpha\,t\,h^2}{k^2}\right) \cdot erfc\left(\frac{x}{\sqrt{4\alpha t}} + \frac{\sqrt{\alpha\,t\,h^2}}{k}\right)\right\}$$
>
> The left-hand side is
>
> $$\frac{T(x,t) - T_i}{T_\infty - T_i} = \frac{-8 - 10}{-13 - 10} = 0.7826$$
>
> Inserting data into above equation, we have
>
> $$-0.782 + erfc(0.409x) - erfc(0.409x + 16.649) \cdot exp(13.63x + 277.21) = 0$$
>
> Solving for x we have x= 0.40358 m = 1.324 ft.

5.2.4   SEMI-INFINITE SOLID: ENERGY PULSE BOUNDARY CONDITION

If the solid initially at T_i is exposed to a short, instantaneous pulse of energy at the surface having an energy flux E_o/A, the temperature at time t>0 is

$$T(x,t) - T_i = \frac{E_o}{A\rho c\sqrt{\pi\alpha t}} \exp\left(\frac{-x^2}{4\alpha t}\right)$$

> ### Example 5.5
>
> In an industrial process, an instantaneous laser pulse of 20 MJ/m^2 is beamed on a semi-infinite stainless steel strip. Before that the strip has a uniform temperature of 60 °C.

(i) Find the temperature at the surface of the strip and at a depth of 2.0 mm after a time of 5 s.

(ii) Find the pulse flux at which the steel would start melting.

The thermophysical properties of stainless steel are $\rho = 8000$ kg/m^3, c $= 450$ J/kg·K, and $\alpha = 4.4 \times 10^{-6}$ m^2/s.

Solution
(i) At x=2mm we have

$$T(x,t) - T_i = \left(\frac{Q_o}{A\rho c \sqrt{\pi \alpha t}} \right) \exp \left(\frac{-x^2}{4\alpha t} \right) = 668.42 \times 0.9555 = 638.72\,^{\circ}\text{C}$$

$$T = 698.72^{\circ}C$$

(ii) The melting temperature of stainless steel is 1375°C. We will continue to increase E_o until we reach surface temperature (x= 0 mm) of 1375°C.

$$E_o = 4.114 \times 10^7 = 41.14 MJ/m^2$$

5.3 TRANSIENT CONDUCTION INSIDE A LARGE SLAB WITH CONVECTIVE COOLING

We now consider the transient heat conduction through a slab with the extent x=-L and x=+L. The slab is suddenly subjected to convective heat transfer with a convective coefficient h. Figure 5.3 shows schematic representation for transient heat flow in a slab with convective boundary conduction at walls. We assume that slab has constant thermal conductivity and has no internal heat generation. The transient heat conduction equation is

$$\alpha \frac{\partial^2 T}{\partial x^2} = \frac{\partial T}{\partial t}$$

We can immediately recognise that the temperature distribution will be symmetrical about x=0, and therefore, the boundary condition equation at the centre of the slab is

$$x = 0 \qquad \frac{\partial T}{\partial x} = 0$$

and at the walls

$$x = \pm L \qquad -k\frac{\partial T}{\partial x} = hA(T - T_\infty)$$

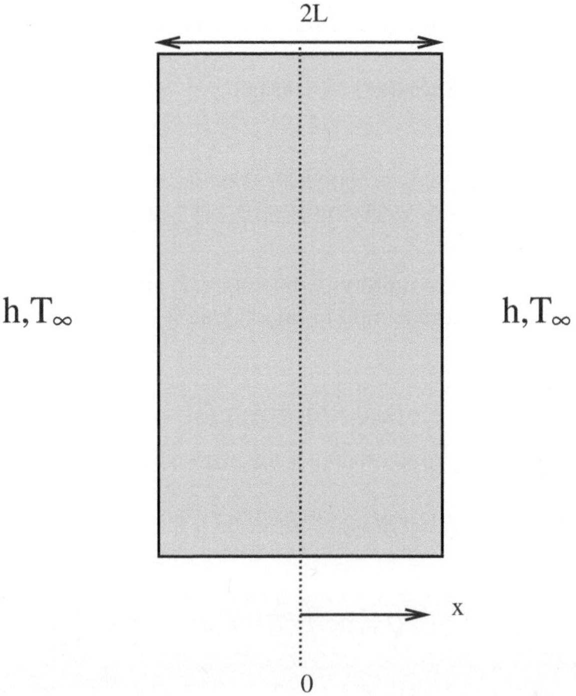

Figure 5.3 Schematic representation for transient heat flow in a slab with convective boundary conduction at walls.

The solution of this problem using separation of variables technique is

$$\frac{T(x,t) - T_\infty}{T_i - T_\infty} = \sum_{n=1}^{\infty} A_n \cos\left(\frac{\lambda_n x}{L}\right) \exp\left(\frac{-\alpha \lambda_n^2 t}{L^2}\right)$$

We introduce the Fourier number into above equation

$$\frac{T(x,t) - T_\infty}{T_i - T_\infty} = \sum_{n=1}^{\infty} A_n \cos\left(\frac{\lambda_n x}{L}\right) \exp\left(-\lambda_n^2 Fo\right)$$

where λ_n are the positive roots (also called eigenvalues) of the transcendental equation:

$$\lambda_n \tan \lambda_n - Bi = 0$$

where Bi is the Biot number. The coefficient A_n can be obtained from equation:

$$A_n = \frac{4 \sin \lambda_n}{2\lambda_n + \sin(2\lambda_n)}$$

> The Maple code for the solution of transcendental equation is
>
> ```
> Bi:=0.1
> lambda := fsolve(lambda*tan(lambda) - Bi)
> An := 4*sin(lambda)/(2*lambda + sin(2*lambda))
> ```
>
> If we take Bi:=0.01 then λ_n:=0.009966755386 and A_n=1.001660844. If we take Bi:=100 then λ_n:= -1.555245129 and A_n=1.273087620.

If the slab was initially at uniform temperature T_i and then after heat transfer it attains temperature T_∞, the reduction in internal energy conversion into heat inside slab is

$$Q_{\max} = \rho c \forall (T_i - T_\infty)$$

where ρ is material density of the slab, c is the material specific heat of the slab, $\forall$ is the volume of the slab.

If the heat transfer process is interrupted, then the decrease in internal energy conversion into heat is

$$Q = \rho c \int_\forall (T_i - T)d\forall$$

Introducing $d\forall = A \cdot dx$ we have

$$Q = \rho cA \int_{-L}^{L} [T_i - T(x,t)]\,dx = 2\rho cA \int_{0}^{L} [T_i - T(x,t)]\,dx$$

and introducing temperature

$$T(x,t) = T_\infty + (T_i - T_\infty) \sum_{n=1}^{\infty} A_n \cos\left(\frac{\lambda_n x}{L}\right) \exp\left(-\lambda_n^2 Fo\right)$$

we have

$$Q = 2\rho cA \left[\int_{0}^{L} (T_i - T_\infty)dx - (T_i - T_\infty) \sum_{n=1}^{\infty} A_n \exp\left(-\lambda_n^2 Fo\right) \int_{0}^{L} \cos\left(\frac{\lambda_n x}{L}\right) dx \right]$$

The integration gives

$$Q = 2\rho cA(T_i - T_\infty)L \left[1 - \sum_{n=1}^{\infty} A_n \exp\left(-\lambda_n^2 Fo\right) \frac{\sin \lambda_n}{\lambda_n} \right]$$

and we can take the ratio of Q with Q_{max}.

The dimensionless heat transfer for slab is

$$\frac{Q}{Q_{\max}} = 1 - \sum_{n=1}^{\infty} A_n \exp\left(-\lambda_n^2 Fo\right) \frac{\sin \lambda_n}{\lambda_n}$$

Example 5.6

A 80cm thick slab at a constant temperature of 20°C is suddenly exposed to fluid having temperature 5°C. The convective heat transfer coefficient is h=20 W/m·K. Find the temperature (i) at the midplane and (ii) at depth of 10cm from wall, after 30 hours of exposure to the fluid. The material properties are ρ=2600 kg/m^3, c= 1300 kJ/kg, and k=16 W/m·K.

Solution
(i) For x=L/2, we have the Biot number

$$Bi = \frac{hx}{k} = \frac{10W/m^2 K \times \left(\frac{40}{100}\right) m}{16W/mK} = 0.1875$$

and the Fourier number is

$$Fo = \left(\frac{\alpha t}{x^2}\right) = 5.68$$

We solve the transcendental equation using Maple:
`lambda := fsolve(Bi*cos(lambda) - lambda*sin(lambda))` This gives λ_n = -0.4199363641.

$$A_n = \frac{4 \sin \lambda_n}{2\lambda_n + \sin(2\lambda_n)} = 1.029$$

The temperature is

$$\frac{T(x,t) - T_\infty}{T_i - T_\infty} = \sum_{n=1}^{\infty} A_n \cos\left(\frac{\lambda_n x}{L}\right) \exp\left(-\lambda_n^2 Fo\right)$$

The RHS is estimated as

$$A_n \cos\left(\frac{\lambda_n x}{L}\right) \exp\left(-\lambda_n^2 Fo\right) = 0.369$$

$$\frac{T(x,t) - T_\infty}{T_i - T_\infty} = 0.369$$

Substituting the temperatures the T at x=L/2 is 10.535°

(i) For x=50cm we have the Biot number

$$Bi = \frac{hx}{k} = \frac{10W/m^2K \times \left(\frac{50}{100}\right)m}{16W/mK} = 0.3125$$

and the Fourier number is

$$Fo = \left(\frac{\alpha t}{x^2}\right) = 2.0449$$

We solve the transcendental equation using Maple:
```
lambda := fsolve(Bi*cos(lambda) - lambda*sin(lambda))
```
This gives $\lambda_n = -0.53151$

$$A_n = \frac{4\sin\lambda_n}{2\lambda_n + \sin(2\lambda_n)} = 1.0467$$

The temperature is

$$\frac{T(x,t) - T_\infty}{T_i - T_\infty} = \sum_{n=1}^{\infty} A_n \cos\left(\frac{\lambda_n x}{L}\right) \exp\left(-\lambda_n^2 Fo\right)$$

The RHS is estimated as

$$A_n \cos\left(\frac{\lambda_n x}{L}\right) \exp\left(-\lambda_n^2 Fo\right) = 0.5307$$

$$\frac{T(x,t) - T_\infty}{T_i - T_\infty} = 0.5307$$

Substituting the temperatures the T at x=50cm is 12.9605°

Example 5.7

A slab of thickness 2L with poor conductivity (see Figure 5.4). The slab was maintained internally at initial temperature of T_o and suddenly immersed in a convective environment with very high convective heat transfer coefficient. The boundaries wall attained temperature of T_w. Assuming that the convection is not influencing the slab due to poor conductivity and find the temperature inside the slab for Fourier numbers Fo = 1, 0.3, 0.1, 0.03, 0.01.

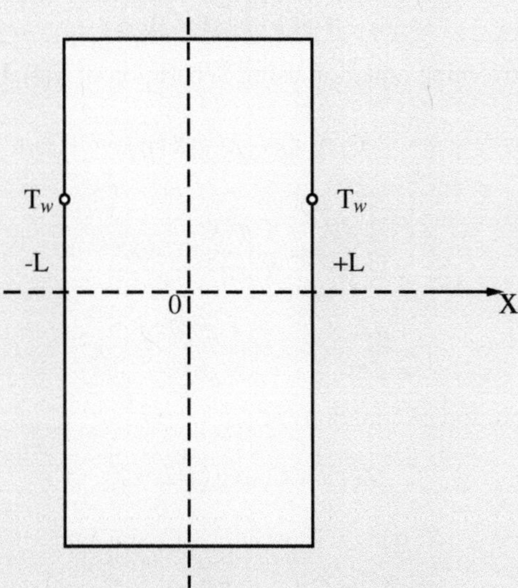

Figure 5.4   A slab with thickness 2L.

Solution

The governing equations for temperature distribution is

$$\frac{\partial T}{\partial t} = \alpha \frac{\partial^2 T}{\partial x^2}$$

The initial and boundary conditions are

$$t = 0 \qquad T = T_o$$

$$x = 0 \qquad \frac{\partial T}{\partial x} = 0$$

$$x = L \qquad T = T_w$$

We dedimensionalise the governing equations and boundary and initial conditions

$$\vartheta = \frac{T - T_w}{T_o - T_w}$$

$$x = \eta L$$

$$\tau = \alpha t / L^2$$

Substituting them into above boundary equations, we have

$$\tau = 0 \qquad \vartheta = 1$$

$$\eta = 0 \qquad \partial \vartheta / \partial \eta = 0$$

$$\eta = 1 \qquad \vartheta = 0$$

We solve the governing equation using Separation of Variable procedure

$$\vartheta(\eta,\tau) = Z(\tau)X(\eta)$$

This gives

$$\frac{1}{Z}\frac{dZ}{d\tau} = \frac{1}{X}\frac{d^2X}{d\eta^2} = -\lambda^2$$

$$\frac{1}{Z}\frac{dZ}{d\tau} = -\lambda^2 \left| \frac{1}{X}\frac{d^2X}{d\eta^2} = -\lambda^2 \right.$$

The separate solution of equations is

$$Z(\tau) = C_1 \exp\left(-\lambda^2\tau\right)$$

$$X(\eta) = C_2\cos(\lambda\eta) + C_1\sin(\lambda\eta)$$

We define new constants $A_1 = C_1C_2$ and $B_1 = C_3C_2$ and equation takes form

$$\vartheta(\eta,\tau) = e^{-\lambda^2\tau}\left(A_1\cos\left(\lambda\eta\right) + B_1\sin\left(\lambda\eta\right)\right)$$

The derivative of dimensionless temperature is

$$e^{-\lambda^2\tau}\left(-A_1\lambda\sin\left(\lambda\eta\right) + B_1\lambda\cos\left(\lambda\eta\right)\right)$$

Substituting $\eta=0$, we obtain

$$1.e^{-1.\lambda^2\tau}B_1\lambda = 0$$

This gives $B_1=0$ as rest of parameters and variables cannot be zero. The reduced form of ϑ equation is

$$\vartheta(\eta,\tau) = A_1 e^{-\lambda^2\tau}\cos\left(\lambda\eta\right)$$

We now apply the boundary condition $\eta=1$:

$$\vartheta(1,\tau) = A_1 e^{-\lambda^2\tau}\cos\left(\lambda\right) = 0$$

which requires that $\cos(\lambda) = 0$, or $\lambda_n = (n+0.5)\pi$ and $n = 0, 1, 2, 3, \ldots$ The λ_n are the eigenvalues for this equation. The solution corresponding to the nth eigenvalue is

$$\vartheta(\eta,\tau) = A_n \exp\left[-\left(n+\frac{1}{2}\right)^2\pi^2\tau\right]\cos\left(n+\frac{1}{2}\right)\pi\eta$$

$$\vartheta(\eta,\tau) = \sum_{n=0}^{\infty} A_n \exp\left[-\left(n+\frac{1}{2}\right)^2\pi^2\tau\right]\cos\left(n+\frac{1}{2}\right)\pi\eta$$

Applying the time condition

$$\vartheta(\eta, \tau = 0) = \sum_{n=0}^{\infty} A_n \exp\left[-\left(n + \frac{1}{2}\right)^2 \pi^2 \tau\right] \cos\left(n + \frac{1}{2}\right) \pi\eta = 1$$

This gives

$$A_n = \frac{2(-1)^n}{\pi\left(n + \frac{1}{2}\right)}$$

$$\vartheta(\eta, \tau) = \sum_{n=0}^{\infty} \left(\frac{2(-1)^n}{\pi\left(n + \frac{1}{2}\right)}\right) \exp\left[-\left(n + \frac{1}{2}\right)^2 \pi^2 \tau\right] \cos\left(n + \frac{1}{2}\right) \pi\eta$$

Inserting τ and η:

$$\frac{T - T_w}{T_o - T_w} = \sum_{n=0}^{\infty} \left(\frac{2(-1)^n}{\pi\left(n + \frac{1}{2}\right)}\right) \exp\left[-\left(n + \frac{1}{2}\right)^2 \pi^2 \left(\frac{\alpha t}{L^2}\right)\right] \cos\left(n + \frac{1}{2}\right) \pi\left(\frac{x}{L}\right)$$

The τ is the Fourier number so we can write

$$\frac{T - T_w}{T_o - T_w} = \sum_{n=0}^{\infty} \left(\frac{2(-1)^n}{\pi\left(n + \frac{1}{2}\right)}\right) \exp\left[-\left(n + \frac{1}{2}\right)^2 \pi^2 Fo\right] \cos\left(n + \frac{1}{2}\right) \pi\left(\frac{x}{L}\right)$$

We can use Maple for plotting the result (see Figure 5.5):

MAPLE CODE

```
L := 1;
  Theta := Sum(2*(-1)^n*exp(-(n + 1/2)^2*3.14^2*Fo)
    *cos((n + 0.5)*Pi*x/L)/((n + 1/2)*3.14),
    n = 0 .. infinity);
plot([seq(Theta, Fo in [1, 0.3, 0.1, 0.03, 0.01])],
  x = -1 .. 1,
legend = [1, 0.3, 0.1, 0.03, 0.01], size = [600, 350]);
```

5.4 TRANSIENT CONDUCTION THROUGH A LONG CYLINDER WITH CONVECTIVE COOLING

We considered now the case of a cylinder in (r, θ, z) coordinates. The initial temperature of the cylinder was T_i and then the cylinder is certainly subjected to the

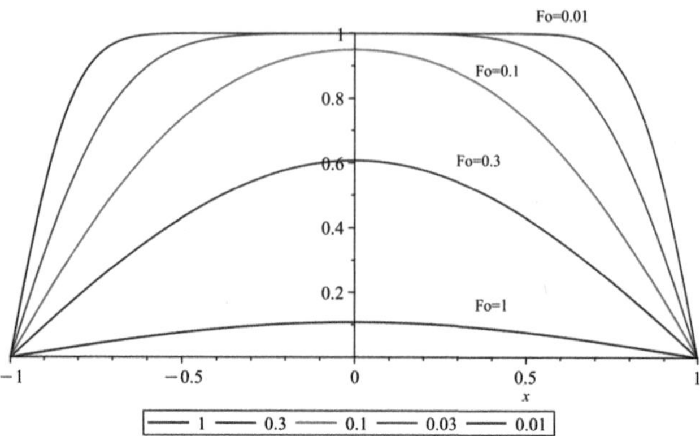

Figure 5.5   Temperature distribution inside the slab placed in highly convective environment at Fourier number Fo=1, 0.3, 0.1, 0.03, 0.01.

convective environment. Assuming that there is no internal energy conversion into heat, and the thermal conductivity is constant, we consider that the temperature distribution is a function of the radial coordinate system only. The conduction heat equation in three-dimensional coordinates is

$$\frac{1}{r}\left[\frac{\partial}{\partial r}\left(kr\frac{\partial T}{\partial r}\right)\right] + \frac{1}{r^2}\left[\frac{\partial}{\partial \theta}\left(k\frac{\partial T}{\partial \theta}\right)\right] + \frac{\partial}{\partial z}\left(k\frac{\partial T}{\partial z}\right) + \tilde{q} = \rho c\frac{\partial T}{\partial t}$$

After assumptions, the reduced form for the one-dimensional transient heat conduction equation is

$$\frac{1}{r}\left[\frac{\partial}{\partial r}\left(r\frac{\partial T}{\partial r}\right)\right] = \frac{1}{\alpha}\frac{\partial T}{\partial t}$$

The temperature distribution is symmetrical at around centre line of the straight cylinder:

at r=0, $\frac{\partial T}{\partial r} = 0$

and the heat fluxes are balancing each other at the cylinder's wall: at $r = r_w$, $-k\frac{\partial T}{\partial r} = h(T - T_\infty)$

Initial condition is t=0, $T(r)=T_i$. With these initial and boundary conditions, we solve the differential equation using Separation of the variable technique and obtain the temperature distribution for any point inside the cylinder as

$$T(r,t) = T_\infty + (T_i - T_\infty)\sum_{n=1}^{\infty} B_n \exp\left(-\frac{\alpha\lambda_n^2 t}{r_w^2}\right)J_o\left(\frac{r\lambda_n}{r_w}\right)$$

In the dimensionless form the temperature distribution is

$$\frac{T(r,t) - T_\infty}{T_i - T_\infty} = \sum_{n=1}^{\infty} B_n \exp\left(-\frac{\alpha \lambda_n^2 t}{r_w^2}\right) J_o\left(\frac{r\lambda_n}{r_w}\right)$$

where J_o is Bessel functions of the first kind. Introducing the Fourier number as $Fo = \alpha t / r_w^2$ we have

$$\frac{T(r,t) - T_\infty}{T_i - T_\infty} = \sum_{n=1}^{\infty} B_n \exp\left(-\lambda_n^2 Fo\right) J_o\left(\frac{r\lambda_n}{r_w}\right)$$

For the eigenvalues of λ_n we only take the positive roots of the following equation

$$Bi J_o(\lambda_n) - \lambda_n J_1(\lambda_n) = 0$$

Here, Bi is Biot number, and J_o and J1 are the Bessel functions of the first kind. The function B_n can be obtained from equation

$$B_n = \frac{2}{\lambda_n}\left[\frac{J_1(\lambda_n)}{J_o^2(\lambda_n) + J_1^2(\lambda_n)}\right]$$

Maple code for λ_n and B_n estimation

We assume a value of λ_n and then solve the Bessel functions and find the corresponding Biot number. Here is the code:

```
lambda := 0.2
BJ0 := evalf(BesselJ(0, lambda));
BJ1 := evalf(BesselJ(1, lambda));
Bi:= fsolve(-Bi*BesselJ(0, lambda)+lambda*BesselJ(1,lambda));
Bn:= 2*BesselJ(1,lambda)/(lambda*(BesselJ(0,lambda)^2+BesselJ(1,lambda)^2))
```

If we take lambda:=0.2 then BJ0 := 0.6433500092 and BJ1 := 0.5118056226, which gives Bi := 0.998 and Bn := 1.2068

For sake of convience, the Zeroth- and first-order Bessel functions of the first kind are tabulated in Table 5.2

You will now investigate the amount of energy exchange from the surface of the cylinder to the ambient. The maximum heat transfer will occur when the slender initially at a temperature T_i is subjected to a convective environment and finally attains the uniform temperature of T_∞.

$$Q_{\max} = \rho c \forall (T_i - T_\infty)$$

The decrease in the internal energy conversion into heat is

$$Q = \rho c \int_\forall (T_i - T)\, d\forall$$

TABLE 5.2

Zeroth- and First-Order Bessel Functions of the First Kind

n	J_0	J_1	n	J_0	J_1
0	1	0	1.5	0.5118	0.5579
0.1	0.9975	0.0499	1.6	0.4554	0.5699
0.2	0.99	0.0995	1.7	0.398	0.5778
0.3	0.9776	0.1483	1.8	0.34	0.5815
0.4	0.9604	0.196	1.9	0.2818	0.5812
0.5	0.9385	0.2423	2	0.2239	0.5767
0.6	0.912	0.2867	2.1	0.1666	0.5683
0.7	0.8812	0.329	2.2	0.1104	0.556
0.8	0.8463	0.3688	2.3	0.0555	0.5399
0.9	0.8075	0.4059	2.4	0.0025	0.5202
1	0.7652	0.44	2.6	0.0968	0.4708
1.1	0.7196	0.4709	2.8	0.185	0.4097
1.2	0.6711	0.4983	3	0.2601	0.3391
1.3	0.6201	0.522	3.2	0.3202	0.2613
1.4	0.5669	0.5419			

where $\forall$ is the volume of the cylinder. If L is the length of the cylinder, the volume of the cylinder is $d\forall = (2\pi rL)\,dr$.

$$Q = \rho c \int_\forall (T_i - T)(2\pi rL)\,dr$$

This leads to the form

$$Q = \left(\rho c L \pi r_w^2\right)(T_i - T_\infty)\left[1 - 2\sum_{n=1}^{\infty} B_n \exp\left(-Fo \cdot \lambda_n^2\right) J_o\left(\frac{r\lambda_n}{r_w}\right)\right]$$

The dimensionless heat transfer for cylinder is

$$\frac{Q}{Q_{max}} = 1 - 2\sum_{n=1}^{\infty} B_n \exp\left(-Fo \cdot \lambda_n^2\right) J_o\left(\frac{r\lambda_n}{r_w}\right)$$

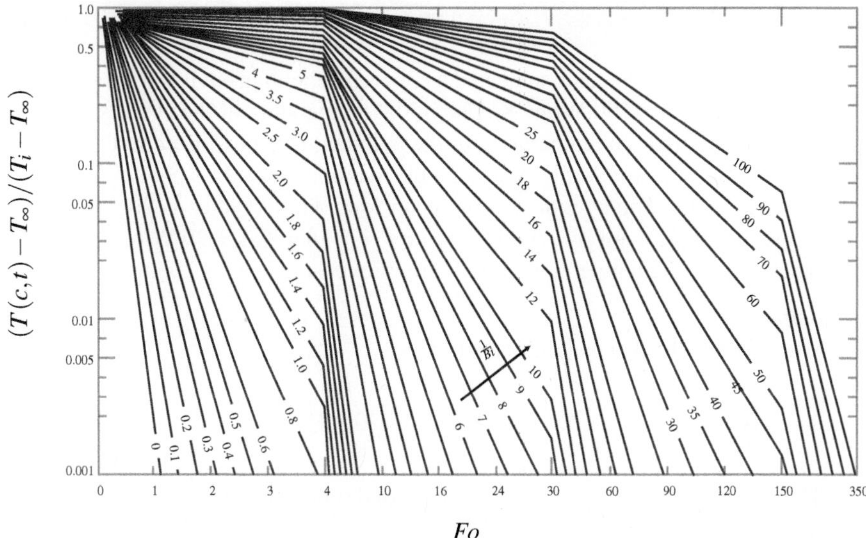

Figure 5.6 Axis temperature for an infinite cylinder of radius r_w subjected to convective cooling (Heisler, 1947). (Printed with permission.)

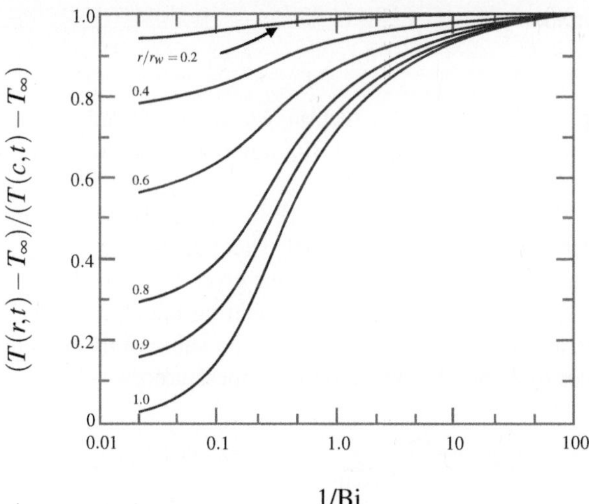

Figure 5.7 Temperature as a function of axis temperature in an infinite cylinder of radius r_w subjected to convective cooling (Heisler 1947). (Printed with permission.)

Heisler (1947) has provided the graphs for the one term approximation which are plotted here in Figures 5.6 and 5.7.

Example 5.8

A solid cylinder of steel is subjected a convective cooling environment where $T_\infty = 5°C$ and h=1100 W/m^2·K. The initial temperature of the steel cylinder was 300°C. Find the temperature of the cylinder, if the exposure to the cold environment was for 9 min. Also, find the maximum heat loss per unit length of the cylinder. The cylinder is one meter long and its radius is 20 cm. The properties of steel are $\rho = 7800$ kg/m^3; c = 490 J/kg·K; k = 39 W/m·K. How much will be the temperature at r=10cm, if exposure span is 20 min?

Solution

We calculate the thermal diffusivity as

$$\alpha = \left(\frac{k}{\rho \cdot c}\right) = 1.020 \times 10^{-5}$$

The characteristic length is $L = \frac{r_w}{2}$ and based on this length the Biot number is

$$Bi = \left(\frac{h \cdot L}{k}\right) = \frac{1100 \cdot 0.1}{39} = 2.82$$

If exposure is for 9 min: Based on the given data the Fourier number at the cylinder wall is

$$Fo = \frac{\alpha t}{L^2} = 0.55$$

Using Maple code given in this section, we adopt trail and error approach for estimation of λ_n until the required Biot number is arrived. This gives
|lambda:= 1.77—
|BJ0:= evalf(BesselJ(0, lambda));— |BJ1:= evalf(BesselJ(1, lambda));—
|lambda2:=fsolve(-BJ0*Biot+ BJ1*lambda)—
We iterate until lambda2=lambda and this gives |Bn:= 1.41—
Alternatively, we can use Table 5.3 for getting the λ_n and B_n. Note that at center, r=0, and $J_0(\lambda_n \cdot r/r_w)$ from Maple the value of this function is unity. Using values of λ_n and B$_n$, we find the temperature:

$$\frac{T(c,t) - T_\infty}{T_i - T_\infty} = \sum_{n=1}^{\infty} B_n \exp\left(-\lambda_n^2 Fo\right) J_0\left(\frac{r\lambda_n}{r_w}\right)$$

The LHS is

$$LHS = \frac{T(c,t) - T_\infty}{T_i - T_\infty} = \frac{Tc}{295} - \frac{1}{59}$$

and after substituting values in RHS of equation for one term we have

$$RHS = B_n \exp\left(-\lambda_n^2 Fo\right) J_o \left(\frac{r\lambda_n}{r_w}\right)$$

$$RHS = 1.411 \times 0.1779430298 \times 1 = 0.2510$$

This gives the equation:

$$\frac{Tc}{295} - 0.2680389445 = 0$$

where c indicates the axial position r=0. This gives T(c,t)= 79.07°C.
The maximum heat transfer from wall of cylinder to surrounding is

$$Q_{max} = (\rho \cdot c \cdot \forall)(T_i - T_\infty)$$

where $\forall$ is the volume of cylinder= $\pi \cdot r_w^2 \cdot L$:

$$\frac{Q_{max}}{L} = \rho \cdot c \cdot (\pi \cdot r_w^2) \cdot (T_i - T_\infty)$$

$$\frac{Q_{max}}{L} = 7800 \cdot 490 \cdot \pi \cdot \left(\frac{20}{100}\right)^2 \cdot (300 - 5) = 1.416 \times 10^8 W/m$$

If the exposure time is 20 min, then the Fourier number at wall is 1.224 and
$J_o(\lambda_n \cdot r/r_w)$ from Maple is
`BesselJ(0, lambda*r/rw) = BesselJ(0, 1.77*0.5) = 0.81357` and
$\exp\left(-\lambda^2 \cdot Fo\right) = 0.02157$

$$\frac{T(r,t) - T_\infty}{T_i - T_\infty} = \sum_{n=1}^{\infty} B_n \exp\left(-\lambda_n^2 Fo\right) J_o \left(\frac{r\lambda_n}{r_w}\right)$$

$$RHS = 1.411 \times 0.0215 \times 0.81357 = 0.02476$$

This gives the equation

$$\frac{Tc}{295} - 0.04171779569 = 0$$

Solving it we have T(r=10 cm, 20 min)= 12.3°C.
We can also use the graphical approach to solve the case of 20 min exposure.
Referring to Figure 5.6 at

$$1/Bi = 0.354 \; and \; Fo = 1.224$$

we have from Figure 5.6

$$\frac{T(c,t) - T_\infty}{T_i - T_\infty} = 0.75$$

for $T_i = 300°C$ and $T_\infty = 5°C$, we have the axis temperature as $T(c,t)=27.12°C$. Now from Figure 5.7 at $r/r_w=0.5$:

$$\frac{T - T_\infty}{T(c,t) - T_\infty} = 0.65$$

This gives temperature at r=10 cm as

$$T = 19.3°C$$

Note that graphical method gives an approximate solution and only good enough to get the ball park figures.

5.5   TRANSIENT HEAT CONDUCTION THROUGH A SPHERE WITH CONVECTIVE COOLING

The heat conduction through spherical coordinates is

$$\frac{1}{r^2}\left[\frac{\partial}{\partial r}\left(kr^2\frac{\partial T}{\partial r}\right)\right] + \frac{1}{r^2\sin\theta}\left[\frac{\partial}{\partial \theta}\left(k\sin\theta\frac{\partial T}{\partial \theta}\right)\right]$$
$$+ \frac{1}{r^2\sin^2\theta}\left[\frac{\partial}{\partial \varphi}\left(k\frac{\partial T}{\partial \varphi}\right)\right] + \tilde{q} = \rho c\frac{\partial T}{\partial t}$$

Assuming one-dimensional heat conduction in radial direction only, with constant thermal conductivity and no internal energy transformation, we have

$$\frac{k}{r^2}\frac{\partial}{\partial r}\left(r^2\frac{\partial T}{\partial r}\right) = \rho c\frac{\partial T}{\partial t}$$

$$\frac{\partial^2 T}{\partial r^2} + \frac{2}{r}\frac{\partial T}{\partial r} = \frac{1}{\alpha}\frac{\partial T}{\partial t}$$

We consider the case of a sphere with an initial temperature of TI. t=0, Ti=0

The sphere is an certainly subjected to a convective environment having a temperature of Tinfi, and the convective heat transfer coefficient h. It is assumed that the temperature distribution inside the sphere it's symmetrical around r=0. Therefore, the bond conditions at the centre of the sphere is r= 0, $\frac{\partial T}{\partial r} = 0$. At the surface of the sphere, the heat fluxes balance is

$$r = r_w, \qquad -k\frac{\partial T}{\partial r} = h(T - T_\infty)$$

TABLE 5.3

One-Term Approximation for Coefficients A_n, B_n, and E_n

Bi	Plate λ_n	A_n	Cylinder λ_n	B_n	Sphere λ_n	E_n
0.01	0.0998	1.0017	0.1412/	1.0025	0.173	1.003
0.02	0.141	1.0033	0.1995	1.005	0.2445	1.006
0.04	0.1987	1.0066	0.2814	1.0099	0.345	1.012
0.06	0.2425	1.0098	0.3438	1.0148	0.4217	1.0179
0.08	0.2791	1.013	0.396	1.0197	0.486	1.0239
0.1	0.3111	1.0161	0.4417	1.0246	0.5423	1.0298
0.2	0.4328	1.0311	0.617	1.0483	0.7593	1.0592
0.3	0.5218	1.045	0.7465	1.0712	0.9208	1.088
0.4	0.5932	1.058	0.8516	1.0931	1.0528	1.1164
0.5	0.6533	1.0701	0.9408	1.1143	1.1656	1.1441
0.6	0.7051	1.0814	1.0184	1.1345	1.2644	1.1713
0.7	0.7506	1.0918	1.0873	1.1539	1.3525	1.1978
0.8	0.791	1.1016	1.149	1.1724	1.432	1.2236
0.9	0.8274	1.1107	1.2048	1.1902	1.5044	1.2488
1	0.8603	1.1191	1.2558	1.2071	1.5708	1.2732
1.5	0.9883	1.1537	1.457	1.2807	1.8364	1.3849
2	1.0769	1.1785	1.5995	1.3384	2.0288	1.4793
3	1.1925	1.2102	1.7887	1.4191	2.2889	1.6227
4	1.2646	1.2287	1.9081	1.4698	2.4556	1.7202
5	1.3138	1.2403	1.9898	1.5029	2.5704	1.787
6	1.3496	1.2479	2.049	1.5253	2.6537	1.8338
7	1.3766	1.2532	2.0937	1.5411	2.7165	1.8673
8	1.3978	1.257	2.1286	1.5526	2.7654	1.892
9	1.4149	1.2598	2.1566	1.5611	2.8044	1.9106
10	1.4289	1.262	2.1795	1.5677	2.8363	1.9249
15	1.4729	1.2676	2.2509	1.58	2.9349	1.963
20	1.4961	1.2699	2.288	1.5919	2.9857	1.9781
30	1.5202	1.2717	2.3261	1.5973	3.0372	1.9898
40	1.5325	1.2723	2.3455	1.5993	3.0632	1.9942
50	1.54	1.2727	2.3572	1.6002	3.0788	1.9962
100	1.5552	1.2731	2.3809	1.6015	3.1102	1.999
∞	1.5708	1.2732	2.4048	1.6021	3.1416	2

with these initial public relations is of the differential equation using the suppression of variable technique, which gives

$$T(r,t) = T_\infty + (T_i - T_\infty) \sum_{n=1}^{\infty} E_n \exp\left(-\frac{\alpha \lambda_n^2 t}{r_w^2}\right) \cdot \sin\left(\frac{\lambda_n r}{r_w}\right) \cdot \left(\frac{r_w}{\lambda_n r}\right)$$

Introducing the Fourier number based on sphere radius ($Fo = \alpha t / r_w^2$) into above equation

$$T(r,t) = T_\infty + (T_i - T_\infty) \sum_{n=1}^{\infty} E_n \exp\left(-\lambda_n^2 Fo\right) \cdot \sin\left(\frac{\lambda_n r}{r_w}\right) \cdot \left(\frac{r_w}{\lambda_n r}\right)$$

The equation gives the temperature at any point in the sphere at a given time. In dimensionless form the equation is

$$\frac{T(r,t) - T_\infty}{(T_i - T_\infty)} = \sum_{n=1}^{\infty} E_n \exp\left(-\lambda_n^2 Fo\right) \cdot \sin\left(\frac{\lambda_n r}{r_w}\right) \cdot \left(\frac{r_w}{\lambda_n r}\right)$$

$$1 - \frac{\lambda_n}{\tan(\lambda_n)} - Bi = 0$$

The λ_n values are the eigenvalues of the transcendental equation

$$1 - \frac{\lambda_n}{\tan(\lambda_n)} - Bi = 0$$

where Bi indicates the Biot number defined as

$$Bi = \left(\frac{h \cdot r_w}{k}\right)$$

Once λ_n is found the function B_n can be found from equation

$$E_n = 4\left[\frac{\sin(\lambda_n) - \lambda_n \cos(\lambda_n)}{2\lambda_n - \sin(2\lambda_n)}\right]$$

The dimensionless heat transfer for sphere case is

$$\frac{Q}{Q_{\max}} = 1 - 3\sum_{n=1}^{\infty} E_n \exp\left(-\lambda_n^2 Fo\right)\left[\frac{\sin \lambda_n - \lambda_n \cos \lambda_n}{\lambda_n^3}\right]$$

where,

$$Q_{\max} = \rho c \forall (T_1 - T_\infty) = \rho c \left(\frac{4}{3}\pi r_w^2\right)(T_1 - T_\infty)$$

Example 5.9

A solid sphere of steel of radius 0.0381 m is subjected a convective cooling environment where $T_\infty = 10°C$ and $h = 50$ W/m$^2 \cdot$K. The initial temperature of the steel sphere was 800°C. The properties of steel are: $\rho = 8000$ kg/m^3; $c = 470$ J/kg$\cdot$K; $k = 41$ W/m$\cdot$K. How long the sphere will take to reach the temperature of 50°C at its surface.

Solution

The given temperature are T_i=800 °C, $T(r=r_w,t)$= 50 °C, $T_\infty = 10$ °C,
This gives us the ratio of temperature differences as

$$\frac{T(r=r_w,t)-T_\infty}{T_i-T_\infty} = \frac{50-10}{800-10} = 0.05063$$

The Fourier number as function of time and Biot number are

$$L = r_w/3$$

$$Fo = \frac{\alpha t}{L^2} = 0.0676t$$

$$Bi = \left(\frac{h \cdot L}{k}\right) = 0.01548$$

Using Table 5.3, or Maple we find the λ_n and function E_n values

```
lambda := 0.2152202272
En := 1.004641685
```

For r=r_w we have

$$\frac{T(r,t)-T_\infty}{(T_i-T_\infty)} = \sum_{n=1}^{\infty} E_n \exp\left(-\lambda_n^2 Fo\right) \cdot \sin\left(\frac{\lambda_n r}{r_w}\right) \cdot \left(\frac{r_w}{\lambda_n r}\right)$$

We get the equation in terms of time, t:

$$0.9969038362 e^{-0.003131516764t} = 0$$

The solution of this equation gives us the time the sphere's surface will take
to drop to the temperature of 50 °C as t= 15.86 min.

Summary of transient results

Slab	$\dfrac{T(x,t)-T_\infty}{T_i-T_\infty} = \sum\limits_{n=1}^{\infty} A_n \exp\left(-\lambda_n^2 Fo\right) \cdot \cos\left(\frac{\lambda_n x}{L}\right)$
Cylinder	$\dfrac{T(r,t)-T_\infty}{T_i-T_\infty} = \sum\limits_{n=1}^{\infty} B_n \exp\left(-\lambda_n^2 Fo\right) \cdot J_o\left(\frac{r\lambda_n}{r_w}\right)$
Sphere	$\dfrac{T(r,t)-T_\infty}{(T_i-T_\infty)} = \sum\limits_{n=1}^{\infty} E_n \exp\left(-\lambda_n^2 Fo\right) \cdot \sin\left(\frac{\lambda_n r}{r_w}\right) \cdot \left(\frac{r_w}{\lambda_n r}\right)$

PROBLEMS

5P-1 A potato having temperature of 20°C is dropped in boiling water at 1 atm. The water temperature is 100°C. Assume that the potato is an ellipsoid. The lengths of all three semi-axes of the ellipsoid are a=5 cm, b=2.2 cm, and c=1 cm, respectively. How long the potato will take till its core temperature is 80°C? The volume of ellipsoid is $4abc/3$.

5P-2 Five Ice cubes at 2°C are dropped into a 0.0002 m^3 glass of water at 20°C. How long the water will take the reach the temperature of 5°C.

5P-3 The spherical steel balls are to be used in a ball bearing are heat treated. Each of the ball has 5 cm diameter and temperature of the ball is 700°C. The ball is quenched in oil at h=400 W/m²·K and T=65°C. Find the time taken by ball to reach the temperature of 200°C, and the heat transfer. The properties of steel are k =55 W/m·K, α=1.35E-5 m²/s.

[Ans: t= 131.4 s, Q=133.32 kW]

5P-4 A chocolate factory is producing the edible chocolate balls of 4 mm diameter. The melting point temperature of chocolate is between 80°F and 90°F. The initial temperature of the chocolate ball is 78°F, and it is desired to cool the balls immediately by submerging them into pool of chilled water at 41°F until the balls have reached temperature of 20°F. How much time the balls will take to reach this temperature. The properties of chocolate are ρ=81.5933 lb/ft³, c_p=0.3824 Btu/lb$_m$·°F, and k=0.15 Btu/h·ft·°F.

[Ans: t= 0.00121 s]

5P-5 In a chemical process industry, certain pellets of 3 mm diameter are produced at temperature of 413 K. It is desired to cool down the pellets till temperature of 330K is attained before storing. An impinging jet is used to cool the pellets at low speeds. The convective environment created by jet gives h=65 W/m²·K and T_{jet}=320 K. If the jet-outlet-to-target wall or pellet distance is 60d, and the mean speed reaching pellets is 0.7 of the jet's outlet velocity, find the jet's outlet speed. The properties of pellet are ρ=450 kg/m³, c_p=3 kJ/kg.K. Assume lumped-capacity method is applicable.

[Ans: Jet's max. velocity = 0.0111 m/s]

5P-6 The material properties of a spherical thermocouple junction (d=9 mm) are ρ=9000 kg/m³, c_p=450 J/kg.K, k=50 W/m·K. Initially the thermocouple was at 50°C and for 10s it is placed in a forced convective environment (h_1=55 W/m²·K, $T_{\infty,1}$=300 °C) and temperature is measured. After that for the next 20s the thermocouple is used to measure the temperature of medium with free convection (h_2=5 W/m²·K, $T_{\infty,2}$=30 °C). Find the temperatures T_{10s} and T_{20s} of the thermocouple junction after 10s and 20s, also how much is the response time of this thermocouple?

[Ans: Response time = 110.45s, T_{10s}=71.639°C, T_{20s} =70.95°C]

5P-7 A mercury based glass thermometer is used to measure the air stream temperature. The convective heat transfer coefficient is 60 W/m$^2\cdot$K. Assuming that the thermometer is like a long cylinder of 3 mm diameter, find the response time of the thermometer and time taken by it to measure the temperature difference which is 1/5th of the initial temperature difference of thermocouple with air stream flow. Take the density 8.6 kg/m^3, and alpha= 0.0164 m^2/h.

[Ans: Response time=24s, t=37.97s]

5P-8 A 80 mm thick Asbestos plate having 0.113 W/m$\cdot$K and α=0.036$\times10^{-5}$ m^2/s, initially at 70°C is suddenly exposed to convective environment of h=50 W/m$^2\cdot$K and temperature 30°C. Assuming constant properties, find the temperature at x=0 (mid-plane) and x=30 mm inside Asbestos, after t=0.07166s.

[Ans: Bi= 17.69, $T_{x=0}$ = 80.76°C, $T_{x=30mm}$=52.3°C]

5P-9 A long cylindrical rod of radius 90 mm just taken out a furnace at 800°C and submerged into an oil bath for quenching. The oil bath temperature is 30°C. The convective heat transfer coefficient is 200 W/m$^2\cdot$K. The thermophysical properties are: k=18 W/m$\cdot$K, α=5.277E-6 m^2/s. Estimate the time taken by the rod center to reach 120°C. Also, find the temperature gradient at the wall.

[Ans: t= 902.92s, dT/dr= 1000 K/m]

5P-10 Gold ingot (k=316 W/m$\cdot$K) heated 1000°C and then quenched in a oil bath at 20°C. Find the time taken by the ingot to drop to T=500°C at the position x=10 cm from the surface. Approximate ingot as a semi-infinite plane plate with α=126.9E-6 m^2/s.

[Ans: t=90.859s]

5P-11 The gas turbine test stand is operated in the close room. The hot exhaust of the turbine is creating temperature of 350°C. The wall of the test section was initially at temperature of 25°C. If the testing was done for 10 hours, then how much is the wall temperature at the wall depth of 100 mm. The wall thermal conductivity and diffusivity is k=0.935 W/m$\cdot$K, α= 4.444$\times 10^{-7}$ m^2/s. Assume semi-infinite solution is applicable.

[Ans: $T_{x=100mm}$=212.24°C]

5P-12 A submarine can be approximated as a cylinder moving in seawater. The submarine has 9 m diameter and 70 m in length. Initially the submarine was at the sea level, where temperature was 20°C and then it descends into depth of 300 m, where the temperature is 4.5°C. The convective environment created by sea water gives h=70 W/m$^2\cdot$K. Find the temperature at the axis of the submarine, if submarine is operated for 9 hours in the same environment. Ignore the transient effects during descent of the submarine.

5P-13 A 5 m^2 wall is experiencing the solar flux daily as expressed by the following relation:

$$q_w = 700 + 10 \cdot \sin\left(\frac{2 \cdot \pi}{24} \cdot t\right)$$

The wall is also exposed to convective environment conditions $T_\infty = 25°C, h_\infty = 10$ W/m·K. Plot the exposed wall temperature for 12 hours duration. Also, how much is the maximum tmeprature experienced by the wall?

REFERENCES

J. P. Holman, Heat Transfer, McGraw-Hill, New York, USA, 1963.

M. P. Heisler, Temperature Charts for Induction and Constant Temperature Heating, Trans. ASMS 69, 227–236, USA, 1947.

6 Heat Conduction Analysis Using the Finite Difference Technique

Most of the analytical solution are restricted to some well-defined geometries. Heat transfer problems can become mathematically too cumbersome to be solved analytically if the thermal conditions are complex. In some materials, the thermal conductivity is varying with temperature, or the surrounding fluid convective heat transfer coefficient might not be constant. Also, in some cases, the radiative flux experienced by surface can be varying. It then become too cumbersome to model these problems analytically and the only way to solve such problems is to use the numerical approaches. In this chapter, we will use the basic finite-difference technique for solution of heat conduction problems. We will use MAPLE to arrive at the solution of set of equations. After finishing this chapter, you will be able to use finite difference method for the solution of 1D and 2D heat conduction problems under steady and unsteady conditions.

The exact analytical models are limited to simple geometrical configurations and are only possible with certain types of boundary conditions, which makes them a restrictive tool. Therefore, it is not always possible to have an analytical solution to a given problem and thus we need to find the solution through numerical solution techniques.

6.1 FINITE DIFFERENCE METHOD

The finite difference approximation is the oldest of the methods applied to obtain numerical solutions of differential equations. Although Finite differences were introduced by Brook Taylor in 1715, it is claimed that the technique was known since the time of Issac Newton. The first application of this technique on practical problems is attributed to Leonhard Euler (1707 - 1783) in 1768. According to this technique, for an arbitrary function u(x), the derivative at point x is defined as

$$\frac{\partial u}{\partial x} = \lim_{\Delta x \to 0} \frac{u(x + \Delta x) - u(x)}{\Delta x} \tag{6.1}$$

If we remove the limit in the above equation, we obtain a finite difference, which explains the name given to this method. For a function u(x), the derivative at point x

DOI: 10.1201/9781003428404-6

is defined by:

$$\frac{\partial u}{\partial x} = \lim_{\Delta x \to 0} \frac{u(x+\Delta x) - u(x)}{\Delta x} \tag{6.2}$$

If Δx is small but finite, the expression on the right-hand side is an approximation to the exact value of $\frac{\partial u}{\partial x}$. The approximation will be improved by reducing Δx, but for any finite value of Δx, an error is introduced, the truncation error, which goes to zero for Δx tending to zero. The power of Δx with which this error tends to zero is called the order of accuracy of the difference approximation and can be obtained from a Taylor series development of u(x+Δx) around point x. We can formulate an expression for u(x+Δx) as

$$u(x+\Delta x) = u(x) + \frac{\partial u}{\partial x}\Delta x + \frac{1}{2}\frac{\partial^2 u}{\partial x^2}(\Delta x)^2 + \frac{1}{3!}\frac{\partial^3 u}{\partial x^3}(\Delta x)^3 + \cdots \tag{6.3}$$

Taylor expansion tells us is that we can find the function at an arbitrary distance far away from point x, if we know the derivatives, at this single point x. This relation can be written as follows:

$$\frac{u(x+\Delta x) - u(x)}{\Delta x} = u_x(x) + \underbrace{\frac{\Delta x}{2}u_{xx}(x) + \frac{(\Delta x)^2}{6}u_{xxx}(x) + \cdots}_{Truncation\ error} \tag{6.4}$$

where u_x, u_{xx}, and u_{xxx} represent first, second, and third-order derivatives. This nomenclature is adopted to reduce the length of equations. The right-hand side of above equation is indeed an approximation to the first derivative u_x in point x. The remaining terms in the r.h.s represent the error associated to this formulation. If we restrict the truncation error to its dominant term, that is to the lower power in Δx, we see that this approximation for u(x) goes to zero like the first power of Δx and is said to be first order in Δx and we write:

$$\frac{u(x+\Delta x) - u(x)}{\Delta x} = u_x(x) + \frac{\Delta x}{2}u_{xx}(x) = u_x(x) + O(\Delta x) \tag{6.5}$$

indicating that the truncation error $O(\Delta x)$ goes to zero like the first power in Δx.

Applying 6.4 and 6.5 we obtain at point i, we obtain the following finite difference approximation for the first derivative $(u_x)i = (\partial u/\partial x)_i$:

$$(u_x)_i = \left(\frac{\partial u}{\partial x}\right)_i = \frac{u_{i+1} - u_i}{\Delta x} - \frac{\Delta x}{2}(u_{xx})_i - \frac{\Delta x^2}{6}(u_{xxx})_i + \cdots \tag{6.6}$$

$$(u_x)_i = \left(\frac{\partial u}{\partial x}\right)_i = \frac{u_{i+1} - u_i}{\Delta x} + O(\Delta x) \tag{6.7}$$

As this formula involves the point (i+1) to the right of point i, it is called the first order forward difference for the first derivative $(u_x)i = (\partial u/\partial x)_i$.

This leads to the relation

$$(u_x)_i = \left(\frac{\partial u}{\partial x}\right)_i = \frac{u_i - u_{i-1}}{\Delta x} + \underbrace{\frac{\Delta x}{2}(u_{xx})_i - \frac{\Delta x^2}{6}(u_{xxx})_i + \cdots}_{Truncation\ error} \tag{6.8}$$

$$(u_x)_i = \left(\frac{\partial u}{\partial x}\right)_i = \frac{u_i - u_{i-1}}{\Delta x} + O(\Delta x) \tag{6.9}$$

With respect to the point $x=x_i$, this formula is the first-order backward difference
for the derivative $(u_x)_i$. Both formulas 6.7 and 6.9 are called one-sided difference
formulas, since they involve points at one side of point i only.

We obtain a second order approximation for the first order as

$$(u_x)_i = \left(\frac{\partial u}{\partial x}\right)_i = \frac{u_{i+1} - u_{i-1}}{2\Delta x} - \frac{\Delta x^2}{6}(u_{xxx})_i + \cdots \tag{6.10}$$

This formula, which involves the points to the left and to the right of point i, is called
a central difference formulation.

We can formulate the FD formulation for the second order derivative $(\frac{\partial^2 u}{\partial x^2})$ by
summing the Taylor series expansions for u_{i+1}, and u_{i-1} as

$$u_{i+1} + u_{i-1} = 2u_i + \left(\frac{\partial^2 u}{\partial x^2}\right)(\Delta x)^2 + \left(\frac{\partial^4 u}{\partial x^4}\right)\frac{(\Delta x)^4}{12} \tag{6.11}$$

Ignoring fourth order and higher order terms and rearranging, we have

$$\left(\frac{\partial^2 u}{\partial x^2}\right) = \frac{u_{i+1} + u_{i-1} - 2u_i}{(\Delta x)^2} + O(\Delta x)^2 \tag{6.12}$$

In above equation, the first term on the right-hand side is a central finite difference
for the second derivative with respect to x evaluated at grid point i from the remaining order-of-magnitude term, we see that this central difference is of second-order
accuracy.

6.2   1D STEADY CONDUCTION

Consider the one-dimensional heat conduction through a small strip exposed to convective and radiative boundary conditions as shown in Figure 6.1.

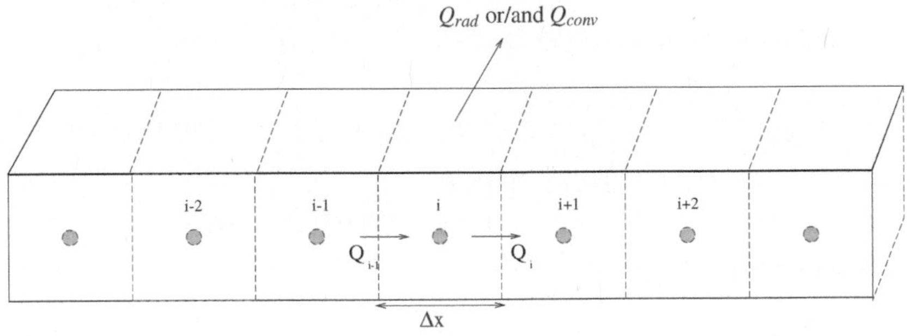

Figure 6.1   Schematic representation of heat flow through node i using finite-difference
approach.

We will apply the first law of thermodynamics on the system for heat flow at node i:

$$Q_{i-1 \to i} + \tilde{Q} = Q_{rad} + Q_{conv} + Q_{i \to i+1}$$

Inserting Fourier-Law of heat conduction, Newton's law of cooling and Stefan-Boltzmann law for gray bodies, we have

$$\left(\frac{kA}{\Delta x}\right)(T_{i-1} - T_i) + \tilde{q}A\Delta x = \varepsilon \sigma A(T_i^4 - T_{amb}^4) + hA(T_i - T_\infty) + \left(\frac{kA}{\Delta x}\right)(T_i - T_{i+1})$$

In case there is no internal heat conversion and radiation we have

$$\left(\frac{kA}{\Delta x}\right)(T_{i-1} - T_i) + \left(\frac{kA}{\Delta x}\right)(T_{i+1} - T_i) + hA(T_\infty - T_i) = 0 \qquad (inner\ nodes)$$

The boundary node heat balance with conduction and convection only is

$$\boxed{\frac{kA\,(T_{i-1} - T_i)}{\Delta x} + \frac{hP\Delta x\,(T_\infty - T_i)}{2} + hA\,(T_\infty - T_i) = 0 \qquad (boundary\ node)} \quad (6.13)$$

Example 6.1

A 15-cm long rectangular pin fin with thermal conductivity k = 645 W/m·K is attached to a wall. The perimeter of the fin is 0.1099 m and cross-sectional area is 0.000961625 m². The ambient air temperature is 50 °C, and the base fin temperature is 200 °C. The convective heat transfer coefficient is 20 W/m²·K. Take 5 nodes and find the temperature inside fin.

Solution
We consider the equidistant grid Δ x = L/(nodes-1)

$$\Delta x = 0.15/4 = 0.0375\ m$$

Boundary nodes i= 1 and i=5.

$$\frac{kA\,(T_{i-1} - T_i)}{\Delta x} + \frac{hP\Delta x\,(T_\infty - T_i)}{2} + hA\,(T_\infty - T_i) = 0 \qquad (boundary\ node)$$

$$\begin{aligned} i &= 1 \quad eq1 := T_1 - 200 = 0 \\ i &= 5 \quad eq5 := 1.1775T_4 - 1.198695T_5 + 0.317925 = 0 \end{aligned}$$

Inner nodes i= 2,3, and 4.

$$\left(\frac{kA}{\Delta x}\right)(T_{i-1} - T_i) + \left(\frac{kA}{\Delta x}\right)(T_{i+1} - T_i) + hA(T_\infty - T_i) = 0 \qquad (inner\ nodes)$$

$$i = 2 \quad eq2 = 236.0298750 - 2.390325T_2 + 1.1775T_3 = 0$$
$$i = 3 \quad eq3 = 1.1775T_2 - 2.390325T_3 + 1.1775T_4 + 0.529875 = 0$$
$$i = 4 \quad eq4 = 1.1775T_3 - 2.390325T_4 + 1.1775T_5 + 0.529875 = 0$$

Solving the equations using Maple `solve` function:
```
solve({eq1, eq2, eq3, eq4, eq5}, [T[1], T[2], T[3], T[4],
T[5]])
```

$$[[T_1 = 200., T_2 = 183.14971, T_3 = 171.34393, T_4 = 164.2284, T_5 = 161.5898]]$$

6.3 2D CONDUCTION

We now consider the heat conduction through two-dimensional region as shown in
Figure 6.2. We consider the three-dimensional heat conduction equation

$$\frac{\partial}{\partial x}\left(k\frac{\partial T}{\partial x}\right) + \frac{\partial}{\partial y}\left(k\frac{\partial T}{\partial y}\right) + \frac{\partial}{\partial z}\left(k\frac{\partial T}{\partial z}\right) + \tilde{q} = \rho c\frac{\partial T}{\partial t}$$

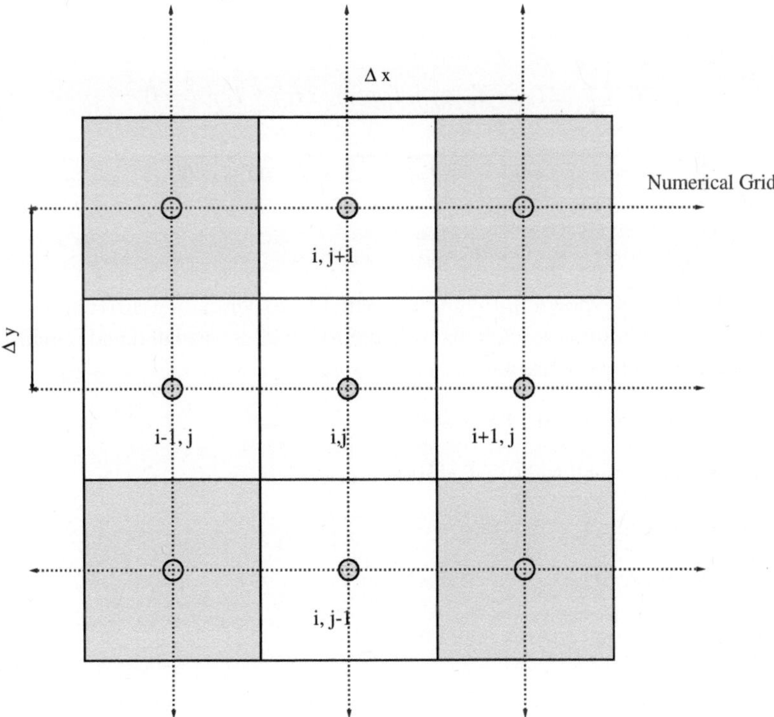

Figure 6.2 Schematic representation of numerical grid using finite-difference approach for
heat conduction problem.

For a steady state, 2D conduction case the equation will be reduced to form

$$\frac{\partial}{\partial x}\left(k\frac{\partial T}{\partial x}\right) + \frac{\partial}{\partial y}\left(k\frac{\partial T}{\partial y}\right) = 0$$

Also, assuming here is no internal heat conversion then

$$\frac{\partial^2 T}{\partial x^2} + \frac{\partial^2 T}{\partial y^2} = 0$$

Using Finite Differences, we can formulate derivatives as

$$\left.\frac{\partial^2 T}{\partial x^2}\right|_{i,j} \equiv \frac{T_{i+1,j} - 2T_{i,j} + T_{i-1,j}}{(\Delta x)^2}$$

$$\left.\frac{\partial^2 T}{\partial y^2}\right|_{i,j} \equiv \frac{T_{i,j+1} - 2T_{i,j} + T_{i,j-1}}{(\Delta y)^2}$$

We have

$$\frac{T_{i+1,j} - 2T_{i,j} + T_{i-1,j}}{(\Delta x)^2} + \frac{T_{i,j+1} - 2T_{i,j} + T_{i,j-1}}{(\Delta y)^2} = 0$$

For a square grid $\Delta x = \Delta y$, we have

$$T_{i+1,j} - 2T_{i,j} + T_{i-1,j} + T_{i,j+1} - 2T_{i,j} + T_{i,j-1} = 0$$

$$\boxed{T_{i+1,j} + T_{i-1,j} + T_{i,j+1} + T_{i,j-1} - 4T_{i,j} = 0} \qquad (6.14)$$

Example 6.2

Consider the slab shown in Figure 6.3, which is experiencing different temperatures at its boundaries. On one side the slab wall is insulated and remaining boundary temperature are

   i. At left wall (x=0) ; T =300 °C;

   ii. At bottom wall (y=0) ; T =100 °C ;

   iii. At top wall; T =600 °C;

   iv. At right wall; $\partial T/\partial y = 0$ (insulated)

Find the temperatures at the nodal position a, b, c and d.

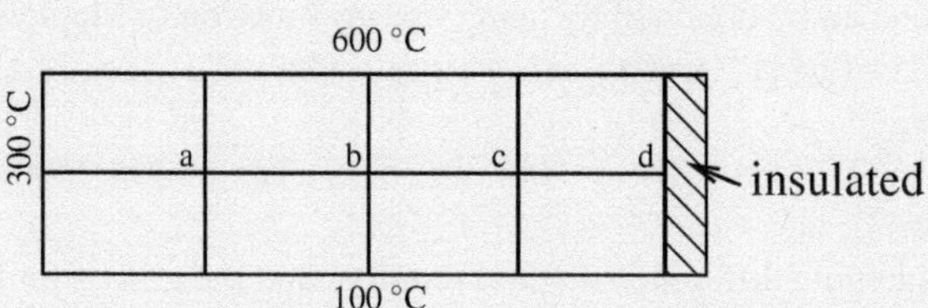

Figure 6.3   The slab is experiencing temperature variation at its boundaries.

Solution The finite-difference formation:

$$\frac{T_{i+1,j} + T_{i-1,j} - 2T_{i,j}}{(\Delta x)^2} + \frac{T_{i,j+1} + T_{i,j-1} - 2T_{i,j}}{(\Delta y)^2} = 0 \qquad (6.15)$$

If Δx and Δy are same we have the equation after rearrangement as:

$$T_{i,j} = \frac{1}{4}\left(T_{i+1,j} + T_{i-1,j} + T_{i,j+1} + T_{i,j-1}\right) \qquad (6.16)$$

This is a very useful relation and one can immediately see that the temperature at a node is depending on the neighboring nodes.

Node a:

$$T_a = \frac{1}{4}(300 + 600 + 100 + T_b)$$

Node b:

$$T_b = \frac{1}{4}(600 + 100 + T_a + T_c)$$

Node c:

$$T_c = \frac{1}{4}(T_b + T_d + 600 + 100)$$

Note that the wall on right-hand side is insulated so there is no gradient in temperature exist at this wall, and we set the temperature same as that on node position c.

Node d:

$$T_d = \frac{1}{4}(T_c + 600 + 100 + T_c)$$

We can write equations in Maple as:

```
eq1 := 300 + 600 + 100 + Tb - 4*Ta;
 eq2 := 600 + 100 + Ta + Tc - 4*Tb;
  eq3 := Tb + Td + 600 + 100 - 4*Tc;
  eq4 := Tc + 600 + 100 + Tc - 4*Td;
```

The solution can be obtained by using `solve` function in Maple:

```
evalf(solve({eq1, eq2, eq3, eq4}, [Ta, Tb, Tc, Td]))
```

This gives

[[Ta = 336.5979381, Tb = 346.3917526, Tc = 348.9690722, Td = 349.4845361]]

Note that the finite difference method requires that the grid be uniform. However, to obtain the solution for temperature, the value of grid spaing is not needed in this particular case.

Matrix-inversion Method We can also solve the equation of equations using the Matrix inversion method. To do so we arrange the equations in format:

$$p_1 T_a + p_2 T_b + p_3 T_c + p_4 T_d = const$$

We can write eq1 as

$$4T_a + (-1)T_b + (0)T_c + (0)T_d = 1000 \tag{6.17}$$

and similarly the remaining equations. We need to define the matrices and their products:

$$\mathbf{A.T} = \mathbf{C} \tag{6.18}$$

where $\mathbf{A}$ is a matrix representing the LHS of the equation 6.17, defined as

$$\mathbf{A} = \begin{bmatrix} 4 & -1 & 0 & 0 \\ -1 & 4 & -1 & 0 \\ 0 & -1 & 4 & -1 \\ 0 & 0 & -2 & 4 \end{bmatrix}$$

and $\mathbf{T}$ represents the matrix of unknown temperatures:

$$T = \begin{bmatrix} Ta \\ Tb \\ Tc \\ Td \end{bmatrix}$$

and, the matrix $\mathbf{C}$ in above equation represent the constant appearing in equations:

$$C = \begin{bmatrix} 1000 \\ 700 \\ 700 \\ 700 \end{bmatrix}$$

Using Maple we now execute the commands

```
with(LinearAlgebra):
  A := Matrix(4, 4, [[4, -1, 0, 0], [-1, 4, -1, 0],
        [0, -1, 4, -1], [0, 0, -2, 4]]);
  T := Vector[column](4, [Ta, Tb, Tc, Td])
  C := Matrix([[1000], [700], [700], [700]])
  Sol:=  B . C
```

The (.) dot in last command can be entered via the full-stop key on the keyboard as set for matrix multiplication in Maple syntax. This gives the solution for temperatures T_a, T_b, T_c and T_d as

[336.59793812749996, 346.39175256, 348.969072151, 349.4845361]

The temperatures values are exhbited in Figure 6.4 at respective nodel positions.

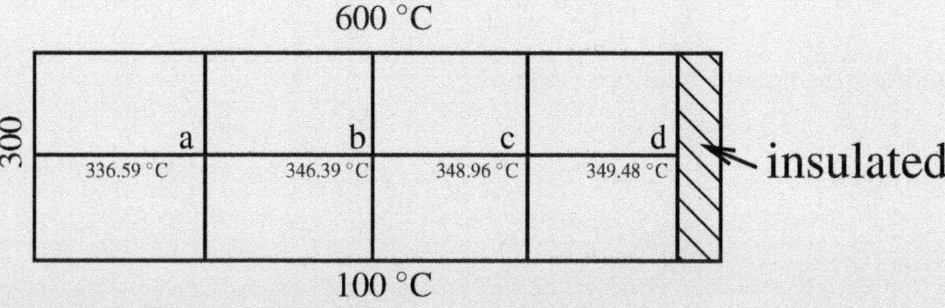

Figure 6.4   The temperatures at nodes inside the slab.

6.4   2D CONDUCTION WITH INTERNAL HEAT CONVERSION

In many engineering problem, there will be internal energy conversion into thermal energy and such problems can be modelled using the finite-difference formulation. Here in this section, we will discuss the two-dimensional heat conduction along with internal heat conversion. In steady-state condition and with constant thermal conductivity, the heat conduction equation can be cast into form

$$\nabla^2 T + \frac{\tilde{q}}{k} = 0 \tag{6.19}$$

This is also known as **Poisson equation of heat conduction**.

$$\frac{\partial^2 T}{\partial x^2} + \frac{\partial^2 T}{\partial y^2} + \frac{\tilde{q}}{k} = 0$$

We transform this PDE into algebraic equation using the finite-difference approach as

$$\frac{T_{i+1,j} - 2T_{i,j} + T_{i-1,j}}{(\Delta x)^2} + \frac{T_{i,j+1} - 2T_{i,j} + T_{i,j-1}}{(\Delta y)^2} + \left(\frac{\tilde{q}}{k}\right) = 0$$

$$\left[\frac{T_{i+1,j} + T_{i-1,j} + T_{i,j+1} + T_{i,j-1} - 4T_{i,j}}{(\Delta x)^2}\right](\Delta x)^2 + \left(\frac{\tilde{q}}{k}\right)(\Delta x)^2 = 0$$

$$\boxed{T_{i+1,j} + T_{i-1,j} + T_{i,j+1} + T_{i,j-1} - 4T_{i,j} + \left(\frac{\tilde{q}}{k}\right)(\Delta x)^2 = 0} \qquad (6.20)$$

Finite Difference Stencils for Common Configurations are provided in Table 6.1.

Example 6.3

A two-dimensional slab of an unknown material is experiencing different levels of temperature at its four boundaries (see Figure 6.5). Find the temperature at discrete points in the slab and assume that grid is not square and has spacing

$$\Delta x = \frac{20}{100}; \Delta y = \frac{10}{100}$$

. Assume no internal heat conversion.

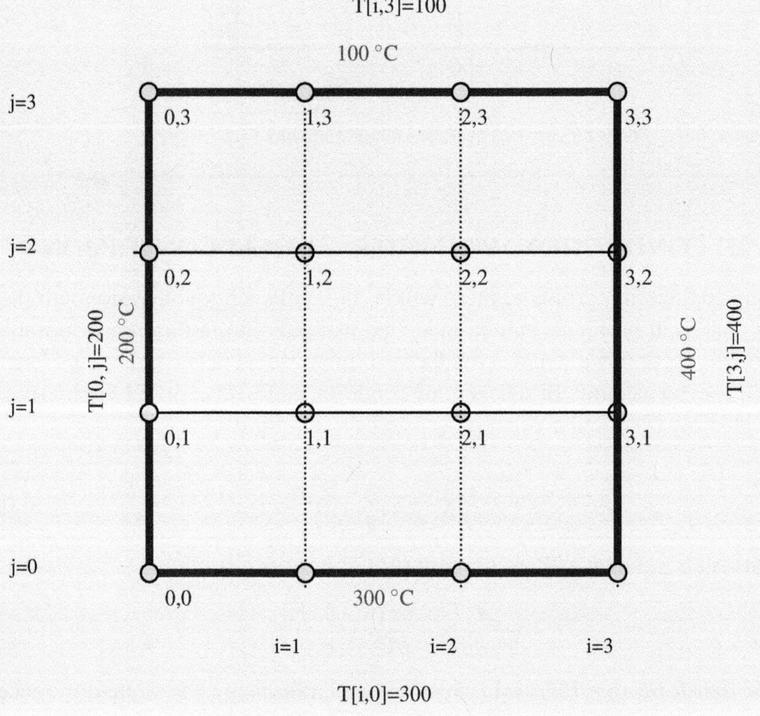

Figure 6.5   The slab along the numerical grid using finite-difference approach for 2D heat conduction problem.

Solution

The applicable equation for the internal nodes

$$T_{i+1,j} + T_{i-1,j} + T_{i,j+1} + T_{i,j-1} - 4T_{i,j} = 0$$

The internal nodes positions in i and j format are

$$i = 1, j = 1$$

$$i = 2, j = 1$$

$$i = 1, j = 2$$

$$i =, j = 2$$

The boundary temperatures are

```
T[i, 0]  := 300;  T[0, j]  := 200;
T[3, j]  := 400;  T[i, 3]  := 100;
```

Since spacing Δx is not same as Δy, we need to define the spacing:

$$\Delta x = \frac{20}{100}; \Delta y = \frac{10}{100}$$

$$
\begin{aligned}
i = 1, j = 1 \quad & eq1 := 25T_{2,1} - 250T_{1,1} + 35000 + 100T_{1,2} = 0 \\
i = 2, j = 1 \quad & eq2 := 40000 - 250T_{2,1} + 25T_{1,1} + 100T_{2,2} = 0 \\
i = 1, j = 2 \quad & eq3 := 25T_{2,2} - 250T_{1,2} + 15000 + 100T_{1,1} = 0 \\
i = 2, j = 2 \quad & eq4 := 20000 - 250T_{2,2} + 25T_{1,2} + 100T_{2,1} = 0
\end{aligned}
$$

We can use Maple to get the solution of equations

```
solve({eq1, eq2, eq3, eq4}, [T[2, 1], T[1, 1], T[2, 2],
T[1, 2]])
```

$$T_{1,1} = 236.483, T_{1,2} = 174.945, T_{2,1} = 265.054, T_{2,2} = 203.516$$

6.5 NUMERICAL TRANSIENT CONDUCTION

The unsteady heat conduction equations is

$$\frac{\partial}{\partial x}\left(k\frac{\partial T}{\partial x}\right) + \frac{\partial}{\partial y}\left(k\frac{\partial T}{\partial y}\right) + \frac{\partial}{\partial z}\left(k\frac{\partial T}{\partial z}\right) + \tilde{q} = \rho c\frac{\partial T}{\partial t}$$

If we consider only one dimensional, unsteady conduction with no internal energy conversion then the applicable equation is

$$\frac{\partial}{\partial x}\left(k\frac{\partial T}{\partial x}\right) = \rho c\frac{\partial T}{\partial t}$$

TABLE 6.1

Finite Difference Stencils for Common Configurations

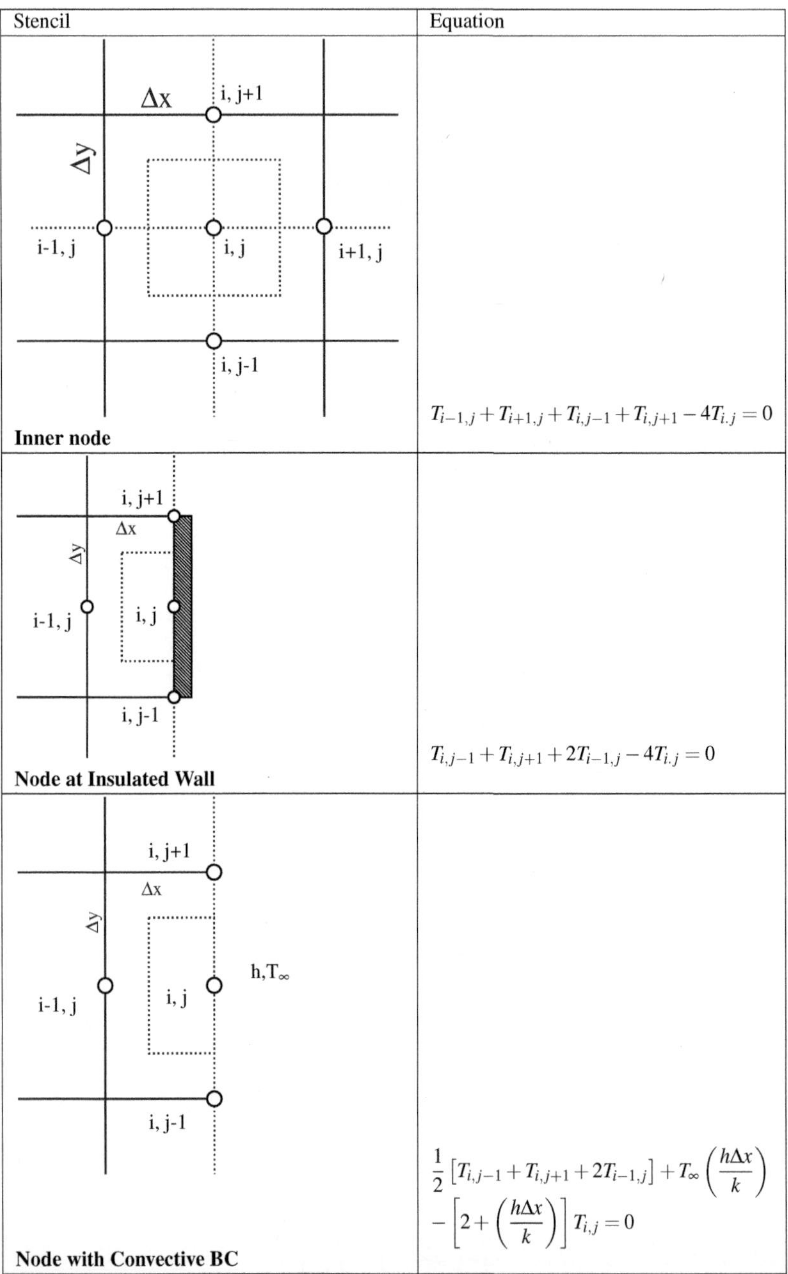

Stencil	Equation
Inner node	$T_{i-1,j} + T_{i+1,j} + T_{i,j-1} + T_{i,j+1} - 4T_{i,j} = 0$
Node at Insulated Wall	$T_{i,j-1} + T_{i,j+1} + 2T_{i-1,j} - 4T_{i,j} = 0$
Node with Convective BC	$\dfrac{1}{2}\left[T_{i,j-1} + T_{i,j+1} + 2T_{i-1,j}\right] + T_{\infty}\left(\dfrac{h\Delta x}{k}\right) - \left[2 + \left(\dfrac{h\Delta x}{k}\right)\right]T_{i,j} = 0$

TABLE 6.2

Finite Difference Stencils for Common Configurations

Stencil	Equation
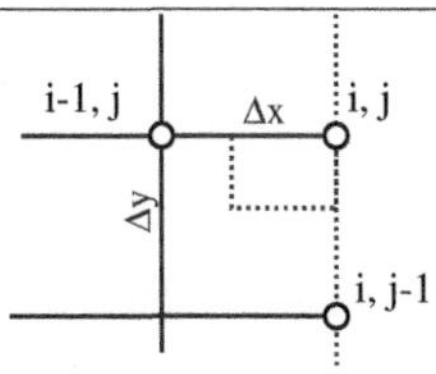 **Insulated Corner**	$T_{i,j-1} + T_{i-1,j} - 2T_{i,j} = 0$
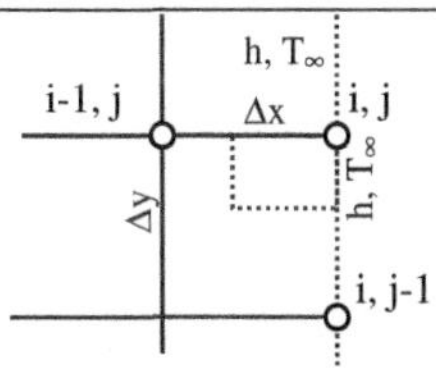 **Corner with Convective BC**	$T_{i,j-1} + T_{i-1,j} + 2T_\infty \left(\dfrac{h\Delta x}{k} \right) - \left[2 + 2 \left(\frac{h\Delta x}{k} \right) \right] T_{i,j} = 0$
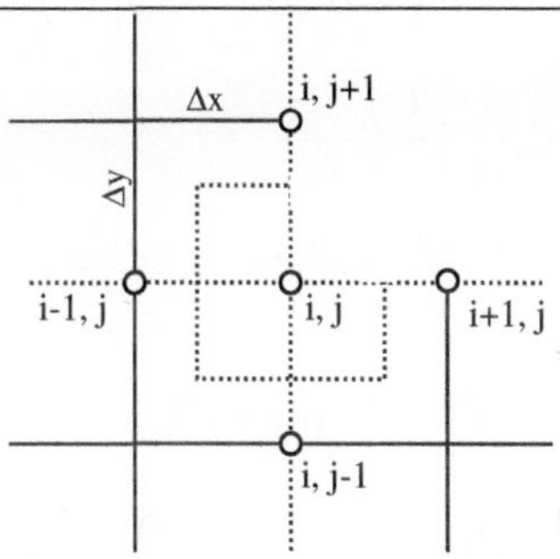 **Interior corner with Convective BC**	$T_{i,j+1} + T_{i+1,j} + 2 \left[T_{i-1,j} + T_{i,j-1} \right] + 2T_\infty \left(\dfrac{h\Delta x}{k} \right) - \left[6 + 2 \left(\frac{h\Delta x}{k} \right) \right] T_{i,j} = 0$
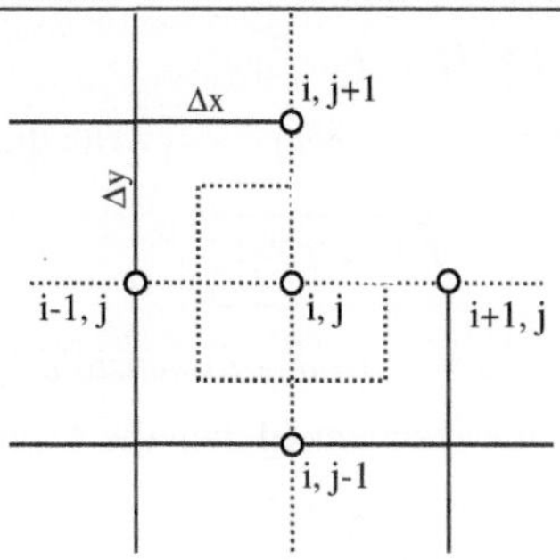 **Interior corner with Insulated BC**	$T_{i,j+1} + T_{i+1,j} + 2 \left[T_{i-1,j} + T_{i,j-1} \right] - 6T_{i,j} = 0$

and for case of constant thermal conductivity

$$\frac{\partial^2 T}{\partial x^2} = \frac{1}{\alpha}\frac{\partial T}{\partial t}$$

We already have created the difference equation for the first- and second-order spatial derivatives in previous section. Here, we will learn how to transform the differential time derivatives into difference form. For this sake, we define time t as

$$t = m \cdot \Delta t$$

where, the parameter m is an integer denoting the time t that has elapsed. This gives the form of difference equation

$$\left.\frac{\partial T}{\partial t}\right|_{i,j} = \frac{T_{i,j}^{m+1} - T_{i,j}^{m}}{\Delta t}$$

The PDE is

$$\frac{\partial^2 T}{\partial x^2} = \frac{1}{\alpha}\frac{\partial T}{\partial t}$$

and the discretised equation for the above equation is

$$\left[\frac{T_{i+1,j}^{m} - 2T_{i,j}^{m} + T_{i-1,j}^{m}}{(\Delta x)^2}\right] = \frac{1}{\alpha}\left[\frac{T_{i,j}^{m+1} - T_{i,j}^{m}}{\Delta t}\right] \qquad (6.21)$$

Solving for $T_{i,j}^{m+1}$

$$T_{i,j}^{m+1} = \left(\frac{\alpha \cdot \Delta t}{(\Delta x)^2}\right)\left[T_{i+1,j}^{m} - 2T_{i,j}^{m} + T_{i-1,j}^{m}\right] + T_{i,j}^{m}$$

$$T_{i,j}^{m+1} = T_{i,j}^{m} + Fo\left[T_{i+1,j}^{m} - 2T_{i,j}^{m} + T_{i-1,j}^{m}\right]$$

where, Fo is grid Fourier number and the equation is an **explicit formulation**.

$$T_{i,j}^{m+1} = T_{i,j}^{m}(1 - 2Fo) + Fo\left[T_{i+1,j}^{m} + T_{i-1,j}^{m}\right]$$

Since we are not considering j direction, we can remove j and rewrite the equation

$$\boxed{T_{i}^{m+1} = T_{i}^{m}(1 - 2Fo) + Fo\left[T_{i+1}^{m} + T_{i-1}^{m}\right]} \qquad (6.22)$$

For stable numerical solution $Fo = \frac{\alpha \cdot \Delta t}{(\Delta x)^2} \leq 0.5 \qquad (one-dimensional)$.

If we include the volumetric heat conversion mechanism then the explicit finite-difference equation is

$$\boxed{T_{i}^{m+1} = T_{i}^{m}(1 - 2Fo) + Fo\left[T_{i+1}^{m} + T_{i-1}^{m} + \frac{\tilde{q}(\Delta x)^2}{k}\right]} \qquad (6.23)$$

6.6 CONVECTIVE BOUNDARY CONDITION

If the convective boundary condition is applied on the wall, then the heat fluxes balance is

$$-k\left.\frac{\partial T}{\partial x}\right|_{x=L} = h\left(T_L - T_\infty\right)$$

Casting this equation into difference equation we have

$$k\left(\frac{T_{i-1}^{m+1} - T_i^{m+1}}{\Delta x}\right) = h\left(T_i^{m+1} - T_\infty\right)$$

Rearranging we have

$$T_i^{m+1} = \frac{\left[T_{i-1}^{m+1} + \left(\frac{h\Delta x}{k}\right)T_\infty\right]}{1 + \left(\frac{h\Delta x}{k}\right)}$$

Another way to estimate the temperature is to apply the first law of thermodynamics

$$k\left(\frac{T_{i-1}^m - T_i^m}{\Delta x}\right) + h\left(T_\infty - T_i^m\right) = \rho c\frac{\Delta x}{2}\left(\frac{T_i^{m+1} - T_i^m}{\Delta t}\right)$$

Rearranging we have

$$T_i^{m+1} = \frac{\alpha\Delta t}{(\Delta x)^2}\left\{\left[\frac{(\Delta x)^2}{\alpha\Delta t} - 2\left(\frac{h\Delta x}{k}\right) - 2\right]T_i^m + 2T_{i-1}^m + 2\left(\frac{h\Delta x}{k}\right)T_\infty\right\}$$

$$\boxed{T_i^{m+1} = Fo\left\{\left[\frac{1}{Fo} - 2\left(\frac{h\Delta x}{k}\right) - 2\right]T_i^m + 2T_{i-1}^m + 2\left(\frac{h\Delta x}{k}\right)T_\infty\right\}} \qquad (6.24)$$

This equation can be arranged as well in the following form

$$T_i^{m+1} = T_i^m(1 - 2\beta_o Fo) + 2FoT_{i-1}^m + 2\gamma_o Fo$$

$$\beta_o = \left[\left(\frac{h\Delta x}{k}\right) + 1\right]$$

$$\gamma_o = \left(\frac{h\Delta x}{k}\right)T_\infty$$

$$Fo = \left(\frac{\alpha\Delta t}{(\Delta x)^2}\right)$$

If internal heat conversion is present at the boundary, then the equation will take the form

$$\boxed{\begin{aligned}
T_i^{m+1} &= T_i^m(1-2\beta_o Fo) + 2Fo T_{i-1}^m + 2\gamma_o Fo + \left(\tfrac{\alpha\Delta t}{k}\right)\widetilde{q} \quad (i=1)\\
T_i^{m+1} &= T_i^m(1-2\beta_o Fo) + 2Fo T_{i+1}^m + 2\gamma_o Fo + \left(\tfrac{\alpha\Delta t}{k}\right)\widetilde{q} \quad (i=0)
\end{aligned}}$$
$$(6.25)$$

For stability, the following condition must hold

$$\frac{(\Delta x)^2}{\alpha\Delta t} \geq 2\left[\left(\frac{h\Delta x}{k}\right)+1\right]$$

or

$$\frac{1}{Fo} \geq 2\left[\left(\frac{h\Delta x}{k}\right)+1\right]$$

Example 6.4

A strip is experiencing temperatures

$$T_0=500, T_1=650, T_2=750, T_3=1000, T_4=750, T_5=650, T_6=500$$

as shown in Figure 6.6. The convective heat transfer coefficient at both sides of wall is 25 W/m^2·K and surrounding temperature is 100 °C. The Properties of the material are: k = 4 W/m · K, ρ = 1200 kg/m^3; c = 920 kg/m^3 and α= k/(ρc) =3.623188405 × 10^{-6} m^2/s.

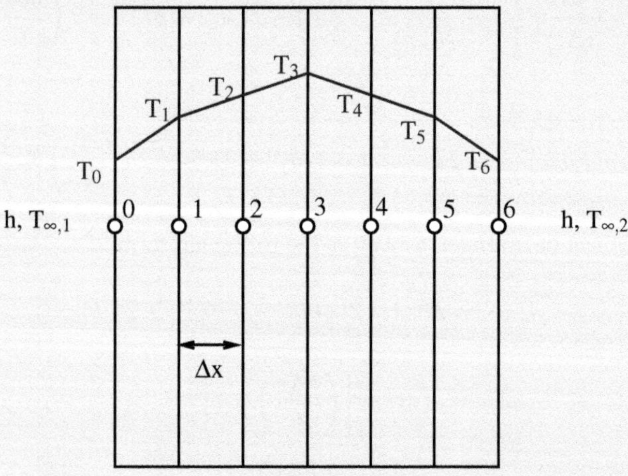

Figure 6.6   Figure example.

Solution We will solve this problem using explicit formulation and for that we must check the stability condition

$$\frac{(\Delta x)^2}{\alpha \Delta t} \geq 2\left[\left(\frac{h\Delta x}{k}\right)+1\right]$$

For h= 15 W/m^{2}K and Δx= 10 cm, we have condition

$$\frac{(\Delta x)^2}{\alpha \Delta t} \geq 2.0075$$

which gives $\Delta t = 849.23$ s $\equiv 850$ s for solution of the problem. The Fourier number is Fo=0.307.

The equation for inner nodes 2,3,4,5 is

$$T_i^{m+1} = T_i^m(1-2Fo) + Fo\left[T_{i+1}^m + T_{i-1}^m\right]$$

We can generate the necessary equations using Maple
```
eq := T[i]*(1 - 2*Fo) + Fo*(T[i + 1] + T[i - 1])
```
For i=1, this gives

$$eq = 0.3840579712 T_i + 0.3079710144 T_{i+1} + 0.3079710144 T_{|i-1|}$$

which represent the RHS of difference equation, and in terms of time increment *m* we have the form

$$T_i^{m+1} = 0.3840579712 T_i^m + 0.3079710144 T_{i+1}^m + 0.3079710144 T_{|i-1|}^m$$

For the boundary nodes 0 and 6, we have

$$T_i^{m+1} = Fo\left\{\left[\frac{1}{Fo} - 2\left(\frac{h\Delta x}{k}\right) - 2\right]T_i^m + 2T_{i-1}^m + 2\left(\frac{h\Delta x}{k}\right)T_\infty\right\}$$

In Maple this can be formulated as
```
eq2 := Fo*((1/Fo - 2*h*'&Delta;x'/k - 2)*T[n]
+ 2*T[abs(n - 1)] + 2*h*'&Delta;x'*Tinfi/k)
```
which represent the RHS of difference equation and the equations for node 0 and 6 are

$$T_i^{m+1} = -0.0009057969562 T_i^m + 0.6159420288 T_{|i-1|}^m + 38.49637680$$

where n=0 and 6 are the nodes at the boundaries. Now we loop in time and compute the temperatures one by one over time increments of 850 s. This leads to following results

m	Δt	T_0	T_1	T_2	T_3	T_4	T_5	T_6
0	0	500	650	750	1000	750	650	500
1	850	438	635	796	846	796	635	438
2	1700	429	624	762	815	762	624	429
3	5100	422	606	736	782	736	606	422
4	20400	412	590	710	754	710	590	412

Example 6.5

The plane wall shown in Figure 6.7 below has an initial uniform temperature of 200°C. The wall is suddenly experiencing internal energy conversion into the heat. The convective boundary conditions are imposed on the sides of the wall as shown in Figure 6.7. The wall has volumetric internal energy conversion into heat as 50 MW/m^3. The material properties are k = 20 W/m · °C, ρ = 7800 kg/m^3, and C = 460 J/kg · °C. Take Δx=2 mm, and find the temperature distribution after time lapse of 1.709s.

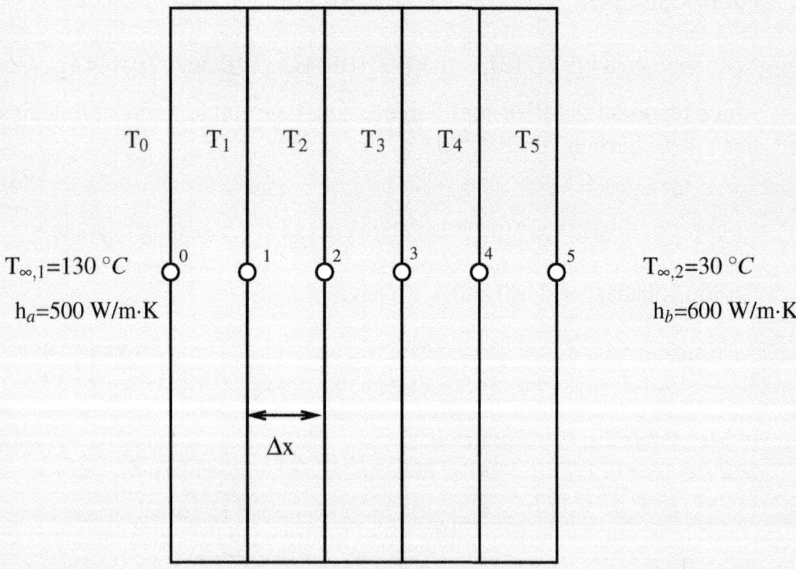

Figure 6.7   Figure example.

Solution
We need to check the stability condition and find the time increment needed for stable solution. We have two different convection conditions at the wall,

so we have two time increments associated with the stability condition

$$\frac{(\Delta x)^2}{\alpha\,(\Delta t_a)} \geq 2\left[\left(\frac{h_a \Delta x}{k}\right)+1\right]$$

$$\frac{(\Delta x)^2}{\alpha\,(\Delta t_b)} \geq 2\left[\left(\frac{h_b \Delta x}{k}\right)+1\right]$$

This gives

$$\Delta t_a = 0.3417142857$$
$$\Delta t_b = 0.3384905660$$

One may choose to use any time increment as the two values are very close to each other, else the smaller value is an obvious choice. We go for

$$\Delta t = \Delta t_b = 0.342s$$

in this solution which gives the Fourier number Fo = 0.476.
We will take 5 nodes to solve this problem using explicit formulation. The applicable difference equation for the internal nodes 1,2,3,4 is

$$T_i^{m+1} = T_i^m(1-2Fo) + Fo\left[T_{i+1}^m + T_{i-1}^m + \frac{\widetilde{q}(\Delta x)^2}{k}\right]$$

In Maple we can formulated the equations using equation
```
T[node]:=0.0476190476*T[i] + 0.4761904762*T[i + 1]
         + 0.4761904762*T[abs(i - 1)] + 4.761904762
```
where node is the number corresponding to the nodes shown in figure. For the boundary nodes 0 and 5 equation is

$$T_i^{m+1} = T_i^m(1-2\beta_{o,a}Fo) + 2FoT_{i+1}^m + 2\gamma_{o,a}Fo + \left(\frac{\alpha\Delta t}{k}\right)\widetilde{q} \quad (i=0)$$
$$T_i^{m+1} = T_i^m(1-2\beta_{o,b}Fo) + 2FoT_{i-1}^m + 2\gamma_{o,b}Fo + \left(\frac{\alpha\Delta t}{k}\right)\widetilde{q} \quad (i=5)$$

where

$$\beta_{o,a} = \left[\left(\frac{h_a \Delta x}{k}\right)+1\right] = 1.05$$
$$\gamma_{o,a} = \left(\frac{h_a \Delta x}{k}\right)T_{\infty,a} = 6.5$$

$$\beta_{o,b} = \left[\left(\frac{h_b \Delta x}{k}\right)+1\right] = 1.06$$
$$\gamma_{o,b} = \left(\frac{h_b \Delta x}{k}\right)T_{\infty,b} = 1.5$$

In Maple we can formulate equations
```
T[0] := 10.95238095 + 0.9523809524*T[1]
T[5] := -0.009523810*T[5] + 0.9523809524*T[4] + 6.190476191
```

The solution of equations in time increment of Δt is as follows

m	Δt	0	1	2	3	4	5
0	0	200	200	200	200	200	200
1	0.342	201.43	204.76	204.76	204.76	204.76	194.76
2	0.683	205.96	207.94	209.52	209.52	204.76	199.35
3	1.025	208.99	212.51	213.53	212.02	209.21	199.30
4	1.367	213.35	216.08	217.09	216.16	210.59	203.54
5	1.709	216.74	220.02	220.93	218.71	214.65	204.82

6.7   1D TRANSIENT CONDUCTION: IMPLICIT FORMULATION

We can also have the implicit formulation for the transient one-dimensional heat conduction problem as

$$\frac{\partial^2 T}{\partial x^2} = \rho c \frac{\partial T}{\partial t}$$

Inserting the finite-difference approximations for the individual derivatives as

$$\left.\frac{\partial^2 T}{\partial x^2}\right|_i \equiv \frac{T_{i+1}^{m+1} - 2T_i^{m+1} + T_{i-1}^{m+1}}{(\Delta x)^2}$$

$$\frac{\partial T}{\partial t} = \frac{T_i^{m+1} - T_i^m}{\Delta t}$$

We get the algebraic equation of the form

$$\frac{T_{i+1}^{m+1} - 2T_i^{m+1} + T_{i-1}^{m+1}}{(\Delta x)^2}$$

$$= \frac{1}{\alpha}\left[\frac{T_i^{m+1} - T_i^m}{\Delta t}\right]$$

Rearranging we get

$$T_i^m = -\left(\frac{\alpha \Delta t}{(\Delta x)^2}\right)\left[T_{i+1}^{m+1} + 2T_i^{m+1} - T_{i-1}^{m+1}\right] + T_i^{m+1}$$

$$T_i^m = T_i^{m+1} - Fo\left[T_{i+1}^{m+1} + 2T_i^{m+1} - T_{i-1}^{m+1}\right]$$

This is an implicit formulation for 1D-conduction problems. This simple implicit scheme is accurate only to $O[\Delta t,(\Delta x)^2]$.

In 1947, an alternative implicit differencing scheme has been proposed by Crank and Nicolson. They proposed to retain the left-hand side of the implicit formulation, along with the inclusion of the arithmetic average of the right-hand sides of the explicit and the implicit formulations for space discretisation. Their proposal leads to the form

$$\frac{1}{2}\left[\left(\frac{T_{i+1}^{m+1}-2T_{i}^{m+1}+T_{i-1}^{m+1}}{(\Delta x)^2}\right)+\left(\frac{T_{i+1}^{m}-2T_{i}^{m}+T_{i-1}^{m}}{(\Delta x)^2}\right)\right]=\frac{1}{\alpha}\left[\frac{T_{i}^{m+1}-T_{i}^{m}}{\Delta t}\right]$$

(6.26)

The Crank–Nicolson formulation is second-order accurate in both space and time.

6.8   2D TRANSIENT CONDUCTION

We can also have the implicit formulation for the transient two-dimensional heat conduction problem by inserting the FD formulations into PDE:

$$\frac{\partial^2 T}{\partial x^2}+\frac{\partial^2 T}{\partial y^2}=\rho c\frac{\partial T}{\partial t}$$

The finite-differences equations are

$$\frac{\partial^2 T}{\partial x^2}\bigg|_{i,j}=\frac{T_{i+1,j}-2T_{i,j}+T_{i-1,j}}{(\Delta x)^2}$$

$$\frac{\partial^2 T}{\partial y^2}\bigg|_{i,j}=\frac{T_{i,j+1}-2T_{i,j}+T_{i,j-1}}{(\Delta y)^2}$$

We get the equation

$$\left[\frac{T_{i+1,j}^{m+1}-2T_{i,j}^{m+1}+T_{i-1,j}^{m+1}}{(\Delta x)^2}\right]+\left[\frac{T_{i,j+1}-2T_{i,j}+T_{i,j-1}}{(\Delta y)^2}\right]=\frac{1}{\alpha}\left(\frac{T_{i,j}^{m+1}-T_{i,j}^{m}}{\Delta t}\right)$$

Rearranging we have

$$T_{i,j}^{m}=-Fo\left[T_{i+1,j}^{m+1}+T_{i-1,j}^{m+1}+T_{i+1,j}^{m+1}+T_{i-1,j}^{m+1}\right]+(1+4Fo)T_{i,j}^{m+1}$$

(6.27)

This scheme is an implicit formulation, and it is a backward-difference formulation as the time derivative moves backward from the time for heat conduction into the node.

In case Δx is not equal to Δy, then we can use the backward-difference or implicit approach and develop the nodal equations

$$-\left(\frac{\alpha\Delta t}{(\Delta x)^2}\right)\left[T_{i+1,j}^{m+1}-2T_{i,j}^{m+1}+T_{i-1,j}^{m+1}\right]-\left(\frac{\alpha\Delta t}{(\Delta y)^2}\right)\left[T_{i,j+1}^{m+1}-2T_{i,j}^{m+1}+T_{i,j-1}^{m+1}\right]$$

$$+T_{i,j}^{m+1}=T_{i,j}^{m}$$

$$T_{i,j}^{m}=T_{i,j}^{m+1}+2Fo_{\Delta x}T_{i,j}^{m+1}2Fo_{\Delta y}T_{i,j}^{m+1}-Fo_{\Delta x}\left[T_{i+1,j}^{m+1}+T_{i-1,j}^{m+1}\right]-Fo_{\Delta y}\left[T_{i,j+1}^{m+1}+T_{i,j-1}^{m+1}\right]$$

(6.28)

where $Fo_{\Delta x}$ and $Fo_{\Delta y}$ are the Fourier numbers associated with x and y grid spacing, respectively. Also, if we include the internal energy conversion term, then equation will take the form

$$T_{i,j}^{m} = T_{i,j}^{m+1}\left[1 + 2Fo_{\Delta x} + 2Fo_{\Delta y}\right] - Fo_{\Delta x}\left[T_{i+1,j}^{m+1} + T_{i-1,j}^{m+1}\right] - Fo_{\Delta y}\left[T_{i,j+1}^{m+1} + T_{i,j-1}^{m+1}\right]$$
$$- \left(\frac{\Delta t \cdot \tilde{q}}{\rho c}\right)$$

$$(6.29)$$

Nodal equations in implicit formulation for transient heat conduction problems

$$T_i^{m+1} = \frac{T_i^m + 2Fo\left[T_{i+1}^{m+1} + BiT_{i-1}^{m+1}\right]}{1 + 2Fo + 2FoBi} \qquad (1D, \; Convective \; boundary \; condition)$$

$$T_i^{m+1} = \frac{T_i^m + 2FoT_{i-1}^{m+1}}{1 + 2Fo} \qquad (1D, \; Insulated \; boundary)$$

$$T_i^{m+1} = \frac{T_i^m + Fo\left[T_{i+1}^{m+1} + T_{i+2}^{m+1} + T_{i+3}^{m+1} + T_{i+4}^{m+1}\right]}{1 + 4Fo} \qquad (2D, \; interior \; node)$$

$$T_i^{m+1} = \frac{T_i^m + 2Fo\left[T_{i+2}^{m+1} + 0.5T_{i+1}^{m+1} + 0.5T_{i+3}^{m+1} + T_{i-1}^{m+1}\right]}{1 + 2Fo(2 + Bi)} \qquad (Convective \; BC)$$

$$T_i^{m+1} = \frac{T_i^m + 2Fo\left[T_{i+1}^{m+1} + T_{i+2}^{m+1} + 2BiT_{i-1}^{m+1}\right]}{1 + 4Fo(1 + Bi)}$$

$$(2D, \; outside \; corner \; node \; with \; convective \; BC)$$

PROBLEMS

6P-1 A rectangular slab as shown in Figure 6P1 is experiencing different temperature at boundaries. Find the steady-state temperatures at the indicated nodal location inside the slab if $\Delta x = \Delta y = 10$ cm, if volumetric heat conversion $\tilde{q} = 0$ W/m^3. The thermal conductivity of slab material is 20 W/m·K.

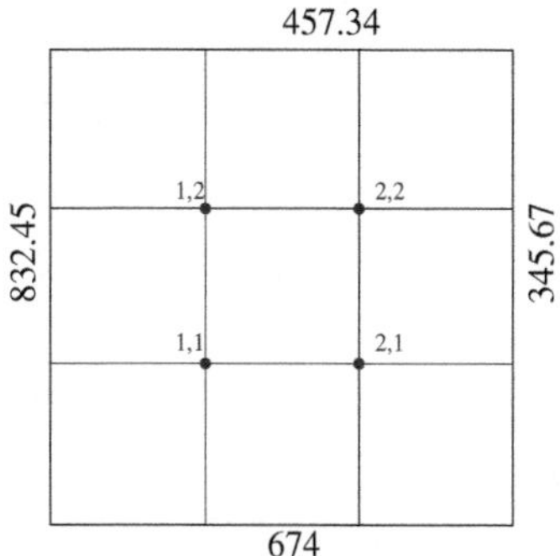

Figure 6P1

[Ans: For $\tilde{q}=0$ W/m^3: $T_{2,1}=542.475, T_{1,1}=661.92, T_{2,2}=488.31, T_{1,2}=607.755$]

6P-2 Repeat Problem 6P1 with $\tilde{q}=100$ W/m^3, and k=150 W/m·K.

[Ans: $T_{2,1}=542.478, T_{1,1}=661.923, T_{2,2}=488.31, T_{1,2}=607.758$]

6P-3 Repeat Problem 6P1 if $\Delta x=\Delta y=500$ cm, and the volumetric heat conversion $\tilde{q}=2000$ W/m^3.

[Ans: $T_{2,1}=1792.475$, $T_{1,1}=1911.92$, $T_{2,2}=1738.31$, $T_{1,2}=1857.755$]

6P-4 A thin strip of thickness δ is exposed to convective and radiative environment on one side. Assuming one-dimensional heat conduction do energy balance and derive an expression for temperature distribution on nodal positions if material is also experiencing volumetric heat generation. Ignore heat transfer on the other side of the strip.

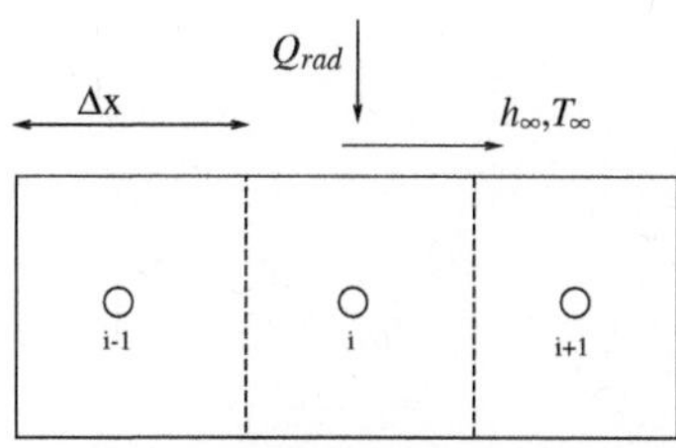

Figure 6P4

6P-5 Figure 6P5 shows the two dimensional region with one wall insulated and one wall exposed to convective environment $h_\infty =25$, $T_\infty= 60$ C. If $\Delta x=100$, $\Delta y=100$ cm and k=80 W/m·K, find the temperature at nodal positions.

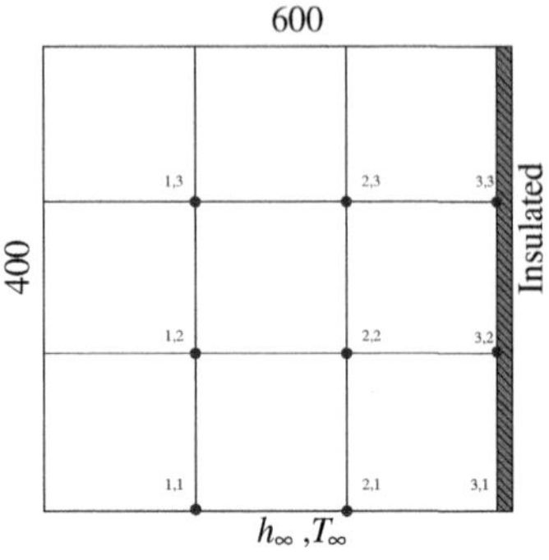

Figure 6P5

REFERENCES

Y. Jaluria and K. E. Torrance. Computational Heat Transfer. New York: Hemisphere, USA, 1986.

G. E. Myers. Analytical Methods in Conduction Heat Transfer. New York: McGraw-Hill, 1971.

M. N. Özisik. Finite Difference Methods in Heat Transfer. Boca Raton, FL: CRC Press, USA, 1994.

7 Fundamentals of Convective Heat Transfer

In this chapter, we will discuss the convection or thermal diffusion along with bulk fluid movement. We will discuss the derivation of mass, momentum and energy equations. After finishing this chapter, one will be able to:

- Derive the differential formulation of the mass continuity, momentum and energy equations.
- Interpret important dimensionless numbers used in convective heat transfer like the Reynolds, Prandtl, Grashof, and Nusselt numbers, and their physical significance.
- Understand the differences in laminar and turbulent heat transfer phenomenon.
- Explain the need and procedure of Reynolds Averaging and Boussinesq approximation.

Convective heat transfer is an important phenomenon of heat transfer and we know that it can be calculated using Newton's law of cooling

$$Q = h \cdot A_w \cdot (T_w - T_\infty)$$

where, A_w is the area of the wall or surface, T_w is wall temperature, and T_∞ is the freestream temperature. The problem is that convective heat transfer coefficient (h) is not a parameter like a thermal conductivity or any other material property. Rather it is based on

$$h = f(Geometry, Flow, Temperature\ Difference, Fluid\ properties)$$

Note that the energy equation for solids or heat conduction equation although describing the energy balance is no longer valid, and therefore, we must formulate the differential form of the energy equation for fluids. In this chapter, we will discuss the kinds of convective heat transfer and also discuss the dimensionless parameters which must be used for the determination of convective heat transfer.

7.1 FORCED CONVECTION AND DIMENSIONLESS PARAMETERS

Forced convection occurs with appreciable fluid speeds. Blowing of wind over objects occurs every day in nature but this blowing of wind over surface would cause the forced convection over the surface.

$$h = f(\rho, \mu, k, c_p, \Delta T, \ell)$$

DOI: 10.1201/9781003428404-7

We convert it into fundamental dimensions of time (t), mass (M), length (L), temperature (θ), and amount of heat (Q). From Newton's law of viscosity, we have

$$Q = h \cdot A \cdot \Delta T$$

$$h = \frac{Q}{A \cdot \Delta T} \approx \frac{Q/t}{L^2 \cdot \theta} \approx \frac{Q}{L^2 \cdot \theta \cdot t}$$

Dimensionally, thermophysical properties are

$$\rho \approx \frac{M}{L^3}, \mu \approx \frac{M}{L \cdot t}, k \approx \frac{Q}{L \cdot \theta \cdot t},$$

$$c_p \approx \frac{Q}{M \cdot \theta}$$

The temperature, time and velocity are

$$\Delta T \approx \theta$$
$$\ell \approx L$$
$$u \approx \frac{L}{t}$$

Using, Rayleigh procedure we have

$$h = C \cdot \mu^a \cdot \rho^b \cdot k^c \cdot c_p^d \cdot \Delta T^e \cdot \ell^f \cdot u^g$$

$$\left(\frac{Q}{L^2 \cdot \theta \cdot t}\right) = C \left(\frac{M}{L \cdot t}\right)^a \left(\frac{M}{L^3}\right)^b \left(\frac{Q}{L \cdot \theta \cdot t}\right)^c \left(\frac{Q}{M \cdot \theta}\right)^d (\theta)^e (L)^f \left(\frac{L}{t}\right)^g$$

After equating the indices in the above fundamental dimensions, one can obtain

$$Q: \quad c + d = 1$$
$$M: \quad a + b - d = 0$$
$$L: \quad -a - 3b - c + f + g = -2$$
$$t: \quad -a - c - g = -1$$
$$\theta: \quad -c - d + e = -1$$

This would lead to

$$h = C \left(\frac{\mu \cdot c_p}{k}\right)^d \left(\frac{k}{\ell}\right) \left(\frac{\rho \cdot u \cdot \ell}{\mu}\right)^d$$

Rearranging we have

$$\left(\frac{h \cdot \ell}{k}\right) = C \left(\frac{\mu \cdot c_p}{k}\right)^d \left(\frac{\rho \cdot u \cdot \ell}{\mu}\right)^d$$

The bracketed parameter on RHS are defined as Reynolds number

$$Re = \left(\frac{\rho \cdot u \cdot \ell}{\mu}\right)$$

and, the Prandtl number

$$Pr = \left(\frac{\mu \cdot c_p}{k}\right)$$

The parameter on the LHS can be defined as Nusselt number

$$Nu = \left(\frac{h \cdot \ell}{k}\right)$$

This lead to form

$$Nu = f(Re, Pr)$$

> ### Pioneers of Heat Transfer
>
> **Osborne Reynolds** born in Belfast in 1842, Ireland, died in 1912 in Watchet in England was an Irish engineer who made a great contribution In the field of hydrodynamics and fluid dynamics at the end of the Nineteenth century.
> **Wilhelm Nusselt** German engineer, born November 25, 1882 in Nuremberg, he studied at Berlin and Munich. He received his doctorate in 1907, and in 1915 he published his work in which he mentioned dimensionless groups, currently known as *Theory of Similarity*. Nusselt was a professor at the University of Karlsruhe from 1920 to 1925, then in Munich until 1957. He lived for 75 years and died in Munich on September 1, 1957.
> **Ludwig Prandtl** born in 1875 in Freising, and died in 1953 in Göttingen, Germany. After his PhD studies at the University of Munich, Prandtl joined the Institute of Fluid Mechanics at the University of Göttingen. He worked mainly on the boundary layer theory.

In Fluid Mechanics and convective heat transfer, Reynolds number is the most often used parameter and it can help us in declaring flow either as laminar or turbulent. The established ranges for the Reynolds numbers for the transition from laminar to turbulent flow are:

$$1000 \leq Re_D \leq 3000 \quad (Free\ jet)$$
$$3 \times 10^5 \leq Re_L \leq 3 \times 10^6 \quad (Flow\ over\ flat\ plate)$$
$$2 \times 10^5 \leq Re_D \leq 3.8 \times 10^5 \quad (Flow\ over\ cylinder)$$
$$3 \times 10^5 \leq Re_D \leq 3.5 \times 10^5 \quad (Flow\ over\ sphere)$$
$$2.3 \times 10^3 \leq Re_D \leq 4 \times 10^3 \quad (Flow\ inside\ pipe)$$
$$500 \leq Re_{Rh} \leq 750 \quad (Open\ Channel\ Flow)$$

where subscript L and D refers to characteristic length, which can be the length of the flat plate or diameter of the cylinder/sphere, respectively. Also, for open channel flow Reynolds number must be computed based on water depth. It would be interesting for reader to learn that we can also use representative time scales to extract dimensionless numbers. The transport of mass, momentum and energy is a time-dependent phenomenon, and it is worthwhile to understand the dynamics of transport mechanism using dimensionless number interpreted via time scales. Lykouds (1990) suggested to relate the time scales or rate of transport of quantities to arrive at the dimensionless number. If we related unsteady inertia force ($\rho \cdot U/t$) and advective inertia force ($\rho \cdot U^2/\ell$), we have the bulk fluid advection time scale:

$$\frac{\rho U}{t} \propto \frac{\rho U^2}{\ell}$$

$$t_{adv} \propto \frac{\ell}{U}$$

We can now write the time scale for momentum and heat transport time scales as

$$t_m = \frac{\ell^2}{\nu}, \text{ and } t_h = \frac{\ell^2}{\alpha}$$

We can relate the time scale for momentum and flow advection and this gives us the Reynolds number:

$$Re = \frac{t_m}{t_{adv}} = \left(\frac{\ell^2}{\nu}\right)\left(\frac{U}{\ell}\right) = \frac{U\ell}{\nu}$$

Prandtl number can be interpreted in number of ways. It can be defined as ratio of momentum diffusivity to thermal diffusivity, i.e. ratio of kinematic viscosity (ν) to thermal diffusivity (α). In forced convection boundary layer flows, Prandtl number is also defined as the square root of ratio of the momentum and thermal boundary layer thicknesses. However, this definition is not universal valid and in the case of free convection for low Prandtl numbers both momentum and thermal boundary layers have similar thicknesses.

Prandtl number is defined as time scale of heat to flow advection scale:

$$Pr = \frac{t_h}{t_m} = \frac{\nu}{\alpha}$$

The representative values of Prandtl number for some fluids are graphically depicted in Figure 7.1.

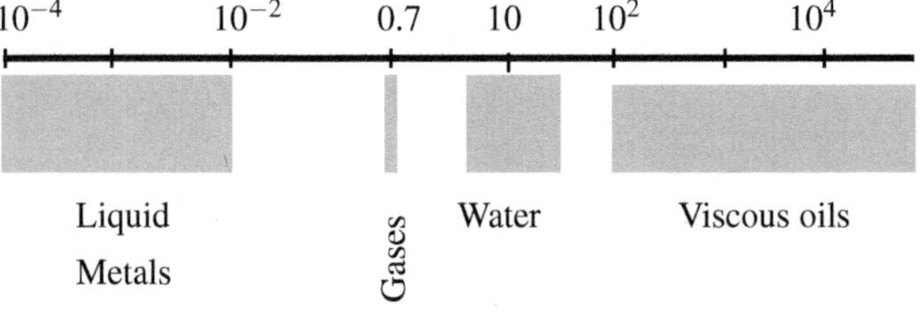

Figure 7.1 Prandtl number ranges for various materials used in thermal engineering.

Since Nusselt number, Reynolds number, and Prandtl number are dimensionless numbers, we can formulate another dimensionless numbers called Stanton Number. It is a ratio of heat transferred to a fluid to the thermal capacity of the fluid. Also, the product of Reynolds and Prandtl number is defined as Peclet number (Pe). As Reynolds and Prandtl numbers are dimensionless, the Peclet number is also a dimensionless number. Phenomenologically it is interpreted as the ratio of convected with the conducted energies in fluid.

Goal: The goal of forced convection heat transfer experiments is to cast the following formulation:

$$Nu = C \cdot Re_D^m \cdot Pr^n$$

where, C is a constant, m and n are experimentally determined indexes. The Nusselt number is defined as

$$Nu = \frac{h \cdot D}{k}$$

7.2 FREE CONVECTION AND DIMENSIONLESS PARAMETERS

In free convection, the density change due to temperature variation is playing a key role. When we heat up the fluid in a gravitational field, the natural convection might occur. The flow speeds associated with natural convection are not appreciable and usually they are of the order of 2 m/s. As appreciable velocity is small in case of a free convection, the Reynolds number does not play a big role, unless it is a case of a mixed convection, where both free and forced convection is happening simultaneously. Overall, natural and free convection related heat transfer coefficients are much smaller than those associated with forced convection. Some representative orders of magnitude for convective heat transfer coefficient in case of forced and free convection are tabulated in Table 7.1

Note that the density variation in fluid can occur due to the composition gradients present in fluid itself, for example, the salt concentration variation in sea water can cause density changes. Also, in atmosphere the moist air rises due to the lower density of the water vapor present in the air-moisture mixture. In rotating machinery, a similar mechanism happens due to the flow induced by a centripetal acceleration of the fluid.

If we relate unsteady inertia force per unit volume with buoyancy force per unit volume we have

$$g\Delta\rho \propto \frac{\rho U}{t}$$

TABLE 7.1

Order of Magnitude for Convective Heat Transfer Coefficient Values

	kcal/h·m^2·°C			W/m^2·K		
Free Convection						
Gases	3	to	20	3	to	23
Water	100	to	600	116	to	698
Boiling	1000	to	20	1163	to	23
Forced Convection						
Gases	10	to	100	12	to	116
Viscous Liquids	50	to	500	58	to	582
Water	500	to	10000	582	to	11630
Condensation	1000	to	100000	1163	to	116300

Also, we can further insert the t=U/L and this shall be

$$t \propto \frac{U}{g(\Delta\rho/\rho)} \equiv \frac{\ell/t}{g(\Delta\rho/\rho)}$$

leading to time taken by the buoyant current to rise or fall as

$$t_{buoyant} \propto \sqrt{\frac{\ell}{g(\Delta\rho/\rho)}}$$

Grashof number It is often defined as the ratio of the buoyant to viscous forces. However, the Grashof number does not involve a velocity which is necessary to define the viscous term. We can define Grashof number as ratio of two time scales:

$$Gr = \left(\frac{t_m}{t_{buoyant}}\right)^2 = \left[\frac{\ell^2/v}{\sqrt{\frac{\ell}{g(\Delta\rho/\rho)}}}\right]^2 = \frac{g(\Delta\rho/\rho)\ell^3}{gv^2}$$

Grashof number is an expression of the relative characteristic times for the diffusion of momentum and the buoyant current time. In case of gases and fluids having moderate Prandtl number, like the liquids with creeping flows it is found that the Grashof number is below 10^4.

Pioneers of Heat Transfer

Franz Grashof was a German engineer, born in 1826 in Dsseldorf, Germany. He leaves school at age of 15, and worked as a mechanic and also take business courses. In 1847 he studied mathematics, physics and machine design at the Technical Institute of Berlin. Between 1849 and 1852, he travelled (Australia) to return to Berlin and continue his studies. He founded the German Society of Engineers (Verein Deutscher Ingenieure, VDI). He was a professor at the University Karlsruhe technique and died in 1893

Dimensionally, the Nusslet number is related with Grashof (Gr) and Prandtl (Pr) number as

$$Nu = f(Gr, Pr)$$

Also, in certain cases the dependence on Prandtl number can also be removed and correlations can be cast into form

$$Nu = f(Gr)$$

Rayleigh number can be defined as ratio of three time scales:

$$Ra = \frac{t_m \cdot t_h}{t_{buoyant}^2} = \frac{(\ell^2/\nu)(\ell^2/\alpha)}{\left(\sqrt{\frac{\ell}{g(\Delta\rho/\rho)}}\right)^2} = \frac{g(\Delta\rho/\rho)\ell^3}{g\nu\alpha}$$

Rayleigh number involves the competition of the harmonic mean between the diffusion times of heat and momentum, and the free fall time of buoyant current generated.

Le Fevre suggested that the product of local Grashof number and Prandtl number $Gr_x Pr$ should be taken as a proper functional parameter for the determination of laminar to turbulent transition in natural convection boundary layers. For plumes and natural convection related flows, the transition to a turbulent boundary layer occurs at a critical value of the Rayleigh number around 10^9.

For Prandtl number is less than unity then the Nusselt number can be related to the product of Grashof number and Prandtl number squared. This parameter is called the Boussinesq number (Bo).

Pioneers of Heat Transfer

Lord Rayleigh was a British physicist, born in Maldon, Essex in 1842. He studied at Trinity College and Cambridge. He lead the Cavendish laboratory as successor to James Clerk Maxwell from 1879 to 1884. Later, he became a professor at the Royal Institute of Great Britain. He was President of the Royal Society from 1905 to 1908. In 1904, he received Nobel prize on the discovery of the inert gas Argon. He died in 1919 in Witham, Essex.

Example 7.1

Air at 10 °R on the side of a vertical plate is heated up due to temperature of the plate at 980 °R. The length of plate is 10 cm. The properties of air at mean temperature of 495 °R:
$\rho = 0.0809$ lb$_m$/ft^3, k = 0.01402 Btu/hr·ft·°R , $c_p = 0.24$ Btu/lb$_m$·°R, $\alpha = 0.685$ ft^2/s, $\nu = 0.0001354$ ft^2/s, Pr = 0.72.

Solution
The thermal expansion coefficient $\beta = 1/495 = 0.0020$ °R^{-1}. Note that β is computed using absolute temperature scale.

The corresponding Rayleigh number is

$$Ra = \frac{g \cdot L^3 \cdot \beta \cdot (T_w - T_a)}{v \cdot \alpha} = \frac{32.2 \cdot (10/100)^3 \cdot 0.0020 \cdot (980 - 10)}{0.0001354 \cdot 0.685}$$
$$= 6.80319 \times 10^5$$

Richardson number or Gradient Richardson number defined as

$$Ri = \frac{g\beta \left(\frac{\partial T}{\partial y} \right)}{\left(\frac{\partial u}{\partial y} \right)^2}$$

is a local parameter denoting the ratio of the buoyancy to inertia forces. This number has a physical significance when the buoyancy effects on the turbulence field.

Goal: The goal of free convection heat transfer experiments is to cast the following formulation:

$$Nu = C \cdot Gr_D^m \cdot Pr^n$$

where, C is a constant, m and n are experimentally determined indexes. The Nusselt number is defined as

$$Nu = \frac{h \cdot D}{k}$$

where, D is a characteristic length

7.3 MASS, MOMENTUM, AND ENTHALPY CONSERVATION EQUATIONS

In this section, we will derive the fundamental mass, momentum, and enthalpy (energy) equation.

7.3.1 CONTINUITY EQUATION

We develop the differential formulation for the continuity equation. Figure 7.2 shows the three-dimensional control volume. For mass conservation, we use the mass balance equations as:

$$[mass\ flow\ rate]_{out} - [mass\ flow\ rate]_{in} = mass\ inside\ CV$$

The control volume has dimensions Δx, Δy and Δz. The velocity and density associated at exit of control volume are indicated by symbol (e). The inlet velocity

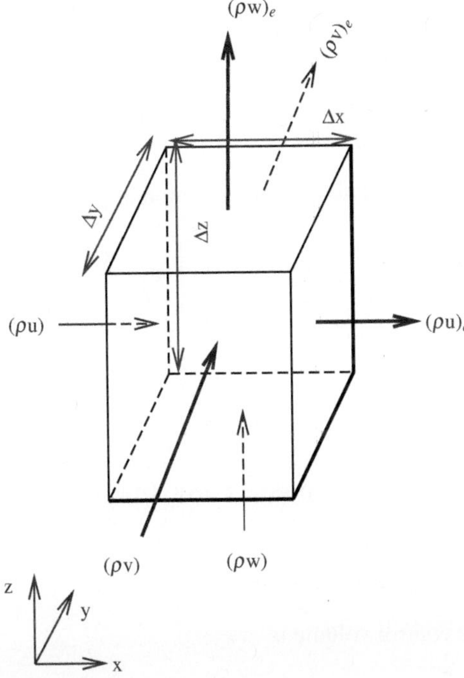

Figure 7.2   Three-dimensional control volume for the derivation of the continuity equation.

and density associated with x, y and z directions are (ρu), (ρv) and (ρw). In order to estimate the out going mass flow rate, we use the Taylor's series:

$$(\rho u)_e = (\rho u) + \frac{\partial(\rho u)}{\partial x}\frac{(\Delta x)^1}{1!} + \frac{\partial^2(\rho u)}{\partial x^2}\frac{(\Delta x)^2}{2!} + \cdots$$

$$(\rho v)_e = (\rho v) + \frac{\partial(\rho v)}{\partial y}\frac{(\Delta y)^1}{1!} + \frac{\partial^2(\rho v)}{\partial y^2}\frac{(\Delta y)^2}{2!} + \cdots$$

$$(\rho w)_e = (\rho w) + \frac{\partial(\rho w)}{\partial z}\frac{(\Delta z)^1}{1!} + \frac{\partial^2(\rho w)}{\partial z^2}\frac{(\Delta z)^2}{2!} + \cdots$$

We ignore the higher order terms and the above expression will be reduced to form

$$(\rho u)_e = (\rho u) + \frac{\partial(\rho u)}{\partial x}\Delta x$$

$$(\rho v)_e = (\rho v) + \frac{\partial(\rho v)}{\partial y}\Delta y$$

$$(\rho w)_e = (\rho w) + \frac{\partial(\rho w)}{\partial z}\Delta z$$

The net mass flow rate in x direction is

$$\Delta\left(\dot{m}_x\right) = \Delta y \Delta z\left[\left(\rho u\right)_e - \left(\rho u\right)\right] = \Delta y \Delta z\left[\left(\rho u\right) + \frac{\partial\left(\rho u\right)}{\partial x}\Delta x - \left(\rho u\right)\right]$$

$$= \Delta y \Delta z\left[\frac{\partial\left(\rho u\right)}{\partial x}\Delta x\right]$$

Similarly, the net mass flow rate in y direction is

$$\Delta\left(\dot{m}_y\right) = \Delta x \Delta z\left[\left(\rho v\right)_e - \left(\rho v\right)\right] = \Delta x \Delta z\left[\left(\rho v\right) + \frac{\partial\left(\rho v\right)}{\partial y}\frac{\left(\Delta y\right)^1}{1!} - \left(\rho v\right)\right]$$

$$= \Delta x \Delta z\left[\frac{\partial\left(\rho v\right)}{\partial y}\Delta y\right]$$

and, the net mass flow rate in z direction is

$$\Delta\left(\dot{m}_z\right) = \Delta x \Delta y\left[\left(\rho w\right)_e - \left(\rho w\right)\right] = \Delta x \Delta y\left[\left(\rho w\right) + \frac{\partial\left(\rho w\right)}{\partial z}\Delta z - \left(\rho w\right)\right]$$

$$= \Delta x \Delta y\left[\frac{\partial\left(\rho w\right)}{\partial z}\Delta z\right]$$

The mass inside control volume is

$$Mass\ inside\ CV = \rho \dot{\forall} = -\frac{\partial}{\partial t}\left(\rho\ \Delta x\ \Delta y\ \Delta z\right)$$

We now substitute the above expressions into mass balance equation

$$\Delta y \Delta z\left[\frac{\partial\left(\rho u\right)}{\partial x}\Delta x\right] + \Delta x \Delta z\left[\frac{\partial\left(\rho v\right)}{\partial y}\Delta y\right] + \Delta x \Delta y\left[\frac{\partial\left(\rho w\right)}{\partial z}\Delta z\right] = -\frac{\partial}{\partial t}\left(\rho\ \Delta x\ \Delta y\ \Delta z\right)$$

This leads to the following form of continuity equation:

$$\boxed{\frac{\partial\left(\rho u\right)}{\partial x} + \frac{\partial\left(\rho v\right)}{\partial y} + \frac{\partial\left(\rho w\right)}{\partial z} = -\frac{\partial\rho}{\partial t}}$$

$$(7.1)$$

Conservation of mass and momentum equations in Cartesian, cylindrical and spherical coordinates

Cartesian coordinates:

$$\frac{\partial\rho}{\partial t} + \frac{\partial\left(\rho u\right)}{\partial x} + \frac{\partial\left(\rho v\right)}{\partial y} + \frac{\partial\left(\rho w\right)}{\partial z} = 0$$

Cylindrical coordinates:

$$\frac{1}{r}\frac{\partial\left(r u_r\right)}{\partial r} + \frac{1}{r}\frac{\partial u_\theta}{\partial \theta} + \frac{\partial u_z}{\partial z} = 0$$

Spherical coordinates:

$$\frac{1}{r}\frac{\partial}{\partial r}\left(r^2 u_r\right) + \frac{1}{\sin\varphi}\left[\frac{\partial}{\partial \varphi}\left(\sin\varphi \cdot u_\varphi\right)\right] + \frac{1}{\sin\varphi}\left[\frac{\partial}{\partial \theta}\left(u_\theta\right)\right] = 0$$

7.3.2 MOMENTUM CONSERVATION OR NAVIER-STOKES EQUATION

Figure 7.3 shows the infinitesimal control volume used for the balance of forces and momentum transfer.

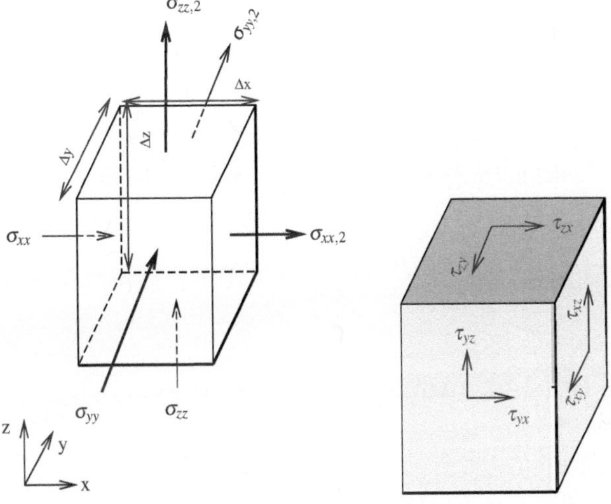

Figure 7.3   Normal and shear stress distribution on an infinitesimal control volume. Normal stresses are represented Greek letter σ and shear stresses are represented by Greek letter τ.

The stresses are different on opposing faces. The number 2 in subscript shows the stress on the opposing faces of the control volume.

Using Taylor series, we can estimate the normal stresses in x, y, and z directions:

$$\sigma_{xx,2} = \sigma_{xx} + \left(\frac{\partial \sigma_{xx}}{\partial x}\right)\Delta x + \left(\frac{\partial^2 \sigma_{xx}}{\partial x^2}\right)\frac{\Delta x^2}{2} + \cdots$$

$$\sigma_{yy,2} = \sigma_{yy} + \left(\frac{\partial \sigma_{yy}}{\partial y}\right)\Delta y + \left(\frac{\partial^2 \sigma_{yy}}{\partial y^2}\right)\frac{\Delta y^2}{2} + \cdots$$

$$\sigma_{zz,2} = \sigma_{zz} + \left(\frac{\partial \sigma_{zz}}{\partial z}\right)\Delta z + \left(\frac{\partial^2 \sigma_{zz}}{\partial z^2}\right)\frac{\Delta z^2}{2} + \cdots$$

We can write the surface stress forces:

$$
\begin{array}{lll}
F_{\sigma x} = (\sigma_{xx,2} - \sigma_{xx})\,\Delta z \Delta y & F_{\tau,yx} = (\tau_{yx,2} - \tau_{yx})\,\Delta x \Delta z & F_{\tau,zx} = (\tau_{zx,2} - \tau_{zx})\,\Delta x \Delta y \\
F_{\sigma y} = (\sigma_{yy,2} - \sigma_{yy})\,\Delta x \Delta z & F_{\tau,xy} = (\tau_{xy,2} - \tau_{xy})\,\Delta y \Delta z & F_{\tau,zy} = (\tau_{zy,2} - \tau_{zy})\,\Delta x \Delta y \\
F_{\sigma z} = (\sigma_{zz,2} - \sigma_{zz})\,\Delta x \Delta y & F_{\tau,xz} = (\tau_{xz,2} - \tau_{xz})\,\Delta y \Delta z & F_{\tau,yz} = (\tau_{yz,2} - \tau_{yz})\,\Delta x \Delta z
\end{array}
$$

From Newton's second law of motion, we have $\sum F_x = m\,a_x$, $\sum F_y = m\,a_y$, and $\sum F_z = m\,a_z$:

For a control volume of size $\forall$ and having fluid with density ρ, the surface and body forces can be written as

$$F_{\sigma,xx} + F_{\tau,yx} + F_{\tau,zx} + F_{b,x} = m\,a_x = \rho \forall a_x$$

$$F_{\sigma,yy} + F_{\tau,xy} + F_{\tau,zy} + F_{b,y} = m\,a_y = \rho \forall a_y$$

$$F_{\sigma,zz} + F_{\tau,xz} + F_{\tau,yz} + F_{b,z} = m\,a_z = \rho \forall a_z$$

Substituting the normal and shear stresses, after neglecting H.O.T we have

$$\rho \forall a_x = (\sigma_{xx,2} - \sigma_{xx})\,\Delta y \Delta z + (\tau_{yx,2} - \tau_{yx})\,\Delta x \Delta z + (\tau_{zx,2} - \tau_{zx})\,\Delta x \Delta y + F_{b,x}$$

$$\rho \forall a_y = (\sigma_{yy,2} - \sigma_{yy})\,\Delta x \Delta z + (\tau_{xy,2} - \tau_{xy})\,\Delta y \Delta z + (\tau_{zy,2} - \tau_{zy})\,\Delta x \Delta y + F_{b,y}$$

$$\rho \forall a_z = (\sigma_{zz,2} - \sigma_{zz})\,\Delta x \Delta y + (\tau_{xz,2} - \tau_{xz})\,\Delta y \Delta z + (\tau_{yz,2} - \tau_{yz})\,\Delta x \Delta z + F_{b,z}$$

We divide the equations with $\forall$:

$$\rho a_x = (\sigma_{xx,2} - \sigma_{xx})\,\Delta z \Delta y + (\tau_{yx,2} - \tau_{yx})\,\Delta x \Delta z + (\tau_{zx,2} - \tau_{zx})\,\Delta x \Delta y + \frac{F_{b,x}}{\forall}$$

We divide the equations with ρ, and this gives accelerations:

$$a_x = \frac{1}{\rho}\left\{ \left(\frac{\partial \sigma_{xx}}{\partial x}\right) + \left(\frac{\partial \tau_{yx}}{\partial y}\right) + \left(\frac{\partial \tau_{zx}}{\partial z}\right) \right\} + f_{b,x}$$

$$a_y = \frac{1}{\rho}\left\{ \left(\frac{\partial \sigma_{yy}}{\partial y}\right) + \left(\frac{\partial \tau_{xy}}{\partial x}\right) + \left(\frac{\partial \tau_{zy}}{\partial z}\right) \right\} + f_{b,y}$$

$$a_z = \frac{1}{\rho}\left\{ \left(\frac{\partial \sigma_{zz}}{\partial z}\right) + \left(\frac{\partial \tau_{xz}}{\partial x}\right) + \left(\frac{\partial \tau_{yz}}{\partial y}\right) \right\} + f_{b,z}$$

where $f_{b,x}$, $f_{b,x}$, and $f_{b,x}$ are body forces per unit mass in x, y, and z directions. These equations are known as *Navier equations*. For incompressible flow, Stoke proposed the constitutive relations which relate stresses with velocity gradients:

$$\tau_{ij} = 2\mu S_{ij} - \frac{2}{3}\mu S_{kk}\delta_{ij}$$

and

$$\sigma_{ij} = -P\delta_{ij} + 2\mu S_{ij} - \frac{2}{3}\mu S_{kk}\delta_{ij}$$

where, δ_{ij} is Kronecker delta, a unit second-order tensor, which is equal to 1 if its indices i and j are equal, and equal to 0 otherwise. The strain-rate and rotation tensors are defined as

$$S_{ij} = \frac{1}{2}\left(\frac{\partial U_i}{\partial x_j} + \frac{\partial U_j}{\partial x_i}\right) \quad Strain-rate\ tensor$$

$$S_{kk} = \frac{1}{2}\left(\frac{\partial U_i}{\partial x_j} - \frac{\partial U_j}{\partial x_i}\right) \quad Rotation\ tensor$$

The S_{ij} tensor describes the local deformation of fluid elements, and the S_{kk} tensor is a measure of the rate of rotation or vorticity in a fluid flow. It describes the antisymmetric part of the velocity gradient tensor, which represents the local rotation of fluid elements. The S_{kk} tensor is also known as the vorticity tensor or the curl tensor.

In vector form, we can write for incompressible flows:

$$\sigma_{xx} = -P + 2\mu\frac{\partial u}{\partial x}, \sigma_{yy} = -P + 2\mu\frac{\partial v}{\partial y}, \sigma_{zz} = -P + 2\mu\frac{\partial w}{\partial z}$$

$$\tau_{xy} = \mu\left(\frac{\partial v}{\partial x} + \frac{\partial u}{\partial y}\right), \tau_{xz} = \mu\left(\frac{\partial w}{\partial x} + \frac{\partial u}{\partial z}\right), \tau_{yz} = \mu\left(\frac{\partial w}{\partial y} + \frac{\partial v}{\partial z}\right)$$

Substituting the Stoke's hypothesis into Navier equation, and introducing the Eulerian formulation for accelerations in x, y and z directions, we have

Navier-Stokes equations in Cartesian coordinates

$$\frac{Du}{Dt} = \frac{\partial u}{\partial t} + u\frac{\partial u}{\partial x} + v\frac{\partial u}{\partial y} + w\frac{\partial u}{\partial z} = -\frac{1}{\rho}\frac{\partial P}{\partial x} + \frac{\mu}{\rho}\nabla^2 u + f_x \qquad (7.2)$$

$$\frac{Dv}{Dt} = \frac{\partial v}{\partial t} + u\frac{\partial v}{\partial x} + v\frac{\partial v}{\partial y} + w\frac{\partial v}{\partial z} = -\frac{1}{\rho}\frac{\partial P}{\partial y} + \frac{\mu}{\rho}\nabla^2 v + f_y \qquad (7.3)$$

$$\frac{Dw}{Dt} = \frac{\partial w}{\partial t} + u\frac{\partial w}{\partial x} + v\frac{\partial w}{\partial y} + w\frac{\partial w}{\partial z} = -\frac{1}{\rho}\frac{\partial P}{\partial z} + \frac{\mu}{\rho}\nabla^2 w + f_z \qquad (7.4)$$

The above set of equations are called the Navier-Stokes equations.

> ### Navier-Stokes equation in cylindrical coordinates
>
> *Navier-Stokes Equations:*
>
> r-direction:
>
> $$\rho\left(\frac{\partial u_r}{\partial t} + u_r\frac{\partial u_r}{\partial r} + \frac{u_\theta}{r}\frac{\partial u_r}{\partial \theta} - \frac{u_\theta^2}{r} + u_z\frac{\partial u_r}{\partial z}\right)$$
>
> $$= -\frac{\partial P}{\partial r} + f_r + \mu\left[\frac{1}{r}\frac{\partial}{\partial r}\left(r\frac{\partial u_r}{\partial r}\right) - \frac{u_r}{r^2} + \frac{1}{r^2}\frac{\partial^2 u_r}{\partial \theta^2} - \frac{2}{r^2}\frac{\partial u_\theta}{\partial \theta} + \frac{\partial^2 u_r}{\partial z^2}\right]$$
>
> θ-direction:
>
> $$\rho\left(\frac{\partial u_\theta}{\partial t} + u_r\frac{\partial u_\theta}{\partial r} + \frac{u_\theta}{r}\frac{\partial u_\theta}{\partial \theta} + \frac{u_r u_\theta}{r} + u_z\frac{\partial u_\theta}{\partial z}\right)$$
>
> $$= -\frac{1}{r}\frac{\partial P}{\partial \theta} + f_\theta + \mu\left[\frac{1}{r}\frac{\partial}{\partial r}\left(r\frac{\partial u_\theta}{\partial r}\right) - \frac{u_\theta}{r^2} + \frac{1}{r^2}\frac{\partial^2 u_\theta}{\partial \theta^2} - \frac{2}{r^2}\frac{\partial u_r}{\partial \theta} + \frac{\partial^2 u_\theta}{\partial z^2}\right]$$
>
> z-direction:
>
> $$\rho\left(\frac{\partial u_z}{\partial t} + u_r\frac{\partial u_z}{\partial r} + \frac{u_\theta}{r}\frac{\partial u_z}{\partial \theta} + u_z\frac{\partial u_z}{\partial z}\right)$$
>
> $$= -\frac{\partial P}{\partial z} + f_z + \mu\left[\frac{1}{r}\frac{\partial}{\partial r}\left(r\frac{\partial u_z}{\partial r}\right) + \frac{1}{r^2}\frac{\partial^2 u_z}{\partial \theta^2} + \frac{\partial^2 u_z}{\partial z^2}\right]$$

7.3.3 ENERGY EQUATION

We are interested in the balance of bulk energy of the fluid. Using Reynolds Transport theorem, we can write the balance of energy as

$$\frac{dN}{dt} = \int_{CS} \eta\rho\mathbf{V}.dA + \frac{\partial}{\partial t}\int_{CV} \eta\rho d\vartheta \tag{7.5}$$

where, N is the conserved quantity, η is N per unit mass. We set N=E, where E is bulk energy of bulk fluid. This helps us in defining energy per unit mass (e). As E= $e \cdot m$, we can related mass with density and volume of the fluid as $e \cdot \rho \cdot \vartheta$. This gives

$$\frac{dE}{dt} = \int_{CS} e\rho\mathbf{V}.dA + \frac{\partial}{\partial t}\int_{CV} e\rho d\vartheta - \sum \dot{Q} - \sum \dot{W} \tag{7.6}$$

The inclusion of work and heat interaction is important, else equation is suffice for heat diffusion analysis without convection currents only. We reconsider equation 7.6 and using the divergence theorem, we replace area integral with volumetric integral as:

$$\sum \dot{Q} + \sum \dot{W} = \int_{CV} \mathbf{div}\,[e\rho\mathbf{V}]\,d\vartheta + \frac{\partial}{\partial t}\int_{CV} e\rho\,d\vartheta \tag{7.7}$$

$$\sum \dot{Q} + \sum \dot{W} = \int_{CV} \left[\mathbf{div}\,[e\rho\mathbf{V}] + \frac{\partial(e\rho)}{\partial t}\right] d\vartheta \tag{7.8}$$

$$\mathbf{div}\,[e\rho\mathbf{V}] + \frac{\partial(e\rho)}{\partial t} = \frac{\dot{Q}}{d\vartheta} + \frac{\dot{W}}{d\vartheta} \tag{7.9}$$

Here

$$\frac{\partial(e\rho)}{\partial t} = \rho\frac{\partial(e)}{\partial t} + e\frac{\partial(\rho)}{\partial t}$$

and

$$\mathbf{div}\,[e\rho\mathbf{V}] = \rho\mathbf{V}.\nabla e + e\nabla.\,[\rho\mathbf{V}]$$

This leads to the new conservation form

$$e\left[\nabla.\,[\rho\mathbf{V}] + \frac{\partial(\rho)}{\partial t}\right] + \rho\left[\mathbf{V}.\nabla e + \frac{\partial(e)}{\partial t}\right] = \frac{\dot{Q}}{d\vartheta} + \frac{\dot{W}}{d\vartheta} \tag{7.10}$$

where the bracketed term:

$$\left[\nabla.\,[\rho\mathbf{V}] + \frac{\partial(\rho)}{\partial t}\right]$$

is continuity equation so its equal to zero. The second term

$$\left[\mathbf{V}.\nabla e + \frac{\partial(e)}{\partial t}\right]$$

can be written as substantial derivative De/Dt. Hence,

$$\rho\frac{De}{Dt} = \frac{\dot{Q}}{d\vartheta} + \frac{\dot{W}}{d\vartheta} \tag{7.11}$$

We focus on the $\dot{Q}/d\vartheta$ term. Heat can enter into system due to thermal diffusion, but also the chemical energy can be converted into thermal energy.

$$\dot{Q} = -\int_{CS} \mathbf{q}.dA + \int_{CV} \tilde{q}\,d\vartheta \tag{7.12}$$

here $\tilde{q}$ is rate of thermal energy conversion in volume (so-called energy generation term) and $\mathbf{q}$ is diffusive flux. Now, using Gauss-Divergence Theorem we convert area integral into volumetric integral:

$$\dot{Q} = \int_{CV} [\tilde{q} - \nabla.\mathbf{q}]\,d\vartheta \tag{7.13}$$

For infinitesimal control volume:

$$\frac{\dot{Q}}{d\vartheta} = \tilde{q} - \nabla.\mathbf{q} \tag{7.14}$$

Introducing this into equation 7.11 we have:

$$\rho\frac{De}{Dt} = \frac{\dot{Q}}{d\vartheta} + \frac{\dot{W}}{d\vartheta} \tag{7.15}$$

$$\rho\frac{De}{Dt} = \tilde{q} - \nabla.\mathbf{q} + \frac{\dot{W}}{d\vartheta} \tag{7.16}$$

$$\frac{\dot{W}}{d\vartheta} = \rho\frac{De}{Dt} - \tilde{q} + \nabla.\mathbf{q} \tag{7.17}$$

For three-dimensional control volume in Cartesian coordinates, we have:

$$\frac{De}{Dt} = \frac{\partial e}{\partial t} + u\frac{\partial e}{\partial x} + v\frac{\partial e}{\partial y} + w\frac{\partial e}{\partial z} \tag{7.18}$$

The expanded from of energy equation 7.17 is

$$\begin{aligned}
\rho\left(\frac{\partial e}{\partial t} + u\frac{\partial e}{\partial x} + v\frac{\partial e}{\partial y} + w\frac{\partial e}{\partial z}\right) \\
+ \left(\frac{\partial q_x}{\partial x} + \frac{\partial q_y}{\partial y} + \frac{\partial q_z}{\partial z}\right) - \tilde{q} \\
= \frac{\dot{W}}{dxdydz}
\end{aligned} \tag{7.19}$$

This equation obscure the information that total energy flux is equal to bulk fluid advection plus the thermal diffusion relative to bulk motion. Therefore, the equation derived so far can only be used for heat conduction or thermal diffusion through mass with no appreciable velocity like air trapped in small area, stationary water, solids, etc. The thermal diffusion form of energy equation developed above is not enough to know the energy change between fluid layers due to fluid bulk movement (i.e. convection or advection) as energy transport is disconnected with stress distribution inside fluid body.

$$\begin{aligned}
\rho\left(\frac{\partial e}{\partial t} + u\frac{\partial e}{\partial x} + v\frac{\partial e}{\partial y} + w\frac{\partial e}{\partial z}\right) \\
+ \left(\frac{\partial q_x}{\partial x} + \frac{\partial q_y}{\partial y} + \frac{\partial q_z}{\partial z}\right) - \tilde{q} \\
= \frac{\dot{W}}{dxdydz}
\end{aligned} \tag{7.20}$$

W is not the shaft work but the work interaction between fluid surrounding and fluid. Dimensionally speaking, $\dot{W}$ here is the rate of work interaction arise due to surrounding surface and body forces. We know that

$$Work = force \cdot displacement$$

$$Work = Stress \cdot area \cdot displacement \approx Stress \cdot Volume$$

Hence,

$$Stress = Work/Volume$$

We can write rate of work per unit volume as

$$\dot{W}/\forall = (Stress.velocity)/length\ dimension$$

$$
\begin{aligned}
\frac{\dot{W}}{dxdydz} = {} & \frac{\partial(u\sigma_{xx})}{\partial x} + \frac{\partial(u\tau_{yx})}{\partial y} + \frac{\partial(u\tau_{zx})}{\partial z} \\
& + \frac{\partial(v\tau_{xy})}{\partial x} + \frac{\partial(v\sigma_{yy})}{\partial y} + \frac{\partial(v\tau_{zy})}{\partial z} \\
& + \frac{\partial(w\tau_{xz})}{\partial x} + \frac{\partial(w\tau_{yz})}{\partial y} + \frac{\partial(w\sigma_{zz})}{\partial z} \\
& + uf_x + vf_y + wf_z
\end{aligned}
\tag{7.21}
$$

$$
\begin{aligned}
\frac{\dot{W}}{dxdydz} = {} & uf_x + vf_y + wf_z + u\frac{\partial(\sigma_{xx})}{\partial x} + \sigma_{xx}\frac{\partial(u)}{\partial x} + u\frac{\partial(\tau_{yx})}{\partial y} + \tau_{yx}\frac{\partial(u)}{\partial y} \\
& + u\frac{\partial(\tau_{zx})}{\partial z} + \tau_{zx}\frac{\partial(u)}{\partial z} + v\frac{\partial(\tau_{xy})}{\partial x} + \tau_{xy}\frac{\partial(v)}{\partial x} \\
& + v\frac{\partial(\sigma_{yy})}{\partial y} + \sigma_{yy}\frac{\partial(v)}{\partial y} + v\frac{\partial(\tau_{zy})}{\partial z} + \tau_{zy}\frac{\partial(v)}{\partial z} \\
& + w\frac{\partial(\tau_{xz})}{\partial x} + \tau_{xz}\frac{\partial(w)}{\partial x} + w\frac{\partial(\tau_{yz})}{\partial y} + \tau_{yz}\frac{\partial(w)}{\partial y} \\
& + w\frac{\partial(\sigma_{zz})}{\partial z} + \sigma_{zz}\frac{\partial(w)}{\partial z}
\end{aligned}
\tag{7.22}
$$

$$\frac{\dot{W}}{dxdydz} = \left[uf_x + u\frac{\partial(\sigma_{xx})}{\partial x} + u\frac{\partial(\tau_{yx})}{\partial y} + u\frac{\partial(\tau_{zx})}{\partial z} \right]$$

$$+ \left[vf_y + v\frac{\partial(\tau_{xy})}{\partial x} + v\frac{\partial(\sigma_{yy})}{\partial y} + v\frac{\partial(\tau_{zy})}{\partial z} \right]$$

$$+ \left[wf_z + w\frac{\partial(\tau_{xz})}{\partial x} + w\frac{\partial(\tau_{yz})}{\partial y} + w\frac{\partial(\sigma_{zz})}{\partial z} \right]$$

$$+ \sigma_{xx}\frac{\partial(u)}{\partial x} + \tau_{yx}\frac{\partial(u)}{\partial y} + \tau_{zx}\frac{\partial(u)}{\partial z}$$

$$+ \tau_{xy}\frac{\partial(v)}{\partial x} + \sigma_{yy}\frac{\partial(v)}{\partial y} + \tau_{zy}\frac{\partial(v)}{\partial z}$$

$$+ \tau_{xz}\frac{\partial(w)}{\partial x} + \tau_{yz}\frac{\partial(w)}{\partial y} + \sigma_{zz}\frac{\partial(w)}{\partial z}$$

$$(7.23)$$

Note that the terms

$$uf_x + u\frac{\partial(\sigma_{xx})}{\partial x} + u\frac{\partial(\tau_{yx})}{\partial y} + u\frac{\partial(\tau_{zx})}{\partial z} \tag{7.24}$$

are actually $\rho u Du/Dt$ and likewise we can write for the bracketed terms in equation 7.23:

$$\rho\frac{DV^2/2}{Dt} = \rho u\frac{Du}{Dt} + \rho v\frac{Dv}{Dt} + \rho w\frac{Dw}{Dt} \tag{7.25}$$

$$\frac{\dot{W}}{dxdydz} = \rho\frac{DV^2/2}{Dt} + \sigma_{xx}\frac{\partial(u)}{\partial x} + \tau_{yx}\frac{\partial(u)}{\partial y} + \tau_{zx}\frac{\partial(u)}{\partial z}$$

$$+ \tau_{xy}\frac{\partial(v)}{\partial x} + \sigma_{yy}\frac{\partial(v)}{\partial y} + \tau_{zy}\frac{\partial(v)}{\partial z}$$

$$+ \tau_{xz}\frac{\partial(w)}{\partial x} + \tau_{yz}\frac{\partial(w)}{\partial y} + \sigma_{zz}\frac{\partial(w)}{\partial z}$$

$$(7.26)$$

We introduced it ino the equation

$$\rho\left(\frac{\partial e}{\partial t} + u\frac{\partial e}{\partial x} + v\frac{\partial e}{\partial y} + w\frac{\partial e}{\partial z} \right)$$

$$+ \left(\frac{\partial q_x}{\partial x} + \frac{\partial q_y}{\partial y} + \frac{\partial q_z}{\partial z} \right) - \tilde{q} \tag{7.27}$$

$$= \frac{\dot{W}}{dxdydz}$$

$$\rho\left(\frac{De}{Dt}\right)+(\nabla.\mathbf{q})-\tilde{q}$$

$$=\rho\frac{DV^2/2}{Dt}+\sigma_{xx}\frac{\partial(u)}{\partial x}+\tau_{yx}\frac{\partial(u)}{\partial y}+\tau_{zx}\frac{\partial(u)}{\partial z}$$
$$+\tau_{xy}\frac{\partial(v)}{\partial x}+\sigma_{yy}\frac{\partial(v)}{\partial y}+\tau_{zy}\frac{\partial(v)}{\partial z}$$
$$+\tau_{xz}\frac{\partial(w)}{\partial x}+\tau_{yz}\frac{\partial(w)}{\partial y}+\sigma_{zz}\frac{\partial(w)}{\partial z}$$

$$(7.28)$$

This equation is also called the **mechanical energy equation**.

In order to obtain thermal energy equation, we introduce I as internal energy per unit mass due to temperature. Note that usually internal energy is indicated by (U) in Thermodynamics literature, but since we take u for x-direction velocity, in order to avoid confusion, I is used to represent internal energy in this section.

$$e = I + \mathbf{V}^2/2 + g\Delta h$$

$$I = e - \mathbf{V}^2/2$$

$$\rho\left(\frac{DI}{Dt}\right)+(\nabla.\mathbf{q})-\tilde{q}$$

$$=\sigma_{xx}\frac{\partial(u)}{\partial x}+\tau_{yx}\frac{\partial(u)}{\partial y}+\tau_{zx}\frac{\partial(u)}{\partial z}$$
$$+\tau_{xy}\frac{\partial(v)}{\partial x}+\sigma_{yy}\frac{\partial(v)}{\partial y}+\tau_{zy}\frac{\partial(v)}{\partial z}$$
$$+\tau_{xz}\frac{\partial(w)}{\partial x}+\tau_{yz}\frac{\partial(w)}{\partial y}+\sigma_{zz}\frac{\partial(w)}{\partial z}$$

$$(7.29)$$

We now insert Stokes Hypothesis into equation, which leads to form

$$\rho\left(\frac{DI}{Dt}\right)+(\nabla.\mathbf{q})-\tilde{q}=-p\nabla.\mathbf{V}+\mu\Phi \qquad (7.30)$$

where Φ is dissipation function and is noramlly not relevant in incompressible flows. In order to simplify further, we replace internal energy (I) with enthalpy (H) as

$$H = I + \frac{P}{\rho}$$

Thus Energy equation takes the form

$$\rho\left(\frac{DH}{Dt}\right)+(\nabla.\mathbf{q})-\tilde{q}=\frac{Dp}{Dt}+\mu\Phi \qquad (7.31)$$

where, Φ is dissipation, defined as

$$\Phi = 2\left[\left(\frac{\partial u}{\partial x}\right)^2 + \left(\frac{\partial v}{\partial y}\right)^2 + \left(\frac{\partial w}{\partial z}\right)^2\right]$$
$$+ \left[\left(\frac{\partial u}{\partial y} + \frac{\partial v}{\partial x}\right)^2 + \left(\frac{\partial v}{\partial z} + \frac{\partial w}{\partial y}\right)^2 + \left(\frac{\partial w}{\partial x} + \frac{\partial u}{\partial z}\right)^2\right]$$
$$- \frac{2}{3}\left[\frac{\partial u}{\partial x} + \frac{\partial v}{\partial y} + \frac{\partial w}{\partial z}\right]^2$$

According to Maxwell relations, thermodynamically:

$$DH = TDs + \left(\frac{1}{\rho}\right)Dp \qquad (7.32)$$

$$\frac{DH}{Dp} = T\frac{Ds}{Dp} + \frac{1}{\rho} \qquad (7.33)$$

also according to Maxwell relations

$$\left(\frac{\partial s}{\partial P}\right)_T = \left(\frac{\partial \vartheta}{\partial T}\right)_P = \frac{1}{\rho^2}\frac{\partial \rho}{\partial T} \qquad (7.34)$$

Also

$$H = c_p T$$

and

$$\beta = -\frac{1}{\rho}\left(\frac{\partial \rho}{\partial T}\right)_p$$

so

$$DH = c_p dT + \frac{1}{\rho}(1 - \beta T)Dp \qquad (7.35)$$

Energy equation takes the form

$$\rho c_p\left(\frac{DT}{Dt}\right) = -(\nabla.\mathbf{q}) + \tilde{q} + \beta T\frac{Dp}{Dt} + \mu\Phi \qquad (7.36)$$

From Fourier Law:

$$\rho c_p\left(\frac{DT}{Dt}\right) = k\left(\nabla^2.T\right) + \tilde{q} + \beta T\frac{Dp}{Dt} + \mu\Phi \qquad (7.37)$$

For constant density fluid β is zero; neglecting Φ we get:

$$\rho c_p\left(\frac{DT}{Dt}\right) = k\left(\nabla^2.T\right) + \tilde{q} \qquad (7.38)$$

This is energy equation we use for convective heat transfer in forced convection problems. The equation actually a form of first law of thermodynamics, which is the fundamental law and it represent a balance and conservation of the sum of energy change, heat transfer, and work transfer (or the interactions of system with surrounding).

Convective Heat Transfer Equation For constant density fluid β is zero; neglecting Φ we get:

$$\rho c_p \left(\frac{DT}{Dt} \right) = k \left(\nabla^2 . T \right) + \tilde{q} \tag{7.39}$$

Energy equation we use for convective heat transfer.

$$\boxed{\rho c_p \left(\frac{\partial T}{\partial t} + u \frac{\partial T}{\partial x} + v \frac{\partial T}{\partial y} + w \frac{\partial T}{\partial z} \right) = k \left(\nabla^2 . T \right) + \tilde{q}} \tag{7.40}$$

Equation 9.36 is valid for incompressible flows only.

Energy equation

Incompressible, constant-property energy equation:

$$\frac{\partial T}{\partial t} + u \frac{\partial T}{\partial x} + v \frac{\partial T}{\partial y} + w \frac{\partial T}{\partial z} = \alpha \nabla^2 T + \frac{\mu}{\rho c_p} \Phi + \frac{\tilde{q}}{\rho c_p}$$

$$\Phi = 2 \left[\left(\frac{\partial u}{\partial x} \right)^2 + \left(\frac{\partial v}{\partial y} \right)^2 + \left(\frac{\partial w}{\partial z} \right)^2 \right]$$

$$+ \left[\left(\frac{\partial u}{\partial y} + \frac{\partial v}{\partial x} \right)^2 + \left(\frac{\partial v}{\partial z} + \frac{\partial w}{\partial y} \right)^2 + \left(\frac{\partial w}{\partial x} + \frac{\partial u}{\partial z} \right)^2 \right]$$

$$- \frac{2}{3} \left[\frac{\partial u}{\partial x} + \frac{\partial v}{\partial y} + \frac{\partial w}{\partial z} \right]^2$$

Cylindrical coordinates (r,θ,z):

$$\rho c_p \left(\frac{\partial T}{\partial t} + u_r \frac{\partial T}{\partial x} + \frac{u_\theta}{r} \frac{\partial T}{\partial \theta} + u_z \frac{\partial T}{\partial z} \right) = \frac{1}{r} \frac{\partial}{\partial r} \left(rk \frac{\partial T}{\partial r} \right) + \frac{1}{r} \frac{\partial}{\partial \theta} \left(k \frac{\partial T}{\partial \theta} \right) + \frac{\partial}{\partial z} \left(k \frac{\partial T}{\partial z} \right)$$

$$+ \beta T \frac{Dp}{Dt} + \nabla \bar{v} : \tau + \mu \Phi$$

Spherical coordinates (r,θ,ϕ):

$$\rho c_p \left(\frac{\partial T}{\partial t} + u_r \frac{\partial T}{\partial x} + \frac{u_\theta}{r} \frac{\partial T}{\partial \theta} + \frac{u_\theta}{r \sin \theta} \frac{\partial T}{\partial \varphi} \right) = \frac{1}{r^2} \frac{\partial}{\partial r} \left(r^2 k \frac{\partial T}{\partial r} \right) + \frac{1}{r \sin \theta} \frac{\partial}{\partial \theta} \left(\frac{k \sin \theta}{r} \frac{\partial T}{\partial \theta} \right)$$

$$+ \frac{1}{r \sin \theta} \frac{\partial}{\partial \varphi} \left(\frac{k}{r \sin \theta} \frac{\partial T}{\partial \varphi} \right) + \beta T \frac{Dp}{Dt} + \nabla \bar{v} : \tau + \mu \Phi$$

Note: Term $\nabla \bar{v} : \tau$ is important in highly compressible flows and it represents the net work done by fluid due to viscous force/volume (force $\times$ deformation)

Tensor form of Conservation Equations

Continuity equation:

$$\frac{\partial \rho}{\partial t} + \frac{\partial \rho u_k}{\partial x_k} = 0$$

Navier-Stokes equation

$$\rho \left[\frac{\partial u_j}{\partial t} + u_i \frac{\partial u_j}{\partial x_j} \right] = \rho f_i - \frac{\partial P}{\partial x_j} + \frac{\partial \tau_{ij}}{\partial x_i}$$

Energy equation:

$$\rho c_p \frac{DT}{Dt} - \rho T \left(\frac{\partial v}{\partial T} \right)_p \left(\frac{Dp}{Dt} \right) = -\frac{\partial}{\partial x_k} \left(-k \frac{\partial T}{\partial x_k} \right) - \Phi_{ij} + \tilde{q} = 0$$

where, $s_{ij} = \frac{1}{2} \left(\frac{\partial u_i}{\partial x_j} + \frac{\partial u_j}{\partial x_i} - \frac{1}{3} \frac{\partial u_k}{\partial x_k} \delta_{ij} \right)$, $\quad \tau_{ij} = 2\mu s_{ij}$, $\Phi_{ij} = \tau_{ij} s_{ij}$

7.4 TURBULENT FLOWS

Turbulence is a three-dimensional time dependent motion in which vortex stretching causes velocity fluctuations. Turbulence is one of the most complex phenomena in nature. Some of the reasons for the complexity are

I- **Intrinsic spatio-temporal randomness:** Turbulent flows are extremely sensitive to small perturbations/disturbances due to which turbulent flows inherently have no repeatability. However, the statistical properties (not only some means, but almost all statistical properties) of turbulent flows are insensitive to disturbances.

1. **Intermittency:** Turbulence can interact with non-turbulent fluid flow and can appear intermittently in time at certain location.

2. **Wide range of non-locally interacting scales:** Turbulent flows are comprised of wide range of scales ranging from hundreds of kilometers to parts of a millimeter.

3. **Continuous self production of vorticity and three-dimensionality:** In turbulence an important role is played by vortex stretching, elongation, spinning, breakup, coalescing, and pairing. Turbulence is a three-dimensional phenomenon which creates and maintains the turbulence vorticity. Note that a two-dimensional vortex cannot be stretched, hence two-dimensional turbulence does not exist in nature.

4. **Strong diffusion:** A source of energy is required to maintain turbulence (gradients of mean velocity, buoyancy, or other external forces). The energy supply

mostly occur at large scales and its dissipation occur at small scales. Turbulent fluctuations cause the stretching and distortion of the fluid elements (containing a lump of property like hot spots of heat, species, momentum, vorticity) until the increase in surface area and the property gradient enables molecular effects to act efficiently. Thus the transport of momentum, heat, species, and mixing is enhanced by several orders of magnitude.

Although turbulence is a very complex flow, fortunately, it can be computed using the purely deterministic equations (Navier-Stokes equations). However, the solution is very sensitive to initial or boundary conditions. The perturbations or disturbances which come from initial conditions, boundary conditions, and external noise can significantly alter the outcome. To study the turbulent flow field, the field variable is treated as random variable and useful information about the behaviour can be obtained by the standard statistical analysis.

7.5 TURBULENT VELOCITY PROFILE AND LAW OF WALL

Although the velocity boundary layer thickness δ is very small region near the wall, experimentally it's been found that it is necessary to distinguish it into different zones in order to facilitate the analytic modeling of the thermal field. These zones are not sharply delineated but merge continuously into one each other.

The layer immediately adjacent to the wall is called viscous sublayer. In this layer, the diffusion of momentum is controlled by molecular viscosity. The velocity variation is almost linear with a constant shear stress at wall. The size of sublayer varies from $\delta_{vsl} = 0.02\delta$ to 0.00005δ for range of Re = $10^5 - 10^6$. Viscous Sub layer is thus a thin region but it plays a crucial role in heat transfer particularly for the flow of high Prandtl number fluids. There are fluctuations in this region so we can no longer call this region laminar sub layer. In the region very near to the wall, the contribution of the molecular transport mechanism is overwhelmingly dominant and we can neglect the contribution of the turbulent transport mechanism. This region is called the laminar sub-layer. The thickness of the laminar sub-layer is known from experiments as $y^+ = 5$. The following equation holds for the viscous sub-layer.

$$\tau = \mu \frac{du}{dy} \tag{7.41}$$

Integrating above Eq. with respect to y^+, we have the following equation:

$$u^+ = y^+ \tag{7.42}$$

where,

$$u^+ = \frac{u}{u_\tau} \qquad y^+ = y\frac{u_\tau}{\nu}$$

turbulent core or fully turbulent region In the region slightly further away from the outer edge of the laminar sub-layer ($y^+ > 30$), where the contribution of the turbulent transport mechanism is dominant, we can neglect the molecular transport mechanism in comparison with the turbulent one. This region is known as the turbulent core or fully turbulent region. Flow in fully turbulent zone is dominated by turbulent mixing. Extensive experimental evidence indicates that the velocity distribution in this layer is of universal form.

From Eqs. we obtain the following equation for this region:

$$\tau = \rho \ell^2 \left(\frac{du}{dy} \right)^2 \tag{7.43}$$

Equation can be approximated by the following equation for turbulent flow near the wall region as

$$\ell = 0.4 * y$$

$$\frac{du^+}{dy^+} = \frac{2.5}{y^+} \tag{7.44}$$

Integrating Eq. with respect to y^+, we have the following equation:

$$u^+ = 2.55 ln y^+ + c \tag{7.45}$$

Wake Region Law of wall does not hold in Wake like region. It is also sensitive to pressure gradient and free stream turbulence.

Laminar sub-layer ($0 \le y^+ \le 5$):

$$u^+ = y^+$$

Buffer layer ($5 \le y^+ \le 30$):

$$u^+ = 5.0 ln y + 3.05$$

Turbulent core ($30 \le y^+$):

$$u^+ = 2.5 ln y^+ 5.5$$

Above equations are referred to as *von Karmans universal velocity distribution law*, and it shows fairly good agreement with observed velocity data, except for data pertaining to the region near the boundary between the buffer layer and the turbulent core (see Figure 7.4). A better fitting of the data is obtained by use of van Driests dumping factor.

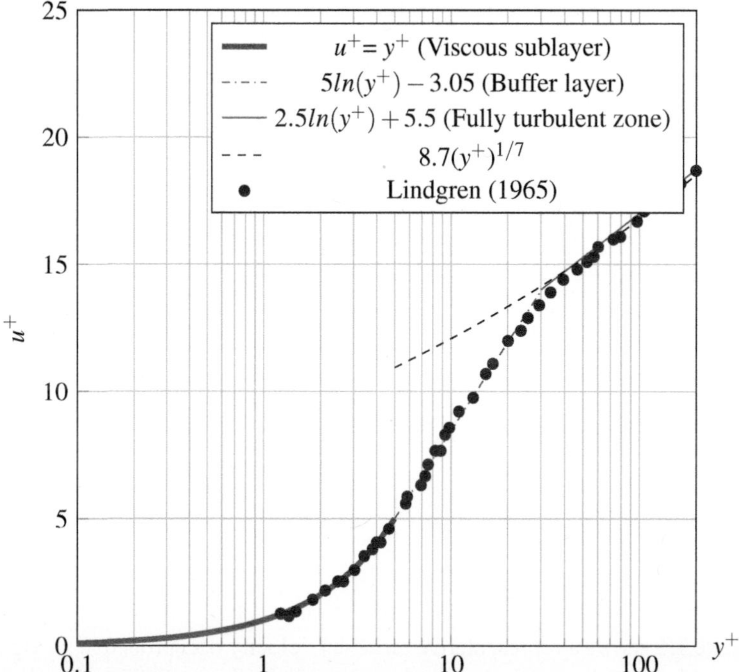

Figure 7.4   Turbulent Law of Wall according to von Karman.

Figure 9.9 shows the velocity distributions as suggested by von Karman. The equation

$$u^+ = 8.7(y^+)^{1/7}$$

matches well with the fully turbulent zone.

7.6   REYNOLDS-AVERAGED NAVIER-STOKES EQUATIONS

Turbulent flow can be predicted through numerical calculations of above equations, provided we use non-dissipative numerical schemes and control time and progress of solution in consideration with Eddy Turn-over Time. Reynolds veraging is defined as

$$< \Psi(x_i,t) > = \lim_{N \to \infty} \frac{1}{N} \sum_{n=1}^{N} \Psi(x_i,t) \tag{7.46}$$

Once the flow becomes statistical stationary (statistics become independent of time), it can be time averaged which is defined as:

$$\overline{\Psi(x_i)} = \lim_{t \to \infty} \int_{-t/2}^{t/2} \Psi(x_i,t)dt \tag{7.47}$$

According to the Reynolds decomposition of turbulence quantities, we have:

$$\psi = \overline{\Psi} + \psi' \tag{7.48}$$

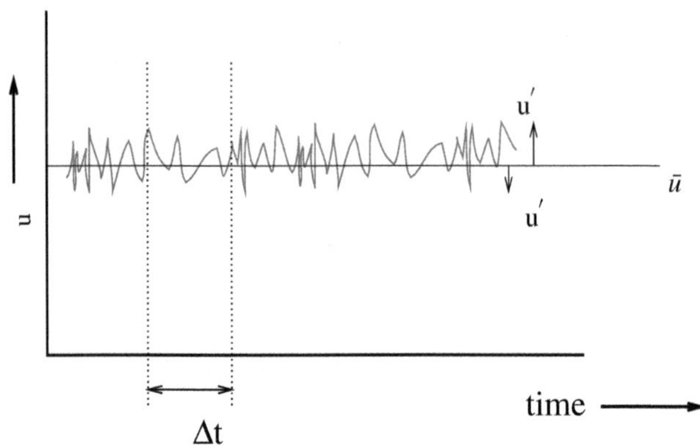

Figure 7.5 A real turbulent velocity data probed at a single point in flow field.

Figure 7.5 shows a representative turbulent velocity data probed at a single point in flow field. According to Reynolds averaging, all the flow quantities like all velocity components temperature, density, and pressure are fluctuating as:

$$u = \bar{u} + u', v = \bar{v} + v', w = \bar{w} + w', p = \bar{p} + p', \rho = \bar{\rho} + \rho', T = \bar{T} + T'$$

We can bifurcate the the velocity $u(t)$ signal into fluctuating and a mean quantity. If the mean is not changing with time then such flow can be locally steady. However, if the mean is changing with time then such flow is considered as unsteady. In turbulent flow, the velocity of the fluid is a continuous random function of time and space. It is a random function cause the instantaneous value of the fluctuating quantity at any point cannot be predicted from the mean value of the flow quantities. The value of the velocity at any point is distributed according to the laws of probability. If the instantaneous velocity is considered the sum of a mean velocity and fluctuating velocity, the distribution of fluctuating velocity follows the Gaussian or normal distribution.

Averaging rules:

Rule i The time-average of all mean quantities is equal to the mean value:

$$\bar{\bar{u}} = \bar{u}$$

Rule ii All fluctuating quantities are zero when time-averaged so

$$\overline{u'} = 0, \overline{v'} = 0, \overline{w'} = 0, \overline{\rho'} = 0, \overline{p'} = 0$$

Rule iii The product of mean and fluctuating quantities will be zero as well.

Rule iv The fluctuating quantities will be zero when time averaged.

Example 7.2

In Reynolds averaging the time-average of the mean quantity is considered same as mean quantity. To understand this assumption we consider an arbitrary temperature field represented by the following function:

$$T(t) = To \cdot \left(\left(1 + 0.5t + 0.01t^2 \right) \right) + 2 \cdot \cos\left(85 \cdot t\right)$$

where To is 293K. Find the arithmetic mean temperature and the time averaged mean temperature in one cycle $2\pi/\omega$.

Solution The temperature field is oscillating bu the temperature is gradually increasing:

$$T(t) = To \cdot \left(\left(1 + 0.5t + 0.01t^2 \right) \right) + 2 \cdot \cos\left(85 \cdot t\right)$$

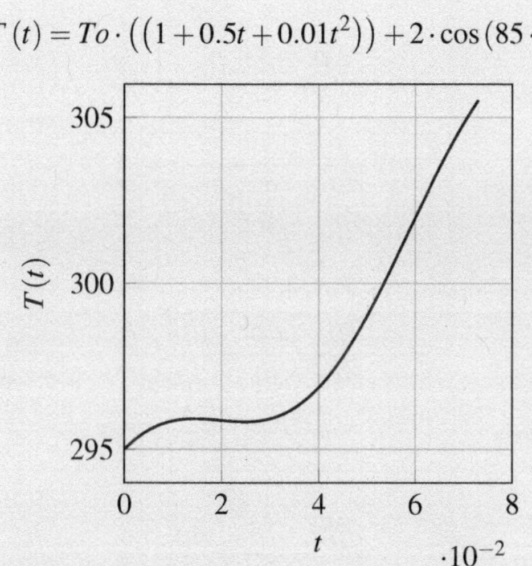

Figure 7.6   Turbulent temperature signal.

Figure 7.6 shows the turbulent temperature signal.
For one cycle we have

$$\text{cycle} = \frac{(2\pi)}{\omega} = 0.073 \, s$$

$$Time \; Average \; of \; T = \frac{1}{0.073} \int_0^{0.073} T(t) \, dt$$

For one cycle, the arithmetic mean temperature is 299.41 K and the time averaged mean temperature is 298.32K. This change is only 0.364% variation and thus the assumption that mean of time averaged quantity is same as mean is justified.

We can apply Reynolds Averaging on Conservation equations, and in this section, Reynolds Averaged Navier-Stokes Equations are developed.

7.6.1   CONTINUITY EQUATION

In this section, we perform Reynolds averaging on Continuity equation 7.1 in tensor notation:

$$\frac{\partial \rho}{\partial t} + \frac{\partial}{\partial x_j}(\rho u_j) = 0 \tag{7.49}$$

As per Reynolds decomposition we divide turbulent quantities into mean and fluctuating components like $\rho = \overline{\rho} + \rho'$ and $u_j = U_j + u'_j$. Here capital letter indicates mean component of turbulent velocity field signal and as per Reynolds averaging we can write $\overline{U_j} = U_j$.

$$\rho = \overline{\rho} + \rho' \tag{7.50}$$

$$\overline{\frac{\partial \rho}{\partial t}} = \overline{\frac{\partial}{\partial t}[\overline{\rho} + \rho']} \tag{7.51}$$

$$\frac{\partial \overline{\rho}}{\partial t} = \frac{\partial \overline{\overline{\rho}}}{\partial t} + \frac{\partial \overline{\rho'}}{\partial t} \tag{7.52}$$

As $\overline{\overline{\rho}} = \overline{\rho}$ we have

$$\frac{\partial \overline{\rho}}{\partial t} = \frac{\partial \overline{\rho}}{\partial t} + \frac{\partial \overline{\rho'}}{\partial t} \tag{7.53}$$

Further we perform operation:

$$u_j \rho = (\overline{U_j} + u'_j)(\overline{\rho} + \rho')$$

$$u_j \rho = \overline{\rho} U_j + \overline{\rho} u'_j + \rho' U_j + \rho' u'_j$$

$$\overline{u_j \rho} = \overline{\overline{\rho} U_j} + \overline{\overline{\rho} u'_j} + \overline{\rho' U_j} + \overline{\rho' u'_j}$$

$$\overline{u_j \rho} = \overline{\overline{\rho} U_j} + \overline{\rho' u'_j}$$

We substitute this result and Eq. 7.53 into Eq. 7.49:

$$\boxed{\frac{\partial \overline{\rho}}{\partial t} + \frac{\partial}{\partial x_j}[\overline{\rho} U_j + \overline{\rho'_j u'_j}] = 0} \tag{7.54}$$

This is RANS Continuity equation valid for unsteady and compressible flows. In case of incompressible flows, ρ is constant and there are no fluctuations in density field so $\rho' = 0$. For that case the applicable continuity equation is:

$$\frac{\partial}{\partial x_j}[\overline{U_j}] = \frac{\partial}{\partial x_j}[U_j] = 0 \tag{7.55}$$

Example 7.3

Convert the tensorial form of the steady state continuity equation into algebraic form.

Solution

As j=1,2,3 only we use tensor summation rule for repeated indices:

$$\frac{\partial U_j}{\partial x_j} = \frac{\partial U_1}{\partial x_1} + \frac{\partial U_2}{\partial x_2} + \frac{\partial U_3}{\partial x_3} \tag{7.56}$$

7.6.2 NAVIER-STOKES EQUATION

Navier's momentum equation in tensor notation is:

$$\rho \frac{\partial u_i}{\partial t} + \rho u_j \frac{\partial u_i}{\partial x_j} = -\frac{\partial p}{\partial x_i} + \frac{\partial \tau_{ji}}{\partial x_j} \tag{7.57}$$

Convective term:

$$\frac{\partial u_i u_j}{\partial x_j} = u_j \frac{\partial}{\partial x_j}[u_i] + u_i \frac{\partial}{\partial x_j}[u_j] \tag{7.58}$$

for incompressible flows

$$\frac{\partial u_i u_j}{\partial x_j} = u_j \frac{\partial}{\partial x_j}[u_i] + u_i \frac{\partial}{\partial x_j}[u_j] \tag{7.59}$$

$$\frac{\partial u_i u_j}{\partial x_j} = u_j \frac{\partial u_i}{\partial x_j} \tag{7.60}$$

We substitute this result into Eq. 7.57:

$$\rho \frac{\partial u_i}{\partial t} + \rho \frac{\partial u_i u_j}{\partial x_j} = -\frac{\partial p}{\partial x_i} + \frac{\partial \tau_{ji}}{\partial x_j} \tag{7.61}$$

The shear stress is described as for incompressible flows $\tau_{ij} = 2\mu S_{ij}$, where S_{ij} is turbulent mean shear strain tensor.

Note that there are symmetric tensors so we have 6 unknowns instead of 9 unknowns.

Doing Reynolds time averaging on whole equation yields:

$$\rho \frac{\partial U_i}{\partial t} + \rho \frac{\partial}{\partial x_j}(\overline{U_i U_j} + \overline{u_i' u_j'}) = -\frac{\partial \overline{p}}{\partial x_i} + \frac{\partial}{\partial x_j}(2\mu S_{ij}) \qquad (7.62)$$

Convective term is brought on RHS and treated as increase in shear stress:

$$\boxed{\rho \frac{\partial U_i}{\partial t} + \rho \frac{\partial}{\partial x_j}(\overline{U_i U_j}) = -\frac{\partial \overline{p}}{\partial x_i} + \frac{\partial}{\partial x_j}(2\mu S_{ij} - \rho \overline{u_i' u_j'})} \qquad (7.63)$$

$$\rho \frac{\partial U_i}{\partial t} + \rho \frac{\partial}{\partial x_j}(\overline{U_i U_j}) = -\frac{\partial \overline{p}}{\partial x_i} + \frac{\partial}{\partial x_j}(\tau_{ij} + \tau_{ij}') \qquad (7.64)$$

The convective fluctuations now treated as turbulent stresses arises due to velocity fluctuations and defined as:

$$\tau_{ij}' = -\rho \begin{bmatrix} \overline{u'u'} & \overline{u'v'} & \overline{u'w'} \\ \overline{v'u'} & \overline{v'v'} & \overline{v'w'} \\ \overline{w'u'} & \overline{w'v'} & \overline{w'w'} \end{bmatrix}$$

Along with u,v,w,p and 6 strain tensors we have total 10 unknowns. However, one mass and three momentum conservation equations are not enough to completely solve the set of equations. To close the problem, we need model to related fluctuating velocity strain tensor with mean velocity components.

7.6.3 ENTHALPY OR ENERGY TRANSFER EQUATION

We start from balance of enthalpy:

$$\frac{\partial(\rho H)}{\partial t} + \frac{\partial(\rho u_j H)}{\partial x_j} = \frac{\partial}{\partial x_j}(u_i \tau_{ij} - q_j) \qquad (7.65)$$

where, $q_j = -k(\partial T / \partial x_j)$. We substitute:

$$H = \overline{H} + H'$$

and

$$u_j = U_j + u_j'$$

also

$$\tau_{ij} = \mu \left(\frac{\partial U_i}{\partial x_j} + \frac{\partial U_j}{\partial x_i} \right) + \frac{2}{3}\delta_{ij}\frac{\partial U_k}{\partial x_k}$$

Taking time averaging and $H = mc_pT$ with substituting NS Eq. into above equation

$$\frac{\partial T}{\partial t} + \frac{\partial \overline{(TU_j)}}{\partial x_j} = \frac{1}{\rho c_p}\left[\frac{\partial \overline{p}}{\partial t} + U_j\frac{\partial \overline{p}}{\partial x_j} + \overline{u'_j\frac{\partial p'}{\partial x_j}}\right]$$
$$+ \frac{\Phi}{\rho c_p} + \frac{\partial}{\partial x_j}\left[\alpha\frac{\partial T}{\partial x_j} - \overline{T'u'_j}\right]$$

We define Reynolds heat flux as:

$$q^R = \rho c_p\overline{T'u'_j}$$

In case of steady, incompressible flows (ignoring pressure fluctuations and viscous dissipation):

$$\boxed{\overline{U_j}\frac{\partial \overline{T}}{\partial x_j} = \frac{\partial}{\partial x_j}\left[\alpha\frac{\partial T}{\partial x_j} - \overline{T'u'_j}\right]} \qquad (7.66)$$

RANS Conservation Equations

RANS conservation equations needed to solve are:

Continuity Equation:

$$\frac{\partial \rho}{\partial t} + \frac{\partial}{\partial x_j}[\rho U_j] = 0 \qquad (7.67)$$

Navier-Stokes Equation:

$$\rho\frac{\partial U_i}{\partial t} + \rho\frac{\partial}{\partial x_j}(\overline{U_iU_j}) = -\frac{\partial \overline{p}}{\partial x_i} + \frac{\partial}{\partial x_j}(2\mu S_{ij} - \rho\overline{u'_iu'_j}) \qquad (7.68)$$

Energy Equation:

$$\frac{\partial \overline{T}}{\partial t} + \overline{U_j}\frac{\partial \overline{T}}{\partial x_j} = \frac{\partial}{\partial x_j}\left[\alpha\frac{\partial \overline{T}}{\partial x_j} - \overline{T'u'_j}\right] \qquad (7.69)$$

7.6.4 BOUSSINESQ APPROXIMATION

Transport phenomena in turbulent flow are much more complicated than those in laminar flow and we cannot obtain any analytical solution of the time-averaged governing equations, unless we describe turbulent shear stress. In most simple terms, turbulent shear stress can be modelled by an approach as proposed by Boussinesq. According to Boussinesq approximation (1877), stress tensors can be related with gradient as:

$$\tau \propto \varepsilon\frac{du}{dy}$$

where, ε is eddy diffusivity for flow. Dimensionally this relationship is very similar to Newton's law of visosity; however, eddy diffusivity (ε) is not a fluid property, rather it based on flow conditions. The turbulent stress and scalar fluxes can be modeled as:

$$\boxed{\tau_{turb} = -\rho(v + \varepsilon_M)\frac{\partial u}{\partial y}} \tag{7.70}$$

Analogous to Boussinesqs model of the turbulent transport of momentum, the turbulent scalar fluxes are assumed to be proportional to the gradient of the transport quantity.

$$\boxed{q_{turb} = -\rho c(\alpha + \varepsilon_H)\frac{\partial T}{\partial y}} \tag{7.71}$$

where, c is specific heat, ρ is density, α is thermal diffusivity, v is kinematic viscosity. The turbulent based parameters ε_M and ε_H are measure experimentally by researchers, and it is found that these parameters are changing in flow. We introduce here a parameter called turbulent Prandtl number defined as

$$Pr_t = \frac{\varepsilon_M}{\varepsilon_H}$$

where ε_M is turbulent eddy diffusivity and ε_H is turbulent thermal diffusivity. Grötzbach found that the turbulent eddy momentum and thermal diffusivities are non-similar in liquid metal (Pr<1) flows. Therefore, Pr_t will be a function of parameters. Even in incompressible flows, we cannot treat turbulent Prandtl number as a constant. For the wall jet boundary layer which is formed after an impinging jet, Uddin (2008) found that turbulent Prandtl number is changing from location to location. The turbulent Prandtl number for radially spreading wall jet is defined as:

$$Pr_t = \frac{\overline{u_r' u_y'}\frac{\partial T}{\partial y}}{\overline{u_y' T'}\frac{\partial U}{\partial y}}$$

Uddin (2008) conducted Large Eddy Simulations and the turbulent Prandtl number is computed at three different radial locations in the wall jet zone.

Figure 7.7 shows the distribution of the turbulent Prandtl number in wall normal direction. As for a channel flow, it is found that near the wall the turbulent Prandtl number reaches a constant value. However, in the near-wall region of a wall jet region the turbulent Prandtl number changes from one radial position to another. One can expect that the turbulence models using constant turbulent Prandtl number might not perform well for complex flow cases.

For the sake of simplicity, we resort to constant turbulent Prandtl number in this text. We can write $\varepsilon_H = \varepsilon_M Pr_t$ and the turbulent heat transfer shall be

$$q_{turb} = -\rho c_p(\alpha + \varepsilon_H)\frac{\partial T}{\partial y}$$

$$q_{turb} = -\rho c_p\left(\frac{v}{Pr} + \frac{\varepsilon_M}{Pr_t}\right)\frac{\partial T}{\partial y}$$

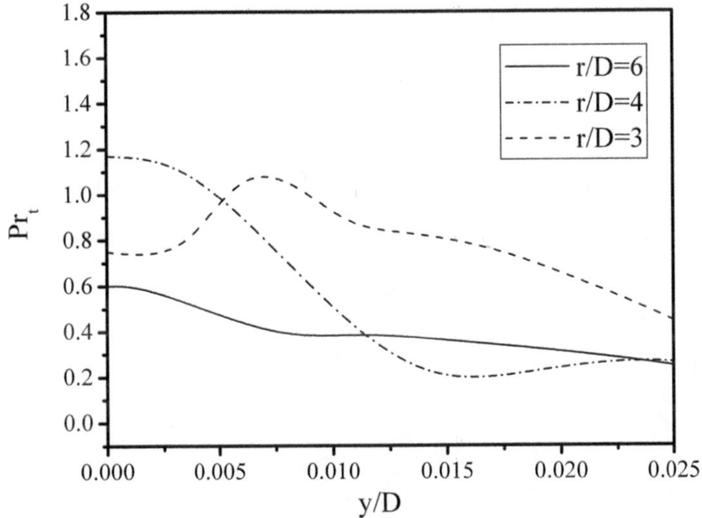

Figure 7.7 Turbulent Prandtl number distribution at r/D=3, 4 and 6 (see Uddin, 2008). y is the wall normal coordinate and r is the radial coordinate.

here can write it in simplified manner as

$$q_{turb} = -\overline{u'_i T'} = \Gamma_t \frac{\partial T}{\partial y}$$

and Γ_t is defined as

$$\Gamma_t = -\rho c_p \left(\frac{v}{Pr} + \frac{\varepsilon_M}{Pr_t} \right)$$

Γ_t is a property of the flow and depends on the local state of the turbulence.

We can write a generalised equation for the turbulent flows. Both the temperature (or enthalpy) and the concentration (or the mass fraction of a chemical species) are scalar quantities and can be represented by a general variable Θ. The time averaged form of the equation governing the transport of Θ can be written as:

$$U_i \frac{\partial \Theta}{\partial x_i} = \frac{\partial}{\partial x_i} \left(\Gamma_\Theta \frac{\partial \Theta}{\partial x_i} - \overline{u'_i \theta'} \right) \tag{7.72}$$

The equation represents a balance between the rates at which the scalar quantity Θ is transported by the mean flow field (advection) and it diffusion by the combined action of molecular and turbulent effects.

PROBLEMS

7P-1 Prove that the following expression has dimensions of rate of work per unit volume.

$$\sigma \cdot u / \ell$$

where, σ, u and ℓ are stress, velocity and length dimensions respectively.

7P-2 What is the difference between Biot number and Nusselt number?

7P-3 The Nusselt number from an experiment for heat transfer by an impinging jet case is tabulated as follows:

r/D	Nu_D
0	139.13
0.271	137.62
0.626	134.58
0.733	127.94
0.884	119.26
1.035	107.51
1.110	102.91
1.283	95.25
1.358	94.75
1.509	94.25
1.809	102.46
1.980	106.56
2.184	107.60
2.366	103.02
2.699	94.86
3.183	83.65

The Reynolds number of flow is $Re_D = \rho U_{jet} D / \mu = 20{,}000$ and fluid is air. Find the constant C and index m, if the data is following the correlation:

$$Nu_D = C \cdot Re^{2/3} Pr^m$$

7P-4 Goldstein and Franchett (1988) have investigated the heat transfer by a circular jet impinging at different oblique angles to a plane surface:

$$\frac{Nu}{Re^{0.7}} = A \cdot \exp\left[-(B + C \cdot \cos\varphi) \cdot \left(\frac{r}{D}\right)^m\right]$$

where D is the pipe diameter, φ is inclination angle, and r is the radial distance on the target wall. If the Reynolds number based on 50 mm pipe diameter is 20000, plot the radial distribution of Nusselt number at $H/D=4$, $\theta=60°$, $A = 0.163$, $B = 0.4$, $C = 0.12$, and $m = 0.75$.

7P-5 The Nusselt number for heated flow through pipe can be expressed as

$$Nu_D = C \cdot Re_D^{0.8} \cdot Pr^{0.4}$$

Find the ratio of convective heat transfer coefficient for air and water: h_w/h_a (i) if the Reynolds number for the two flows is same, and (ii) if the mean velocity and internal diameter are the same.

REFERENCES

R. D. Blevin. Applied Fluid Dynamics Handbook. New York: Van Nostrand Reinhold, USA, 1984.

N. Kasagi, Y. Tomita, and A. Kuroda, J. Heat Transfer, Vol. 114, pp. 598–606, 1992.

N. Uddin, Turbulence modeling of complex flows in CFD, PhD Thesis, Universitaet Stuttgart, Germany, 2008.

R. J. Goldstein and M. E. Franchett, Heat transfer from a flat surface to an oblique impinging jet, Journal of Heat Transfer, vol 110, pp. 84–90, 1988.

8 Forced Convection: Internal Flows

Convection can be classified as external or internal convective flow, depending on whether the fluid is forced to flow over a surface or in a pipe or duct. We discuss the importance of hydrodynamic and thermal entrance lengths in internal flows. For laminar flows inside the pipe, the boundary conditions play an important role. Traditionally the constant temperature and constant wall heat-flux boundary conditions are used as they are the simplest to implement experimentally. It has been found that for turbulent flow this sharp distinction in boundary conditions is not needed. After finishing this chapter one will be able to:

- Calculate the heat transfer in entrance regions of pipe.
- Calculate the heat transfer for both laminar and turbulent flows.
- Calculate the heat transfer for transition flows.
- Calculate the heat transfer in coiled tube.

The character of flow in a round pipe depends on four variables fluid density ρ, fluid viscosity μ, pipe diameter D (or d), and average velocity of flow $\bar{u}$. Osborne Reynolds has showed through his famous dye experiment that flow has three distinct characters. When he injected the dye in the flow, the dye did not diffuse, and he named this flow layered flow. In another type of flow, the dye completely diffused into the flow, and he called this flow a sinuous flow. The latter is now called turbulent flow, and the former is called laminar flow. Between these two types of flows, there is another type of flow called the transition flow. In this famous Reynolds experiment, Osborne Reynolds employed the streakline visualisation, as he injected a colored dye into water which was flowing in pipe at different speeds. In turbulent flow, the diffusion of dye was strong and streakline is immediately lost into diffusing mixture of water and dye. However, in laminar flow the streakline was maintained and can be visualised through a naked eye. Here what is crucial for us is the dimensionless number called the Reynolds number.

$$Re_D = \frac{\rho \bar{u} D}{\mu}$$

$$Re_D = \frac{D \bar{u}}{\nu}$$

DOI: 10.1201/9781003428404-8

where mean velocity can be estimated as:

$$\bar{u} = \frac{1}{A}\int_A u_z(r)\,dA \tag{8.1}$$

It has been found that the flow has lower and upper critical Reynolds number:

$$2300 < \mathrm{Re}_D < 4000$$

In practical calculations, we consider the step transition and treat flow as either laminar or turbulent. So if

$$\mathrm{Re}_D < 2300 \quad \textit{Laminar}$$
$$\mathrm{Re}_D > 2300 \quad \textit{Turbulent}$$

In many industrial applications, the pipes are not circular and may have different cross sections as shown in Figure 8.1. In such cases, the correlations and charts available in this chapter are still applicable if we calculate the hydraulic diameter as:

$$D_h = \frac{4A}{WP}$$

where D_h is called the hydraulic diameter, A is the cross section of the non-circular duct, and WP is the wetted perimeter, i.e., the perimeter touched by the fluid. In half-filled liquid ducts, this has to be calculated with care. However, in the case of gases, the wetted perimeter will be the complete perimeter in the cross section, as gas will spread out and touch all the walls. Hydraulic diameters of some cross-sections are shown in Table 8.1.

Hydrodynamic Entrance Length　　　　Fully Developed Flow

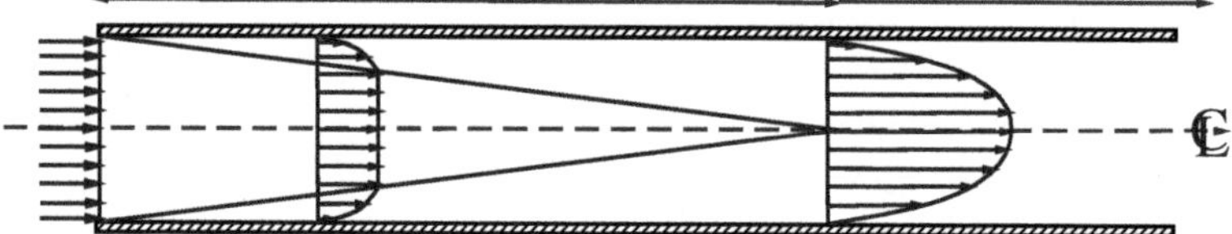

Figure 8.1　　Hydrodynamic entrance length in pipe flow.

Note that the use of hydraulic diameter for the sharp-edged cross sections like triangular, square, and cusped ducts may lead to significantly unacceptable errors of the order of 35% in turbulent flows friction factors if determined from the circular pipe flow correlations. The error is not substantial for cross sections that do not have sharp edges. Also, the hydraulic diameter concept gives acceptable results for turbulent flow, but it is not a very reliable approach for laminar flow. Therefore, it is advised to consult the handbooks for sharp-edged cross sections for improved and accurate results.

TABLE 8.1

Hydraulic Diameter of Some Cross Sections

Name	Cross section	Area	Hydraulic diameter
Circle	D or d	$A = \frac{\pi}{4}D^2$	D or d
Semicircle	2R	$A = \frac{\pi}{2}R^2$	$D_h = \frac{2\pi R}{\pi+2}$
Sector	R, θ	$A = \frac{\theta}{2}R^2$	$D_h = \frac{2\theta R}{\theta+2}$
Trapezoid	a, b, c, θ	$A = \frac{b}{2}(a+c)$	$D_h = \frac{2b(a+c)}{a+b+c}$
Isosceles Triangle	a, a, θ	$A = \frac{a^2}{2}\sin\theta$	$D_h = \frac{a\sin\theta}{1+\sin\left(\frac{\theta}{2}\right)}$

8.1 HYDRODYNAMIC ENTRANCE LENGTH IN LAMINAR FLOWS

We now consider the fluid entering a circular pipe or a duct at a uniform velocity (see Figure 8.1). The in contact with the stationary wall will attain zero velocity. The layer next to the surface comes to rest due to wall friction. As the mass flow is always conserved, the flow in the core region will be sped up. This also happens as the boundary layer develops on the wall, and the fluid in the inner core and close to the center starts squeezed and sped up. Therefore, there are two regions in the entrance region: the flow velocity and the flow near the wall, where flow is experiencing the boundary layer phenomenon. The zone from the pipe or duct inlet to where the boundary layer combines at the centerline is called the hydrodynamic entrance zone, and the length of this inlet zone is called the hydrodynamic entrance length.

Once the flow becomes fully developed, the velocity profile will become independent of the length of pipe or duct and mathematically we can write

$$\frac{du}{dx} = 0$$

Laminar Flow Hydrodynamic Entrance Length Estimation Following are some of the correlation in Literature:

$$\left(\frac{L_{hyd}}{D}\right)_{lam} \simeq 0.065 \mathrm{Re}_D \qquad \begin{array}{l} Boussinesq \\ Nikuradse \end{array}$$

$$\left(\frac{L_{hyd}}{D}\right)_{lam} \simeq 0.06 \mathrm{Re}_D \qquad Asao\ et\ al.$$

Chen et al. (1973) have derived the entrance length for pipe and channel flow valid in range of 1<Re<2000.

$$\left(\frac{L_{hyd}}{D}\right)_{lam} \simeq 0.061 \mathrm{Re} + \frac{0.72}{0.04 \mathrm{Re} + 1} \qquad (pipe)$$

$$\left(\frac{L_{hyd}}{D}\right)_{lam} \simeq 0.053 \mathrm{Re} + \frac{0.79}{0.04 \mathrm{Re} + 1} \qquad (channel)$$

Example 8.1

Air at 80 °C is flowing through a 50 mm diameter pipe at speed of 1 m/s. Estimate the hydrodynamic entrance length using different correlations, if the wall temperature is constant.

Solution From property data for air, we have the properties at mean temperature of 80 ^{o}C.

$$\mu_w = 0.0000235\ kg/m.s, \quad v = 0.00002094\ m^2/s, \quad \mu = 0.0000210\ kg/m.s$$
$$k = 0.03 W/m.K, \quad \beta = 0.00283\ K^{-1}, \quad Pr = 0.708$$

The Reynolds number of the flow is

$$\mathrm{Re} = \frac{u \cdot D}{v} = 1432.66$$

The flow is laminar. We now use the correlations to estimate the hydrodynamic entrance length for laminar flow.

Boussinesq-Nikuradse correlation:

$$\left(\frac{L_{hyd}}{D}\right)_{lam} \simeq 0.065 Re_D$$

$$L_{hyd} = 2.793m$$

Asao et al. correlation:

$$\left(\frac{L_{hyd}}{D}\right)_{lam} \simeq 0.06 Re_D$$

$$L_{hyd} = 2.578m$$

Chen correlation:

$$\left(\frac{L_{hyd}}{D}\right)_{lam} \simeq 0.061 Re + \frac{0.72}{0.04 Re + 1}$$

$$L_{hyd} = 2.622m$$

As can be observed, all these correlations predict the hydrodynamic entrance length value close to each other. To be on safe side, we can take the highest value and hence for practical calculation, we take the value of entrance length as

$$L_{hyd} = 2.793m$$

8.2 HEAT TRANSFER IN LAMINAR ENTRANCE LENGTH

In laminar flows we need to make distinction between the boundary conditions. The two most widely used boundary conditions are constant wall temperature and constant heat flux boundary conditions. Following are some of the useful relations for estimation of Nusselt number in the thermal entrance region.

Constant wall temperature condition
Based on Graetz (1883) und Nusselt (1910) derivations, Shah (1975) proposed the correlation in terms of the ratio of thermal entrance length with hydrodynamic entrance length for constant wall temperature condition as

$$\frac{L_{th}}{L_{hyd}} = \frac{0.0335 Pr}{0.056 + \frac{0.6}{Re(1+0.035Re)}} \tag{8.2}$$

This relationship shows that for a flow with sufficiently large Reynolds number and for fluid with Prandtl numbers $Pr >> 1$ (viscous liquids) the thermal inlet length will be larger than the hydrodynamic entrance length.

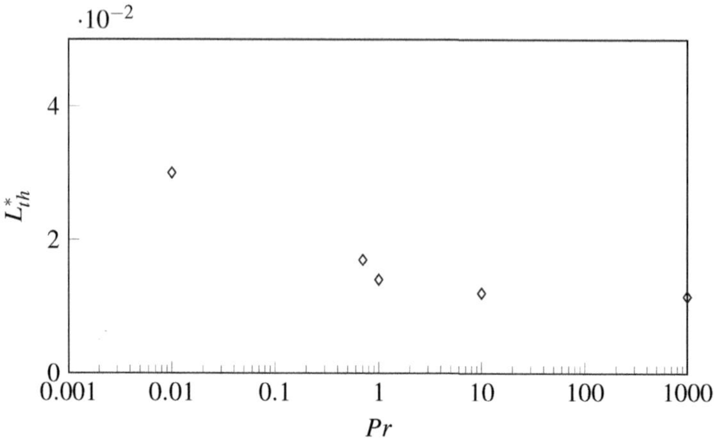

Figure 8.2   Thermal entrance lengths for different fluids.

Figure 8.2 shows the dimensionless thermal entrance lengths for different fluids, which is defined as

$$L_{th}^{*} = \frac{L_{th}}{D_h \cdot \mathrm{Re}_D \cdot Pr}$$

The crude assumption of $L_{th}^{*} = 0.05$ is not a valid approximation for all kind of fluids.

Example 8.2

Consider the case of previous example and estimate the thermal entrance length for laminar flow. At what pipe length would the flow be both thermally and hydrodynamically developed?

Solution We have already computed the Reynolds number as

$$Re = 1432.66$$

The hydrodynamic entrance length from previous example is

$$L_{hyd} = 2.793m$$

Using the equation 8.2, we estimate the thermal entrance length:

$$\frac{L_{th}}{L_{hyd}} = \frac{0.0335Pr}{0.056 + \frac{0.6}{Re(1+0.035Re)}}$$

This gives

$$L_{th} = 1.1830m$$

We notice here that

$$L_{hyd} > L_{th}$$

Hence, the flow is fully developed i.e. both thermally and hydrodynamically developed once it reaches the pipe length of 2.8 m. For 50 mm diameter pipe, we have for air:

$$Lth^* = \frac{Lth}{D \cdot Re \cdot Pr} = 0.0199$$

This shows that taking $Lth^* == 0.05$ is not a good estimate for air flow.

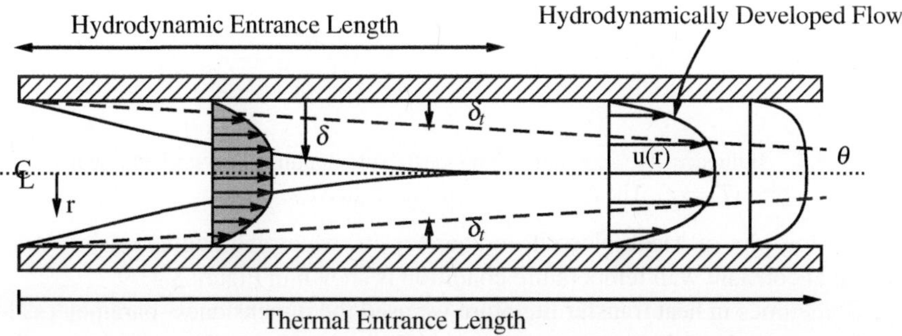

Figure 8.3 Momentum and thermal boundary layer development inside the pipe entrance length.

Figure 8.3 shows the velocity boundary layer development in the entrance region of the pipe (δ). The flow is fully developed once the boundary layer merge at the core of the pipe. In Figure 8.3 the thermal boundary layer (δ_t) also formed, but it is still developing as the length required is longer than depicted in this figure. The dimensionless temperature $\theta = (T_w - T)/(T_w - T_b)$, where, T_b is a bulk mean temperature and it will be discussed later in this chapter.

Hausen (1959) analytically solved the problem with inclusion of Prandtl number

$$Nu_{m,T} = 3.657 + \frac{0.19(x^+/Pr)^{-0.8}}{1 + 0.117(x^+/Pr)^{-0.467}} \tag{8.3}$$

where m indicate the mean value and T corresponds to constant wall temperature boundary condition. We define here dimensionless distance x^+ from entrance

$$x^+ = \frac{x}{D_h Re}$$

As $x^+ \to \infty$ the flow would reach to fully developed solution.

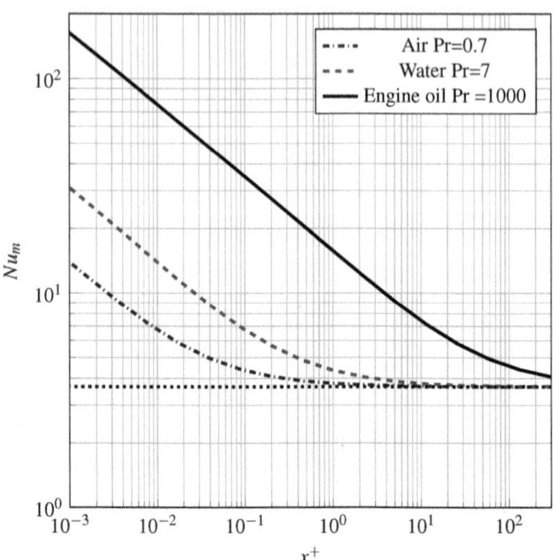

Figure 8.4   Influence of Prandtl number on mean Nusselt number versus location in entrance length of a pipe ($T_w = c$). The dotted line corresponds to Nu_m =3.657.

log-log plot of Mean Nusselt number for air, water and engine oil in entrance region at constant wall temperature condition is shown in Figure 8.4.

Sometimes in heat transfer literature we used the dimensioness parameter called the Graetz number which is named after the German engineer Gustav Graetz, defined as:

$$Grz = \frac{Re \cdot Pr}{(L/D_h)^2}$$

A low value of Graetz number indicates that the thermal diffusion is dominant over the momentum diffusion, and the fluid temperature distribution is almost uniform across the pipe or channel cross-section. A high Graetz number indicates that the momentum diffusion is dominant over the thermal diffusion, and the fluid temperature distribution is highly non-uniform across the pipe or channel cross-section.

Example 8.3

Air at 80°C is flowing through a metallic tube having internal diameter of 20 mm at speed of 2 m/s. The tube is 500 mm in length. Estimate the Nusselt number for this case if tube is maintained at constant temperature of 100°C. Also, calculate the Graetz number for this flow.

Solution From property data for air we have the properties at mean temperature of 80 °C.

$\mu_w = 0.0000235 \ kg/m.s, \ \ v = 0.00002094 \ m^2/s, \ \ \mu = 0.0000210 \ kg/m.s$
$k = 0.03 W/m.K, \ \ \beta = 0.00283 \ K^{-1}, \ \ Pr = 0.708$

The Reynolds number of the flow is

$$Re = \frac{u \cdot D}{\nu} = 1910.21$$

The flow is laminar. We estimate the hydrodynamic entrance length using Bossinesq-Nikuradse correlation

$$L_{hyd} = 0.065 \cdot Re \cdot D = 2.483$$

and thermal entrance length from correlation

$$L_{th} = \frac{0.0335 \cdot Pr \cdot L_{hyd}}{0.056 + \frac{0.6}{Re \cdot (1 + 0.035 \cdot Re)}} = 1.0516 \text{m}$$

The actual length of tube is $\ell = 0.5$ m and thus

$$\ell < L_{th} < L_{hyd}$$

Indicating that the tube length is not enough to let the flow fully develop. In this case we use the correlation

$$Nu_{m,T} = 3.657 + \frac{0.19(x^+/Pr)^{-0.8}}{1 + 0.117(x^+/Pr)^{-0.467}}$$

where,

$$x^+ = \frac{x}{D_h Re} = 0.01163$$

and the mean Nusselt number value is found to be

$$Nu_{m,T} = 6.48$$

The Graetz number (Grz) is defined as

$$Grz = \frac{Re \cdot Pr}{(L/D_h)^2} = 2.139 \times 10^6$$

Edwards et al. (1979) proposed a correlation for mean Nusselt number in both entrance and fully developed zone as

$$Nu_m = 3.66 + \frac{0.065 \cdot \left(\frac{d}{L} \cdot Re \cdot Pr\right)}{1 + 0.04 \cdot \left(\frac{d}{L} \cdot Re \cdot Pr\right)^{\frac{2}{3}}} \tag{8.4}$$

Edwards et al. (1979) proposed a correlation for the mean Nusselt number in case flow between parallel plates as

$$Nu_{m,T} = 7.54 + \frac{0.03 \cdot \left(\frac{d}{L} \cdot Re \cdot Pr\right)}{1 + 0.016 \cdot \left(\frac{d}{L} \cdot Re \cdot Pr\right)^{\frac{2}{3}}} \qquad Re \le 2800 \qquad\qquad (8.5)$$

Example 8.4

Engine oil flows with mass flow rate of 0.009 kg/s, moves through a 0.5 cm internal diameter tube (d). The tube length (L) is 20 m with wall maintained at 300 K. The bulk inlet oil temperature is 377 K. Find the average heat transfer coefficient. The properties of engine oil are

$k = 0.137 \frac{W}{m \cdot K}$, $\mu = 0.0189$ Pa·s, $c_p = 2200$ J/kg·K
$Pr = 300$, $\mu_w = 0.503$;

Solution We compute the flow Reynolds number as

$$Re = \frac{4 \cdot \dot{m}}{\pi \cdot d \cdot \mu} = 121.26$$

As Re< 2300, the flow is laminar. We now compute the hydrodynamic and thermal entrance lengths:

$$L_{hyd} = 0.065 \cdot Re \cdot d = 0.0394 m$$

$$L_{th} = 0.0335 \cdot Pr \cdot L_{hyd}/(0.056 + 0.6/(Re \cdot (1 + 0.035 \cdot Re))) = 6.955 m$$

This shows that flow is not thermal developed in 34.7% of the tube length. We use the Edward et al relation to estimate the mean Nusselt number for the complete length of the tube:

$$Nu_{m,T} = 3.66 + \frac{0.065 \cdot \left(\frac{d}{L} \cdot Re \cdot Pr\right)}{1 + 0.04 \cdot \left(\frac{d}{L} \cdot Re \cdot Pr\right)^{\frac{2}{3}}} = 4.163$$

This gives convective heat transfer coefficient as

$$h = 114.077 \frac{W}{m^2 \cdot K}$$

Another way to estimate the Nusselt number is to use Hausen correlation:

$$x^+ = 32.98$$

$$Nu_{m,T} = 3.657 + \frac{0.19 \cdot \left(\frac{x^+}{Pr}\right)^{-0.8}}{1 + 0.117 \cdot \left(\frac{x^+}{Pr}\right)^{-0.467}}$$

$$Nu_m = 4.49$$

$$h = 123.3 \frac{W}{m^2 \cdot K}$$

Hausen correlation gives 7.89% over-prediction for convective heat transfer coefficient compared with Edwards et al correlation.

Constant wall heat flux condition

The mean Nusselt number in entrance region can be estimated from Shah und London (1978) relation:

$$Nu_{m,H} = \begin{cases} 1.953(x^+/Pr)^{-1/3} & x^* \le 0.03 \\ 4.364 + \frac{0.0722}{x^+/Pr} & x^* \ge 0.03 \end{cases} \tag{8.6}$$

where,

$$x^* = \frac{x}{D_h \cdot Re \cdot Pr}$$

and

$$x^+ = \frac{x}{D_h \cdot Re}$$

m indicate the mean value and H corresponds to constant wall heat-flux boundary condition.

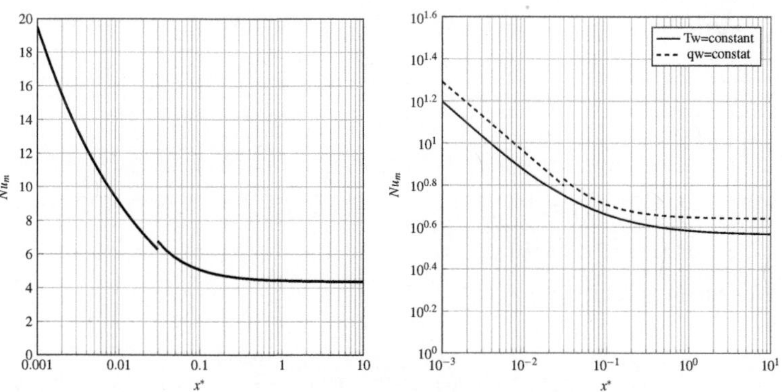

Figure 8.5   (Right) Mean Nusselt number versus location in entrance length of a pipe $(q''_w = c)$. (Left) Thermal entrance length at constant wall temperature and constant heat flux conditions for air.

Figure 8.5 shows that the Nu values for constant wall heat flux is bit higher than the constant wall temperature conditions.

If the wall and the fluid temperatures are a lot different, then accounting the variation of viscosity with temperature will give better prediction for heat transfer coefficient. Sieder and Tate (1936) proposed a correlation for the average Nusselt number in case of developing laminar flow:

$$Nu_{m,T} = 1.86 \left(\frac{d}{L} \cdot Re \cdot Pr \right)^{1/3} \left(\frac{\mu_b}{\mu_w} \right)^{0.14}$$

8.3 HYDRODYNAMICALLY DEVELOPED LAMINAR FLOW THROUGH CIRCULAR PIPE

In case of fluid flowing isothermally through a smooth tube at small velocity and with no turbulence introduced at the entrance of the pipe, the flow shall be called laminar. In this flow if a thin stream of colored dye is injected at the axis of the pipe/tube, the dye will not diffuse throughout the pipe length, instead it but will remain moving at the center of the pipe for a considerable distance. This proves that the radial velocity in the pipe is very small for laminar flows. The laminar flow consists of thin concentric tubular filaments sliding along the axis of the tube. The fluid at the larger radii will be moving more slowly than those with smaller radii near the center. Thus, there will be a paraboloid type of velocity profile.

$$u = \frac{p_o - p_L}{4L\mu} \left[r_o^2 - r^2 \right] \tag{8.7}$$

where, For $z = 0$, $p = p_o$ and at $z = L$, $p = p_L$. The velocity distribution at a given section of the tube is parabolic. The mean velocity u_m may be defined by

$$
\begin{aligned}
\pi r_o^2 u_m &= \int_0^{r_o} u 2\pi r \, dr = \int_0^{r_o} \frac{1}{4L\mu} (p_o - p_L)[r_o^2 - r^2] 2\pi r \, dr \\
&= \frac{\pi}{2L\mu} [p_o - p_L] \left[\frac{r_o^2 r^2}{2} - \frac{r^4}{4} \right]_o^{r_o} \\
&= \frac{\pi}{2L\mu} [p_o - p_L] \left[\frac{r_o^4}{2} - \frac{r_o^4}{4} \right] \\
&= \frac{\pi r_o^4}{8L\mu} [p_o - p_L]
\end{aligned}
\tag{8.8}
$$

This gives mean velocity as

$$u_m = \frac{r_o^2}{8L\mu} [p_o - p_L] = \frac{u_o}{2} \tag{8.9}$$

The above equation states that the mean velocity in a pipe is equal to one half the maximum velocity at the center of the tube. Solving for $\frac{1}{4L\mu}[p_o - p_L]$ in the above equation, and substituting it in the velocity distribution a more convenient expressions for the velocity field can be obtained:

$$u = 2u_m \left[1 - \frac{r^2}{r_o^2} \right] = u_o \left[1 - \frac{r^2}{r_o^2} \right] \tag{8.10}$$

Pressure drop in pipe

Consider a length dz of tubular filament experiencing the pressure drop in the length dz as $\dfrac{\partial p}{\partial z}dz$. As the fluid flow is steady, the shear stress in the fluid layers at any radius r is balanced by the pressure drop in length dz multiplied by the cross sectional area of all of the filaments contained in the pipe of radius r.

$$\left(\pi r^2 \frac{dp}{dz}\right)dz = (2\pi r \cdot \tau_w)\,dz$$

$$\pi r^2 \frac{dp}{dz}dz = 2\pi r\left(\mu \frac{du}{dr}\right)dz \tag{8.11}$$

Since p is a function of z and u is a function of r alone, we can reduce the equation 8.11 and integrate it.

$$u = \frac{1}{2\mu}\int_r^{r_o} r\frac{dp}{dz}dr = \frac{1}{2\mu}\frac{dp}{dz}\int_r^{r_o} r\,dr$$

or

$$u = -\frac{1}{4\mu}\frac{dp}{dz}[r_o^2 - r^2] \tag{8.12}$$

Integration between the limits $p = p_o$ and $p = p_L$, $z = 0$ and $z = L$ gives

$$u = \frac{r_o^2 - r^2}{4\mu L}[p_o - p_L]$$

The laminar fully developed velocity profile indicates that the laminar flow is directly proportional to the fourth power of the radius and inversely proportional to the length. This is in accordance with experimental evidence given by Poiseuille in 1840, and therefore, this flow is also called Hagen-Poiseuille flow. The friction factor (f) can be computed for laminar flow.

$$\Delta p = p_o - p_L = 8L\mu\frac{u_m}{r_0^2} = \frac{32L\mu}{d^2}u_m \tag{8.13}$$

Since $(p_o - p_L)$ is the work per unit volume to overcome friction we can write

$$f = 64\left[\frac{\mu}{\rho D u_m}\right] \tag{8.14}$$

The quantity $\rho D u_m/\mu$ is known as Reynolds number (R_e). Thus for laminar flows we have

$$f = 64/Re \tag{8.15}$$

Experimentally it is confirmed that flow cannot be turbulent before a critical value of R_e which is found to be near 2300 in a pipe. In normal engineering practices the flow having Reynolds number higher than this value is considered as either transitional flow or a turbulent flow.

There are two definitions of friction factors used in literature. One is named after American architect and hydraulic engineer John Thomas Fanning (1837-1911) and another one is named after French engineer Henry Philibert Gaspard Darcy (1803-1858).

Fanning Friction Factor

$$C_f = \frac{\tau_w}{\frac{1}{2}\rho u^2}$$

Darcy Friction Factor

$$f = \frac{\Delta p}{\frac{1}{2}\rho u^2}\left(\frac{D}{\ell}\right)$$

8.4 THERMALLY AND HYDRODYNAMICALLY DEVELOPED LAMINAR FLOW IN PIPE

For the analysis of heat transferred to the fluid moving inside a heated pipe, it is customary to define the bulk mean temperature or mixing cup temperature. Figure 8.6 depicts the case of fluid movement inside a pipe with heated walls. Note that the we are considering here the simple cases of constant wall temperature and constant heat flux boundary conditions. The bulk mean temperature associated with inlet is indicated as T_{b1}, and the bulk mean temperature associated with pipe outlet is indicated in figure as T_{b1}.

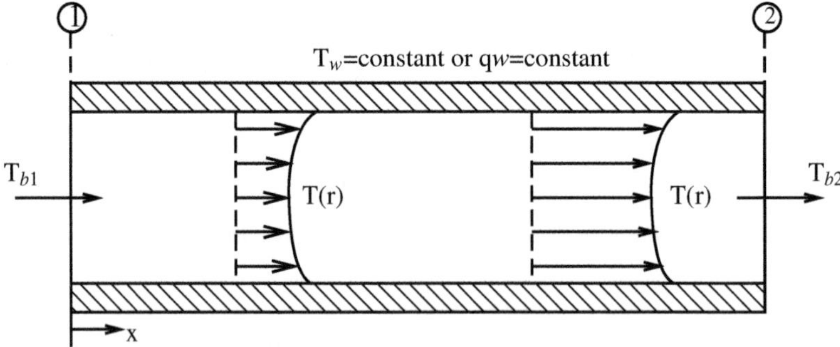

Figure 8.6 The thermally developed flow of a cold fluid in a heated pipe would experience rise in temperature as pipe wall would constantly heat up the fluid inside it and temperature distribution will change from location to location.

At any translational location in streamwise direction, we can do the balance of energy as

$$\int_{A_c} \rho u c_p T \ dA = c_p \dot{m} T_b$$

where, A_c is the cross-sectional area of pipe, c_p is specific heat and u is the mean velocity.

This gives the definition of bulk mean temperature as

$$T_b = T_m = \frac{\int_{A_c} \rho u c_p T \ dA}{c_p \dot{m}}$$

In laboratory, such temperatures can be measured by extracting fluid at a certain location and mixing the fluid, hence, bulk fluid temperature is also called the mixing cup temperature. In heat transfer literature both T_b and T_m are used to indicate bulk-mean or mixing-cup temperature. Note that the radial distribution of temperature field is constantly changing and we can no longer take $\partial T / \partial x = 0$ as a thermally developed flow condition. In convective heat transfer, we define the thermally developed flow based on temperature difference as

$$\frac{\partial}{\partial x}\left(\frac{T_w - T(r,x)}{T_w - T_b(x)}\right) = 0$$

We consider the bracketed term on LHS as θ:

$$\theta = \frac{T_w - T(r,x)}{T_w - T_b(x)}$$

and we can notice that derivative of θ with respect r must also be independent of streamwise position, leading to result:

$$\frac{\partial}{\partial r}\left(\frac{T_w - T(r,x)}{T_w - T_b(x)}\right)_{r=r_o} = \frac{-\left(\frac{\partial T}{\partial r}\right)_{r=r_o}}{T_w - T_b}$$

Comparing Fourier's law and Newton's law of cooling at wall we have

$$\dot{q} = h(T_w - T_b) = -k\left(\frac{\partial T}{\partial r}\right)_{r=r_o}$$

$$h = \frac{-k\left(\frac{\partial T}{\partial r}\right)_{r=r_o}}{(T_w - T_b)}$$

Since RHS is independent of streamwise location, we can infer that for pipe or duct flows, convective heat transfer coefficient is independent of x. It is a great simplification, as now we can treat convection heat transfer coefficient as a constant for the fully developed flow.

8.5 FULLY DEVELOPED LAMINAR FLOW THROUGH A CIRCULAR TUBE WITH VOLUMETRIC HEAT CONVERSION

We now consider the case of fully developed Laminar flow through a circular tube with uniform q_w applied at the pipe wall. Also, energy is being liberated uniformly in fluid as it flows through pipe at a rate of $\tilde{q}$ per unit volume (W/m^3). We assume that properties remain constant and velocity field is independent of temperature field and internal heat generation (W/m^3).

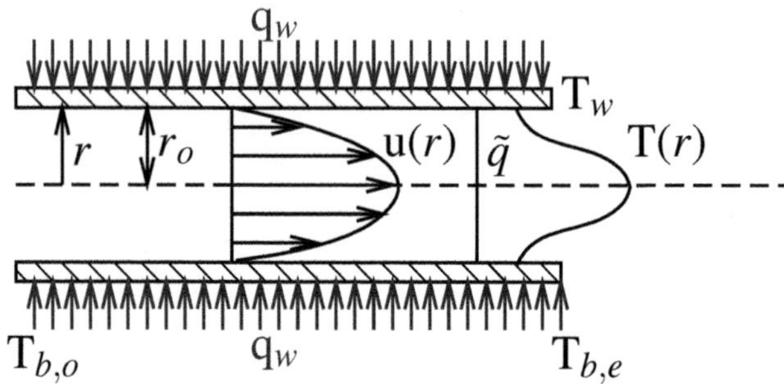

Figure 8.7   The thermally developed flow with constant wall heat flux condition. Note that the flow is experiencing the radial variation in temperature field.

Figure 8.7 shows the case of both hydrodynamically and thermally developed flow.

For laminar flow the velocity profile can be taken from Hagen-Poisuielle flow:

$$\frac{u}{u_o} = \left[1 - \left(\frac{r}{r_o}\right)^2\right]$$

where u_o is the centerline velocity, and r_o is radius at pipe or tube wall. This is the parabolic velocity profile for fully developed laminar flow. The energy equation is

$$u\frac{\partial T}{\partial x} + \cancel{u_r}\frac{\partial T}{\partial r} + \frac{\cancel{u_\theta}}{r}\frac{\partial T}{\partial \theta} = \frac{k}{\rho cp}\left(\frac{\partial^2 T}{\partial r^2} + \frac{1}{r}\frac{\partial T}{\partial r} + \frac{1}{r^2}\frac{\partial^2 T}{\partial \theta^2} + \frac{\partial^2 T}{\partial x^2}\right) + \frac{\tilde{q}}{\rho c_p}$$

$$u\frac{\partial T}{\partial x} = \frac{k}{\rho c_p}\left(\frac{\partial^2 T}{\partial r^2} + \frac{1}{r}\frac{\partial T}{\partial r}\right) + \frac{\tilde{q}}{\rho cp}$$

For for fully developed Thermal field, we can set the derivative of temperature in dimensionless form as zero as:

$$\frac{\partial}{\partial x}\left(\frac{T_w - T}{T_w - T_o}\right) = 0$$

where, T_o is bulk inlet temperature and T_w is the pipe wall temperature. With a little algebra we can prove that

$$= \frac{(T_w - T_o)\left(\frac{\partial T_w}{\partial x} - \frac{\partial T}{\partial x}\right) - (T_w - T)\left(\frac{\partial T_w}{\partial x} - \frac{\partial T_o}{\partial x}\right)}{(T_w - T_o)^2}$$

$$= \frac{1}{(T_w - T_o)}\left(\frac{\partial T_w}{\partial x} - \frac{\partial T}{\partial x}\right) - \left(\frac{T_w - T}{(T_w - T_o)^2}\right)\left(\frac{\partial T_w}{\partial x} - \frac{\partial T_0}{\partial x}\right)$$

$$= (T_w - T_o)\left(\frac{\partial T_w}{\partial x} - \frac{\partial T}{\partial x}\right) - (T_w - T)\left(\frac{\partial T_w}{\partial x} - \frac{\partial T_o}{\partial x}\right)$$

So,

$$\frac{\partial T}{\partial x} = \frac{\partial T_w}{\partial x} - \left(\frac{T_w - T}{T_w - T_o}\right)\left(\frac{\partial T_w}{\partial x} - \frac{\partial T_o}{\partial x}\right) \tag{8.16}$$

Since heat flux at wall is specified q_w = constant Fourier law of heat conduction dictates that:

$$q_w = +k\left(\frac{\partial T}{\partial r}\right)_{r=r_o}$$

The sign in taken as positive since r is measured from center to wall. The dimensionless temperature will be the function of radial coordinate so we can define a function $G(r/r_o)$ as

$$\frac{T_w - T}{T_w - T_o} = G\left(\frac{r}{r_o}\right)$$

Differentiation would lead to

$$\frac{dG}{d(r/r_o)} = \frac{-1}{(T_w - T_o)}\frac{\partial T}{\partial(r/r_o)} \approx \frac{-r_o}{(T_w - T_o)}\frac{\partial T}{\partial r}$$

hence,

$$\frac{\partial T}{\partial r} = \frac{(T_w - T_o)}{r_o}\frac{dG}{d(r/r_o)}$$

and

$$q_w = -\frac{k(T_w - T_o)}{r_o}\frac{dG}{d(r/r_o)}\bigg|_{r=r_o} = \text{Constant}$$

This shows that $T_w - T_o$ is a constant.

$$\frac{dT_w}{dx} = \frac{dT_o}{dx}$$

$$u\frac{\partial T}{\partial x} = \frac{k}{\rho c_p}\left(\frac{\partial^2 T}{\partial r^2} + \frac{1}{r}\frac{\partial T}{\partial r}\right) + \frac{\tilde{q}}{\rho c_p} \tag{8.17}$$

We can write energy equation as

$$\frac{\partial T}{\partial x} = \frac{dT_w}{dx}$$

since $\dfrac{dT_w}{dx} = \dfrac{dT_o}{dx}$ we have

$$u\frac{dT_w}{dx} = \frac{k}{\rho cp}\left(\frac{\partial^2 T}{\partial r^2} + \frac{1}{r}\frac{\partial T}{\partial r}\right) + \frac{\tilde{q}}{\rho c_p}$$

$$u\frac{dT_w}{dx} = \frac{k}{\rho cp}\left(\frac{1}{r}\frac{\partial}{\partial r}\left(r\frac{\partial T}{\partial r}\right)\right) + \frac{\tilde{q}}{\rho c_p}$$

From Hagen-Poiseuille flow we have

$$\frac{u}{u_o} = \left[1 - \left(\frac{r}{r_o}\right)^2\right]$$

where, $r^* = r/r_o$

$$u_o\left[1 - \left(\frac{r}{r_o}\right)^2\right]\frac{dT_w}{dx} = \frac{k}{\rho c_p}\left(\frac{r_o}{r_o^3(r/r_o)}\frac{\partial}{\partial(r/r_o)}\left((r/r_o)\frac{\partial T}{\partial(r/r_o)}\right)\right) + \frac{\tilde{q}}{\rho c_p}$$

Multiplying both sides with $r_o^2 r^*$ we have

$$u_o\left[1 - r^{*2}\right]\frac{dT_w}{dx}r_o^2 r^* = \frac{k}{\rho c_p}\left(\frac{\partial}{\partial r^*}\left(r^*\frac{\partial T}{\partial r^*}\right)\right) + \frac{\tilde{q}}{\rho c_p}r_o^2 \cdot \frac{r}{r_o}$$

Multiplying both sides with $\dfrac{\rho c_p}{k}$:

$$u_o\frac{\rho c_p}{k}\left[r^* - r^{*3}\right]\frac{dT_w}{dx}r_o^2 = \frac{\partial}{\partial r^*}\left(r^*\frac{\partial T}{\partial r^*}\right) + r_o^2\frac{\rho c_p}{k}\cdot\frac{\tilde{q}r^*}{\rho c_p}$$

$$Pr = \frac{c_p\mu}{k} \qquad \text{and } \alpha = \frac{k}{\rho c_p} \qquad \text{and } Pr = \frac{\nu}{\alpha}$$

$$\frac{Pr}{\nu}u_o r_o^2\left[r^* - r^{*3}\right]\frac{dT_w}{dx} - r_o^2 r^*\frac{\tilde{q}}{k} = \frac{\partial}{\partial r^*}\left(r^*\frac{\partial T}{\partial r^*}\right)$$

Boundary conditions:

i At $r = 0$ $\quad \dfrac{\partial T}{\partial r} = 0,$

ii $r = r_o$ $\quad T = T_w.$

We now simplify the algebra by combining the constants together and calling them C_1 and C_2.

$$\frac{Pr}{v} u_o r_o^2 \frac{dT_w}{dx_\sim} = \text{Constant} = C_1$$

$$r_o^2 \frac{\dot{q}}{k} = \text{Constant} = C_2$$

$$C_1 \left[r^* - r^{*3} \right] - r^* C_2 = \frac{\partial}{\partial r^*} \left(r^* \frac{\partial T}{\partial r^*} \right)$$

We now integrate this equation

$$C_1 \left[\frac{r^{*2}}{2} - \frac{r^{*4}}{4} \right] - C_2 \frac{r^{*2}}{2} + A = r^* \frac{\partial T}{\partial r^*}$$

where, A is a constant of integration.

$$C_1 \left[\frac{r^*}{2} - \frac{r^{*3}}{4} \right] - \frac{C_2}{2} r^* + \frac{A}{r^*} = \frac{\partial T}{\partial r^*}$$

from boundary conditions: $\dfrac{\partial T}{\partial r} = 0$ at $r = 0$. This makes the constant of integration as zero. At $r^* = 0$ we have $\dfrac{\partial T}{\partial r^*} = 0$. This gives A=0.

We now integrate again

$$T = C_1 \left[\frac{r^{*2}}{4} - \frac{1}{4} \frac{r^{*4}}{4} \right] - \frac{C_2}{2} \frac{r^{*2}}{2} + B$$

where, B is constant of integration.

$$T = C_1 \left[\frac{r^{*2}}{4} - \frac{r^{*4}}{16} \right] - C_2 \frac{r^{*2}}{4} + B$$

at $r = r_o$ (or $r^* = 1$), $\quad T = T_w$.

$$T_w = C_1 \left[\frac{1}{4} - \frac{1}{16} \right] - C_2 \frac{1}{4} + B$$

$$T_w = C_1 \left[\frac{4-1}{16} \right] - \frac{C_2}{4} + B$$

$$B = -\frac{3C_1}{16} + \frac{C_2}{4} + T_w$$

hence,

$$T = C_1 \left[-\frac{3}{16} + \frac{r^{*2}}{4} - \frac{r^{*4}}{16} \right] - C_2 \left[\frac{r^{*2}}{4} - \frac{1}{4} \right] + T_w$$

$$T = T_w - C_1 \left[\frac{3}{16} - \frac{r^{*2}}{4} + \frac{r^{*4}}{16} \right] + \frac{C_2}{4} \left[1 - r^{*2} \right] \tag{8.18}$$

Since,

$$q_w = k \left. \frac{\partial T}{\partial r} \right|_{r=r_o} = \frac{k}{r_o} \left. \frac{\partial T}{\partial r^*} \right|_{r^*=1}$$

$$\frac{\partial T}{\partial r^*} = C_1 \left[\frac{r^*}{2} - \frac{r^{*3}}{4} \right] - C_2 \frac{r^*}{2}$$

$$\left. \frac{\partial T}{\partial r^*} \right|_{r^*=1} = C_1 \left[\frac{1}{2} - \frac{1}{4} \right] - \frac{C_2}{2} = \frac{C_1}{4} - \frac{C_2}{2}$$

We obtain heat flux as

$$q_w = \frac{k}{r_o} \left[\frac{Pr}{v} u_o r_o^2 \frac{dT_w}{dx} \frac{1}{4} - \frac{r_o^2 \tilde{q}}{2k} \right]$$

By little rearrangement of above equation, we find a new expression for C_1 in terms of heat flux as

$$C_1 = \frac{Pr}{v} u_o r_o^2 \frac{dT_w}{dx} = 4 \left[\frac{q_w r_o}{k} + \frac{\tilde{q} r_o^2}{2k} \right] \tag{8.19}$$

We will now estimate the mean temperature of the laminar flows as:

$$T_m = \frac{1}{\dot{m}} \int_A \rho u T \, dA = \frac{1}{\pi r_o^2 \cdot \rho u_m} \int_0^{r_o} \rho u T 2\pi r \, dr$$

$$= \int \frac{2u_o}{r_o^2 u_m} \left\{ T_w - C_1 \left[\frac{3}{16} - \frac{r^{*2}}{4} + \frac{r^{*4}}{16} \right] + \frac{C_2}{4} \left[1 - r^{*2} \right] \right\} \left[1 - r^{*2} \right] r \, dr$$

$$= \int_0^{r_o} \frac{4}{r_o^2} \left\{ T_w - C_1 \left[\frac{3}{16} - \frac{r^{*2}}{4} + \frac{r^{*4}}{16} \right] + \frac{C_2}{4} \left[1 - r^{*2} \right] \right\} \left[1 - r^{*2} \right] r \, dr$$

$$\approx \int_0^1 4 \left\{ T_w - C_1 \left[\frac{3}{16} - \frac{r^{*2}}{4} + \frac{r^{*4}}{16} \right] + \frac{C_2}{4} \left[1 - r^{*2} \right] \right\} \left[1 - r^{*2} \right] r^* dr^*$$

$$T_m = \left\{ \frac{T_w}{4} - \frac{11}{384} C_1 + \frac{C_2}{24} \right\} 4$$

$$T_m = T_w - \frac{11}{96} C_1 + \frac{C_2}{6}$$

This gives us the difference between the wall and bulk mean temperature as

$$T_w - T_m = \frac{11}{96} C_1 - \frac{C_2}{6} \tag{8.20}$$

Now substituting new expressions for C_1 and C_2 as

$$C_1 = 4q_w\frac{r_o}{k} + 4\frac{\tilde{q}r_o^2}{2k}$$

$$C_2 = \frac{\tilde{q}r_o^2}{k}$$

We arrive at result:

$$T_w - T_m = \frac{11}{96}\left[4q_w\frac{r_o}{k} + \frac{4}{2}\frac{\tilde{q}r_o^2}{k}\right] - \frac{1}{6}\frac{\tilde{q}r_o^2}{k}$$

$$T_w - T_m = \frac{11}{24}\frac{q_w r_o}{k} + \frac{1}{16}\frac{\tilde{q}r_o^2}{k}$$

Invoking, Newton's law of cooling, the expression for convective heat transfer coefficient is

$$h = \frac{q_w}{T_w - T_m} = \frac{q_w}{\frac{11}{24}\frac{q_w r_o}{k} + \frac{1}{16}\frac{\tilde{q}r_o^2}{k}}$$

$$h = \frac{1}{\frac{11}{24}\left(\frac{r_o}{k}\right) + \frac{1}{16}\frac{\tilde{q}r_o^2}{q_w k}} = \frac{1}{\left(\frac{r_o}{k}\right)\left[\frac{11}{24} + \frac{r_o}{16}\left(\frac{\tilde{q}}{q_w}\right)\right]}$$

The Nusselt number based on diameter is

$$Nu_D = h\left(\frac{2r_o}{k}\right) = \frac{1}{\frac{11}{24}\left(\frac{r_o}{k}\right)\cdot\left(\frac{k}{2r_o}\right) + \frac{1}{16}\cdot\frac{r_o^2}{k}\left(\frac{\tilde{q}}{q_w}\right)\left(\frac{k}{2r_o}\right)}$$

$$\boxed{Nu_D = \frac{1}{\frac{11}{48} + \frac{r_o}{32}\left(\frac{\tilde{q}}{q_w}\right)}}$$

If there is no internal energy conversion into thermal energy i.e. $\tilde{q}=0$, we have

$$\boxed{Nu_{D,H} = \frac{h\cdot D}{k} = \frac{48}{11} = 4.363} \tag{8.21}$$

This is the Nusselt number for the case of constant wall heat flux. For the constant wall temperature conditions the Nusselt number can be computed from

$$\boxed{Nu_{D,T} = \frac{h\cdot D}{k} = 3.66} \tag{8.22}$$

Example 8.5

Consider fully-developed laminar internal flow of a fluid having thermal conductivity value of 6 W/m·K. The inner diameter of tube is 25 mm and the fluid is experiencing volumetric energy conversion of 20 W/m^3. Find the difference between wall and mean temperature, if heat flux of 1000 W/m^2 is applied on the walls.

Solution We use the equation developed in this section:

$$T_w - T_m = \frac{11}{24}\frac{q_w r_o}{k} + \frac{1}{16}\frac{\tilde{q} r_o^2}{k}$$

$$T_w - T_m = \frac{11}{24}\left(\frac{1000 \times 0.0125}{6}\right) + \frac{1}{16}\left(\frac{20 \times (0.0125)^2}{6}\right)$$

$$T_w - T_m = 0.95489 \text{ K}$$

Shah and London (1975) tabulate Nusselt number for fully developed laminar flow through different flow cross-sections. The Nusselt number based on hydraulic diameter are calculated by Shah and London have reported the Nusselt numbers for for top and bottom wall at constant wall heat flux and constant wall temperature boundary conditions. Their results are listed here in Table 8.2.

TABLE 8.2
Nusselt Number for Constant Lower Wall Heat Flux (H,1), Constant Upper Wall Heat Flux (H,2), and Constant Wall Temperature Boundary Conditions for Various Cross Sections

Cross-section ($L/D_h > 100$)		$Nu_{H,1}$	$Nu_{H,2}$	Nu_T	Re_h
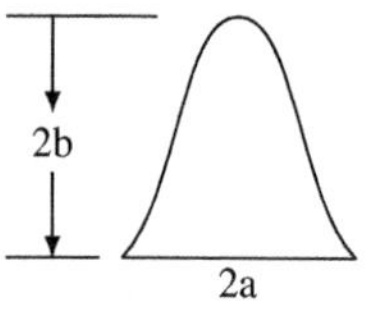	$\frac{2b}{2a} = \frac{\sqrt{3}}{2}$	3.014	1.474	2.39*	12.630
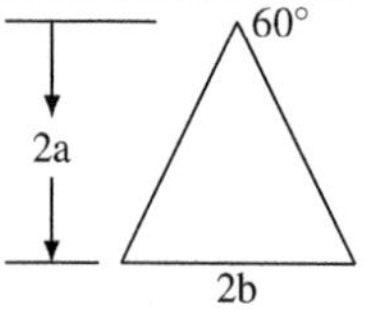	$\frac{2b}{2a} = \frac{\sqrt{3}}{2}$	3.111	1.892	2.47	13.333
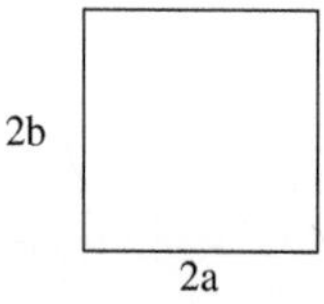	$\frac{2b}{2a} = 1$	3.608	3.091	2.976	14.227

(hexagon)	4.002	3.862	3.34*	15.054
$2b$ (rectangle) $2a$, $\dfrac{2b}{2a}=\dfrac{1}{2}$	4.123	3.017	3.391	15.548
(circle)	4.364	4.364	3.657	16.000
$2b$ (ellipse) $2a$, $\dfrac{2b}{2a}=.9$	5.099	4.35*	3.66	18.700
$2b$ (rectangle) $2a$, $\dfrac{2b}{2a}=\dfrac{1}{4}$	5.331	2.930	4.439	18.233
$2b$ (rectangle) $2a$, $\dfrac{2b}{2a}=\dfrac{1}{8}$	6.490	2.904	5.597	20.585
$\dfrac{2b}{2a}=0$	8.235	8.235	7.541	24.000
$\dfrac{b}{a}=0$, insulated	5.385	-	4.861	24.000

Example 8.6

Develop an expression for temperature distribution in case of fluid heated inside a duct with constant wall heat flux (q_w). The bulk inlet temperature is $T_{b,1}$ and the bulk outlet temperature is $T_{b,2}$. The perimeter of duct is P and corss-sectional area of duct is A.

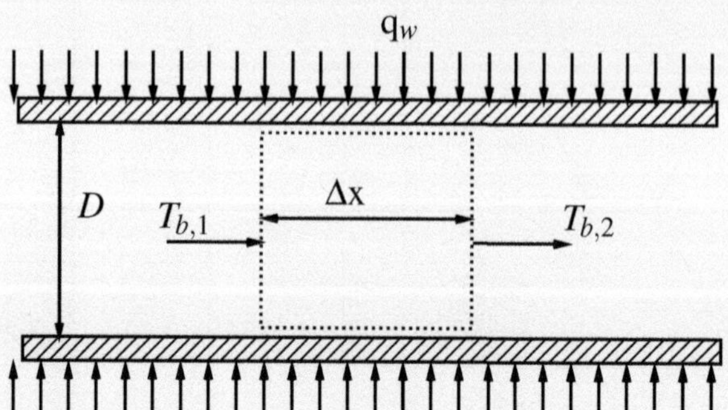

Figure 8.8   Control volume used for heat transfer analysis at constant heat flux conditions.

Solution Considering the control volume shown in Figure 8.8, we estimate the bulk temperature at the exit of control volume. The temperature $T_{b,2}$ can be estimated using Taylor series:

$$T_{b,2} = \left(T_{b,1} + \frac{dT_b}{dx}\Delta x \right)$$

We use the bulk heat transfer relation:

$$Q = \dot{m}\, c_p \left[\left(T_{b,1} + \frac{dT_b}{dx}\Delta x \right) - T_{b,1} \right] = q_w \cdot dA$$

$$\dot{m}\, c_p \left[\frac{dT_b}{dx}\Delta x \right] = q_w \cdot P \cdot \Delta x$$

where, P is the perimeter. Rearrangement gives

$$\dot{m}\, c_p \left[\frac{dT_b}{dx} \right] = q_w \cdot P$$

Since heat-flux, perimeter, area, and mass flow rate are constant, we can see that gradient of temperature is also constant.

$$\frac{dT_b}{dx} = \frac{q_w \cdot P}{\dot{m} \cdot c_p} = const$$

Integrating above equation, we have the temperature distribution:

$$\boxed{T_{b,2} = \left(\frac{q_w \cdot P}{\dot{m} \cdot c_p} \right) \Delta x + T_{b,1}}$$

8.6 THE BULK ENERGY BALANCE IN A PIPE AT CONSTANT WALL TEMPERATURE

We will noe develop an expression for temperature distribution and heat transfer in case of fluid heated inside a duct with constant wall temperature (T_w). The bulk inlet temperature is $T_{b,1}$ and the bulk outlet temperature is $T_{b,2}$. The perimeter of duct is P and cross-sectional area of duct is A. The area of pipe wall is A_w.

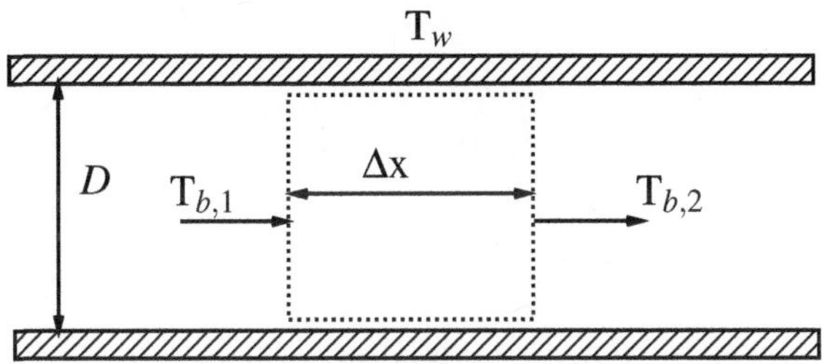

Figure 8.9   Control volume used for heat transfer analysis at constant wall temperature condition.

Considering the control volume shown in Figure 8.9, we do the balance of energy using the Newton's law of cooling:

$$Q = \dot{m}\, c_p \left[\left(T_b + \frac{dT_b}{dx}\Delta x \right) - T_b \right] = h \cdot P \cdot \Delta x \cdot (T_w - T_b)$$

here, P is perimeter and h is convective heat transfer coefficient.

$$\dot{m}\, c_p \frac{dT_b}{dx}\Delta x = h \cdot P \cdot \Delta x \cdot (T_w - T_b)$$

$$\dot{m}\, c_p \frac{dT_b}{dx} = h \cdot P \cdot (T_w - T_b)$$

$$\frac{dT_b}{(T_w - T_b)} = \frac{h \cdot P}{\dot{m}\, c_p} \cdot dx$$

For internal flows, h, A, $\dot{m}$ and perimeter are constant so we set them as a constant.

$$\beta = \frac{h \cdot P}{\dot{m}\, c_p}$$

$$\frac{dT_b}{(T_w - T_b)} = \beta \cdot dx$$

We do substitution

$$\vartheta = T_w - T_b$$
$$d\vartheta = -dT_b$$

This helps in casting the equation

$$\frac{-dT_b}{(T_w - T_b)} = \beta \cdot dx$$

$$\frac{d\vartheta}{\vartheta} = -\beta \cdot dx$$

Integrating on both sides

$$\int_{\vartheta,1}^{\vartheta,2} \frac{d\vartheta}{\vartheta} = -\beta \int_{x_1}^{x_2} dx$$

$$\ln \vartheta_2 - \ln \vartheta_1 = -\beta (x_2 - x_1)$$

$$\ln(T_w - T_{b,2}) = -\beta (x_2 - x_1) + \ln(T_w - T_{b,1})$$

$$\ln\left(\frac{T_w - T_{b,2}}{T_w - T_{b,1}}\right) = -\beta \cdot (x_2 - x_1)$$

This gives

$$\boxed{\frac{T_w - T_{b,2}}{T_w - T_{b,1}} = \exp\left[-\left(\frac{h \cdot P}{\dot{m}\, c_p}\right) \cdot (x_2 - x_1)\right]} \tag{8.23}$$

We can also cast this equation in another form by inserting $P = \pi D$ and $\dot{m} = \rho \cdot A \cdot u$ and this gives

$$\ln\left(\frac{T_w - T_{b2}}{T_w - T_{b1}}\right) = -\left(\frac{h \cdot P}{\dot{m} \cdot c_p}\right)\Delta x = -\left(\frac{h \cdot \pi D}{\rho \cdot u \cdot \frac{\pi}{4} \cdot D^2 \cdot c_p}\right)\Delta x$$

$$\ln\left(\frac{T_w - T_{b2}}{T_w - T_{b1}}\right) = -\left(\frac{4h}{\rho \cdot u \cdot D \cdot c_p}\right)\Delta x$$

where, u is the mean velocity. Since $\alpha = k/\rho c_p$, we can substitute this in above equation:

$$\ln\left(\frac{T_w - T_{b2}}{T_w - T_{b1}}\right) = -\left(\frac{4\alpha h}{k \cdot u \cdot D}\right)\Delta x$$

$$\boxed{\left(\frac{T_w - T_{b2}}{T_w - T_{b1}}\right) = \exp\left[-\left(\frac{4\alpha h}{k \cdot u \cdot D}\right)\Delta x\right]} \tag{8.24}$$

We now rearrange the above equation and write the mean velocity as

$$\boxed{u = \frac{-\left(\frac{4\alpha h}{k \cdot D}\right)\Delta x}{\ln\left(\frac{T_w - T_{b2}}{T_w - T_{b1}}\right)}} \tag{8.25}$$

From bulk temperatures we can calculate the heat transfer as

$$Q = \dot{m} \cdot c_p (T_{b2} - T_{b1})$$

Adding and subtracting T_w we have

$$Q = \rho \cdot u \cdot \frac{\pi}{4} \cdot D^2 \cdot c_p \left[(T_w - T_{b1}) - (T_w - T_{b2}) \right]$$

Inserting mean velocity in above equation we get

$$Q = \rho \left(-\left(\frac{\alpha h}{k} \right) \Delta x \right) \cdot \pi \cdot D \cdot c_p \frac{\left[(T_w - T_{b1}) - (T_w - T_{b2}) \right]}{\ln \left(\frac{T_w - T_{b2}}{T_w - T_{b1}} \right)}$$

and

$$Q = -h \cdot A_w \left\{ \frac{\left[(T_w - T_{b1}) - (T_w - T_{b2}) \right]}{\ln \left(\frac{T_w - T_{b2}}{T_w - T_{b1}} \right)} \right\}$$

$$Q = h \cdot A_w \left\{ \frac{\left[(T_w - T_{b2}) - (T_w - T_{b1}) \right]}{\ln \left(\frac{T_w - T_{b2}}{T_w - T_{b1}} \right)} \right\}$$

where, A_w is the wall area. The bracketed term above equation is known as log mean temperature difference (LMTD) in heat transfer literature.

$$LMTD = \frac{\left[(T_w - T_{b2}) - (T_w - T_{b1}) \right]}{\ln \left(\frac{T_w - T_{b2}}{T_w - T_{b1}} \right)} \qquad (8.26)$$

and finally the heat transfer equation is

$$\boxed{Q = h \cdot A_w \cdot LMTD} \qquad (8.27)$$

8.7 HYDRODYNAMIC ENTRANCE LENGTH IN TURBULENT FLOWS

Turbulent boundary layer development in the entrance region of the pipe takes longer than the laminar boundary length development inside the entrance region. The rough estimate is that the turbulent entrance length is from 70 to 100 times the pipe diameter. Following are some of the correlation in Literature:

Turbulent Flow Entrance Lengths Estimation

$$\left(\frac{L_e}{D} \right)_{turb} \simeq 0.693 \mathrm{Re}_D^{1/4} \qquad Latzko \ (1944)$$

$$\left(\frac{L_e}{D} \right)_{turb} \simeq 1.359 \mathrm{Re}_D^{1/4} \qquad Zhi-qing(1982)$$

$$\left(\frac{L_e}{D}\right)_{turb} \simeq 4.4\mathrm{Re}_D^{1/6} \qquad Nikuradse(1933)$$

$$\left(\frac{L_e}{D}\right)_{turb} \simeq 14.2\log_{10}\mathrm{Re} - 46 \quad \mathrm{Re} > 10,000 \quad Bowlus \ \& \ Brighton(1968)$$

For engineering estimates the turbulent flow can be considered as fully developed if

$$\left(\frac{L_t}{D}\right)_{turb} > 10$$

8.8 HYDRODYNAMICALLY DEVELOPED TURBULENT FLOW THROUGH PIPE

The correlation for friction factor for turbulent flow in a smooth pipe are reported by many researchers.

Blasius correlation:

$$f = \frac{0.3164}{\mathrm{Re}^{1/4}} \qquad (\mathrm{Re} = 4 \times 10^3 - 2 \times 10^5)$$

Nikuradse correlation I

$$f = 0.0032 + \frac{0.221}{\mathrm{Re}^{0.237}} \qquad (\mathrm{Re} = 10^5 - 3 \times 10^6)$$

Karman-Nikuradse correlation

$$f = \frac{1}{\left[2\log_{10}(\mathrm{Re}\sqrt{f}) - 0.81\right]^2} \qquad (\mathrm{Re} = 3 \times 10^3 - 3 \times 10^6)$$

Itaya correlation

$$f = \frac{0.314}{0.7 - 1.65\log_{10}(\mathrm{Re}_D) + (\log_{10}\mathrm{Re})^2}$$

Nikuradse correlation II

$$\frac{1}{\sqrt{f}} = 4\log(\mathrm{Re}\sqrt{f}) - 0.4$$

McAdams correlation

$$f \simeq \frac{0.184}{\text{Re}^{1/5}} \qquad (\text{Re} = 3 \times 10^4 - 10^6)$$

Bhatti and Shah correlations

$$C_f = A + \frac{B}{\text{Re}_D^{1/m}}$$

Flow	A	B	m
Laminar	0	16	1
Transitional Flow	0.0054	2.30×10^{-8}	-0.666
Turbulent Flow	1.28×10^{-3}	0.1143	3.2154

Techo, Tickner, and James Correlation

$$\frac{1}{\sqrt{C_f}} = 1.7372 \ln \left[\frac{\text{Re}}{1.964 \ln(\text{Re}) - 3.8215} \right]$$

This correlation is defined using the Fanning's definition of the friction factor but it covers the wide range of Reynolds number.

The pipe is considered smooth as long as the roughness is submerged in the viscous sublayer. If the wall roughness is large and it protrude out of the viscous sublayer, pipe is considered as rough. The flow in pipe can be divided into three different flow regimes:

i **Hydraulically Smooth:** In this regime, the roughness is fully submerged inside the viscous sublayer and thus roughness is not influencing the flow.

ii **Hydraulically Rough:** In this regime, the roughness is not influencing the flow and friction factor is solely the function of Reynolds number.

iii **Transitional Roughness:** In this regime, the roughness is influencing the flow and friction factor is function of both Reynolds number and roughness ratio. This regime is also sometimes referred as transitional flow and care should be exercised to avoid confusion to mix it up with laminar-to-turbulent transition.

Table 8.3 lists the wall roughness for different materials. Some of the widely used Darcy Friction Factors for turbulent flow through rough pipes are:

Colebrook-White correlation

$$\frac{1}{\sqrt{f}} = 1.74 - 2\log_{10} \left[\frac{2\varepsilon}{D} + \frac{18.7}{\text{Re}\sqrt{f}} \right]$$

TABLE 8.3

Roughness (ε) for Some Pipe Materials

Description	Condition	ε (mm)
Seamless metal	New drawn Copper, brass or lead	0.0015-0.01
	New drawn Aluminium	0.015-0.06
	Commercial steel – light rust	0.02-0.1
	Commercial steel – mildly rusted	0.15-1
	Commercial Steel – heavily rusted	>5
Welded Steel	New, smooth, enemaled surface	0.006-0.06
	Very poor condition	10-3.5
Iron	New wrought	0.045
	Cast	0.1-1
	Cast, corroded	2-4
Sheet metal	Ducts with smooth joints	0.02-0.1
Concrete	Smooth	0.025-0.18
	Rough	2.5-9
Rock	Tunnel	600
Glass/Plastic		0.0015-0.01
Rubber	Smooth tube	0.006-0.07
Ceramic		1.4

Jain correlation (1976)

$$f = \left[1.14 - 2\log_{10}\left(\frac{\varepsilon}{D} + \frac{21.25}{\text{Re}^{0.9}} \right) \right]^{-2}$$

Haaland correlation (1983) gave a correlation which can wide range of flow types:

$$\frac{1}{\sqrt{f}} = \frac{-1.8}{n}\log\left[\left(\frac{6.9}{\text{Re}} \right)^n + \left(\frac{\varepsilon}{3.75D} \right)^{1.11n} \right]$$

where n= 3 is suitable for natural gas pipelines and n=1 is good for abrupt transition cases.

The pressure loss (or major loss) in a pipe, tube or duct can be calculated using an empirical equation:

$$\Delta p = f \left(\frac{\ell}{D} \right) \left(\frac{\rho \bar{u}^2}{2} \right)$$

The equation is named after French engineer Henry Philibert Gaspard Darcy and German engineer Julius Ludwig Weisbach. Here ℓ is the pipe length, D is the diameter of the pipe, $\bar{u}$ is the mean velocity of the flow. The quantity Δp is also called the major loss in literature. The minor contribution comes from the bends, elbows, fittings, expansions, and contraction, etc.

Figure 8.10 shows the Moody diagram for turbulent flow in rough pipes. The smooth pipe line is along the $\varepsilon = 0.00001$ D. For smooth pipe McAdams and Blasisus correlations are used. For rough pipes Jain's correlation is used. Note that in original Moody diagram for the transitional flow Colebrook-White correlation is used.

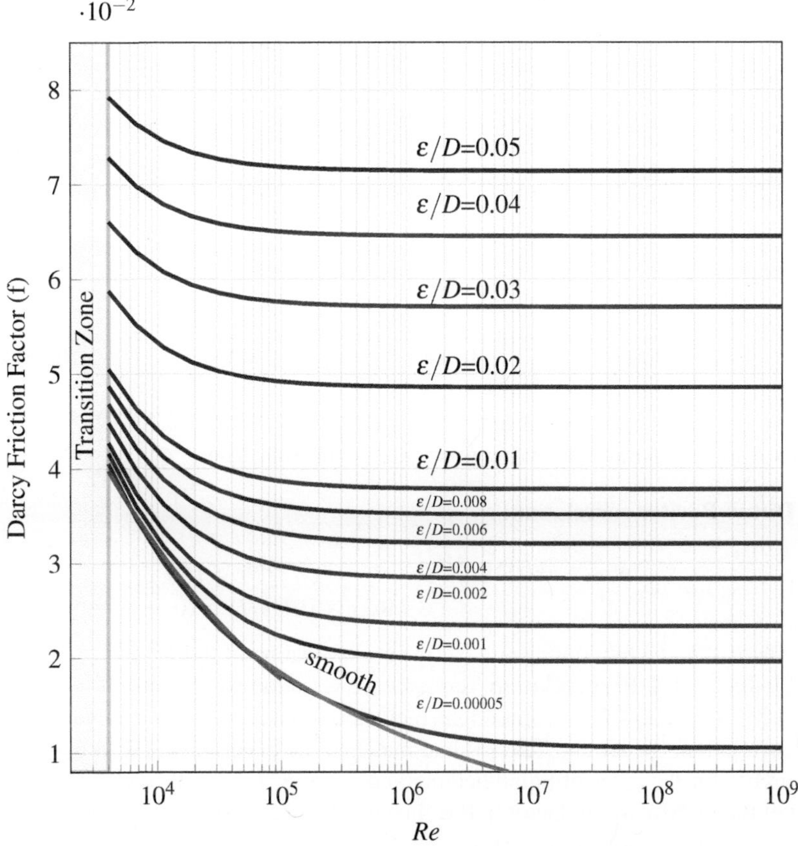

Figure 8.10   The Moody diagram for turbulent flow in rough pipes. The smooth pipe line is along the $\varepsilon = 0.00001$D. For smooth pipe McAdams and Blasisus correlations are used. For rough pipes Jain's correlation is used.

8.9   CONVECTIVE HEAT TRANSFER IN TURBULENT PIPE FLOW

For laminar flows, it is necessary to distinguish between the boundary conditions as the governing equations and results are different for each case. However, it is not necessary to distinguish between the boundary conditions of constant wall temperature and constant wall heat flux for heat transfer in a turbulent flow. Figure 8.11 shows the mean Nusselt number plotted along Reynolds number of the gas flowing in a pipe.

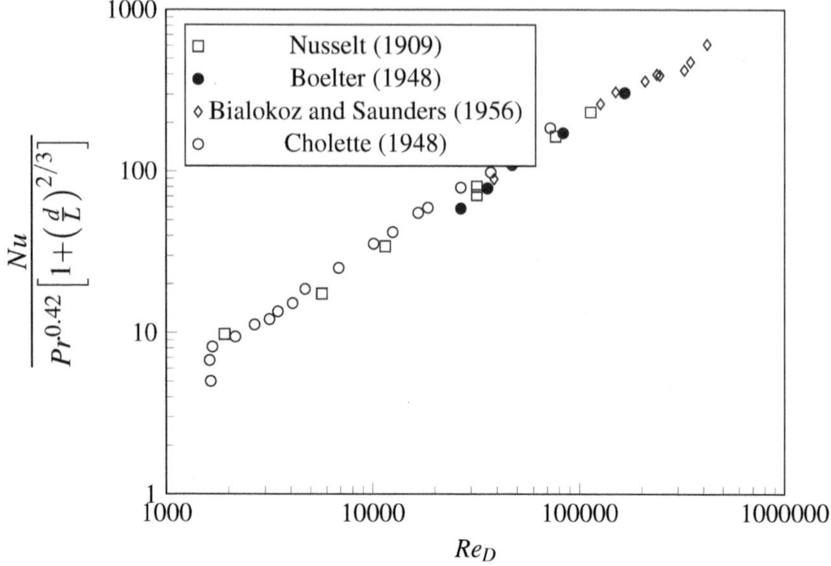

Figure 8.11   Mean Nusselt number versus Re_D for different gases.

Dittus-Boelter correlation:

$$Nu_D = 0.023 Re_D^{4/5} Pr^n \qquad \begin{array}{c} 0.6 \leq Pr \leq 160 \\ Re_D \geq 10^4, \ell/D \geq 10 \end{array} \tag{8.28}$$

where, n is 0.3 for cooling and 0.4 in case of heating of fluid.

Sieder and Tate correlation Temperature variation would cause the change in viscosity. Sieder and Tate proposed the inclusion of viscosity ratio into Nusselt number correlation. Sieder and Tate [3] had matched the experimental data by different researcher for the cases of liquid heating and cooling. They indicated that the inclusion of the viscosity ratio factor in the Nu correlation can help in eliminating the use of different exponents for the Prandtl number.

Figure 8.12 shows the influence of viscosity on the temperature distribution inside pipe.

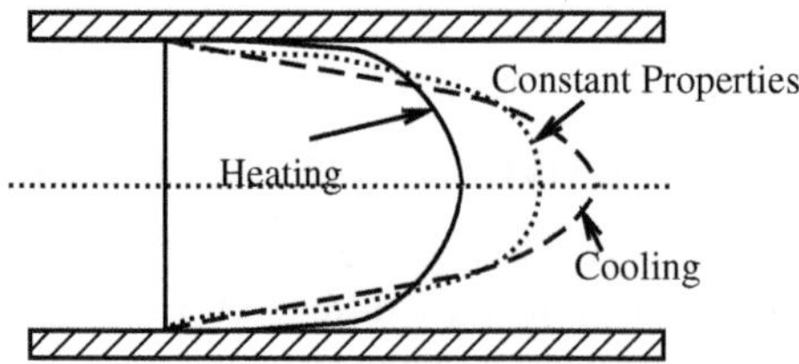

Figure 8.12   The velocity profiles with heating and cooling in a fully developed laminar pipe flow compared with flow profile with constant fluid properties condition.

$$Nu_D = 0.027\mathrm{Re}_D^{4/5}Pr^{1/3}\left(\frac{\mu}{\mu_w}\right)^{0.14} \qquad \begin{array}{l} 0.7 \leq Pr \leq 16700 \\ \mathrm{Re}_D \geq 10^4, \ell/D \geq 10 \end{array} \tag{8.29}$$

In above correlation, all properties except viscosity at wall, are evaluated at mean temperature. The Nusselt number predicted by the above equations can be $\pm 25\%$.

Hausen correlation: In 1959, Hausen proposed heat transfer correlation for turbulent flow in tubes.

$$Nu_D = 0.037(\mathrm{Re}^{0.75} - 180)Pr^{0.42}\left[1 + \left(\frac{d}{\ell}\right)^{2/3}\right]\left(\frac{\mu_b}{\mu_w}\right)^{0.14} \tag{8.30}$$

This correlations includes the influence of developing hydraulic and thermal boundary layers from the beginning of the pipe.

Gnielinski correlation: Gnielinski (1975,1976,1984) proposed the following equation, which he arrived at after modification in the Petukhov's correlation:

$$Nu_D = \frac{\left(\frac{f}{8}\right)Pr\mathrm{Re}}{1 + 12.7(Pr^{2/3} - 1)\sqrt{\left(\frac{f}{8}\right)}}\left[1 + \left(\frac{d}{\ell}\right)^{2/3}\right] \tag{8.31}$$

the correlation is valid for range

$$10^4 \leq \mathrm{Re}_D \leq 10^6$$

$$0.6 \leq Pr \leq 1000$$

where, Konakov (1946) correlation can be used for friction factor (f):

$$f = \frac{1}{[1.8 \cdot \log_{10}(\mathrm{Re}) - 1.5]^2}$$

Gnielinski (2013) further improved it and proposed the following equation:

$$Nu_D = \frac{(f/8)(\mathrm{Re} - 1000)Pr}{1 + 12.7(Pr^{2/3} - 1)\sqrt{(f/8)}}\left[1 + \left(\frac{D}{\ell}\right)^{2/3}\right] \cdot K \tag{8.32}$$

$$K = \left(\frac{Pr_b}{Pr_w}\right)^{0.14} \quad for \quad liquids$$

$$K = \left(\frac{T_b}{T_w}\right)^m \quad for \quad air \quad m = 0.45$$

Seban and Shimazaki correlation: For liquid metals above correlations do not give satisfactory results. Seban and Shimazaki proposed the following correlation for Peclet number greater than and equal to 100.

$$Nu_D = 5 + 0.025Pe_D^{0.8} \qquad T_w = constant \tag{8.33}$$

Example 8.7

Air enters a circular aluminium tube at 80°C with mass flow rate of 0.2 kg/s. The tube is 10 m long with 500 mm as an internal diameter. What will be the exit temperature if tube wall is maintained at 70°C. The properties of air are

$$\rho = 0.999 \ \text{kg/m}^3, \ k = 0.0295 \frac{W}{m \cdot K}, \ \nu = 0.7154 \ m^2/s, \ c_p$$
$$= 1008 \ \text{J/kg} \cdot \text{K}, \ Pr = 0.712$$

Solution

We calculate the cross-sectional area of of tube and compute the velocity of the air as

$$A = \frac{\pi}{4} \cdot d^2 = 0.19634 \text{m}^2$$

$$u = \frac{\dot{m}}{\rho \cdot A} = \frac{0.2}{0.999 \times 0.19634} = 1.019 \ m/s$$

We now compute the Reynolds number

$$Re = \frac{u \cdot d}{\nu} = 2.43 \times 10^4 > 2300$$

hence flow is turbulent. We use Dittus-Boelter correlation to estimate the Nusselt number:

$$Nu_{DB} = 0.023 \cdot Re^{0.8} \cdot Pr^{0.3} = 67.08$$

$$h = 3.961 \frac{W}{m^2 \cdot K}$$

As the wall is maintained at constant temperature, we use the equation 8.23:

$$\frac{(T_w - T_e)}{(T_w - T_i)} = \exp\left(-\frac{h \cdot (\pi \cdot d) \cdot L}{\dot{m} \cdot c_p}\right) = \exp\left(-\frac{3.96 \cdot (\pi \cdot 0.5) \cdot 10}{(0.2) \cdot (1008)}\right) = 0.7344$$

where, T_i=80°C is bulk inlet temperature, T_e is bulk exit temperature, and T_w= 70°C is the wall temperature. Solving this we have the exit temperature as

$$T_e = 77.344°C$$

We estimate the skin friction coefficient:

$$f = \frac{1}{(1.8 \cdot \log_{10}(2E4) - 1.5)^2} = 0.02566$$

We now use Gnielinski correlation to estimate the Nusselt number:

$$Nu_G = \frac{\left(\frac{f}{8}\right)\cdot(Rey-1000)\cdot Pr}{1+12.7\cdot\left(\frac{f}{8}\right)^{\frac{1}{2}}\cdot\left(Pr^{\frac{2}{3}}-1\right)}\cdot\left(1+\left(\frac{d}{L}\right)^{\frac{2}{3}}\right)\cdot\left(\frac{T_i}{T_w}\right)^{0.45} = 75.34$$

$$h = 4.45\frac{W}{m^2\cdot K}$$

As the wall is maintained at constant temperature, we use the equation 8.23:

$$\frac{(T_w-T_e)}{(T_w-T_i)} = \exp\left(-\frac{h\cdot(\pi\cdot d)\cdot L}{\dot{m}\cdot c_p}\right) = \exp\left(-\frac{4.45\cdot(\pi\cdot 0.5)\cdot 10}{(0.2)\cdot(1008)}\right) = 0.7069$$

where, T_i=80°C is bulk inlet temperature, T_e is bulk exit temperature, and T_w= 70°C is the wall temperature. Solving this we have the exit temperature as

$$T_e = 77.069°C$$

8.10   MOMENTUM AND HEAT TRANSFER ANALOGIES

Many earlier researchers before 1960s, have proposed that in turbulent flow a good estimate of convective heat transfer can be arrived at by assuming an analogy between heat and momentum transfer. One of the earliest proposals is attributed to Reynolds, which is defined as

$$St = C_f/2 \tag{8.34}$$

Prandtl proposed the following analogy:

$$St = \frac{\left(\frac{C_f}{2}\right)}{\left[1+5\sqrt{\frac{C_f}{2}}(Pr-1)\right]} \tag{8.35}$$

Von Karman (1939) derived Stanton number expression by consideration of the respective thermal resistances in laminar sublayer, a buffer layer, and a turbulent outer region:

$$St = \frac{C_f/2}{1+5\sqrt{C_f/2}\left[(Pr-1)+\ln\left(1+\frac{5}{6}(Pr-1)\right)\right]} \tag{8.36}$$

Colburn (1933) proposed an analogy as

$$St = \frac{C_f/2}{Pr^{2/3}} \tag{8.37}$$

It is important to note that these analogies have basis in experimental correlation and analysis of heat transfer and flow fields and the results are applicable if the ratio of energy thickness to momentum thickness, is essentially constant in the streamwise direction at a value close to unity. In case of the nozzle flows, these analogies are not applicable.

Pioneers of Heat Transfer

Allan Philip Colburn born in Madison, Wisconsin, USA in 1904. He studied chemistry at University of Wisconsin, and in 1929 he obtained his doctorate. He worked at the University of Delaware from 1938 until his death in 1955.

8.11   HEAT TRANSFER IN TRANSITION ZONE

Gnielinski (2013) proposed the following correlation for the transition heat transfer inside pipe or tube with properties estimated at the bulk mean inlet and outlet temperatures:

$$Nu_{m,trans} = (1 - \gamma)Nu_{m,Lam,2300} + \gamma Nu_{m,Turb,4000} \tag{8.38}$$

where, γ is defined as

$$\gamma = \frac{Re - 2300}{4000 - 2300} \tag{8.39}$$

where, the Reynolds number (Re) is the transitional Reynolds number of the flow.

The $Nu_{m,Lam,2300}$ can be computed for laminar flow based on constant wall temperature and heat-flux boundary conditions as

Constant heat-flux condition:

$$Nu_{m,H,2300} = \sqrt[3]{83.326 + (Nu_{m,H,1} - 0.6)^3 + (Nu_{m,H,2})^3} \tag{8.40}$$

$$Nu_{m,H,1} = 1.953 \left[2300 \cdot Pr \left(\frac{d}{\ell}\right)\right]^{1/3} \tag{8.41}$$

$$Nu_{m,H,2} = 0.924 Pr^{1/3} \left[2300 \cdot Pr \left(\frac{d}{\ell}\right)\right]^{1/2} \tag{8.42}$$

Note that Constant wall temperature condition:

$$Nu_{m,T,2300} = \sqrt[3]{49.371 + (Nu_{m,T,1} - 0.7)^3 + (Nu_{m,T,2})^3} \tag{8.43}$$

$$Nu_{m,T,1} = 1.615 \left[2300 \cdot Pr \left(\frac{d}{\ell} \right) \right]^{1/3} \qquad (8.44)$$

$$Nu_{m,T,2} = \left(\frac{2}{1 + 22 \cdot Pr} \right)^{1/6} \left[2300 \cdot Pr \left(\frac{d}{\ell} \right) \right]^{1/2} \qquad (8.45)$$

Note that in above equations for laminar flow, the quantities are computed at Re=2300. The turbulent flow can be computed from Gnielinski correlation:

$$Nu_{m,Turb,4000} = \frac{\left(\frac{f}{8} \right) Pr(4000 - 1000)}{1 + 12.7(Pr^{2/3} - 1)\sqrt{\left(\frac{f}{8} \right)}} \left[1 + \left(\frac{d}{\ell} \right)^{2/3} \right] \qquad (8.46)$$

where,

$$f = \frac{1}{(1.8\log_{10} 4000 - 1.5)^2} \qquad (8.47)$$

Note that $Nu_{m,Turb}$ is computed at 4000.

Example 8.8

Water with an inlet temperature of 10 °C flows in a tube with an inner diameter of d = 3 mm. The length of pipe is 1000 mm. The speed of water is 0.7 m/s. At the outer surface of pipe the steam is condensing and we can assume it as a constant wall temperature of 100 °C. Find the convective heat transfer coefficient, if the water outlet temperature is 62 °C ? Use Gnielinski correlations for transitional flows.

Solution We take the properties at the mean temperature of 36 °C:
v= 0.711 × 10^{-6} m^2/s; k = 0.623 W/m·K; ρ = 993.68 kg/m^3; c$_p$ = 4179 J/kg·K; Pr = 4.735; Pr$_w$ = 1.76;
The Reynolds number of the flow is

$$Re = \frac{u \cdot d}{v} = 2953.58$$

We compute the laminar flow related equations at constant wall temperature:

$$Nu_{m,T,1} = 1.615 \left[2300 \cdot Pr \left(\frac{d}{\ell} \right) \right]^{1/3} = 5.162$$

$$Nu_{m,T,2} = \left(\frac{2}{1 + 22 \cdot Pr} \right)^{1/6} \left[2300 \cdot Pr \left(\frac{d}{\ell} \right) \right]^{1/2} = 2.953$$

$$Nu_{m,T,2300} = \sqrt[3]{49.371 + (Nu_{m,T,1} - 0.7)^3 + (Nu_{m,T,2})^3} = 5.473$$

We compute the turbulent flow related equations:

$$f = \left(\frac{1}{(1.8 \cdot \log 10\,(4000) - 1.5)^2} \right) = 0.04026$$

$$Nu_{m,Turb,4000} = \frac{\left(\frac{f}{8}\right) \cdot Pr \cdot (4000 - 1000) \cdot \left(1 + \left(\frac{d}{L}\right)^{\frac{2}{3}}\right)}{1 + 12.7 \cdot \left(Pr^{\frac{2}{3}} - 1\right) \cdot \left(\frac{f}{8}\right)^{0.5}} = 27.647$$

$$\gamma = (Re - 2300)/(4000 - 2300) = 0.3844$$

$$Nu_{Trans} = (1 - \gamma)Nu_{m,Lam,2300} + \gamma Nu_{m,Turb,4000} = 13.99$$

$$K = \left(\frac{Pr}{Pr_w} \right)^{0.111} = 1.116$$

$$Nu_{Trans,corr} = Nu_{Trans} \cdot K = 15.624$$

$$h = Nu_{Trans,corr} \cdot k/d = 3244.65\,\frac{W}{m \cdot K}$$

Note that the flow Reynolds number is only used in the estimation of γ parameter. Using equation 8.27 we can estimate the heat transfer:

$$Q = h \cdot A_w \cdot LMTD$$

where, A_w is the heated wall area. We compute LMTD = 60.309 °C and this gives the heat transfer as

$$Q = (3244.65\,\frac{W}{m^2 \cdot K}) \cdot (0.00942m^2) \cdot (60.309°C) = 1843.329W$$

Levenspiel (2014) proposed the correlation for the transition range:

$$Nu_D = 0.116 Pr^{1/3} \left[Re_D^{2/3} - 125 \right] \left[1 + \left(\frac{d}{\ell}\right)^{2/3} \right] \left(\frac{\mu}{\mu_w}\right)^{0.14} \qquad 2100 < Re_D < 10{,}000$$

8.12 HEAT TRANSFER AND PRESSURE LOSS IN A COILED TUBE

In order to increase heat transfer, the flow is sometimes moved through a coiled tube. This leads to the increase in the secondary velocities and thus overall heat transfer would be increased. Figure 8.13 shows that due to the centrifugal forces, a secondary flow will be generated which creates the counter rotating flow in the crosssection of the pipe. The large-scale vortices appear perpendicular to the axial-flow direction and will enhance the heat transport.

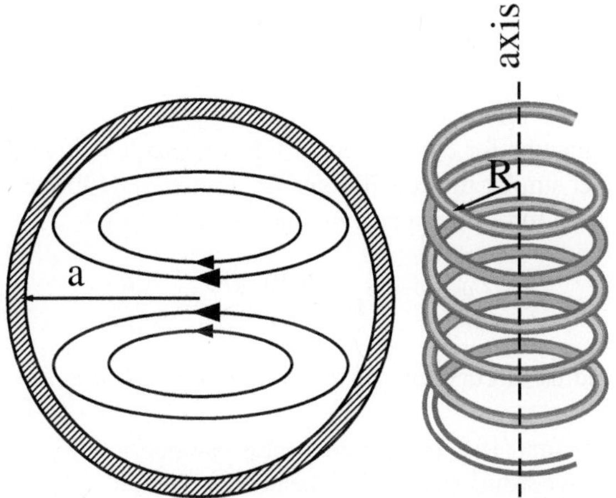

Figure 8.13 Schematic representation of secondary flow in a pipe. The geomtry of the helical coiled tube.

However, there is an associated pressure loss which can be expressed using the Dean number, defined as

$$De = \text{Re}\sqrt{\left(\frac{a}{R}\right)}$$

where, a is the radius of the coiled tube, and R is the radius of curvature of the coil itself.

Srinivasan and Nandapukar (1970) proposed the following correlation for friction factor in coiled tubes compare with straight tubes

$$\frac{f_c}{f_s} = \begin{cases} 1 & De < 30 \\ 0.419De^{0.275} & 30 < De < 300 \\ 0.1125De^{0.5} & De > 300 \end{cases}$$

This is valid for the range $7 < R/a < 104$.

For the coiled tube, we can no longer use $\text{Re}_{crit} = 2300$, instead the upper critical Reynolds number must also be calculated as

$$\text{Re}_{crit} = 2100\left[1 + 12\left(\frac{R_{min}}{a}\right)^{-0.5}\right]$$

where, R_{min} is the minimum radius of curvature of a coil.

For heat transfer through the coiled helical pipe, Schimdt (1967) proposed the correlation:

$$\frac{Nu_{coil}}{Nu_o} = 1 + 3.6\left(\frac{a}{R}\right)^{0.8}\left[1 - \left(\frac{a}{R}\right)\right] \qquad \begin{array}{l} 5 < R/a < 84 \\ 2 \times 10^4 < \text{Re} < 1.5 \times 10^5 \end{array}$$

where, Nu_o is Nusselt number inside straight pipe. Pratt (1947) recommended the correlation for water flowing in helical coil at low Reynolds number

$$\frac{Nu_{coil}}{Nu_o} = 1 + 3.4\left(\frac{a}{R}\right) \qquad 1.5 \times 10^3 < \text{Re} < 2 \times 10^4$$

Note that above correlations are for the thermally developed flow. In general, the thermal entrance lengths for curved/helical tubes is much shorted than that for the straight tubes under similar flow conditions. Acharya et al. (1964) recommended the following correlation for estimation of thermal entrance length:

$$\frac{L_{th}}{d_{coiled}} = 1.66 + \frac{0.102 Re^{1/4}}{Pr^{0.5}}\sqrt{\frac{2R}{d}}$$

where, R is the radius of curvature of the coil, and $d = 2 \cdot a$ is the diameter of the tube.

Mori and Nakayama (1967) proposed the correlation for fluids with Prandtl number close to unity. According to them the mean Nusselt number based on diameter of coil curvature can be estimated as

$$Nu_{m,2R} = \frac{Pr \cdot Re_D^{4/5}}{26.2(Pr^{2/3} - 0.074)}\left(\frac{D}{d_c}\right)^{1/10}\left[1 + 0.098\left\{Re_D\left(\frac{D}{d_c}\right)^2\right\}^{1/5}\right]$$

and for liquids their proposed correlation is

$$Nu_{m,2R} = \frac{Pr^{0.4} \cdot Re_D^{5/6}}{41}\left(\frac{D}{d_c}\right)^{1/12}\left[1 + 0.61\left\{Re_D\left(\frac{D}{d_c}\right)^{2.5}\right\}^{1/6}\right]$$

D is twice the radius of curvature, and d_c=2a is the diameter of the coiled tube.

Example 8.9

An organic fluid (kinematic viscosity v=3E-06 m^2/s, k= 3 W/m·K, and Pr=12) is flowing in a coiled smooth tube of diameter 10 cm. The constant radius of curvature of coil is 80 cm and the total length of the coiled tube is 50 m. If the volume flow rate is $\dot{V}$=0.03 m^3/s, Find the head loss for flow through the coiled tube and heat transfer if organic fluid is cooled from 500K to 300K. The coiled tube is maintained at temperature of 298K.

Solution We calculate the velocity of flow through the tube as

$$u = \frac{\dot{V}}{A} = \frac{0.03}{\frac{\pi}{4}\left(\frac{10}{100}\right)^2} = 3.82 m/s$$

$$\text{Re} = \frac{u.d}{v} = \frac{3.82 \times 0.1}{3 \times 10^{-6}} = 1.27 \times 10^5$$

$$Re_{crit} = 2100 \left[1 + 12\left(\frac{R}{a}\right)^{-0.5}\right] = 8400$$

As Re>Re$_{crit}$, flow is turbulent.
We now calculate the Dean number as

$$De = Re\sqrt{\frac{a}{R}} = 1.27 \times 10^5 \sqrt{\frac{5}{80}} = 31847$$

Note that De >300.
For smooth tube the friction factor can be calculated from Techo et. al. relation

$$f_s = \left(\frac{1}{\ln\left(\frac{Rey}{1.96 \cdot \ln(Rey) - 3.8215}\right) \cdot 1.7372}\right)^2 = 0.00427$$

For coiled tube friction factor f_c is

$$f_c = f_s.0.1125 \cdot De^{0.5} = 0.0859$$

$$h_f = \frac{f_c \cdot L}{d} \cdot \frac{u^2}{2 \cdot 9.81} = 31.98\text{m}$$

We compute R/a = 8 so the Nusselt number for smooth pipe is

$$Nu_o = 0.023 \cdot Re^{0.8} \cdot Pr^{0.3} = 586.847$$

the Nusselt number for coiled pipe as per Schmidt correlation is

$$Nuc = Nu_o \cdot \left(1 + 3.6 \cdot \left(\frac{a}{R}\right)^{0.8} \cdot \left(1 - \left(\frac{a}{R}\right)\right)\right) = 937.08$$

The convective heat transfer coefficient is

$$h = 14056.28 W/m^2 \cdot K$$

The area of wall is $A_w = 2 \cdot \pi \cdot r \cdot (1) = \pi/5 m^2$. The heat loss from organic fluid is

$$Q = h \cdot A_w \cdot (T_w - T_m) = 14056.28 \times (\pi/5) \times (298 - 400) = -900846.2057 W$$

8.13   THERMAL PERFORMANCE INDEX

Often in engineering devices, the surface area is increased by adding ribs on the surface of duct (see Figure 8.14). The flow separation and reattachment on the surface increases the heat transfer. However, this also leads to added pressure loss.

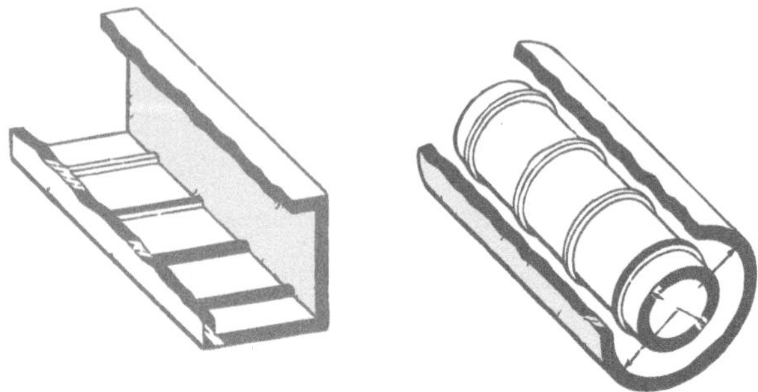

Figure 8.14   Heat transfer in an annulus and square duct region can be augmented by at ribs on the surface.

To simultaneously account for increased heat transfer and pressure loss, the researchers have proposed a parameter called Thermal Performance Index (η), defined as

$$\eta = \frac{(Nu/Nu_o)}{\sqrt[3]{(f/f_o)}}$$

where,
Nu_o is Nusselt number in a plane duct or pipe,
Nu is Nusselt number in a ribbed duct or pipe,
f_o is the Darcy friction factor in a plane duct or pipe without ribs, and
f is the Darcy friction factor in a ribbed plane duct or pipe, and

The augmentation in heat transfer can also be accounted by calculating thermal enhancement factor (TEF) which is defined as the ratio of the heat transfer coefficient of an augmented surface, h to that of a smooth surface, h_o, at an equal pumping power and expressed as

$$TEF = \left(\frac{h}{h_o}\right)_{pp} = \left(\frac{Nu}{Nu_o}\right)_{pp}$$

where, subscript pp stand for same pumping power.

Example 8.10

Find the thermal performance index for air flow in a heated ribbed duct having Reynolds number 6000. The hydraulic diameter of the duct is 0.2804 m and the friction factor for the ribbed duct is 0.04. The duct has roughness elements of 0.15 mm. The Nusselt number for the ribbed duct is 30. Find the thermal performance index for the ribbed duct. Use Swamee Jain (1976) correlation for estimation of friction factor for the duct without ribs.

$$f_o = \frac{1.325}{\left[\ln\left(\frac{\varepsilon}{3.7D_h} + \frac{5.75}{Re_D^{0.9}}\right)\right]^2}$$

valid for $Re_D > 2320$. The Prandtl number for air is 0.7.

Solution We use Swamee Jain (1976) correlation for estimation of the Darcy friction factor (f_o) in a plane duct without ribs

$$f_o = \frac{1.325}{\left[\ln\left(\frac{\varepsilon}{3.7D_h} + \frac{5.75}{Re_D^{0.9}}\right)\right]^2}$$

$$f_o = 0.03657$$

We use the Dittus-Boelter correlation for estimation of Nusselt number of unribbed duct:

$$Nu_0 = 0.023 \cdot Re_D^{0.8} \cdot Pr^{0.4} = 21$$

The thermal performance index is

$$\eta = \frac{(Nu/Nu_o)}{\sqrt[3]{(f/f_o)}} = \frac{(30/21)}{\sqrt[3]{(0.04/0.0365)}} = 1.38$$

It means that for same pumping power we would have more enhanced heat transfer in a ribbed duct.

8.14 NANOFLUID THROUGH PIPES

Vajjha et al. (2010) proposed the relation for the friction factor:

$$f = 0.316Re^{-0.25}\left(\frac{\rho_{nf}}{\rho_f}\right)^{0.707}\left(\frac{\mu_{nf}}{\mu_f}\right)^{0.108} \qquad \begin{array}{l} (for\ \varphi = 0.1\%\ SiO_2) \\ 3000 < Re < 16000 \end{array}$$

Xuan and Li (2003) using Cu-H_2O solution proposed a correlation:

$$Nu_D = 0.0059Re^{0.9238}Pr^{0.4}\left[1 + 7.628\varphi^{0.6886}Pe^{0.001}\right] \qquad \left\{ \begin{array}{l} 10,000 < Re < 25,000 \\ \varphi \le 2\% \end{array} \right.$$

Suresh et al. (2012) proposed the relation for the friction factor:

$$f = 26.4Re^{-0.8737}(1+\varphi)^{156.23} \qquad \begin{array}{l} (for\ \varphi = 0.1\%\ Al_2O_3 - Cu) \\ Re < 2300 \end{array}$$

Maiga et al. (2005) proposed the relation for the Nusselt number:

$$Nu = 0.086Re^{0.55}Pr^{0.5} \quad (q_w = Const, Re \leq 1000, \ Al_2O_3 \ \varphi = 10\%)$$

$$Nu = 0.067Re^{0.71}Pr^{0.35} + 0.0005Re \quad (T_w = Const, Re \leq 1000, Al_2O_3 \ \varphi = 10\%)$$

Maiga et al. (2006) proposed the relation for the Nusselt number:

$$Nu = 0.085Re^{0.71}Pr^{0.35} \quad (10^4 \leq Re \leq 5 \times 10^5, Al_2O_3 \ \varphi = 10\%)$$

Duangthongsuk and Wongwises (2010) proposed the correlation applicable on TiO_2-water nanofluids:

$$Nu_D = 0.074Re^{0.707}Pr^{0.385}\varphi^{0.074} \quad \left\{ \begin{array}{c} 3000 \leq Re \leq 18,000 \\ \varphi \leq 1\% \end{array} \right.$$

Vajjha et al (2015) experimented with aluminum oxide Al_2O_3 and copper oxide CuO nanoparticles which were suspended in ethylene glycol or water mixture for the heat transfer in an automotive radiator tubes and proposed a correlation:

$$Nu_D = 0.023Re^{0.8}Pr^{0.3}(1+0.1771\varphi^{0.1465}) \quad \left\{ \begin{array}{c} 3000 \leq Re \leq 8,000 \\ \varphi \leq 0.06, 1.988 \leq Pr \leq 13.44 \end{array} \right.$$

Esfe et al. (2021) experimented with water as the base fluid, and MgO as nanoparticles and proposed a correlation:

$$Nu_D = 0.0129Re^{0.9363}Pr^{0.3167}\varphi^{0.0932} \quad \left\{ \begin{array}{c} 0.005 < \varphi < 0.02 \\ 6000 \leq Re \leq 32000 \end{array} \right.$$

Correlation Sheet for Internal Flows

Laminar Flow:

Heat transfer in Thermal Entrance Length:

$$Nu_{m,T} = 3.657 + \frac{0.19(x^+/Pr)^{-0.8}}{1+0.117(x^+/Pr)^{-0.467}}$$

$$Nu_{m,H} = \left\{ \begin{array}{ll} 1.953(x^+/Pr)^{-1/3} & x^* \leq 0.03 \\ 4.364 + \frac{0.0722}{x^+/Pr} & x^* \geq 0.03 \end{array} \right.$$

$$x^* = x/D_h \cdot Re \cdot Pr \ and \ x^+ = x/D_h \cdot Re$$

Heat transfer for laminar flow in fully developed zone:

$$Nu_{D,H} = 4.363 \ (q_w = const) \ and \ Nu_{D,T} = 3.66 \ (T_w = const)$$

Turbulent Flow:
Dittus-Boelter correlation:

$$Nu_D = 0.023 \mathrm{Re}_D^{4/5} Pr^n \qquad 0.6 \leq Pr \leq 160$$
$$\mathrm{Re}_D \geq 10^4, \ell/D \geq 10$$

Sieder and Tate correlation:

$$Nu_D = 0.027 \mathrm{Re}_D^{4/5} Pr^{1/3} \left(\frac{\mu}{\mu_w}\right)^{0.14} \qquad 0.7 \leq Pr \leq 16700$$
$$\mathrm{Re}_D \geq 10^4, \ell/D \geq 10$$

Hausen correlation:

$$Nu_D = 0.037(\mathrm{Re}^{0.75} - 180)Pr^{0.42}\left[1 + \left(\frac{D}{\ell}\right)^{2/3}\right]\left(\frac{\mu_b}{\mu_w}\right)^{0.14}$$

Seban and Shimazaki correlation for liquid metals:

$$Nu_D = 5 + 0.025 Pe_D^{0.8} \qquad T_w = constant$$

Sleicher and Rouse (1975) correlations for liquid metals applicable in range $10^4 < \mathrm{Re} < 10^6$:

$$Nu_D = 4.8 + 0.0156 \mathrm{Re}^{0.85} Pr_w^{0.93} \quad (T_w = const)$$
$$Nu_D = 6.3 + 0.0167 \mathrm{Re}^{0.85} Pr_w^{0.93} \quad (q_w = const)$$

Transition Flow:
Gnielinski Correlation:

$$Nu_{m,trans} = (1 - \gamma)Nu_{m,Lam,2300} + \gamma Nu_{m,Turb,4000}$$

$$\gamma = \frac{Re - 2300}{4000 - 2300}$$

$$Nu_{m,H,2300} = \sqrt[3]{83.326 + (Nu_{m,H,1} - 0.6)^3 + (Nu_{m,H,2})^3} \quad (q_w = c)$$

$$Nu_{m,H,1} = 1.953\left[2300 \cdot Pr\left(\frac{d}{\ell}\right)\right]^{1/3} \quad Nu_{m,H,1} = 1.953\left[2300 \cdot Pr\left(\frac{d}{\ell}\right)\right]^{1/3}$$

$$Nu_{m,H,2} = 0.924 Pr^{1/3}\left[2300 \cdot Pr\left(\frac{d}{\ell}\right)\right]^{1/2}$$

$$Nu_{m,T,1} = 1.615\left[2300 \cdot Pr\left(\frac{d}{\ell}\right)\right]^{1/3}$$

$$Nu_{m,T,2} = \left(\frac{2}{1+22 \cdot Pr}\right)^{1/6}\left[2300 \cdot Pr\left(\frac{d}{\ell}\right)\right]^{1/2}$$

$$Nu_{m,Turb,4000} = \frac{\left(\frac{f}{8}\right)Pr(4000-1000)}{1+12.7(Pr^{2/3}-1)\sqrt{\left(\frac{f}{8}\right)}}\left[1+\left(\frac{d}{\ell}\right)^{2/3}\right]$$

$$f = \frac{1}{(1.8\log_{10}4000 - 1.5)^2}$$

Heat Transfer and Pressure Loss in a Coiled Tube

Schmidt correlation:

$$\frac{Nu_{coil}}{Nu_o} = 1+3.6\left(\frac{a}{R}\right)^{0.8}\left[1-\left(\frac{a}{R}\right)\right] \qquad \begin{array}{c} 5 < R/a < 84 \\ 2\times10^4 < \mathrm{Re} < 1.5\times10^5 \end{array}$$

Pratt correlation:

$$\frac{Nu_{coil}}{Nu_o} = 1+3.4\left(\frac{a}{R}\right) \qquad 1.5\times10^3 < \mathrm{Re} < 2\times10^4$$

Thermal Performance Index

$$\eta = \frac{(Nu/Nu_o)}{\sqrt[3]{(f/f_o)}}$$

PROBLEMS

8P-1 Engine oil at 20 °C is moving through a pipe having internal diameter 30 cm and length 100 m. The pipe wall is maintained at 0 °C. If the flow speed is 3 m/s, find the heat transfer from oil to the pipe wall. The properties of engine oil are

$$\rho = 888 \ kg/m^3$$
$$v = 901 \times 10^{-6} \ m^2/s$$
$$k = 0.145 \ W/m.K$$
$$c_p = 1880 \ J/kg.°C$$
$$Pr = 10400$$

8P-2 The bulk inlet temperature of fluid in a pipe is 10°C and the bulk exit temperature of fluid is 65°C. The pipe is held at constant wall temperature of 100°C. Find the log mean temperature.

8P-3 Figure 8.15 shows the schematic representation of a counter flow heat exchanger. The fluid is being heated inside the pipe (d=0.09 ft) and the outer walls of pipe are exposed to steam condensing at 240°F. The inlet and outlet temperatures of fluid are 60°F and 150°F. The length of heat exchanger is 30 ft. The mass flow rate of heated fluid is 2 lb_m/s. Assuming fully-developed flow and ignoring the conduction losses, find the convective heat transferred to the fluid. The thermophysical properties of fluid are

$\rho = 61.9 \; lb_m/ft^3, \quad k = 0.364 \; Btu/hr \cdot ft \cdot °R, \quad c_p = 0.9980 \; Btu/lb_m \cdot °R$
$\alpha = 5.86E - 3 \; ft^2/hr, \quad Pr = 4.34.$

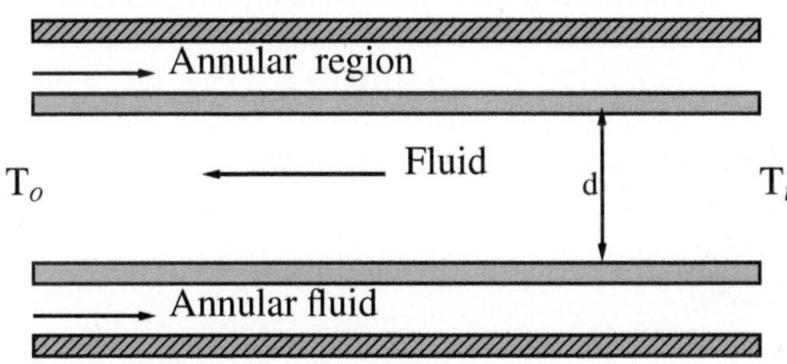

Figure 8.15 The counter flow heat exchanger for problem 8P-3.

8P-4 Compare heat-transfer coefficients for air flowing at an average temperature of 550 K, and at a velocity of 4.46 m/s in a 25.4 cm diameter duct using (1) the Prandtl–Taylor analogy, (2) Von Karman analogy, and (3) Colburn analogy. Use Techo, Tickner and James Correlation to estimate the skin-friction factor.

8P-5 Water at bulk inlet temperature of 50°C enters a pipe of 30 mm diameter. The pipe length is 5 m and it's wall is maintained at constant temperature of -10°C. Estimate the outlet temperature of water, if speed of water is (i) 0.07 m/s, (ii) 1.5 m/s. Ignore entrance lengths effects, and consider the flow as both thermally and hydrodynamically developed flow.

8P-6 Water at bulk inlet temperature of 30°C enters a pipe of 30 mm diameter. The pipe length is 10 m and it's wall is maintained at constant heat-flux of 1000 W/m². Estimate the outlet temperature of water, if speed of water is (i) 0.01 m/s, (ii) 2.5 m/s. Ignore entrance lengths effects, and consider the flow as both thermally and hydrodynamically developed flow.

8P-7 A 100 liter tank, full of oil at 400 K has a coiled tube inside it. The coiled tube contains an organic fluid which is to be heated from 200K to 340 K. The inside diameter of the coiled tube is 1 cm, having 10 turns of 0.5 m diameter. If the tank is well insulated, estimate the heat transferred to the organic fluid.

8P-8 Liquid metals are used in nuclear reactors owing to their good heat transfer characteristics. Liquid sodium flows at speed to 5 m/s in a long, 1 cm-I.D. tube. Calculate the Nusselt number and heat transfer coefficient if the pipe wall is maintained at (i) a uniform wall temperature and (ii) a uniform wall heat flux q_w.

8P-9 Helium flows at 2.0×10^{-3} kg/s in a 10 mm-I.D., 5 m long tube which is wound in a coil. The coil radius of curvature is 100 mm. The Helium enters at

330 K. The coil is maintained at constant wall temperature of 20 K. Find the exit temperature.

8P-10 Air at Reynolds number of 3000 is flowing through 70 mm-I.D, 10 m long pipe. The air enters at 300 K and exit at 500 K. If pipe wall is at 800 K Find the heat transferred to the air.

8P-11 It is planned to replace the heat transfer fluid in a 70 mm-I.D, 100 m long pipe from water to a nanofluid comprised of water-Al_2O_3 with φ=10%. The bulk inlet temperature in both cases is 300K and pipe wall temperature is maintained at 350 K. The mass flow rate is 5 kg/s irrespective of the fluid used. Assuming Dittus-Boelter correlation for regular fluids presented in this chapter is valid for nanofluids, find the bulk outlet temperature. Take the thermophysical properties of nanofluid as ρ_{nf}=1297 kg/m^3, k_{nf}=3.55170, μ_{nf}= 0.0010678868 Pa·s, cp_{nf}= 3849.1 J/kg·K. The thermophysical properties of pure water are k_f=0.597, ρ_f=998 kg/m^3, μ_f=1×10^{-3} Pa·s, c_p=4182 J/kg·K. Compare the result with Maiga correlation. What is the percentage increase in temperature with the use of nanofluid?

[Ans: $T_{b,out,f}$=367.56 K, $T_{b,out,nf,DB}$= 386.14 K (5.05%), $T_{b,out,nf,Maiga}$= 389.14 K (5.87 %)]

8P-12 Assuming that the entrance lengths correlations presented in this chapter are valid for nanofluids. Investigate the influence of volumetric concentration φ on the hydrodynamic and thermal entrance lengths if the water-Al_2O_3 based nanofluid is entering a smooth pipe.

8P-13 Water at 20 °C is being heated in a straight pipe with the uniform heat flux of $q_w = 10^4$ W/m^2, and the local difference between the wall temperature and the mean temperature of the stream is 4 °C. and verify that the flow is turbulent. The inside diameter of pipe is 2.3 cm. Assume both the flow and temperature fields are fully developed. Calculate the mass flow rate of the water stream.

[Ans: $\dot{m} = 0.2327 kg/s$]

8P-14 The cold water stream with $\dot{m}$=10 tons/h is pumped downward through the annular space. The outer diameter $d_o = 23$ cm and inner diameter $d_i = 14$ cm. Calculate the temperature difference between the outer wall of the annulus and the mean temperature of the water stream. The bulk mean temperature of the stream is 80 °C, and the bulk mean temperature gradient dT_b/dx is 200 °C/km. Take C_f=0.005.

[Ans: ΔT=7.79°C]

8P-15 Stephan (1959) proposed the following relation for the estimation of mean Nusselt number in laminar entrance length region:

$$Nu_{D,T} = 3.66 + \frac{0.0667(1/x^*)^{1.333}}{1+0.1Pr(1/x^+)^{0.83}}$$

The correlation is valid for constant wall temperature condition. Compare Stephan correlation with the Hausen (1959) correlation by plotting the correlation for water flow (ν=10^{-6} m^2/s, Pr=7) in a 50 mm pipe at speed of 0.02 m/s.

8P-16 Stephan (1959) proposed the following relation for the estimation of mean Nusselt number in laminar entrance length region:

$$Nu_{D,H} = 4.36 + \frac{0.01(1/x^*)^{1.333}}{1 + 0.0226pr^{0.155}(1/x^+)^{0.83}}$$

The correlation is valid for constant wall heat-flux condition. Compare Stephan correlation with the Hausen (1959) correlation by plotting the correlation for water flow (v=10^{-6} m^2/s, Pr=7) in a 40 mm pipe at speed of 0.02 m/s.

REFERENCES

O. Reynolds. On the Experimental Investigation of the Circumstances Which Determine Whether the Motion of Water Shall Be Direct or Sinuous, and the Law of Resistance in Parallel Channels. Philosophical Transactions of the Royal Society of London 174 (1883), pp. 935–982. UK, 1883.

S. Kakac and Y. Yerner, Convective heat transfer, 2nd edn. CRC Press, New York, USA, 1995.

M. S. Bhatti and R. K. Shah. Turbulent and Transition Flow Convective Heat Transfer in Ducts. In Handbook of Single-Phase Convective Heat Transfer, ed. S. Kakaç, R. K. Shah, and W. Aung. New York: Wiley Interscience, USA, 1987.

R. K. Shah, Thermal entry length solutions for the circular tube and parallel plates. Proc. Natl. Heat Mass Transfer Conf., 3rd, Indian Inst. Techno., Bombay, Vol. I, Pap. no HMT-11-75, 1975

F. W. Dittus and L. M. K. Boelter. University of California Publications on Engineering 2, pp. 433, USA, 1930.

L. F. Moody. Friction Factors for Pipe Flows. Transactions of the ASME 66, pp. 671–684, USA, 1944.

T. H. Von Karman, The Analogy Between Fluid Friction and Heat Transfer, Trans ASME, v. 61, Nov. USA, 1939.

H. Hausen, Neue Gleichungen für die Wärmeüübertragung bei freier und erzwungener Strömung, Allg. Wärmetechnik 9 (4/5), 75–79, 1959.

E. N. Sieder and G. E. Tate, Heat transfer and pressure drop of liquids in tubes, Ind. Eng. Chem. 28, 1429-1435, 1936.

A. P. Colburn, A method of correlating forced convection heat-transfer data and comparison with fluid friction, Int. J. Heat Mass Transfer 7 (1964), 1359–1384. (Reprinted from Trans. Amer. Inst. Chem. Eng. Vol. 29, 1933.

V. Gnielinski, Neue Gleichungen fr den Wärme- und Stoffaustausch in turbulent durchströmten Rohren und Kanälen, Forsch. Ing. Wes. 41 (1), 8–16, 1975. Englisch Translation: New equations for heat and mass transfer in turbulent pipe and channel flow, Int. Chem. Eng. 16 (2), 359–368, 1976.

V. Gnielinski, On heat transfer in tubes, International Journal of Heat and Mass Transfer 63, 134–140, 2013.

V. Gnielinski, Wärmeübertragung bei erzwungener Konvektion: Durchströmte Rohre, VDI-Wärmeatlas, 12th ed., VDI, Springer, 2019.

O. Levenspiel Engineering Flow and Heat Exchange, 3 ed. Springer, 2014.

D. K. Edwards, V. E. Denny, and A. F. Mills, Transfer Processes, 2nd ed., Hemisphere, Washington, D.C., 1979.

P. K. Konakov, Eine neue Formel für den Reibungskoeffizienten glatter Rohre (Orig. Russ), Berichte der Akademie der Wissenschaften der UdSSR L1 (7), 503–506. A new equation for the frictional resistance in smooth tubes, 1946.

W. Nußelt, Der Wärmeübergang in Rohrleitungen, Forsch.-Arb. Ing.-Wes. Nr. 89. Berlin: VOI-Verlag, 1909.

L. M. K. Boelter, G. Young, and H. W. Iversen, An investigation of aircraft heaters, Part XXVII, Distribution of heat-transfer rate in the entrance section of a circular tube. Nat. Adv. Comm. Aeron. NACA-TN 1451. Washington 1948.

A. Cholette, Local and average coefficients for air flowing inside tubes, Chem. Engng. Progr. 44, 1948.

J. E. Bialokoz and O. A. Saunders, Heat transfer in pipe flow at high speeds, Prec. Instn. Mech. Engrs. 170 Nr. 12 S. 389/99, 1956.

E. F. Schimdt, Warmeubergang und Druckverlust in Rohrschlangen, Chem. Ing. Tech. Vol 39, 1967.

N. H. Pratt, The heat transfer in reaction tank cooled by mean of a coil, Trans. Inst Chem Engg, vol 25, 1947.

N. Acharya, M. Sen, and H. C. Chang, Thermal Entrance Length and Nusselt Numbers in Coiled Tubes, Int. J. Heat Mass Transfer, 37(2), 336–340, 1994.

Y. Mori and W. Nakayama, Study on Forced Convective Heat Transfer in Curved Pipes (3rd Report, Theoretical Analysis Under the Condition of Uniform Wall Temperature and Practical Formulae), International Journal of Heat and Mass Transfer, vol. 10, pp. 681–695, 1967.

S. Suresh, K. P. Venkitaraj, P. Selvakumar, and M. Chandrasekar, Effect of Al_2O_3-Cu/water hybrid nanofluid in heat transfer. Experimental Thermal and Fluid Science, 38:54–60, 2012.

R. S. Vajjha, D. K. Das, and D. P. Kulkarni, Development of new correlations for convective heat transfer and friction factor in turbulent regime for nano- fluids. International Journal of Heat and Mass Transfer 53:4607–4618, 2010.

S. Suresh, M. Chandrasekar, and S. C. Sekhar, Experimental studies on heat transfer and friction factor characteristics of CuO/water nanofluid under turbulent flow in a helically dimpled tube. Experimental Thermal and Fluid Science 35:542–549, 2011.

S. E. B. Maiga, S. J. Palm, C. T. Nguyen, G. Roy, and N. Galanis, Heat transfer enhancement by using nanofluids in forced convection flows. International Journal of Heat and Fluid Flow, 26:530–546, 2005.

S. E. B. Maiga, C. T. Nguyen, N. Galanis, G. Roy, T. Mare, and M. Coqueux, Heat transfer enhancement in turbulent tube flow using Al 2O 3 nanoparticle suspension. International Journal Numerical Methods Heat Fluid Flow 16:275–292, 2006.

K. Stephan, Waermeubergang und Druckabfall bei nicht ausgebildeter Laminarstroemnung in Rohren und in ebenen Spalten, Chem. Ing. Tech. 31, 773–778, 1959.

Xuan, Y., and Li, Q. Investigation on convective heat transfer and flow features of nanofluids. Journal of Heat Transfer, 125(1) (2003), 151–155.

Ravikanth S. Vajjha, Debendra K. Das, Dustin R. Ray, Development of new correlations for the Nusselt number and the friction factor under turbulent flow of nanofluids in flat tubes, International Journal of Heat and Mass Transfer 80 (2015) 353–367.

Duangthongsuk, W., and Wongwises, S. (2010). An experimental study on the heat transfer performance and pressure drop of TiO2-water nanofluids flowing under a turbulent flow regime. International Journal of Heat and Mass Transfer, 53(1–3), 334–344.

Hemmat Esfe, M., Abbasian Arani, A. A., Rezaee, M. (2021). Experimental thermal analysis of a turbulent nano enriched water flow in a circular tube. Physica A: Statistical Mechanics and Its Applications, 580.

9 Forced Convection along Flat Surface

This chapter is devoted to the convective heat transfer in external flows. We will discuss the momentum and thermal boundary layer development over a flat surface. We will discuss both differential and integral formulations based models for analysis of thermal boundary layer flows. After finishing this chapter, one should be able to:

- Use the Blasius-Polhausen solution for laminar thermal boundary layer flows over a flat surface in order to estimate the heat transfer.
- Drive integral formulation of an energy equation for flat plate flows.
- Compute heat transfer using correlations for both laminar and turbulent boundary layer flows.

When hot fluid moves over a cold surface, the heat would flow from fluid towards the surface and two distinct boundary layers will form. In one boundary layer, the momentum will diffuse predominantly in vertical direction and velocity will vary from zero to free stream value. The thickness of the momentum boundary layer is indicated by Greek symbol δ and at the edge of boundary layer the velocity will be u=0.99U_∞. The thermal boundary layer is indicated by δ_t and it is defined as the region in which the temperature difference ratio

$$(T_w - T)/(T_w - T_\infty)$$

varies from zero to the value of 0.99. The two boundary layers are depicted schematically in Figure 9.1.

The surface will be conducting heat to fluid or vice-versa depending on the temperatures. Due to viscous nature of all fluids the fluid layer adjacent to the wall will have same speed as that of the surface as per no-slip condition. The applicable equation for the heat conduction through fluid is Fourier law and we can write heat-flux as

$$q_w = -k_f \left(\frac{dT}{dy} \right)_w \tag{9.1}$$

where, q_w is amount of heat (Q) per unit area , and y is wall normal direction. The subscript f indicate the fluid and w indicates the wall thermal conditions. The heat inside thermal boundary layer will flow as per the Newton's law of cooling:

$$q_w = h(T_w - T_\infty) \tag{9.2}$$

DOI: 10.1201/9781003428404-9

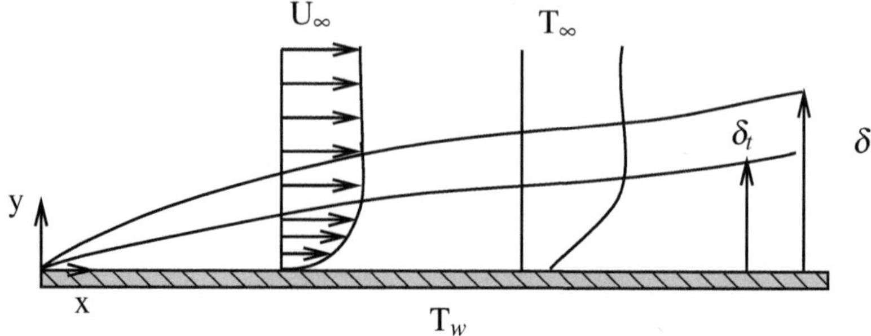

Figure 9.1   Momentum and thermal boundary layer development over an isothermal flat surface for fluids with Prandtl number close to 7.

Comparing equation 9.1 and 9.2, we have

$$h = \frac{-k_f \left(\frac{dT}{dy}\right)_{y=0}}{(T_w - T_\infty)} \tag{9.3}$$

where, h is convective heat transfer coefficient. Note that the temperature will progressively change as the thermal boundary layer grows over the surface and convective heat transfer coefficient will be the function of streamwise coordinate i.e. $h = f(x)$. The surface averaged mean convective heat transfer coefficient is defined as

$$\bar{h} = h_m = \frac{1}{A_w} \int_{A_w} h \, dA$$

where over-line $\overline{(.)}$ and subscript m both used in heat transfer literature to indicate the surface averaged values. The corresponding Nusselt number are defined as

$$\overline{Nu} = \bar{h} \cdot L/k$$

or

$$Nu_m = h_m \cdot L/k$$

depending on the nomenclature used.

9.1   METHOD OF DIFFERENTIAL SIMILARITY

Obtaining a dimensionless number from mere balance of forces or rates is purely an intuitive process. There may be still an ambiguity present and more refined procedures are adopted to remove it. One of the method of extracting the right dimensionless parameters is the method of differential similarity. In this approach the partial differential equations are analysed based on the order of magnitude of terms and balance between the terms can provide us the significant dimensionless numbers.

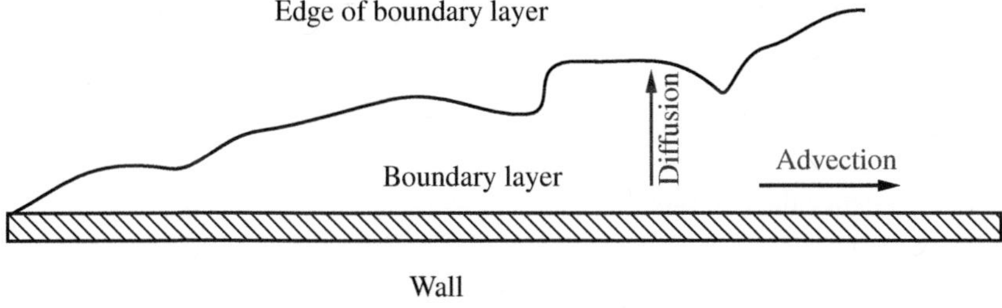

Figure 9.2   The dominant directions of diffusion and advection in a boundary layer.

Two mechanisms control the boundary layer (i) advection and (ii) diffusion. An advection (or isothermal convection) process is the movement of bulk fluid movement, whereas diffusion is develops perpendicularly to advection phenomenon. Figure 9.2 shows the schematic representation of dominant direction of advection and diffusion in a boundary layer. We can investigate the interplay between the two phenomenon using the time scales as

$$\textit{Diffusion} \qquad t_{diff} = \frac{\delta^2}{\mu/\rho}$$

$$\textit{Advection} \qquad t_{adv} = \frac{x}{U_\infty}$$

With balance between the time scales, we have

$$t_{diff} \sim t_{adv}$$

$$\frac{\delta}{x} \sim \frac{1}{\sqrt{\mathrm{Re}_x}} \tag{9.4}$$

This shows that boundary layer thickness is related with flow's Reynolds number. Likewise, when we inspect the energy equation with constant thermophysical properties:

$$\frac{\partial T}{\partial t} + u\frac{\partial T}{\partial x} + v\frac{\partial T}{\partial y} = \alpha\left(\frac{\partial^2 T}{\partial x^2} + \frac{\partial^2 T}{\partial y^2}\right) = 0$$

$$\underbrace{\left[\frac{\partial T}{\partial t} - \alpha\left(\frac{\partial^2 T}{\partial y^2}\right)\right]}_{\textit{Thermal Diffusion}} + \underbrace{\left[u\frac{\partial T}{\partial x} + v\frac{\partial T}{\partial y}\right]}_{\textit{Thermal Advection}} = 0$$

we can infer that there is an interplay between thermal advection and diffusion phenomenon inside boundary layer. The time scale associated with diffusion mechanism of heat transfer in thermal boundary layers is

$$\textit{Thermal Diffusion} \qquad t_{T,diff} = \frac{\delta_t^2}{\alpha}$$

We assume that the rate of bulk fluid contribution to heat transfer is same as that of fluid movement rate

$$\textit{Thermal advection} \qquad t_{T,adv} = \frac{x}{U_\infty}$$

If we balance the two rates

$$t_{T,diff} \sim t_{T,adv}$$

$$\frac{\delta_t^{\,2}}{\alpha} \sim \frac{x}{U_\infty} \tag{9.5}$$

Taking the ratios of δ_t and δ from equations 9.4 and 9.5. This would lead to the following relationship for fluid with constant thermophysical properties:

$$\frac{\delta_t}{\delta} \approx \frac{1}{Pr^{1/2}}$$

For air and water (moderate Prandtl number fluids) n can be taken as unity and this gives us $\delta \equiv \delta_t$. For liquid metals, Prandtl number is less than unity leading to case where $\delta_t \gg \delta$. For engine oils, Prandtl number is order of 1000 or more, and thus $\delta \gg \delta_t$.

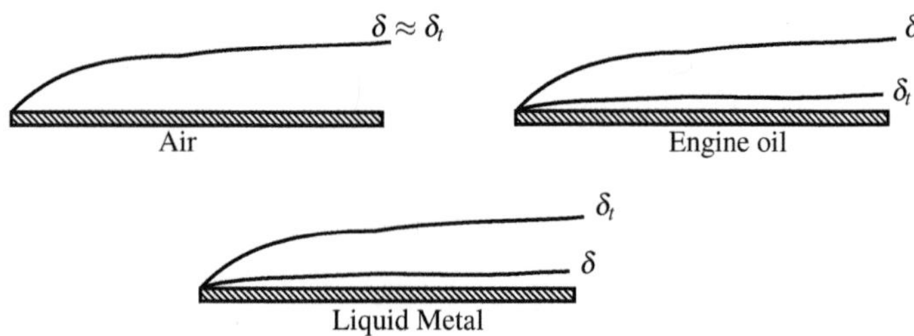

Figure 9.3   Schematic presentation of boundary layer thicknesses for air, engine oil and liquid metals.

Figure 9.3 shows the schematic representation of thicknesses of momentum and thermal boundary layers for fluids with different Prandtl numbers.

9.2 SOME IMPORTANT RELATIONS FOR BOUNDARY LAYER FLOWS

Figure 9.4 shows the transition from laminar to turbulent flow. The flow's Reynolds number over a flat plate is defined as

$$Re = \frac{\rho \cdot U_\infty \cdot x}{\mu}$$

where, x is the distance from leading edge and U_∞ is the free stream velocity. The flow will go into transitional flow as disturbuances appear in flow and no longer damped by viscosity. The lower critical Reynolds number Re_1 is 3×10^5, and the upper critical Reynolds number Re_2 is 5×10^5.

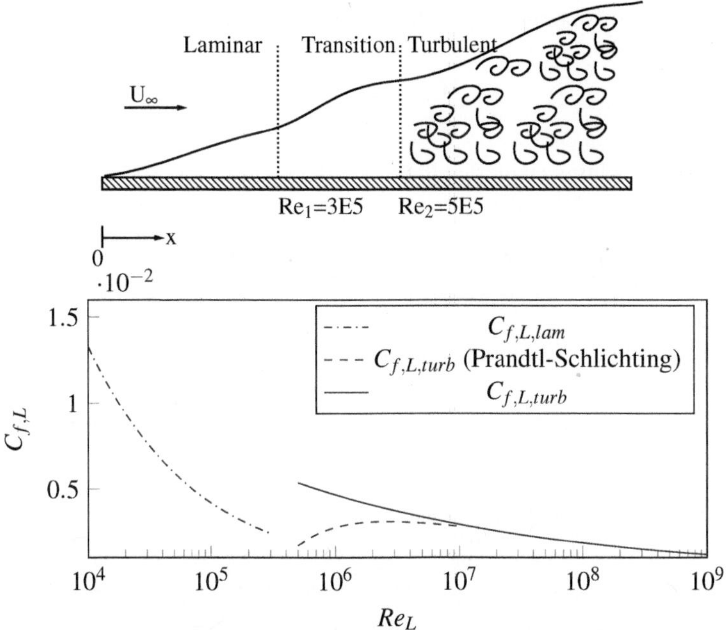

Figure 9.4   Mean skin-friction coefficient ($C_{f,L}$) for flow over a flat plate.

The laminar boundary layer thickness can be found from Blasius solution

$$\left(\frac{\delta}{x}\right)_{laminar} = \frac{5}{\sqrt{\mathrm{Re}_x}} \tag{9.6}$$

The local friction coefficient for laminar can be estimated from correlated

$$C_{f,x} = \frac{0.664}{\sqrt{Re_x}} \tag{9.7}$$

The total skin-friction drag coefficient for laminar flow is

$$C_{f,L,lam} = \frac{1.328}{\sqrt{\mathrm{Re}_L}} \tag{9.8}$$

The turbulent boundary layer thickness can be found from integral solution

$$\left(\frac{\delta}{x}\right)_{turbulent} = 0.37 Re_x^{-1/5} \tag{9.9}$$

The local friction coefficient for turbulent can be estimated from correlated

$$C_{f,x} = \frac{0.0592}{Re_x^{1/5}} \tag{9.10}$$

The total skin-friction drag coefficient for turbulent flow is

$$C_{f,L,turb} = \frac{0.074}{Re_L^{1/5}}$$

(9.11)

Prandtl-Schlichting suggested the following relation for estimation of total skin-friction drag on the flat plat:

$$C_{f,L} = \frac{0.455}{[\log(Re_L)]^{2.58}} - \frac{1700}{Re_L} \qquad 10^7 < Re < 10^{11}$$

(9.12)

Figure 9.4 shows the plot of equations 9.8, 9.11, and 9.12.

9.3 BLASIUS-POLHAUSEN SOLUTION FOR LAMINAR BOUNDARY LAYER FLOWS (PR≈1)

We use energy equation with no viscous dissipation:

$$\rho c_p \left(\frac{\partial T}{\partial t} + u\frac{\partial T}{\partial x} + v\frac{\partial T}{\partial y} \right) = k \left(\frac{\partial^2 T}{\partial x^2} + \frac{\partial^2 T}{\partial y^2} \right) + \dot{q}$$

(9.13)

As $\alpha = k/\rho c_p$ we have

$$u\frac{\partial T}{\partial x} + v\frac{\partial T}{\partial y} = \alpha \left(\frac{\partial^2 T}{\partial y^2} \right)$$

(9.14)

We introduce variable θ as $(T_w - T)/(T_w - T_\infty)$ which leads to

$$u\frac{\partial \theta}{\partial x} + v\frac{\partial \theta}{\partial y} = \alpha \left(\frac{\partial^2 \theta}{\partial y^2} \right)$$

(9.15)

subject to boundary conditions:

at y=0: $\theta = 0$

at y= δ_t or ∞: $\theta = 1$

at x = 0 : $\theta = 1$

We already know the hydrodynamic counter-part of the solution as Blasius Solution.

$$2f''' + ff'' = 0$$

$$u\frac{\partial u}{\partial x} + v\frac{\partial u}{\partial y} = -\frac{1}{\rho}\frac{\partial P}{\partial x} + v \left(\frac{\partial^2 u}{\partial y^2} + \frac{\partial^2 u}{\partial x^2} \right)$$

(9.16)

$$u\frac{\partial v}{\partial x} + v\frac{\partial v}{\partial y} = -\frac{1}{\rho}\frac{\partial P}{\partial y} + v \left(\frac{\partial^2 v}{\partial y^2} + \frac{\partial^2 v}{\partial x^2} \right)$$

(9.17)

With the boundary conditions:

u(x,0) = v(x,0)=0 (no-slip condition)

u(x,∞) = U_∞ (matching condition)

u(0,y)=U_y (initial condition)

We introduce the dimensionless similarity variable

$$\eta = \left(\frac{U_\infty}{vx}\right)^{1/2} y = \frac{y}{\sqrt{vx/U_\infty}}$$

and the stream function

$$\psi = (vxU_\infty)^{1/2} f(\eta)$$

into momentum equation, leading to the Blasius equation:

$$\boxed{2f''' + ff'' = 0} \tag{9.18}$$

The boundary conditions for theflow field are f, $f' = 0$ at y=0 , $f' \to 1$ as $\eta \to \infty$. The energy equation 9.15 will take the form

$$\boxed{\theta'' + \frac{Pr}{2} f\theta' = 0} \tag{9.19}$$

The boundary conditions for the thermal field are : $\theta(0)=0, \theta(\infty)=1$.

MAPLE CODE

```
>N:=10; Pr:=1;
>eq1:= (10-9*lambda1)*2*diff(f(eta),eta,eta,eta)+
((f(eta)*diff(f(eta),eta,eta)))=0;
>eq2:=diff(theta(eta),eta,eta)+
(Pr/2)*f(eta)*diff(theta(eta),eta)=0;

>bc1:= f(0)=0,D(f)(0)=0,D(f)(N)=1;
>bc2:=theta(0)=0,theta(N)=1;
>A1:=dsolve({eq1,eq2,bc1,bc2},
numeric, continuation=lambda1,output=
array([seq(i,i=0..N,0.5)]));

>A2:=dsolve({eq1 ,bc1},numeric,
continuation=lambda1, output=array([seq(i,i=0..N,0.5)]));

>with(plots):
```

```
>odeplot(A1,[[eta,f(eta),color=green],[eta,diff(f(eta),eta),
color=red],[eta,diff(f(eta),eta,eta), color=blue]],0..N);

>odeplot(A1,[[eta,theta(eta),color=green],
[eta,diff(theta(eta),eta), color=red]],0..N);
```

TABLE 9.1

Numerical Solution of Polhausen and Blasius Equations

η	$f(\eta)$	$f'(\eta)$	$f''(\eta)$	$\theta(\eta)$	$\theta'(\eta)$
0	0.0000	0.0000	0.3321	0.0000	**0.3321**
0.5	0.0415	0.1659	0.3309	0.1659	0.3309
1	0.1656	0.3298	0.3230	0.3298	0.3230
1.5	0.3701	0.4868	0.3026	0.4868	0.3026
2	0.6500	0.6298	0.2668	0.6298	0.2668
2.5	0.9963	0.7513	0.2174	0.7513	0.2174
3	1.3968	0.8460	0.1614	0.8460	0.1614
3.5	1.8377	0.9130	0.1078	0.9130	0.1078
4	2.3057	0.9555	0.0642	0.9555	0.0642
4.5	2.7901	0.9795	0.0340	0.9795	0.0340
5	3.2833	0.9915	0.0159	0.9915	0.0159
5.5	3.7806	0.9969	0.0066	0.9969	0.0066
6	4.2796	0.9990	0.0024	0.9990	0.0024
6.5	4.7793	0.9997	0.0008	0.9997	0.0008
7	5.2792	0.9999	0.0002	0.9999	0.0002
7.5	5.7792	1.0000	0.0001	1.0000	0.0001
8	6.2792	1.0000	0.0000	1.0000	0.0000

Table 9.1 lists the numerical solution of the equations.

For case of fluids Prandtl number close to unity, the viscous and thermal boundary layers are similar and we can assume that $\theta = \theta(\eta)$ so we apply boundary condition $\theta(0) = 0$ and $\theta(\infty) = 1$.

$$q_w = -k \left(\frac{\partial T}{\partial y} \right)_{y=0}$$

$$q_w = -k(T_\infty - T_w) \left(\frac{\partial \theta}{\partial y} \right)_{y=0} = k(T_w - T_\infty) \left(\frac{\partial \theta}{\partial \eta} \cdot \frac{\partial \eta}{\partial y} \right)_{\eta=0}$$

$$q_w = \frac{k(T_w - T_\infty)\theta'(0)}{\sqrt{vx/U_\infty}}$$

we know

$$h_x = q_w / (T_w - T_\infty)$$

so

$$Nu_x = \frac{xh_x}{k} = \frac{\theta'(0)}{\sqrt{v/xU_\infty}} = \theta'(0) \cdot \sqrt{Re_x} \qquad (9.20)$$

We introduce dimensionless parameter Stanton number defined as St=Nu/Re.Pr. For Pr= 1 fluids Nu is expressed as:

$$Nu_x = \frac{xh_x}{k} = 0.332 \cdot \sqrt{Re_x} \qquad (9.21)$$

experimentally it is found that for fluids with Pr=0.15 to 15 the correlations valid is:

$$\boxed{Nu_x = 0.332 \cdot \sqrt{Re_x} \cdot Pr^{1/3}} \qquad (9.22)$$

Example 9.1

The local Nusselt number on the flat plate for laminar boundary layer is given by equation 9.22. Find an expression for the mean Nusselt number at the end of plate of length L.

Solution We use the local Nusselt number relation and write the local heat transfer coefficient as

$$h_{x,lam} = \frac{0.332 \cdot \sqrt{\left(\frac{U \cdot x}{v}\right)} \cdot Pr^{\frac{1}{3}} \cdot k}{x}$$

We now integrate this expression over the entrie length of the plate to obtain the mean heat transfer coefficient:

$$h_{m,lam} = \frac{1}{L} \int_0^L h_{x,lam}\, dx$$

$$= \frac{0.664\sqrt{\frac{UL}{v}}Pr^{1/3}k}{L}$$

substituting the definition of Reynolds number, we have

$$h_{m,lam} = \frac{0.664\sqrt{Re_L}Pr^{1/3}k}{L}$$

leading to the mean Nusselt number expression:

$$Nu_{m,lam} = \frac{h_{m,lam} \cdot L}{k} = 0.664\sqrt{Re_L}\sqrt[3]{Pr}$$

The mean Nusselt number we obtained in previous example is

$$\boxed{Nu_L = 0.664 \cdot \sqrt{Re_x} \cdot Pr^{1/3}}$$

(9.23)

In terms of Stanton number, we we may write

$$\boxed{St_L . Pr^{2/3} = 0.332 Re_L^{-1/2}}$$

(9.24)

This solution is known as *Polhausen Solution* in the heat transfer literature.

Figure 9.5 shows the mean Nusselt number distribution for laminar flow over a flat plate.

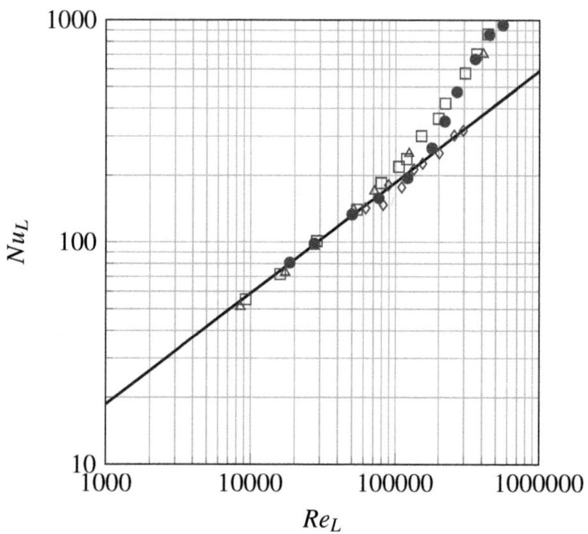

Figure 9.5 Mean Nusselt number (Nu_L) for laminar flow vs. Reynolds number (Re_L) over a flat plate for air. (Data taken from Edward and Furber (1956).)

Example 9.2

Reynolds has suggested that heat and momentum are analogous phenomenon with the relationship of the form

$$St_x = \left(\frac{C_f}{2} \right)$$

Using this analogy derive an expression for the Colburn factor (j_H) in terms of heat transfer parameters, where (j_H) is defined as

$$j_H = \frac{Cf}{2}$$

Solution
When Pr = 1, and pressure gradient is zero (dP/dx = 0) then for the case of flow over a flat plate, the velocity profile is very close to temperature profile. In this case the gradient at wall is also similar and an analogy is proposed that

$$St_x = \left(\frac{C_f}{2}\right)$$

For laminar flow over a flat plate from Blasius solution, we have $C_f = 0.664 Re_x$ and from Polhausen solution we have $Nu_x = 0.332.\sqrt{Re_x}.Pr^{1/3}$ so Stanton number relation can be expressed as

$$St = \frac{Nu}{Re.Pr} = \frac{0.332.Re_x^{1/2}.Pr^{1/3}}{Re.Pr}$$

$$St = 0.332/Re_x^{1/2}Pr^{2/3}$$

$$St.Pr^{2/3} = 0.332 Re_x^{-1/2}$$

We note that the right hand side can be cast into skin friction coefficient relation:

$$\boxed{St.Pr^{2/3} = \frac{Cf}{2} = j_H} \tag{9.25}$$

Here j_H is the Colburn j-factor.

9.4 INTEGRAL FORMULATION FOR THERMAL BOUNDARY LAYER FLOWS

We will now develop an integral energy equation for a thermal boundary layer over a flat plate. For sake of analysis the control volume is shown in Figure 9.6.

Energy entering the control volume

$$E_{in} = \int_0^\ell \rho c_p u T\, dy \tag{9.26}$$

where, ℓ is height of control volume. We use Taylor series to estimate the energy leaving the control volume at x+Δ x:

$$\int_0^\ell \rho c_p u T\, dy + \frac{d}{dx}\left[\int_0^\ell \rho c_p u T\, dy\right] dx \tag{9.27}$$

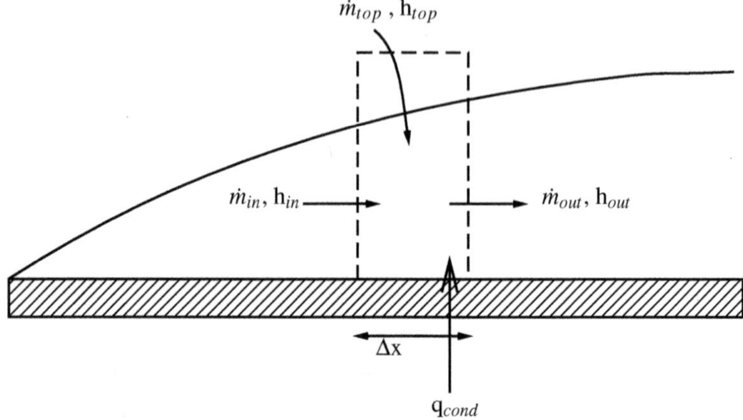

Figure 9.6   Control volume used for integral energy equation. Here h is enthalpy and $\dot{m}$ is mass flow rate.

Thus the net energy convected is:

$$E_{net} = \frac{d}{dx}\left[\int_0^{\ell} \rho c_p u T \, dy\right] dx \tag{9.28}$$

Net mass inflow entering from the top face of control volume can be obtained by doing the balance of mass flow rate, which gives

$$\dot{m}_{top} = \frac{d}{dx}\left[\int_0^{\ell} \rho u \, dy\right] dx$$

so thermal energy or enthalpy entering from top is $E_{top} = \dot{m}_{top} c_p T_{\infty}$, which will take form

$$E_{top} = c_p T_{\infty} \frac{d}{dx}\left[\int_0^{\ell} \rho u \, dy\right] dx$$

Energy from wall (E_{wall}) is the heat conducted from the surface is

$$q_{cond} = -k\Delta x \left(\frac{\partial T}{\partial y}\right)_{y=0} \tag{9.29}$$

We do the energy balance for the whole control volume:

$$E_{net-conv} = E_{wall} + E_{top} \tag{9.30}$$

$$-k\Delta x \left(\frac{\partial T}{\partial y}\right)_{y=0} + c_p T_{\infty} \frac{d}{dx}\left[\int_0^{\ell} \rho u \, dy\right] dx = \frac{d}{dx}\left[\int_0^{\ell} \rho u c_p T \, dy\right] dx \tag{9.31}$$

$$-k\left(\frac{\partial T}{\partial y}\right)_{y=0} = \frac{d}{dx}\left[\int_0^{\ell} \rho u c_p T \, dy\right] - c_p T_{\infty}\frac{d}{dx}\left[\int_0^{\ell} \rho u \, dy\right] \tag{9.32}$$

The extend of the control volume (ℓ) in the wall normal direction is up till the edge of the thermal boundary layer, so $\ell=\delta_t$:

$$-k\left(\frac{\partial T}{\partial y}\right)_{y=0} = \frac{d}{dx}\left[\int_0^\ell \rho u c_p T\, dy\right] - c_p T_\infty \frac{d}{dx}\left[\int_0^\ell \rho u\, dy\right] \qquad (9.33)$$

For thermal Boundary layer with thickness δ_t and constant fluid properties:

$$-k\left(\frac{\partial T}{\partial y}\right)_{y=0} = \rho c_p \frac{d}{dx}\left[\int_0^{\delta_t} u(T - T_\infty)\, dy\right] \qquad (9.34)$$

$$\boxed{-\alpha\left(\frac{\partial T}{\partial y}\right)_{y=0} = \frac{d}{dx}\left[\int_0^{\delta_t} u(T - T_\infty)\, dy\right]} \qquad (9.35)$$

The equation is called integral energy equation as it provides the balance of energy in integral formulation for the thermal boundary layer flows.

9.5 HEAT TRANSFER BY LAMINAR BOUNDARY LAYER WITH UNHEATED STARTING LENGTH

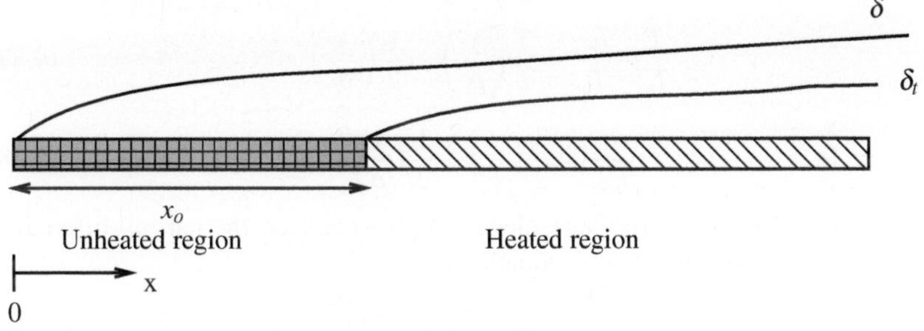

Figure 9.7　Schematic representation of boundary layer formation over a flat plate with an unheated starting length region of length x_0.

Figure 9.7 shows the boundary layer formation over a flat plate with an unheated starting length region of length x_0. To solve this problem we take the following assumption:

Assumptions:

1 constant thermophysical properties;

2 constant wall temperature;

3 no viscous dissipation ;

4 Steady ;

5 Incompressible flows

The applicable boundary conditions are
  Boundary conditions

i at y=0: $T = T_w$

ii at y= δ_t: $\partial T/\partial y = 0$

iii at y= δ_t: $T = T_\infty$

iv at y= 0: $\partial^2 T/\partial y^2 = 0$

The last boundary condition is invoked by applying Energy Equation at wall y=0:

$$\rho c_p \left(\cancel{\frac{\partial T}{\partial t}} + u\frac{\partial T}{\partial x} + v\frac{\partial T}{\partial y} \right) = k \left(\cancel{\frac{\partial^2 T}{\partial x^2}} + \frac{\partial^2 T}{\partial y^2} \right) + \cancel{\dot{q}} \tag{9.36}$$

For integral solutions as per procedure, we need to assume a polynomial which best fit with experimental or analytical data. To do so we use the following *reasonable polynomials*, i.e. the assumed velocity and dimensionless temperature distributions which matches well with the experimental data:

$$\frac{T - T_w}{T_\infty - T_w} = \frac{3}{2}\left(\frac{y}{\delta_t}\right) - \frac{1}{2}\left(\frac{y}{\delta_t}\right)^3 \tag{9.37}$$

$$\frac{u}{U_\infty} = \frac{3}{2}\left(\frac{y}{\delta}\right) - \frac{1}{2}\left(\frac{y}{\delta}\right)^3 \tag{9.38}$$

We introduce the similarity variables in order to convert the partial differential equations into ordinary differential equations:

$$\eta = y/\delta$$

and

$$\eta_t = y/\delta_t$$

and cast the velocity ($\bar{u} = u/U_\infty$) and temperature ($\theta = (T-T_w)/(T_\infty-T_w)$) profiles into form

$$\theta = \frac{3}{2}\eta_t - \frac{1}{2}\eta_t^3 \tag{9.39}$$

$$\bar{u} = \frac{3}{2}\eta - \frac{1}{2}\eta^3 \tag{9.40}$$

The dimensionless Velocity and temperature profiles are plotted in Figure 9.8.

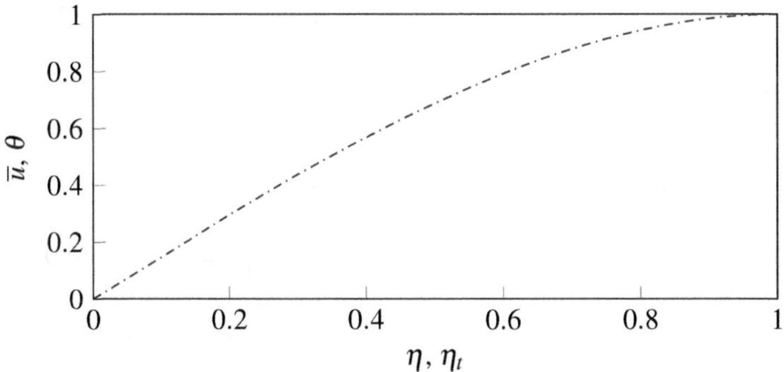

Figure 9.8   The *reasonable polynomial* for flow and thermal fields in case of laminar boundary layer over a flat plate.

We can calculate the temperature gradient as

$$\frac{\partial \theta}{\partial \eta_t} = \frac{3}{2} - \frac{3\eta_t^2}{2} \tag{9.41}$$

$$\left(\frac{\partial \theta}{\partial \eta_t}\right)_{\eta_t=0} = \frac{3}{2} \tag{9.42}$$

$$T - T_\infty = (\theta - 1)(T_\infty - T_w) \tag{9.43}$$

We differentiate T with y:

$$\frac{\partial T}{\partial y} = -(T_w - T_\infty)\frac{\partial \theta}{\partial y} \tag{9.44}$$

$$\frac{\partial T}{\partial y} = -(T_w - T_\infty)\frac{\partial \theta}{\delta_t \partial (y/\delta_t)} \tag{9.45}$$

$$\frac{\partial T}{\partial y} = -\frac{(T_w - T_\infty)}{\delta_t}\frac{\partial \theta}{\partial \eta_t} \tag{9.46}$$

Substitute 9.43 and 9.46 into equation 9.35:

$$\boxed{-\alpha\left(\frac{\partial T}{\partial y}\right)_{y=0} = \frac{d}{dx}\left[\int_0^{\delta_t} u(T - T_\infty)dy\right]}$$

$$T - T_\infty = (\theta - 1)(T_\infty - T_w)$$

$$\frac{\partial T}{\partial y} = \frac{(T_\infty - T_w)}{\delta_t}\frac{\partial \theta}{\partial \eta_t}$$

$$-\alpha \frac{(T_\infty - T_w)}{\delta_t} \frac{\partial \theta}{\partial \eta_t} = \frac{d}{dx}\left[\int_0^{\delta_t} u\left[(\theta - 1)(T_\infty - T_w)\right]dy\right]$$

$$-\alpha \frac{1}{\delta_t} \frac{\partial \theta}{\partial \eta_t} = \frac{d}{dx}\left[\int_0^{\delta_t} u\left[(\theta - 1)\right]dy\right] \tag{9.47}$$

Multiply and divide both sides by U_∞ and change the variable of integration we get:

$$-\alpha \frac{1}{U_\infty \delta_t} \frac{\partial \theta}{\partial \eta_t} = \frac{d}{dx}\left[\int_0^1 \frac{u}{U_\infty}\left[(\theta - 1)\right]\delta_t d(y/\delta_t)\right] \tag{9.48}$$

$$\alpha \frac{1}{U_\infty \delta_t}\left(\frac{\partial \theta}{\partial \eta_t}\bigg|_{\eta_t=0}\right) = \frac{d}{dx}\left[\int_0^1 \bar{u}\delta_t\left[(1 - \theta)\right]d\eta_t\right] \tag{9.49}$$

$$\alpha \frac{1}{U_\infty \delta_t}\left(\frac{3}{2}\right) = \frac{d}{dx}\left[\int_0^1 \bar{u}\delta_t\left[(1 - \theta)\right]d\eta_t\right] \tag{9.50}$$

Consider integral term on RHS of Eq. 9.56:

$$Integral \Rightarrow \int_0^1 \bar{u}\delta_t\left[(1 - \theta)\right]d\eta_t = \delta_t\left[\int_0^1 k(\eta_t)d\eta_t\right] \tag{9.51}$$

Substitute polynomials into above equation and neglect H.O.T.

$$\theta = \frac{3}{2}\eta_t - \frac{1}{2}\eta_t^3$$

$$\bar{u} = \frac{3}{2}\eta - \frac{1}{2}\eta^3$$

Assume $\delta_t \approx \varphi\delta$ or $\eta = \varphi.\eta_t$ then we get:

This gives:

$$LHS - integral = -(3/280) * \varphi^3 + (3/20) * \varphi; \tag{9.52}$$

Ignore φ^3 and higher order terms:

$$Integral \Rightarrow \int_0^1 \bar{u}\delta_t\left[(1 - \theta)\right]d\eta_t = \frac{3}{20}\delta_t\varphi \tag{9.53}$$

$$LHS \approx \frac{d}{dx}\left(\delta_t\frac{3}{20}\varphi\right) \tag{9.54}$$

$$RHS = \alpha \cdot (1.5 - 1.5 \cdot \eta_t^2)/(\delta_t \cdot Uo) \tag{9.55}$$

We get Eq. 9.56 in form:

$$\alpha \frac{1}{U_\infty \delta_t}\left(\frac{3}{2}\right) = \frac{d}{dx}\left[\delta_t \frac{3}{20}\varphi\right] \tag{9.56}$$

$$\frac{\alpha}{\delta_t U_\infty} = \frac{d}{dx}\left[\frac{\varphi \delta_t}{10}\right] \tag{9.57}$$

We can write in terms of δ by replacing $\delta_t = \varphi\delta$:

$$\frac{d}{dx}(\varphi^2 \delta) = \frac{10\alpha}{U_\infty}\frac{1}{\delta\varphi} \tag{9.58}$$

We differentiate with x using rules:

$$\varphi^2 \frac{d\delta}{dx} + 2\delta\varphi\frac{d\varphi}{dx} = \frac{10\alpha}{U_\infty}\frac{1}{\delta\varphi} \tag{9.59}$$

Since

$$\frac{\delta}{x} = \frac{4.64}{\sqrt{Re_x}} = \underbrace{\left(4.64\sqrt{v/U_\infty}\right)}_{constant}x^{-1/2} \equiv a\frac{1}{\sqrt{x}} \tag{9.60}$$

$$\frac{d\delta}{dx} = a\frac{d}{dx}\left[x^{1/2}\right] = \frac{a}{2\sqrt{x}} \tag{9.61}$$

We substitute result from Eq. 9.61 into Eq. 9.59 and after some rearrangement Eq. takes form

$$\varphi^3 + \frac{4}{3}x\frac{d^3\varphi}{dx^3} = \frac{0.9289}{Pr} \tag{9.62}$$

as at $x=x_o \rightarrow \varphi = 0$, we have

$$\boxed{\frac{\delta_t}{\delta} = \frac{0.976}{Pr^{1/3}}\left[1 - \left(\frac{x}{x_o}\right)^{-3/4}\right]^{1/3}} \tag{9.63}$$

As

$$q = -k\left(\frac{dT}{dy}\right)_{y=0}$$

so

$$q = \frac{3}{2}\frac{k}{\delta_t}(T_w - T_\infty) = \frac{3}{2}\frac{k}{\delta}(T_w - T_\infty)\left[\frac{\sqrt[3]{Pr}}{0.976}\left[1 - \left(\frac{x}{x_o}\right)^{-3/4}\right]^{-1/3}\right] \tag{9.64}$$

The approximate solution with heated starting length: is:

$$Nu_x = 0.332 Pr^{1/3} Re_x^{1/2} \left[1 - \left(\frac{x}{x_o} \right)^{-3/4} \right]^{-1/3} \tag{9.65}$$

and for no heated starting length case it reduces to form of exact solution. The exact solution without heated starting length:

$$Nu_x = 0.332 . \sqrt{Re_x} . Pr^{1/3} \tag{9.66}$$

This proved that integral solutions can be used to solve complex heat transfer situations.

Example 9.3

Air is flowing over at flat plate at freestream temperature of 15°C and velocity U_∞=30 m/s. The unheated starting length is 10 cm and the length of the plate is 10 m. The properties of air are k=0.025 W/m·K, $v=1.5\times10^{-5}$ m²/s, Pr=0.7. Find the local heat transfer coefficient at the end of the laminar boundary layer region.

Solution
We first find the location of transition by taking critical Reynolds number as 3×10^5. The location of transition is

$$x_{transition} = \frac{3 \times 10^5 \cdot v}{U} = 15cm > 10cm$$

This shows that up till x=15 cm the boundary layer is laminar and it completely cover the unheated starting length region. Hence, we can use the Nusselt number correlation developed in this section:

$$Nu_{x,lam} = 0.332 Pr^{1/3} Re_x^{1/2} \left[1 - \left(\frac{x}{x_o} \right)^{-3/4} \right]^{-1/3}$$

Substituting, Re_x=3×10^5, x=$x_{transition}$ = 15 cm, and x_o = 10 cm, we have the Nusselt number at the onset of transition:

$$Nu_{x,lam} = 252.259$$

The corresponding convective heat transfer coefficient is

$$h_{x,lam} = \frac{Nu_{x,lam} \cdot k}{x_{transition}} = 42.04 \frac{W}{m^2 \cdot K}$$

9.6 TURBULENT BOUNDARY LAYER FLOWS

For flow over a flat surface it is found that velocity field is following the profile of the form

$$\frac{u}{U_\infty} = \left(\frac{y}{\delta}\right)^{1/7}$$

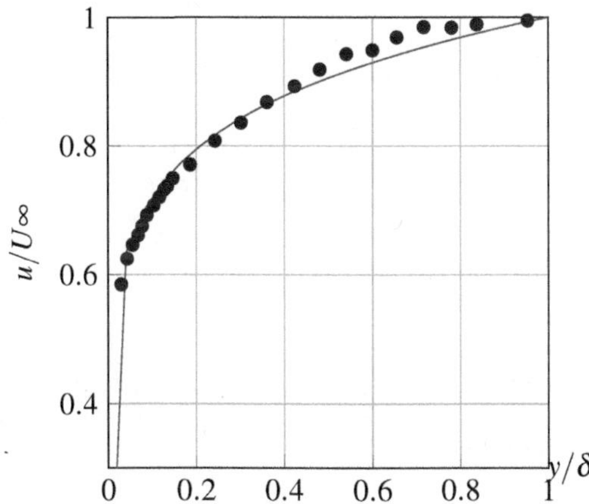

Figure 9.9 Turbulent 1/7th velocity profile at Re=9.5×10⁵. (Data from Edwards and Furber 1956.)

We assume 1/7th velocity and temperature profiles as *reasonable* expression for the case of turbulent flow over a flat plate with unheated starting length (see Figure 9.9):

$$\frac{u}{U_\infty} = \left(\frac{y}{\delta}\right)^{1/7}$$

$$\frac{T - T_w}{T_\infty - T_w} = \left(\frac{y}{\delta_t}\right)^{1/7}$$

Also, we assume shear stress distribution as:

$$\tau / \tau_w = \left[1 - \left(\frac{y}{\delta}\right)^{9/7}\right]$$

In turbulent flow, in the region near the wall, the contribution of the turbulent transport mechanism far outweighs that of the molecular mechanism. O. Reynolds was the first to apply the concept of analogy to the prediction of turbulent transport phenomena. We use the equation 7.70 and 7.71 for the derivation of the solution:

$$\tau = \rho(v + \varepsilon_M)\frac{du}{dy} \tag{9.67}$$

$$q = \rho c_p(\alpha + \varepsilon_H)\frac{dT}{dy} \tag{9.68}$$

where $Pr_t = \varepsilon_M/\varepsilon_H$, and for the case of $Pr_t \approx Pr = 1$, we assume that

$$(v + \varepsilon_M) = (\alpha + \varepsilon_H) \tag{9.69}$$

We modify the RHS of equation 9.69 by substituting equation 7.70:

$$(v + \varepsilon_M) = \frac{\tau/\rho}{du/dy} \tag{9.70}$$

and, we modify the LHS of equation 9.69 by substituting equation 7.71:

$$(\alpha + \varepsilon_H) = \frac{q/(\rho c_p)}{dT/dy} \tag{9.71}$$

We further rearrange the equations by multiplying and dividing with U_∞ and τ_w, leading to more simplification for the shear stress as

$$\frac{\tau/\rho}{du/dy} = \frac{U_\infty^2}{U_\infty^2} \frac{\tau}{\rho} \frac{\tau_w}{\tau_w} \frac{1}{du/dy} = U_\infty \frac{\tau_w}{\rho U_\infty^2} \frac{\tau}{\tau_w} \frac{1}{du/dy} \tag{9.72}$$

From 1/7th velocity profile, we have

$$u = U_\infty \left(\frac{y}{\delta}\right)^{1/7} = U_\infty \eta^{1/7}$$

Differentiating the profile with y, we have

$$du/dy = U_\infty \frac{d}{d\eta}(\eta^{1/7}) \cdot \frac{d\eta}{dy} = \frac{U_\infty}{7\delta\eta^{6/7}}$$

Since $\eta = y/\delta$, the derivative of η is $d\eta/dy = 1/\delta$. By inserting these results into shear stress expression (equation 9.72), we have

$$\frac{\tau/\rho}{du/dy} = U_\infty^2 \frac{\tau_w}{\rho U_\infty^2} \frac{\tau}{\tau_w} \frac{1}{du/dy} = U_\infty^2 \left(\frac{C_f}{2}\right)\left(\frac{\tau}{\tau_w}\right) \frac{7\delta\eta^{6/7}}{U_\infty} \tag{9.73}$$

We will now introduce the dimensionless shear stress profile in above equation

$$\frac{\tau/\rho}{du/dy} = 7U_\infty\delta\eta^{6/7}\left(\frac{C_f}{2}\right)\left(\frac{\tau}{\tau_w}\right) = 7U_\infty\delta\eta^{6/7}\left(\frac{C_f}{2}\right)\left[1 - \left(\frac{y}{\delta}\right)^{9/7}\right] \tag{9.74}$$

Hence

$$(v + \varepsilon_M) = \frac{\tau/\rho}{du/dy} = 7U_\infty\delta\eta^{6/7}\left(\frac{C_f}{2}\right)\left[1 - \left(\frac{y}{\delta}\right)^{9/7}\right] \tag{9.75}$$

Now, since $(v + \varepsilon_M) = (\alpha + \varepsilon_H)$ as per equation 9.69, the turbulent heat transfer equation takes the form

$$\frac{q_y}{\rho c_p} = (\alpha + \varepsilon_H)\frac{dT}{dy} = \left\{7U_\infty\delta\eta^{6/7}\left(\frac{C_f}{2}\right)\left[1 - \left(\frac{y}{\delta}\right)^{9/7}\right]\right\}\frac{dT}{dy} \tag{9.76}$$

At $y = 0$ we have the wall:

$$\frac{q_w}{\rho c_p} = (\alpha + \varepsilon_H)\frac{dT}{dy} = 7U_\infty \delta \eta^{6/7}\left(\frac{C_f}{2}\right)\frac{dT}{dy} \tag{9.77}$$

Like velocity profile, we assume the dimensionless temperature distribution to be following 1/7th power law. This gives

$$T - T_\infty = \left(\frac{y}{\delta}\right)^{1/7}\Delta T$$

$$\frac{dT}{dy} = \frac{1}{7}\frac{1}{\delta_t^{1/7}}y^{-6/7}\Delta T$$

where, $\Delta T = (T_\infty - T_w)$. We now introduce the temperature gradient into turbulent heat transfer equation which we have developed:

$$\frac{q_w}{\rho c_p} = (\alpha + \varepsilon_H)\frac{dT}{dy} = \left(7U_\infty \delta \eta^{6/7}\right)\left(\frac{C_f}{2}\right)\frac{dT}{dy} \tag{9.78}$$

$$\frac{-q_w}{\rho c_p} = (7U_\infty \delta \eta^{6/7})\left(\frac{C_f}{2}\right)\frac{1}{7}\frac{1}{\delta_t^{1/7}}y^{-6/7}\Delta T \tag{9.79}$$

The negative sign appear with q_w because they way we defined ΔT.

$$\frac{-q_w}{\rho c_p U_\infty} = (\delta\left(\frac{y}{\delta}\right)^{6/7})\left(\frac{C_f}{2}\right)\frac{1}{\delta_t^{1/7}}y^{-6/7}\Delta T \tag{9.80}$$

$$\frac{-q_w}{\rho c_p U_\infty} = \frac{\delta^{1/7}}{\delta_t^{1/7}}\Delta T\left(\frac{C_f}{2}\right) \tag{9.81}$$

$$\boxed{\frac{q_w}{\rho c_p U_\infty (T_w - T_\infty)} = \left(\frac{\delta}{\delta_t}\right)^{1/7}\left(\frac{C_f}{2}\right)} \tag{9.82}$$

According to integral energy equation

$$\frac{d}{dx}\left[\int_0^{\delta_t} u(T - T_\infty)dy\right] = \frac{q_w}{\rho c_p} \tag{9.83}$$

$$U_\infty\frac{d}{dx}\left[\int_0^{\delta_t}\frac{u}{U_\infty}(\theta - 1)(T_\infty - T_w)dy\right] = \frac{q_w}{\rho c_p} \tag{9.84}$$

$$U_\infty\frac{d}{dx}\left[\int_0^{\delta_t}\frac{u}{U_\infty}(1 - \theta)(T_w - T_\infty)dy\right] = \frac{q_w}{\rho c_p} \tag{9.85}$$

$$\frac{d}{dx}\left[\int_0^1\left(\frac{u}{U_\infty}\right)(1 - \theta)\delta_t d(y/\delta_t)\right] = \frac{q_w}{\rho c_p U_\infty (T_w - T_\infty)} \tag{9.86}$$

$$\frac{d}{dx}\left[\delta_t \int_0^1 \left(\frac{u}{U_\infty}\right)(1-\theta)d\eta_t\right] = \frac{q_w}{\rho c_p U_\infty (T_w - T_\infty)} \tag{9.87}$$

We now substitute Eq.9.82 into RHS Eq. 9.87:

$$\frac{d}{dx}\left[\frac{7}{72}\delta_t\left(\frac{\delta_t}{\delta}\right)^{1/7}\right] = \left(\frac{\delta}{\delta_t}\right)^{1/7}\left(\frac{C_f}{2}\right) \tag{9.88}$$

Setting $\zeta = \delta_t/\delta$ we have

$$\zeta^{9/7} = 0.998 + C.x^{-9/10} \tag{9.89}$$

At $x_o = 0$ we have $\zeta = 0$ From equation 9.82 we have the local Stanton number:

$$St_x = \frac{q_w}{\rho c_p U_o (T_w - T_o)} = \left[1 - \left(\frac{x_o}{x}\right)^{9/10}\right]^{-1/9}(C_f/2) \tag{9.90}$$

$$St_x = \frac{Nu}{Re \cdot Pr} = \left(\frac{C_f}{2}\right)\left[1 - \left(\frac{x_o}{x}\right)^{9/10}\right]^{-1/9} \tag{9.91}$$

Substituting skin-friction coefficient $C_f/2 = 0.0291/Re_x^{1/5}$ and boundary layer thickness $\delta/x = 0.37/Re_x^{1/5}$, we have

$$C_{f,x} = 0.059 Re_x^{-1/5}$$

We have the Nusselt number

$$Nu_{x,turb} = \frac{0.0295 Re^{0.8} Pr^{1/3}}{\left[1 - \left(\frac{x_o}{x}\right)^{9/10}\right]^{1/9}} \tag{9.92}$$

Burmeister (1993) suggested a more convenient approximation for the above equation

$$\frac{Nu}{Re.Pr^{0.6}} = \frac{0.0291}{Re_x^{1/5}}\left[1 - \left(\frac{x_o}{x}\right)\right]^{-0.12} \tag{9.93}$$

valid for $5 \times 10^5 \le Re \le 10^7$. Kays et al. (2004) have reported that the Stanton number for turbulent flow over flat plate without any heated starting length as

$$St_{turb,x} = \frac{0.0287}{Re_x^{1/5} Pr^{0.6}} \tag{9.94}$$

which can be arranged for Nusselt number as

$$Nu_{x,turb} = 0.0287 \cdot Re_x^{4/5} \cdot Pr^{0.4} \tag{9.95}$$

Figure 9.10 shows the local Nusselt number for the laminar and turbulent flow over a flat plate.

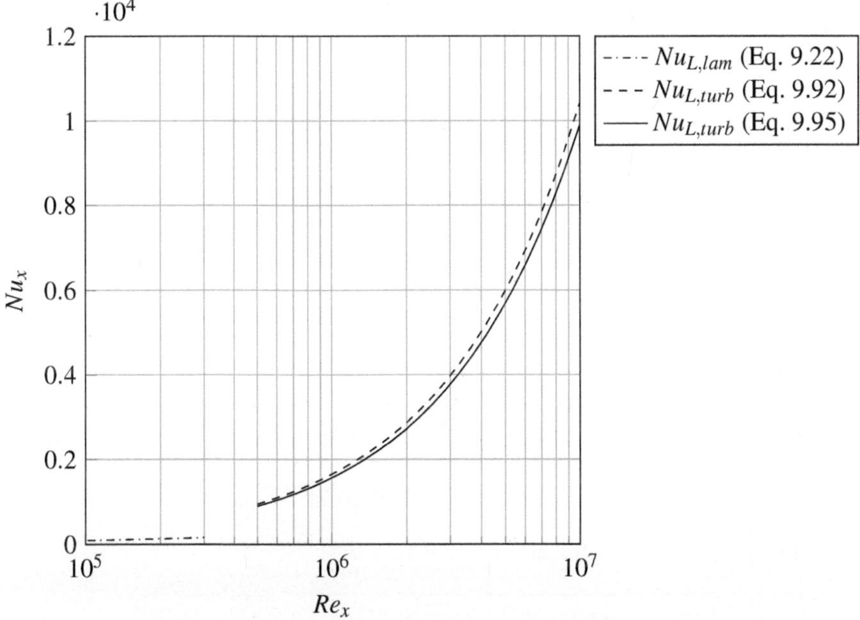

Figure 9.10   Local Nusselt number for the laminar and turbulent flow over a flat plate.

9.7   COMBINED LAMINAR AND TURBULENT FLOW OVER A PLATE

We have already developed the Nusselt number relation for laminar flow over the flat plate as

$$Nu_{x,lam} = 0.332 Re_x^{1/2} Pr^{0.333}$$

For turbulent flow over flat plate, we use the equation reported in previous section:

$$Nu_{x,turb} = 0.0287 Re_x^{4/5} Pr^{0.4}$$

Using these relations and taking lower critical Reynolds number ($Re_c = 2 \times 10^5$) as point of transition, we can arrive at mean Nusselt number defined as

$$Nu_m = \frac{h_m \cdot L}{k}$$

where, the mean convective heat transfer coefficient (h_m) is defined as

$$h_{mean} = \frac{1}{L} \left[\int_0^{x_{tran}} h_{x,lam} dx + \int_{x_{tran}}^{L} h_{x,turb} dx \right] \tag{9.96}$$

where, $x_{tran} = 2E5 \times v/U$ and U_∞ is the freestream velocity. We insert the local convective heat transfer coefficients:

$$h_{x,lam} = \frac{0.332\sqrt{\frac{xU}{v}}Pr^{0.333}k}{x}$$

$$h_{x,turb} = \frac{0.0287\left(\frac{xU}{v}\right)^{0.8}Pr^{0.4}k}{x}$$

On integration we arrive at expression:

$$h_{mean} = \frac{k}{L}\left[296.94Pr^{0.333} - 624.62Pr^{2/5} + 0.03587\left(\frac{U \cdot L}{v}\right)^{4/5}Pr^{2/5}\right]$$

where, $(U \cdot L/v)$ is Re_L. We notice that the middle term can be cast into critical Reynolds number Re_c as it is set to 2×10^5:

$$Nu_m = 624.62 \cdot (0.7)^{2/5} \approx 0.036 \cdot (2E5)^{\frac{4}{5}} \cdot (0.7)^{0.4} = 0.036 \cdot Re_c^{\frac{4}{5}} \cdot (0.7)^{0.4}$$

This leads to the simplification:

$$\boxed{Nu_m = (0.036Re_L^{4/5} - 0.036Re_c^{4/5})Pr^{2/5} + 296.94Pr^{1/3}} \tag{9.97}$$

Now since $Re_{0.8}^c = 17411.01 \approx 17400$, we write mean Nusselt number for the entire plate as

$$Nu_{m,L} = 0.036Pr^{0.43}(Re_L^{0.8} - 17400) + 297Pr^{1/3} \tag{9.98}$$

We further adjust the term $297Pr^{1/3}$ by introducing an approximation

$$297Pr^{1/3} \approx 297Pr^{0.43} \qquad Pr \sim 1$$

which leads to a simplification

$$\boxed{Nu_m = 0.036Pr^{0.43}\left[Re_L^{0.8} - 10300\right]} \tag{9.99}$$

Some researcher have derived the equation for Nusselt number as

$$\boxed{Nu_m = 0.036Pr^{0.43}\left[Re_L^{0.8} - 9400\right]} \tag{9.100}$$

As can be seen in Figure 9.11 the solution proposed by these relations have a validity from $Re \geq 2 \times 10^5$. Also, some had proposed the relation:

$$Nu_m = (0.037Re_L^{0.8} - 871)Pr^{1/3} \quad 5E5 \leq Re \leq 1E7,\ 0.6 \leq Pr \leq 60 \tag{9.101}$$

However, the plot in Figure 9.11 shows that this correlation is only giving acceptable results when the range is set to $2 \times 10^6 \leq Re \leq 1 \times 10^7$.

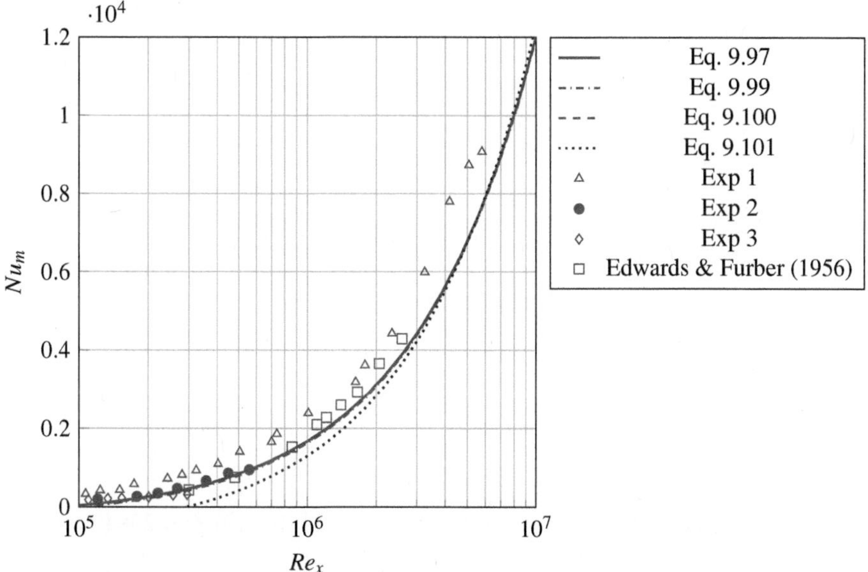

Figure 9.11 Averaged Nusselt number (Nu_m) vs Reynolds number (Re_L) over a flat plate for air. ○ and △ represent laminar flow experimental data, whereas, the symbol ◇ represents turbulent flow experimental data.

The figure shows that equations 9.97, 9.99, and 9.100 are giving same level of Nusselt number. Zhukauskas and Ambrazyavichyus (1978) suggested to include the viscosity ratio for the improved result

$$Nu_m = 0.036 Pr^{0.43} \left[Re^{0.8} - 9400 \right] \left(\frac{\mu_\infty}{\mu_w} \right)^{1/4}$$

If laminar boundary layer is covering small surface area compared to turbulent boundary layer, we can ignore the laminar boundary layer completely specially for large Reynolds number flows and the above equation is reduced to form

$$Nu_m = 0.036 Pr^{0.43} Re^{0.8} \left(\frac{\mu_\infty}{\mu_w} \right)^{1/4}$$

Example 9.4

Air is flowing over a 1.5m×7m flat plate at speed of 9 m/s. The wall temperature (T_w) is 140°C, and free stream temperature (T_∞) is 20°C. Find the rate of heat transfer from the flat plate.

Solution The properties of air at mean temperature of 80°C are

$$k = 0.02953 W/m \cdot K, Pr = 0.7154, v = 0.00002097 m^2/s$$

We compute the Reynolds number:

$$Re_L = \frac{u \cdot L}{v} = \frac{9m/s \times 7m}{0.00002097} = 3 \times 10^6 > 5 \times 10^5$$

The flow is turbulent. The Nusselt number is computed for the whole plate as

$$Nu_m = 0.036 \cdot Pr^{0.43} \cdot (Re_L^{0.8} - 9400) = 4448.88$$

This gives the convective heat transfer coefficient as

$$h_m = \frac{Nu_L \cdot k}{L} = 18.76 \ \frac{W}{m^2 \cdot K}$$

The heat transfer from the surface is

$$Q = h_m \cdot A \cdot (T_w - T_\infty) = 18.76 \frac{W}{m^2 \cdot K} \times 10.5 m^2 \times (140 - 20)^\circ C = 23647.60 W$$

9.8   HEAT TRANSFER FROM FLAT PLATE AT CONSTANT WALL HEAT-FLUX

In the previous section, we considered the laminar heat transfer from an isothermal flat plate. In many industrial applications we encounter situations where wall heat flux is maintained at a constant level and then the wall is experiencing the variation in the temperature. In such cases we take $T_w = f(x)$, where x is the streamwise boundary layer development direction. For the case of wall maintained at the constant-heat-flux, the local Nusselt number for laminar flow can be obtained from relation:

$$\boxed{Nu_x = 0.453 Re_x^{1/2} Pr^{1/3} \qquad (q_w = const)} \qquad (9.102)$$

which is valid for fluid with Prandtl number greater than 0.6. The mean temperature difference is

$$\overline{T_w - T_\infty} = \frac{1}{L} \int_0^L (T_w - T_\infty)\, dx = \frac{1}{L} \int_0^L \left(\frac{q_w x}{k Nu_x} \right) dx$$

$$\overline{T_w - T_\infty} = \frac{q_w L / k}{0.6795 Re_L^{0.5} Pr^{0.333}}$$

For the case of turbulent flow, the applicable correlation is

$$\boxed{Nu_x = 0.0308 Re_x^{4/5} Pr^{0.6} \qquad (q_w = const)} \qquad (9.103)$$

which is valid for $0.6 \leq \text{Pr} \leq 60$.

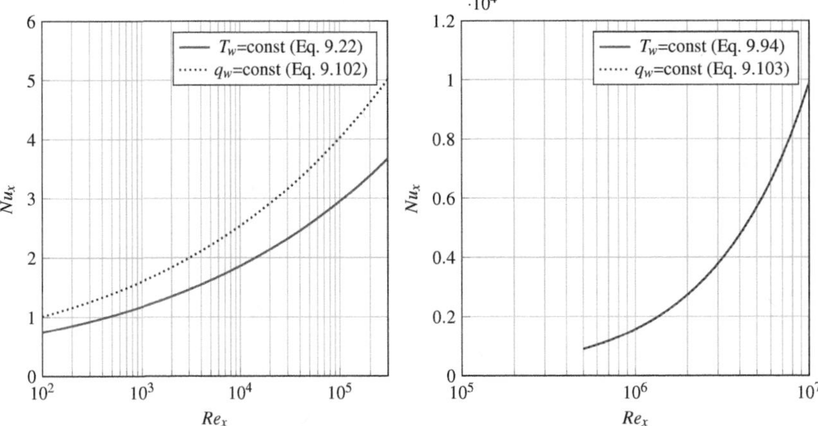

Figure 9.12   (Right) Local Nusselt number versus Reynolds number for laminar flow for constant temperature and constant heat-flux conditions (Pr=0.7). (Left) Local Nusselt number vs. Reynolds number for turbulent flow for constant temperature and constant heat-flux conditions (Pr=0.7).

Figure 9.12 shows that boundary conditions have a strong influence in case of laminar flow, whereas, the Nusselt number in case of turbulent flow is independent of the wall boundary conditions. In case of unheated starting length, for laminar flow the correlation is

$$Nu_x = 0.453 \text{Re}_x^{1/2} Pr^{1/3} \left[1 - \left(\frac{x_o}{x} \right)^{3/4} \right]^{-1/3} \qquad (q_w = const)$$

and for turbulent flow the correlation is

$$Nu_x = 0.0308 \text{Re}_x^{4/5} Pr^{0.6} \left[1 - \left(\frac{x_o}{x} \right)^{9/10} \right]^{-1/9} \qquad (q_w = const)$$

where, x_o is the unheated starting length.

Example 9.5

The cooling fluid moving at 10 ft/s with a free-stream temperature of 300 °R over a series of electronic devices which can be approximated as a flat surface of length 1.25 ft. The device surfaces are experiencing temperature variation with temperature reaching as high as 800 °R. The electronic device is dissipating constant heat flux of 300 Btu/hr $\cdot ft^2$. The properties of cooling fluid at mean temperature are

$$\rho = 0.0812 lb_m/ft^3, \quad c_p = 0.2918 Btu/(lb_m \cdot {}^\circ R),$$
$$k = 0.01546 Btu/(hr.ft \cdot {}^\circ R), v = 17.07 \times 10^{-5} \ ft^2/s$$
$$Pr = 0.709$$

Solution We calculate the Reynolds number over the flat surface as

$$Re = \frac{u \cdot L}{v} = \frac{(10)(1.25)}{17.07 \times 10^{-5}} = 73227.88 < 3 \times 10^5$$

We assume that the Nu relation for the heat transfer with constant heat-flux condition (equation 9.102) is applicable here

$$Nu = 0.453 \cdot Re^{0.5} \cdot Pr^{0.333} = 109.320$$

Using Newton's law of cooling, we can write

$$\Delta T = \frac{\frac{q_w \cdot L}{k}}{0.453 \cdot Re^{0.5} \cdot Pr^{0.333}} = 222.745 \, {}^\circ R$$

Since

$$\Delta T = T_w - T_\infty$$

the temperature at the end of plate is

$$T_w = 300 + 222.745 = 522.74 \, {}^\circ R$$

Note that this solution is based on a crude approximation of constant heat-flux conditions. In reality, the flux will be varing over the surface of device and due to small gaps between the devices the boundary layer would trigger into transition.

Correlation Sheet for Flat plate Flows

Laminar flow over a flat plate

$$\left(\frac{\delta}{x}\right)_{laminar} = \frac{5}{\sqrt{Re_x}}$$

$$C_{f,x} = \frac{0.664}{\sqrt{Re_x}} \quad \text{and} \quad C_{f,L,lam} = \frac{1.328}{\sqrt{Re_L}}$$

Heat transfer in laminar flow over a flat plate ($T_w = $ const)

$$
\left.
\begin{array}{l}
Nu_{x,lam} = 0.332 \cdot \sqrt{Re_x} \cdot Pr^{1/3} \\
Nu_{L,lam} = 0.664 \cdot \sqrt{Re_x} \cdot Pr^{1/3}
\end{array}
\right\} \quad 0.5 \leq Pr \leq 10, \ Re < 3 \times 10^5
$$

$$
Nu_{x,lam} = 0.339 \cdot \sqrt{Re_x} \cdot Pr^{1/3} \qquad Pr > 10
$$

$$
Nu_{x,lam} = 0.565 \cdot \sqrt{Re_x \cdot Pr} \qquad Pr \leq 1
$$

Turbulent flow over a flat plate

$$
\left(\frac{\delta}{x} \right)_{turbulent} = 0.37 Re_x^{-1/5}
$$

$$
C_{f,x} = \frac{0.0592}{Re_x^{1/5}} \ \text{and} \ C_{f,L,turb} = \frac{0.074}{Re_L^{1/5}}
$$

Heat transfer in turbulent flow over a flat plate ($T_w = $ const)

$$
Nu_{x,turb} = 0.0287 Re_x^{4/5} Pr^{0.4} \qquad Re > 5 \times 10^5
$$

Combined laminar and turbulent flow over a flat plate ($T_w = $ const)

$$
Nu_{m,lam-turb} = 0.036 Pr^{0.43} \left[Re^{0.8} - 9400 \right]
$$

Flow over a flat plate with unheated starting length ($T_w = $ const)

$$
Nu_{x,lam} = 0.332 Pr^{1/3} Re_x^{1/2} \left[1 - \left(\frac{x_o}{x} \right)^{3/4} \right]^{-1/3} \qquad Re < 3 \times 10^5
$$

$$
Nu_{x,turb} = 0.0295 Re^{0.8} Pr^{1/3} \left[1 - \left(\frac{x_o}{x} \right)^{9/10} \right]^{-1/9} \qquad Re > 5 \times 10^5
$$

Flow over a plate with constant heat flux condition ($q_w = $ const)

$$
Nu_x = 0.453 Re_x^{1/2} Pr^{1/3} \left[1 - \left(\frac{x_o}{x} \right)^{3/4} \right]^{-1/3} \qquad Re < 3 \times 10^5
$$

$$
Nu_x = 0.0308 Re_x^{4/5} Pr^{0.6} \left[1 - \left(\frac{x_o}{x} \right)^{9/10} \right]^{-1/9} \qquad Re > 3 \times 10^5
$$

PROBLEMS

9P-1 Air at a free stream temperature of $T_\infty = 30\ °C$ is moving over a flat plate of length L=6 m. The plate is maintained at temperature of $T_w = 90\ °C$. The temperatures inside the thermal boundary layer can be expressed as $T = 30 + 60 \cdot \exp(-500 \cdot y \cdot x)$, where x and y are in meters. Find the local convective heat transfer coefficient (h_x) and the mean convective heat transfer coefficient (h_L) over the plate.

9P-2 In an industrial application, air at 2 °C is flowing over a surface of 1 m length. The heating of plate is started after x_o cm such that the wall temperature is kept constant at 40 °C. Find the heat transfer from the surface to air if speed of air is 0.5 m/s. Find (i) how much is Nusselt number if x_o is 30 cm? (ii) How much is Nusselt number if x_o is increased to 50 cm? (iii) What is the thermal boundary layer thickness at the end of the surface if x_o is 10 cm?

9P-3 Hydrogen is flowing at $T_\infty=200\ °C$ over a 2 m flat surface at velocity of 50 m/s. The plate is maintained at a constant wall temperature $T_\infty=300\ °C$. The properties of hydrogen at mean temperature of 250 °C are: $\rho= 0.09819\ kg/m^3$, $c_p = 14059$ J/kg·K, $\nu= 80.64 \times10^{-6}\ m^2/s$, k= 0.1561 W/m·K, $\alpha= 1.130\times10^{-4}\ m^2/s$, Pr 0.713.

9P-4 The local convective heat transfer over a surface is described by the relation:

$$h(x) = \frac{68.34}{x^{\frac{1}{5}}}$$

Find the mean convective heat transfer coefficient.

9P-5 The radial distribution of convective heat transfer due to jet impingement on a target wall is expressed as

$$h(r) = 5 + 100 \cdot r^2 + \frac{0.0011}{r^4}$$

for $0.5 \leq r/D \leq 2$. Find the radial mean convective heat transfer coefficient.

9P-6 Refrigerant 134a is flowing over a flat heated plate at free stream velocity of 40 m/s and temperature -43°C. The plated is maintained at constant temperature of 97°C. (i) Sketch momentum and thermal boundary layers along with velocity and temperature distributions inside respective boundary layers, (ii) Find heat transfer from plate using Reynolds and Colbourn analogy, (iii) Find heat transfer from plate using appropriate correlation.

9P-7 In an industrial application, air at 2°C is flowing over a surface of 1 m length. The heating of plate is started after x_o such that the wall temperature is kept constant at 40°C. Find the heat transfer from the surface to air if speed of air is 0.5 m/s. Find (i) How much is Nusselt number if x_o is 30 cm? (ii) How much is Nusselt number if x_o is increased to 50 cm? (iii) What is the thermal boundary layer thickness at the end of the surface if x_o is 10 cm?

9P-8 The wing of an airplane is manufactured from aluminium. At an altitude of 2000 m the wing is exposed to solar insolation of 800 W/m^2. The wing has a chord of 7-m length and it experiencing free stream speed of 300 m/s. Find the heat transfer from the wing to air. Take assumptions to sove this problem.

9P-9 Air is moving over an electronic circuit board of length 20 cm at temperature of $T_\infty=25°$C. The convective cooling creates h=20W/m^2·K. Due to protuding devices the boundary layer immediately turned into turbulent flow. If the circuit board is dissipating 300 W and temperature of the circuit board should not exceed 70°C, what speed is best suited to achieve this?

9P-10 Wind blows at speed of 0.2 m/s parallel to a flat roof surface of length 10 m. The area of the surface is 0.04 m^2. The surface temperature is 40 °C, and the free stream temperature of the air is 20 °C. Find the total drag force experienced by the surface and, also estimate the total heat transfer rate.

9P-11 Water at speed of 0.05 m/s enters in a duct of 20 cm × 30 cm at 20 °C. Assume that boundary layer is small and flow over the duct walls can be treated as a flat plate boundary layer flow. (a) Find the local heat transfer coefficient at position 1 m. (b) find the total heat transfer. The duct is maintained at constant temperature of 50 °C.

REFERENCES

H. Blasius. The Boundary Layers in Fluids with Little Friction (in German). Z. Math. Phys., 56, 1 (1908); pp. 1–37; English translation in National Advisory Committee for Aeronautics Technical Memo No. 1256, February 1950.

H. Blasius H (1908) Grenzschichten in flüssigkeiten mit kleiner Reibung. Z Angew Math Phy 56: 1–37. English version: Blasius H (1950) The boundary layers in fluids with little fiction, NACA-TM-1256, 1908.

L. Prandtl, Investigations on Turbulent Flow, ZAMM 5, 136, Germany, 1925.

E. Pohlhausen, Der Wärmeaustausch zwischen festen körpern und flüssigkeiten mit kleiner reibung und kleiner wärmeleitung. Z Angew Math Mech 1(2):115–121, 1921.

P. S. Klebanoff, Characteristics of Turbulence in a Boundary Layer with Zero Pressure Gradient, NACA Report 1247, USA, 1955.

E. R. Van Driest, On Turbulent Flow Near a Wall, J. Aero. Sci. 23, 1007, 1956.

V. S. Arpaci, P.S. Larsen, Convection Heat Transfer, Prentice-Hall, Inc., Englewood Cliffs, NJ, USA, 1984.

L. C. Burmesiter, Convective Heat and Mass Transfer, John Wiley, New York, 2004.

W. Kays, M. Crawford, B. Weigand, Convective Heat and Mass Transfer, Mc Graw-Hill, New York, 2004.

10 Forced Convection on Curved Surfaces and on an impact

Heat transfer over curved surfaces is a kind of a complex flow, as extra strains are generated in flow due to curvature. The flow will undergo boundary layer separation and wake region would have a profound influence over the heat transfer. We will discuss the drag experience by the flow and general flow features to understand the complex dynamics of the flow. We will learn about different correlations to be used for the estimation of Nusselt number for flow over cylinder and spheres. We will learn about heat transfer in case of cross flow over tube banks in aligned and staggered configurations. After finishing this chapter, one should be able to:

- Compute heat transfer by a cylinder or sphere surface.
- Compute heat transfer for flow through a tube bank.

Heat transfer over curved surfaces is bit complex as compared to flow over flat surface (see Figure 10.1). The flow over curved surface involves the flow acceleration and separation. The pressure gradient cannot be taken as zero and formulation of an analytical model is no longer a simple task. The fluid moves over the curved surface with acceleration due to presence of the favourable pressure gradient ($dU/dx > 0$ when $dp/dx < 0$), and it attains increased velocity when $dp/dx = 0$. Further downstream, the flow would decelerate due to the adverse pressure gradient ($dU/dx < 0$ when $dp/dx > 0$). Eventually, the flow would separate and a wake region will be formed.

Due to these issues, the researchers resort to experiments to understand the flow related drag and heat transfer in details. The part of the drag that arises due to surface friction is called the skin friction drag, whereas the drag that occurs because of the object's shape, which causes the pressure difference across a body, is called the pressure drag. In certain cases, the vortex shedding also induces a drag in the object, called the induced drag. If the object is experiencing a shockwave, it will be an added drag component, called the wave drag.

The bodies with extensive surface area experience considerable frictional drag. The friction drag is minimal for the flat surface that is normal to the flow, whereas the pressure drag is minimum for the surface parallel to the flow. In specific industrial and practical applications, the object's shape cannot be changed. However, by

DOI: 10.1201/9781003428404-10

designing the wall/surface, the surface drag or skin friction drag can be significantly reduced. Experimentally it has been found that different vortex structures form over the cylinder when free-stream flow passes over it. Different flow behaviour according to the Reynolds number, based on the cylinder diameter and the free-stream velocity are outlined below:

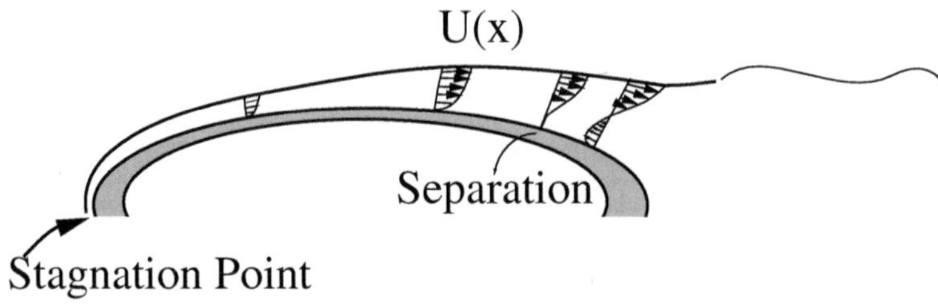

Figure 10.1   Flow over curved surfaces.

Classification of different flow regimes for flow over a circular cylinder (see Zdravkovich (1997)		
Flow	Regime	Re range
Laminar	No separation	0 to 4 - 5
	Closed wake	4 - 5 to 30 - 50
	Periodic wake	30 - 50 to 180 - 200
Transition in Wake	Far wake transition	180 - 200 to 220 - 250
	Near wake transition	220 - 250 to 350 - 400
Subcritical transition	Lower	350 - 400 to 1E3 - 2E3
	Intermediate	1E3 - 2E3 to 2E4 - 4E4
	Upper	2E4 - 4E4 to 1E5 - 2E5
Transition in Boundary layer	Precritical	1E5 - 2E5 to 3E5 - 3.4E5
	Single bubble	3E5- 3.4E5 to 3.8E5 - 4E5
	Two bubble	3.8E5 - 4E5 to 5E5 - 1E6
	Supercritical	5E5- 1E6 to 3.5E6 - 6E6
	Postcritical	3.5E6 - 6E6

The table lists different flow regimes which may occur for flow over a cylinder, as stipulated by Zdravkovich (1997).

We may divide the flow cylinder into several regimes. Some of the flow regimes are depicted schematically in Figure 10.2.

Regime I (Creeping flow) [Re< 1] In this regime the Reynolds number is small being less than one. We know that the Reynolds number is the ratio of inertia forces to viscous forces so we could say that the viscous forces predominate in this flow. The flow pattern in this case is almost symmetric over the cylinder and the wake will be free from oscillation. This is one type of the creeping flow.

Regime II [10 < Re < 1000] As a Reynolds numbers increased beyond 10 to 1000, small eddies are formed near the rear stagnation point and they will grow larger as a Reynolds number is increased. The pattern of flow in wake is called the von Karman vortex street or vortex trail. The wake in this regime is unsteady but patriotic.

Regime III [1000 < Re < 5 $\times 10^5$] In this third regime the point of separation stabilises about $80°$ from the forward stagnation point. The wake is no longer characterised by large eddies, but the flow remains unsteady in wake. The boundary layer near the stagnation point to the point of separation is laminar and the shear stress in this interval is appreciable only in a thin layer near the surface. The drag coefficient value levels out to almost constant value of one.

Regime IV [Re $\approx 5\times10^5$] Reynolds number near 5×10^5 the drag coefficient suddenly decreases to value of 0.3. The point of separation will move past $90°$. Close to the stagnation point on the cylinder, the boundary layer will quickly go into the transition and flow is predominantly controlled by inertia forces as the boundary layer traverse over the remaining cylinder body.

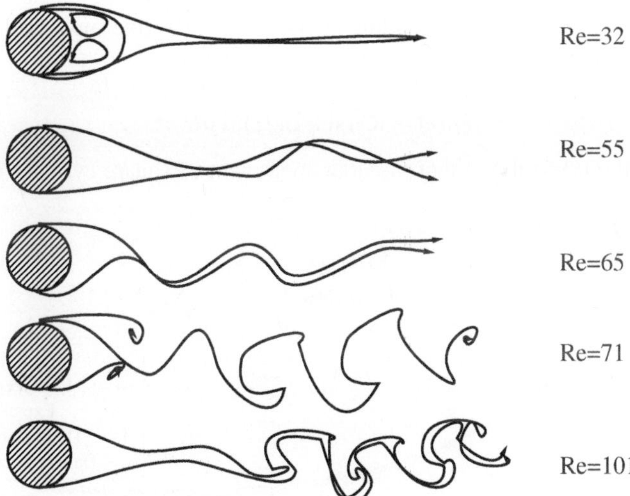

Figure 10.2   The vortex shedding behind cylinder at different Reynolds numbers.

Figure 10.2 shows the vortex shedding behind cylinder at Reynolds numbers from Re_D=55-101.

Definition (*Drag and Lift Forces*): Drag Force:

$$F_D = C_D \underbrace{\left(\frac{1}{2}\rho V^2\right)}_{Dynamic\ Pressure} A_{projected}$$

Lift Force:

$$F_L = C_L \underbrace{\left(\frac{1}{2} \rho V^2 \right)}_{Dynamic\ Pressure} A_{planform}$$

Wadell (1934) and Dallavalle (1948) proposed the equation for drag coefficient calculation:

$$C_D = \left(0.632 + \frac{4.8}{\sqrt{Re_D}} \right)^2$$

Darby proposed the equation for drag coefficient calculation:

$$C_D = \left(1.05 + \frac{1.9}{\sqrt{Re_D}} \right)^2 \qquad Re \leq 2 \times 10^5$$

Figure 10.3 shows the drag coefficients for cylinders and other column shaped bodies.

In case of periodic vortex shedding after Re=3.5×10^6 (the super critical regime), the drag force could also oscillate in sinusoidal manner:

$$F_D(t) = [C_D \sin(2\pi f t)] \tfrac{1}{2}\rho U^2 A$$

Lyn et al. (1995) found that the drag coefficient of square cylinder is related to circular cylinder as:

$$\frac{C_{Dsquare}}{C_{Dcircular}} \simeq 1.7$$

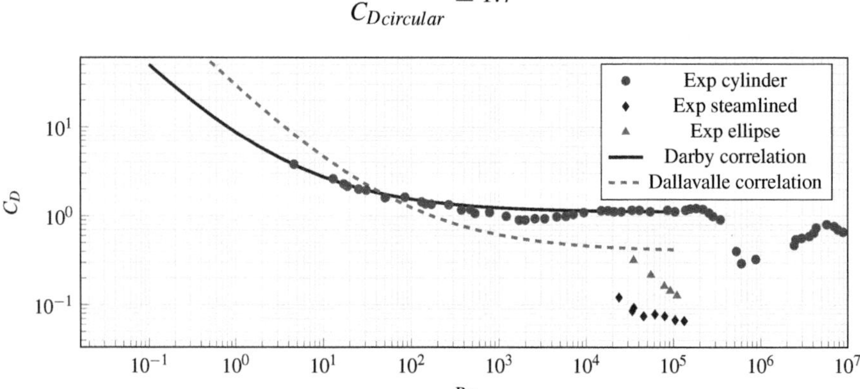

Figure 10.3   Drag coefficients for cylinders and other column shaped bodies.

The drag over a sphere is an interesting inquiry which has attracted the attention of many researchers over the year. Flow over sphere has been divided into regimes:

$$Re \sim 24 \quad (Laminar\ Flow\ Separation)$$
$$Re \le 2 \times 10^5\ (Subcritical\ Flow)$$
$$2 \times 10^5 \le Re \le 4 \times 10^5\ (Critical\ Flow)$$
$$4 \times 10^5 \le Re \le 10^6\ (Supercritical\ Flow)$$
$$Re \ge 10^6\ (Transcritical\ Flow)$$

Figure 10.4 shows the drag coefficient over the sphere.

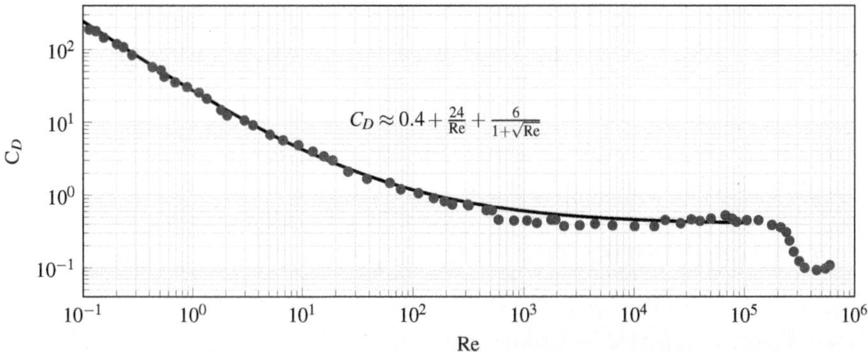

Figure 10.4 Comparison of experiment and empirical formulas for drag coefficients of a smooth sphere.

The following curve-fit formula for the (laminar-flow) data is provided by White (1993) and plotted along with experimental data.

$$C_D \approx 0.4 + \frac{24}{Re} + \frac{6}{1 + \sqrt{Re}}$$

It is valid for the range $0 \le Re \le 2 \times 10^5$.

The accuracy is $\pm\ 10\ \%$ up to the drag crisis, $Re_D \approx 250{,}000$, where the boundary layer on the sphere will become turbulent and the wake will be thin. This will lead to sudden drag reduction, and it may occur earlier with rough surfaces and with fluctuating free-stream velocity conditions.

In tube bundles, the vortex shedding of one tube will directly effects the vortex shedding frequency of the subsequent tubes. The range of Strouhal numbers is between 1 and 1.6. In case the vortex shedding frequency matches to the bundle's Eigen-frequency or natural frequency due to resonance, the bundle will experience severe oscillations and may damage.

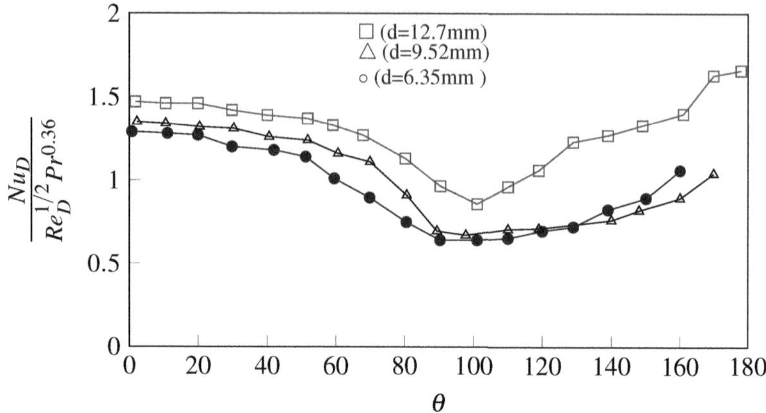

Figure 10.5 Nu_D over cylinders of three different cross-sections at Re=10,000. (Data from Perkins and Leppert (1964).)

10.1   CORRELATIONS FOR HEAT TRANSFER OVER A CYLINDER

Figure 10.5 shows the Nusselt number distribution over cylindrical objects of different shapes as reported by Perkins and Leppert (1964). We focus here on flow and heat transfer from cylindrical surface. The Reynolds number of the flow is 10,000 ($<$ 2×10^5) and thus the flow is in laminar regime. The expected boundary layer separation is around angle 82° measured from stagnation point (see Uddin 2022) and from Figure 10.5 it is clear that after this angle the heat transfer is also changed drastically. Thus the flow has a huge influence over the heat transfer.

Martinelli et al. (1943) have proposed the correlation for Nusselt number distribution over the surface as a function of angle β measured from stagnation point.

$$Nu_D(\beta) = \frac{h(\beta)D}{k} = 1.05 Pr^{1/3} \sqrt{\mathrm{Re}_D} \left[1 - \left(\frac{\beta}{90} \right)^3 \right] \qquad 0 \leq \beta \leq 80° \qquad (10.1)$$

Knudsen and Katz (1958) proposed the following relation for Nusselt number distribution over cylinder:

$$Nu_D = 1.14 Pr^{0.4} \mathrm{Re}_D^{0.5} \left[1 - \left(\frac{\beta}{90} \right)^3 \right] \qquad 0 \leq \beta \leq 80° \qquad (10.2)$$

Figure 10.6 shows that correlation 10.2 is giving better predictions for Nusselt number in the laminar flow regime. Morgan (1975) proposed the Nusselt number correlation based on the cylinder diameter:

$$Nu_D = m.\mathrm{Re}_D^p \qquad (10.3)$$

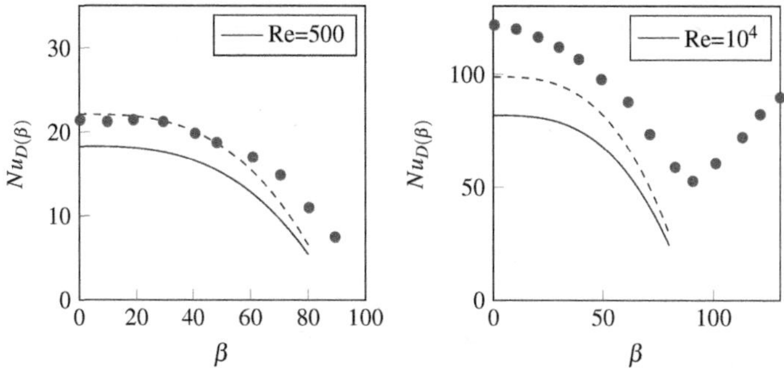

Figure 10.6   Nu distribution with cylinder angle (β) measured from stagnation point at Re=500 and 10,000 (Žukauskas data). Compared with Martinelli et al. (Eq. 10.1), and Knudsen and Katz (Eq. 10.2) proposals.

where Re_D Reynolds number based on cylinder diameter and the upstream normal velocity. The m and p are listed in Table 10.1

Nusselt number predictions for flow of air over cylinder is plotted in Figure 10.7.

Pasternak and Gauvin (1960) proposed to use a new length scale ℓ defined as the surface area of a cylinder divided by the perimeter wetted by fluid. For long cylinder with diameter d, the perimeter length ℓ is

$$\ell = \frac{A_s}{WP} = \lim_{L \to \infty} \frac{\pi.d.L}{2(d+L)} = \lim_{L \to \infty} \frac{\pi.dL}{2L\left[\frac{d}{L}+1\right]} = \frac{\pi.d}{2} \tag{10.4}$$

TABLE 10.1

Coefficients to be Used with Equation 10.3, Air Flow Normal to a Circular Cylinder [Morgan (1975)]

Re_D	m	p
0.0001 - 0.004	0.437	0.0895
0.004 - 0.09	0.565	0.136
0.09 - 1	0.8	0.28
1 - 35	0.795	0.384
35 - 5,000	0.583	0.471
5,000 - 50,000	0.148	0.633
50,000 - 200,000	0.0208	0.814

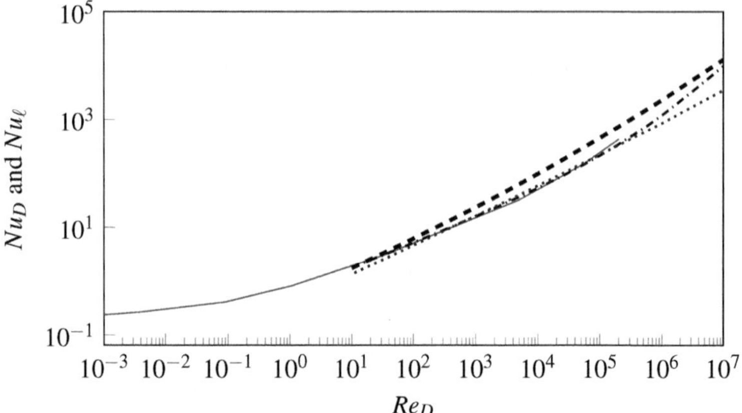

Figure 10.7   Nusselt number predictions for flow of air (Pr=0.7) over cylinder according to Morgan correlation 10.3 (solid line), Gnielinski correlation (1975) (dashed line), and Churchill and Bernstein (1977) dash-dotted line, Collis and Williams (1959) dotted line.

Gnielinski (1975) has proposed the following correlation for Nu predictions spanning over both laminar and turbulent regimes:

$$Nu_{o,\ell} = 0.3 + \sqrt{Nu^2_{\ell,lam} + Nu^2_{\ell,turb}} \tag{10.5}$$

$$Nu_{\ell,lam} = 0.664(Pr)^{1/3}\sqrt{\mathrm{Re}_\ell} \tag{10.6}$$

$$Nu_{\ell,turb} = \frac{0.037\, Pr\, \mathrm{Re}_\ell^{0.8}}{1 + 2.334\mathrm{Re}_\ell^{-0.1}(Pr^{2/3}-1)} \tag{10.7}$$

where,

$$Nu_\ell = \frac{h.\ell}{k}$$

$$\mathrm{Re}_\ell = \frac{u.\ell}{\nu} \qquad 10 < \mathrm{Re}_\ell < 10^7$$

and

$$0.6 < Pr < 1000$$

Churchill and Bernstein (1977) have proposed a correlation for a wide range of Pr:

$$\overline{Nu_D} = 0.3 + \frac{0.62Pr^{1/3}\sqrt{\mathrm{Re}_D}}{\sqrt[4]{1+\left(\frac{0.4}{Pr^{2/3}}\right)}}\left[1+\left(\frac{\mathrm{Re}_D}{282000}\right)^{5/8}\right]^{4/5} \qquad Pe \geq 0.2 \tag{10.8}$$

For flow Reynolds numbers in the range of 0.02 to 44 with negligible temperature differences, Collis and Williams (1959) presented the correlation:

$$Nu = 0.35\sqrt{Re_D} + 0.052\sqrt[3]{Re_D^2} \tag{10.9}$$

valid for $Re_D > 10$ and valid for air only.

The above correlations were mainly meant for isothermal conditions. Following correlations were developed for iso-flux boundary condition:

Sarma and Sukhatme (1977) Correlation:

$$Nu_D = 0.62Re_D^{0.505} \qquad 1200 < Re_D < 4700 \tag{10.10}$$

Žukauskas and Žiugžda (1985) Correlation:

$$Nu_D = 0.29Re_D^{0.6} \qquad 10^3 < Re_D < 10^5 \tag{10.11}$$

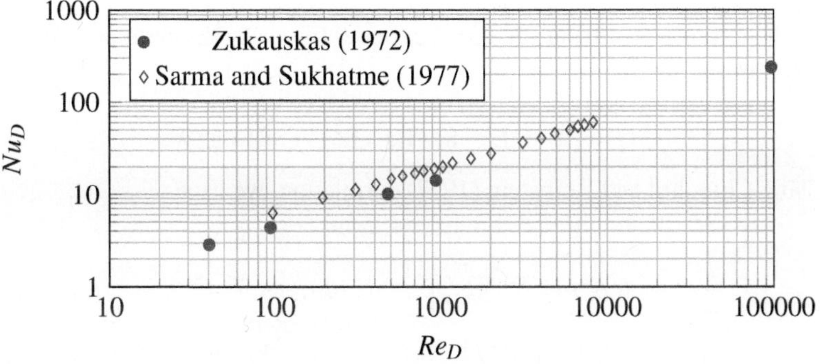

Figure 10.8 Mean Nusselt number distribution over cylinder with Reynolds number for isothermal (circle) and isoflux boundary condition (diamond).

Nusselt number predictions for flow of air over cylinder is plotted in Figure 10.8.

Example 10.1

Atmospheric Air is moving over a 5.5 cm diameter cylinder at a speed of 60 m/s. The length of cylinder is 10 cm. Calculate the heat loss from cylinder, if the free-stream and cylinder temperatures are 35°C and 150°C.

Solution We use the properties at the mean temperature of 92.5 °C:
$\rho = 0.966 \ kg/m^3, \mu = 2.14 \times 10^{-5} \ Pa.s,$
$k = 0.0312 \ W/m \cdot K, Pr = 0.695$
Given in problem statement the cylinder dimensions and speed are

$$d = 0.055 \ m, u = 60 m/s, L = 0.10 m$$

We calculate the Reynolds number

$$Re = \frac{\rho \cdot u \cdot d}{\mu} = 1.489 \times 10^5$$

Using Morgan correlation and Table 10.1 we have

$$m = 0.0208$$

$$p = 0.814$$

$$Nu = m \cdot Re^p = 338.02$$

The convective heat transfer coefficient is

$$h = \frac{Nu \cdot k}{d} = 191.75 W/m^2 \cdot K$$

and heat transferred from surface of cylinder (ignoring the circular cross sections) is

$$Q = (\pi \cdot d \cdot L \cdot) \cdot h \cdot (T_w - T_\infty) = 381.022 W$$

Using Churchill and Bernstein (1977) correlation, we have

$$\overline{Nu}_D = 0.3 + \frac{0.62 Pr^{1/3} \sqrt{Re_D}}{\sqrt[4]{1 + \left(\frac{0.4}{Pr^{2/3}}\right)}} \left[1 + \left(\frac{Re_D}{282000}\right)^{5/8} \right]^{4/5} \qquad Pe \geq 0.2$$

$$Nu_D = 288.687$$

The convective heat transfer coefficient is

$$h = \frac{Nu \cdot k}{d} = 163.764 W/m^2 \cdot K$$

and heat transferred from surface of cylinder (ignoring the circular cross sections) is

$$Q = (\pi \cdot d \cdot L) \cdot h \cdot (T_w - T_\infty) = 325.40 W$$

Example 10.2

Atmospheric Air is moving over a 10 cm diameter cylinder at a speed of 20 m/s. The length of cylinder is 10 cm. Calculate the temperature of cylinder surface, if the free-stream temperature is 92.5 °C, and cylinder is maintained at constant heat flux of 800 W/m^2.

Solution We use the properties at the mean temperature of 92.5 °C:

$$\rho = 0.966 \ kg/m^3, \mu = 2.14 \times 10^{-5} \ Pa.s, k = 0.0312 \ W/m \cdot K, Pr = 0.695$$

Given in problem statement the cylinder dimensions and speed are
$d = 0.1 \ m, u = 20 m/s, L = 0.10 m$
We calculate the Reynolds number

$$Re = \frac{\rho \cdot u \cdot d}{\mu} = 9.0280 \times 10^4$$

Using Žukauskas and Žiugžda (1985) Correlation, we have

$$Nu_D = 0.29 Re_D^{0.6} \qquad 10^3 < Re_D < 10^5$$

$$Nu_D = 272.74$$

$$h = \frac{Nu \cdot k}{d} = 85.09 W/m^2 \cdot K$$

$$q_w = \frac{Q}{A} = h(T_w - T_\infty)$$

$$T_w = \frac{q_w}{h} + T_\infty = 44.401 \ ^\circ C$$

10.2 CORRELATIONS FOR HEAT TRANSFER OVER A SPHERE

Figure 10.9 shows the local Nusselt number along the surface of a sphere.

Experimental investigations indicate that for heat transfer from sphere there is a sudden rise in the heat transfer coefficient beyond a critical Reynolds number of 2.9×10^5.

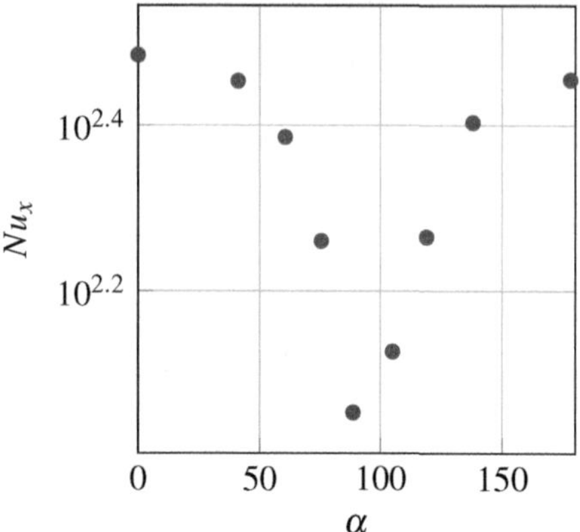

Figure 10.9   Local Nu vs. angle on a sphere (data according to Aufdermauer and Joss 1967).

Kramers (1946) proposed following correlation for heat transfer predictions:

$$Nu_{sphere} = 1 + 1.3Pr^{0.915}Re^{0.5} \qquad \begin{array}{c} 0.4 < Re < 2100 \\ 0.71 < Pr < 380 \end{array}$$

Yuge (1960) proposed correlations for air:

$$Nu_{sphere} = 2 + 0.493Re^{0.5} \qquad 10 < Re < 1.8E3$$
$$Nu_{sphere} = 2 + 0.3Re^{0.57} \qquad 1.8E3 < Re < 1.5E5$$

Whitaker (1972) proposed the following correlation that is reported in many textbooks:

$$\overline{Nu}_D = 2 + \sqrt[4]{\left(\frac{\mu}{\mu_s}\right)}\left[0.4\sqrt{Re_D} + 0.06\sqrt[3]{Re_D^2}\right]Pr^{0.4}$$

valid for

$$0.71 \leq Pr \leq 380$$

$$3.5 \leq Re_D \leq 76000$$

$$1 \leq \left(\frac{\mu}{\mu_s}\right) \leq 3.2$$

However, the Nu predicted from this correlation can have maximum deviation with reported experimental data as high as 30% (see Churchill and Bernstein (1977)).

Eastop and Smith (1972) developed a correlation for heat transfer by air based on theoretical considerations of laminar boundary layers for the stagnation zone of the sphere and experimental data for the rear half of the sphere where the flow has been separated. Their correlation for Nusselt number and fluid as air is:

$$Nu = 0.42\sqrt{Re_D} + 0.0035Re_D^{0.92}$$

valid for $3000 < Re_D < 10^5$.

Ranz and Marshall (1952) proposed a correlation for heat and mass transfer from spheres which are freely falling as liquid drops as:

$$\overline{Nu}_D = 2 + 0.6 Pr^{1/3}\sqrt{Re_D}$$

Will, Kruyt, and Venner (2017) suggested the correlation for heat transfer from sphere as

$$Nu_D = A \cdot \sqrt{Re_D} + B \cdot Re_D \} \qquad 7800 \le Re \le 2.9E5$$

where,

$$A = 0.493 \pm 0.015$$
$$B = 0.0011 \times (1 \pm 0.035)$$

Example 10.3

Atmospheric Air at 27°C moves around a sphere of 10 mm-diameter. The speed of air is 6 m/s. The sphere surface temperature is held contant at 77 °C. Find the heat transferred from sphere is the atmospheric air.

Solution

We use the properties at temperature of 27°C:
$$\mu = 1.8462 \times 10^{-5} \ Pa.s, \ \mu_w = 2.075 \times 10^{-5} \ Pa.s,$$
$$v = 15.7 \times 10^{-6} m^2/s, k = 0.0262 \ W/m \cdot K, Pr = 0.709$$

$$d = 0.1 \ \text{m}, u = 20 \text{m/s}, A_w = 4\pi r^2 = 0.000452 \text{m}^2$$

The Reynolds number is

$$Re = \frac{\rho \cdot u \cdot d}{\mu} = 3824.09$$

We use the Whitaker (1972) correlation:

$$\overline{Nu}_D = 2 + \sqrt[4]{\left(\frac{\mu}{\mu_s}\right)} \left[0.4\sqrt{Re_D} + 0.06 \ \sqrt[3]{Re_D^2} \right] Pr^{0.4}$$

$$Nu = 2 + \left(\frac{\mu}{\mu_w}\right)^{\frac{1}{4}} \cdot Pr^{0.4} \cdot \left(0.4 \cdot \sqrt{Re} + 0.06 \cdot \sqrt[3]{Re^2} \right) = 35.32$$

$$h = \frac{Nu \cdot k}{d} = 92.69 W/m^2 \cdot K$$

$$Q = h \cdot A_w \cdot (T_w - T_\infty) = 1.456 W$$

From Eastop and Smith (1972) correlation we have

$$Nu = 0.42\sqrt{Re_D} + 0.0035 Re_D^{0.92}$$

$$Nu = 32.89$$

$$h = \frac{Nu \cdot k}{d} = 86.305 W/m^2 \cdot K$$

$$Q = h \cdot A_w \cdot (T_w - T_\infty) = 1.355 W$$

Yuge correlations for air:

$$Nu_{sphere} = 2 + 0.493 Re^{0.5} \qquad 10 < Re < 1.8E3$$

$$Nu = 32.486$$

$$h = \frac{Nu \cdot k}{d} = 85.24 W/m^2 \cdot K$$

Comparison of convective heat transfer coefficient

Correlation	h $W/m^2 \cdot$K
Whitaker (1972)	92.69
Eastop and Smith (1972)	86.3
Yuge (1960)	85.24

Figure 10.10 shows the Yuge and Eastop and Smith (1972) correlations are giving results close to each other.

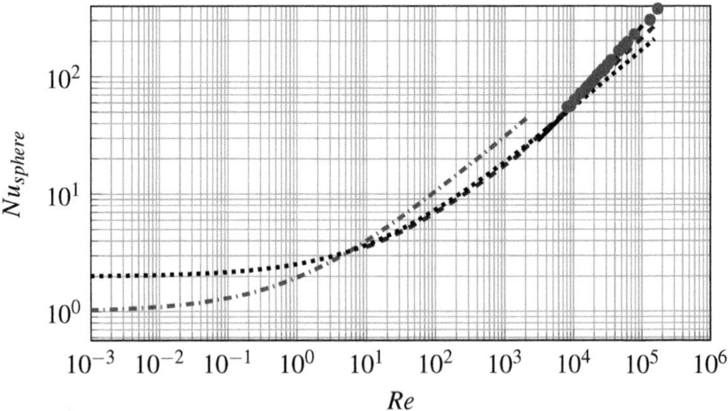

Figure 10.10   Heat transfer from a sphere. Kramers correlation (dash dotted line), Eastop and Smith correlation (solid line), Yuge correlation (dashed line), Ranz and Marshall (dotted line); experimental data from Churchill and Bernstein (1977).

10.3 FLOW OVER TUBE BANKS

Flow over tube banks is very complex, and it has applications in heat exchangers, boilers, air conditioners (see Figure 10.11). Also, the tube banks can be found in filtration and food processing plants. Owing to their importance, flow and heat transfer over tube banks has been the focus of investigations. Zukauskas (1972) found that a small displacement between adjacent tubes in a transverse row may result in a significant change in the pressure gradient. Also, the velocity distribution inside tube banks is influenced by a change in the boundary layer separation and wake region.

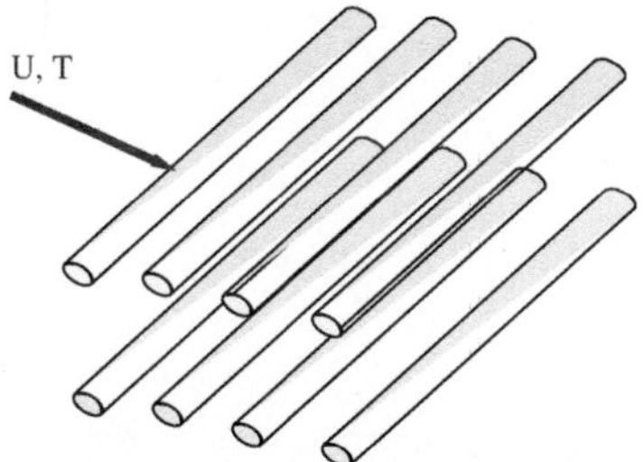

Figure 10.11   Schematic diagram of a tube bank in a cross flow.

In Figure 10.12 S_T and S_L are transverse and longitudinal pitch, respectively. The ratio of pitches with diameter is defined as dimensionless transverse pitch (a) and longitudinal pitch (b) parameters:

$$a = \frac{S_T}{D}$$

$$b = \frac{S_L}{D}$$

Most commonly used tube banks arrangements are

1. Inline square (a = b)

2. Rotated square (a = 2b), and

3. Equilateral triangle (a = 2b/$\sqrt{(3)}$)

If the product of tube banks parameter $a \times b < 1.25^2$, then such tube banks are considered as compact arrangements. Whereas, those having $a \times b > 4$ are considered as widely spaced tube banks (Beale 1997). The pressure distribution in the tube bank is much affected by the longitudinal pitch. Whereas, the velocity distribution is much influenced by the transverse pitch.

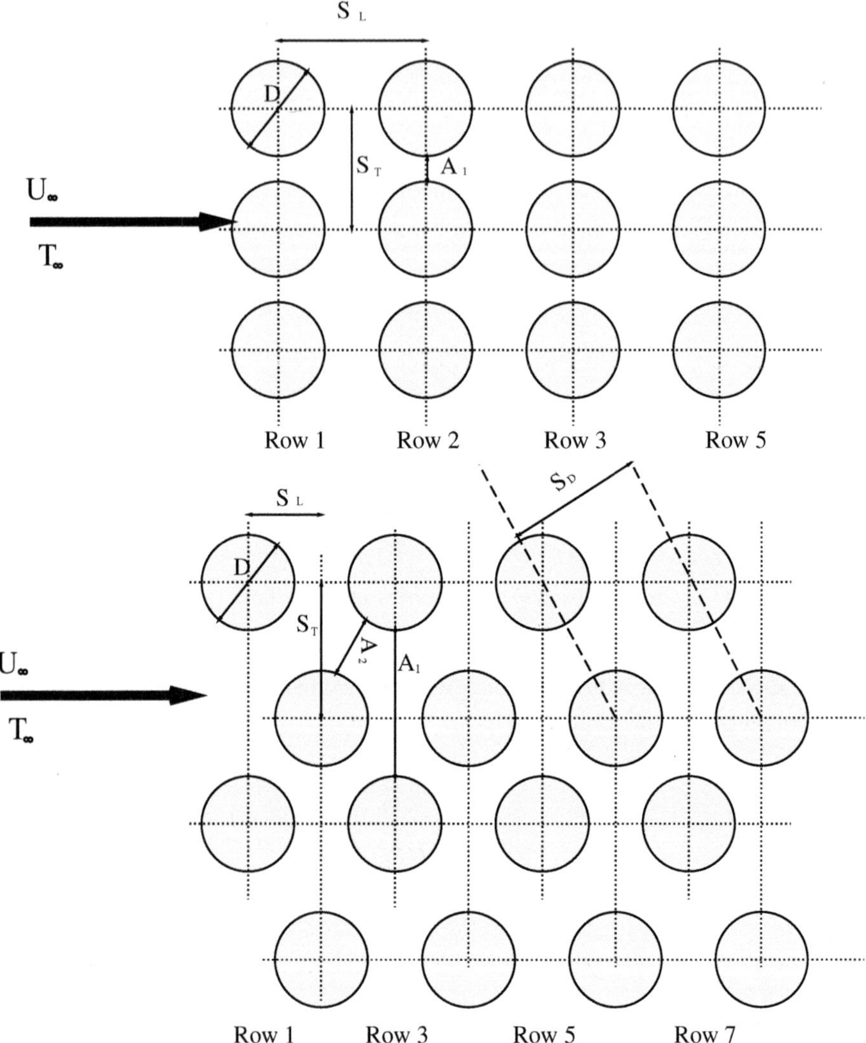

Figure 10.12 Tube bank arrangements for inline and staggered configurations.

10.3.1 CORRELATIONS FOR TUBE BANK

The heat transfer distribution between the tubes is strongly depending on the flow pattern inside the tubes. The flow between tube banks is mostly laminar when Re $< 10^3$. With respect to the Reynolds numbers, there are three distinguishable flow regimes in tube banks:

i. Laminar flow regime at $\mathrm{Re} < 10^3$,

ii. Mixed or sub-critical flow regime at 5×10^2 to 2×10^5, and

iii. Turbulent or critical flow regime at $\mathrm{Re} > 2 \times 10^5$.

The Nusselt number for the flow over tube banks can be a function of several parameters like

$$Nu = f\left(\mathrm{Re}, Pr, S_\mathrm{T}, S_L\right)$$

Colburn Correlation: In 1933, Colburn (1933) derived an equation for cross flow heat transfer in staggered banks, which is valid for flow ovwe tube banks with more than ten rows and Reynolds numbers ranging from 10 to 40000.

$$Nu = 0.33\mathrm{Re}^{0.6} Pr^{\frac{1}{3}}$$

Grimson Correlation: Grimson (1937) linked the experimental findings of Huge (1937) and Pierson (1937) for tubes positioned in an inline and staggered manner for the same number of tubes as mentioned above. The correlation only is valid for air.

$$Nu = C\mathrm{Re}^n$$

The maximum velocity is

$$U_{\max} = \begin{cases} U_{in}\left(\dfrac{S_T}{S_T - D}\right) & \textit{inline arrangement} \\[2em] U_{in}\left(\dfrac{S_T}{\sqrt{[(S_T/2)^2 + S_L^2]} - D}\right) & \textit{staggered arrangement} \end{cases}$$

Grimson (1937) used the experimental measurements of Pierson (1937) and Huge (1937) and introducing the dimensionless transverse and longitudinal pitches, he proposed relation

$$Nu_{10} = C\mathrm{Re}^n_{D,max} Pr^{1/3}$$

where, C and n can be taken from the table 10.2. Note that Reynolds number is based on maximum velocity. Table also lists the factor to be multiplied with Nu_{10} if number of rows are less than 10.

The temperature of the fluid will increase as it flows through the tube bank. We can calculate the mass flow rate through the tubes as

$$\dot{m} = \rho_{in} U_{in} S_T n_{rows}$$

The balance of heat transfer is

$$Q = \dot{m} \cdot c_p \cdot (T_{\infty,2} - T_{\infty,1}) = h \cdot A \cdot [T_w - (T_{\infty,2} + T_{\infty,1})/2]$$

TABLE 10.2

Parameters for Grimson Correlation

Factor to be multiplied with Nu_{10} if rows are less than 10 (Note: N is no. of rows)

N	1	2	3	4	5	6	7	8	9	10
in-line	0.64	0.8	0.87	0.9	0.92	0.94	0.96	0.98	0.99	1
staggered	0.68	0.75	0.83	0.89	0.92	0.95	0.97	0.98	0.99	1

Values of C and n to be used in Grimson correlation for Nusselt number (inline tube banks of 10 or more rows)

Inline arrangement $b=S_L/D$	$S_T/D=1.25$ C	n	1.5 C	n	2 C	n	3 C	n
1.25	0.386	0.592	0.305	0.608	0.111	0.704	0.0703	0.752
1.5	0.407	0.586	0.278	0.62	0.112	0.702	0.0753	0.744
2	0.464	0.57	0.332	0.602	0.254	0.632	0.22	0.648
3	0.322	0.601	0.396	0.584	0.415	0.581	0.317	0.608

Values of C and n to be used in Grimson correlation for Nusselt number (staggered tube banks of 10 or more rows)

Staggered arrangement $b=S_L/D$ $b=S_L/D$	$S_T/D=1.25$ C	n	1.5 C	n	2 C	n	3 C	n
0.6	-	-	-	-	-	-	0.236	0.636
0.9	-	-	-	-	0.495	0.571	0.445	0.581
1	-	-	0.552	0.558	-	-	-	-
1.125	-	-	-	-	0.531	0.565	0.575	0.56
1.25	0.575	0.556	0.561	0.554	0.576	0.556	0.579	0.562
1.5	0.501	0.568	0.511	0.562	0.502	0.568	0.542	0.568
2	0.448	0.572	0.462	0.568	0.535	0.556	0.498	0.57
3	0.344	0.592	0.395	0.58	0.488	0.562	0.467	0.574

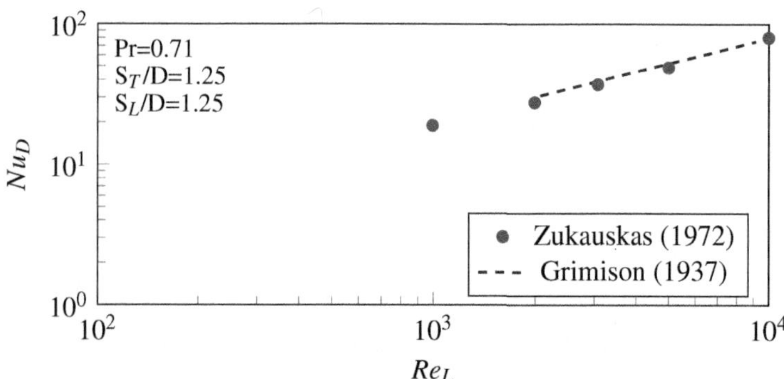

Figure 10.13 Mean Nusselt numbers for an in-line arrangement.

Figure 10.13 shows the mean Nusselt numbers for an in-line arrangement as predicted by Grimison correlation, compared with experimental data.

Example 10.4

Atmospheric air at $T_{\infty,1} = 10\ ^\circ C$ flows across a bank of tubes with in-line arrangement of 4 rows (perpendicular to the flow). The tube walls are maintained at $T_w = 65\ ^\circ C$. The velocity of air before it enters the tube bank is 6 m/s. The diameter (d) of the tubes is 0.0254 m. The normal and parallel spacings for the flow are 0.0508 m. There are 16 rows parallel to the flow. Find the total heat transfer per unit length for the tube bank and the exit air temperature. Use Grimson correlation.

Solution

The properties at the mean temperature at (65+10)/2=37.5°C:
$\mu = 1.894 \times 10^{-5}\ Pa.s,\ \rho = 1.137\ kg/m^3,$
$k = 0.027\ W/m \cdot K, Pr = 0.706, c_p = 1007\ J/kg \cdot K$
The rows in streamwise direction are 4 i.e. less than 5. We use the Grimson correlation and data tables:

$$S_T = S_L = 0.0508m$$

The flow will accelerate through the tube bank and velocity will increase, which is computed as

$$U_{max} = U_{in} \cdot \left(\frac{S_T}{S_T - d} \right) = (6\ m/s) \cdot \left(\frac{0.0508}{0.0508 - 0.0254} \right) = 12\ m/s$$

The Reynolds number based on maximum velocity is

$$Re = \frac{\rho \cdot U_{max} \cdot d}{\mu} = 18297.65$$

The Nusselt number as per Grimson correlation is

$$Nu_{10} = C \cdot Rey^n \cdot Pr^{0.333}$$

As we have only four rows in flow direction, we do not need to use Table 10.2. As $S_T/d = S_L/d = 2$ we have from Table 10.2:

$$C = 0.254; n = 0.632$$

The Nusselt number is 111.768, and the convective heat transfer coefficient is

$$h = \frac{Nu_{10} \cdot k}{d} = 118.80 W/m^2 K$$

As number of rows in-line $n_{rows} = 4$ and perpendicular to flow $n_{depth} = 16$, the area of tube bank per unit length to be used in Grimson correlation is

$$A = \pi \cdot d \cdot (1) \cdot n_{rows} \cdot n_{depth} = 5.106 m^2$$

$$Q = \dot{m} \cdot c_p \cdot (T_{\infty,2} - T_{\infty,1}) = h \cdot A \cdot [T_w - (T_{\infty,2} + T_{\infty,1})/2]$$

We can formulate the following equation using T_w and $T_{\infty,1}$, h, and area per unit length as

$$\dot{m} \cdot c_p \cdot (T_{\infty,2} - T_{\infty,1}) - h \cdot A \cdot [T_w - (T_{\infty,2} + T_{\infty,1})/2] = 0$$

Solving this equation for $T_{\infty,2}$, we have the solution:

$$T_{\infty,2} = 15.66°C$$

Using this the heat transfer per unit length is

$$Q = 31651.797W$$

Zukauskas Correlation

Zukauskas (1972) proposed the following correlation for Nusselt number, if rows are more than 20.

$$Nu_{20} = C.Re^{m}_{D,\max} \cdot Pr^{0.36} \left(\frac{Pr}{Pr_w} \right)^{1/4} \cdot F \qquad \left\{ \begin{array}{c} 10 < Re_{D,\max} < 10^6 \\ 0.7 < Pr < 500 \end{array} \right.$$

Table 10.3 lists the factor (F) to be multiplied with Nu_{20} if number of rows are less than 20.

Example 10.5

The tube bank consists of 20 tubes with 5 rows in flow direction and 4 rows in the transverse direction. The tubes are arranged in an in-line arrangement. Each of the tubes is consist of $d=50$ mm diameter and has length of 0.7 m. Air enters the tube bank at the speed of 2 m/s. The inlet and wall temperatures are 30 °C and 150 °C. Find the rate of heat transfer, if $S_T = 0.07$ m, and $S_L = 0.06$ m.

Solution
We use the properties at $T_{in}=30$ °C:

$$k = 0.0259W/m \cdot K, \rho = 1.164kg/m^3, v = 0.00001608m^2/s;$$
$$c_p = 1007J/kg \cdot K, Pr = 0.728, Pr_w = 0.709;$$

TABLE 10.3

Constants and Factors for Zukauskas Correlation

Constants of Zukauskas Correlation for Nusselt number over the tube bank
(Note: SC means single cylinder)

Configuration	$Re_{D,max}$	C	m
Aligned	10-1E2	0.8	0.4
Aligned	1E2-1E3	SC	SC
Aligned	1E3-2E5	0.27	0.63
($S_T/S_L > 0.7$)			
Aligned	2E5-2E6	0.021	0.84
Staggered	10-1E2	0.9	0.4
Staggered	1E2-1E3	SC	SC
Staggered	1E3-2E5	$0.35(S_T/S_L)^{1/5}$	0.6
($S_T/S_L < 2$)			
Staggered	1E3-2E5	0.4	0.6
($S_T/S_L > 2$)			
Staggered	2E5-2E6	0.022	0.84

Factor (F) to be multiplied with Nu_{20} if rows are less than 20

N	2	3	4	5	6	8	10	16	20
In-line	0.7	0.8	0.9	0.92	0.94	0.97	0.98	0.99	1
Staggered	0.77	0.84	0.89	0.92	0.94	0.97	0.98	0.99	1

The geometry of the inline arrangement is

$$d = 0.05m, L = 0.7m, S_T = 0.07m, S_L = 0.06m, S_T/d = 1.4, S_L/d = 1.2$$

The inlet speed is 2 m/s, but the flow will accelerate through the tube bank and velocity will increase, which is computed as

$$U_{max} = U_{in} \cdot \left(\frac{S_T}{S_T - d} \right) = 7 m/s$$

The Reynolds number based on maximum velocity is

$$Re = \frac{\rho \cdot U_{max} \cdot d}{\mu} = 21766.16 = 1.17 \times 10^4$$

The Nusselt number as per Zukauskas correlation is

$$Nu_{20} = C \cdot Re^m \cdot Pr^{0.36} \cdot \left(\frac{Pr}{Pr_w} \right)^{\frac{1}{4}} F$$

From Table 10.3

$$C = 0.254; m = 0.632$$

and also from Table 10.3 F=0.92.

$$Nu_{20} = 0.27 \cdot Re^{0.63} \cdot Pr^{0.36} \cdot \left(\frac{Pr}{Pr_w}\right)^{\frac{1}{4}}$$

For 5 rows the Nusselt number is

$$Nu = Nu_{20} \cdot F = 120.55$$

$$h = \frac{Nu \cdot k}{d} = 62.4W/m^2 \cdot K$$

$$A = \pi \cdot d \cdot (L) \cdot n_{rows} \cdot n_{depth} = 2.199m^2$$

The heat transfer is

$$Q = h \cdot A \cdot (T_w - T_\infty) = 16479.79W$$

Hausen Correlation: Hausen (1983) presented an empirical calculation for the factor F. His proposal for inline tube banks arrangement is

$$Nu = 0.34FRe^{0.61}Pr^{0.31}$$

$$F = 1 + \left(b + \frac{7.17}{b} - 6.52\right) \cdot \left[\frac{0.266}{(a-0.8)^2} - 0.12\right]\sqrt{\frac{1000}{Re}}$$

where,

$$a = S_T/D, b = S_L/D$$

and for staggered tube banks, the correlation and correction factor is

$$Nu = 0.35FRe^{0.57}Pr^{0.31}$$

$$F = 1 + 0.1b + \frac{0.34}{a}$$

Example 10.6

Repeat Problem 10.6 and find heat transfer using Hausen correlation.

Solution

From the data

$$a = \frac{S_T}{d} = 1.4; b = \frac{S_L}{d} = 1.2$$

For Re $=21766.16$, this gives

$$F = 1 + \left(b + \frac{7.17}{b} - 6.52\right) \cdot \left[\frac{0.266}{(a-0.8)^2} - 0.12\right] \sqrt{\frac{1000}{\text{Re}}} = 1.0868$$

$$Nu_{Hausen} = 0.34 \cdot F \cdot Rey^{0.61} \cdot Pr^{0.31} = 148.24$$

Based on this Nusselt number, the heat transfer is

$$Q = 20.264 kW$$

Gnielinski Correlation: Gnielinski (1978) proposed the relation based on a geometry parameter ψ defined as

$$\psi = \begin{cases} 1 - \frac{\pi}{4b} & for \quad a \geq 1 \\ 1 - \frac{\pi}{4ab} & for \quad a < 1 \end{cases}$$

where, $a = S_T/d$, and $b = S_L/d$. The Reynolds number based on this geometry parameter is

$$\text{Re}_\psi = \frac{U_b . \ell_{bank}}{v . \psi}$$

where the length scale is related to the flow over one row of tubes as:

$$\ell_{bank} = \frac{\pi}{2} D$$

U_b is the bulk velocity between the tubes, estimated as

$$U_b = \frac{\dot{m}}{\rho \cdot S}$$

and S is area facing the flow S $= n_{depth} \cdot S_L \cdot L$, and n_{depth} are tubes per unit rows. In this procedure several intermediary Nusselt numbers are computed:

$$Nu_o = 0.3 + \sqrt{Nu_{lam}^2 + Nu_{turb}^2}$$

$$Nu_{lam} = 0.664 \sqrt{\text{Re}_\psi} \sqrt[3]{Pr}$$

$$Nu_{turb} = \frac{0.037 Pr \text{Re}_\psi^{0.8}}{1 + 2.443 \text{Re}_\psi^{-0.1}(Pr^{2/3} - 1)} \qquad \begin{cases} 10 < \text{Re}_\psi < 10^6 \\ 0.6 < Pr < 10^3 \end{cases}$$

The Nusselt number with the effect of property variation for a single row of a tube bank is

$$Nu_{o,bank} = Nu_o \left(\frac{Pr_m}{Pr_w}\right)^{0.25}$$

The row arrangement factor F is

$$F_{inline} = 1 + \frac{0.7}{\psi^{3/2}} \left[\frac{a/b - 0.3}{(a/b + 0.7)^2} \right]$$

$$F_{stagg} = 1 + \frac{2}{3a}$$

which should be used with the following equation of Nu for n number of rows

$$Nu_{bank} = Nu_{o,bank} \left[\frac{1 + (n-1)F}{n} \right]$$

The Nusselt number is defined as

$$Nu_{bank} = \frac{h \ell_{bank}}{k}$$

The heat transfer can be computed as

$$q''_w = h \cdot LMTD$$

where log-mean-temperature is defined as

$$LMTD = \frac{(T_w - T_{in}) - (T_w - T_{out})}{\ln \left(\frac{T_w - T_{in}}{T_w - T_{out}} \right)}$$

Example 10.7

The water at 103 kg/s enters a tube bank consisting of 5 rows in flow direction and 6 tubes/rows with inlet temperature of 20 °C. The tube walls have temperature of 100 °C. The tubes are arranged in an inline arrangement with geometry configuration as

$$d = 0.022; S_T = 0.024; S_L = 0.032, n = 6, n_{depth} = 10$$

where d is diameter of a tube and n_{rows} is number of rows. If the length of a tube is 2.5 m, find the amount of heat transfer in watts using Gnielinski correlation.

Solution

Gnielinski recommended to use the properties at mean temperature of $(T_{in} + T_{out})/2$. In given problem T_{in} = 20°C, but T_{out} is unknown. As a first step, we assume T_{out}=29.7 °C. Hence, the properties are
k = 0.606 W/m · K, ρ = 997 kg/m^3. v = 0.897e-6 m^2/s, c_p = 4179 J/kg · K, Pr = 6.163, Pr$_w$ = 1.75.
Using the data given in problem statement, we calculate the area for the flow as

$$S = n_{depth} \cdot S_L \cdot L = 10(0.032)(2.5) = 0.8 m^2$$

The bulk inlet velocity is computed as

$$U_b = \frac{\dot{m}}{\rho \cdot S} = \frac{103}{997 \times 0.8} = 0.129 m/s$$

We compute the dimensionless translational (a) and transverse (b) distances as

$$a = S_T/d = 1.090, b = S_L/d = 1.454$$

We compute the characteristic length and ψ function:

$$\psi = 1 - \frac{\pi}{4b} = 0.46$$
$$\ell_{bank} = \pi \cdot d/2 = 0.03454$$

This gives the Reynolds number:

$$\text{Re}_\psi = \frac{U_b.\ell_{bank}}{v.\psi} = 10,809$$

We will now compute the various Nusselt numbers as given in Gnielinski procedure:

$$Nu_{lam} = 0.664\sqrt{\text{Re}_\psi}\sqrt[3]{Pr} = 126.49$$

$$Nu_{turb} = \frac{0.037 Pr \text{Re}_\psi^{0.8}}{1 + 2.443\text{Re}_\psi^{-0.1}(Pr^{2/3} - 1)} \quad \begin{cases} 10 < \text{Re}_\psi < 10^6 \\ 0.6 < Pr < 10^3 \end{cases}$$

$$Nu_{turb} = 117.44$$

$$Nu_o = 0.3 + \sqrt{Nu_{lam}^2 + Nu_{turb}^2} = 172.90$$

$$Nu_{o,bank} = Nu_o \left(\frac{Pr_m}{Pr_w}\right)^{0.25} = 236.86$$

$$F_{inline} = 1 + \frac{0.7}{\psi^{3/2}}\left[\frac{a/b - 0.3}{(a/b + 0.7)^2}\right] = 1.480$$

$$Nu_{bank} = Nu_{o,bank}\left[\frac{1 + (n_{rows} - 1)F_{inline}}{n_{rows}}\right] = 331.63$$

$$A = \pi \cdot d \cdot L \cdot (n_{rows} \cdot n_{depth}) = 10.36 \ \text{m}^2$$

The log mean temperature difference is

$$LMTD = \left(\frac{(T_w - T_{in}) - (T_w - T_{out})}{\ln\left(\frac{(T_w - T_{in})}{(T_w - T_{out})}\right)}\right) = 75.097°C$$

The heat transfer as per Newton's law of cooling:

$$Q = h \cdot A \cdot LMTD = 4.530 \times 10^6 W$$

We compute the outlet temperature using the bulk energy equation:

$$T_{out} = \frac{Q}{\dot{m} \cdot c_p} + T_{in} = 30.5°C$$

As $T_{out} = 30.5 \approx 29$, we can either take the above calculations as a solution or do one more iteration to arrive at more closer agreement for outlet temperature.

10.4 HEAT TRANSFER BY IMPINGING JET

Impinging jets, known for their ability to efficiently transfer heat and mass, are utilised in various industrial applications. These jets are instrumental in creating the lift force required for Vertical Take-Off and Landing (VTOL) aircraft by directing the hot exhaust from the jet engine towards the ground. Their applications extend to areas such as the vertical take-off and landing of aircraft and rockets, as well as the de-icing of aircraft.

The dynamics of flow and heat within an impinging jet are highly intricate, encompassing phenomena such as jet stagnation, ring vortex impact, flow acceleration, and subsequent deceleration. Similar to other types of flows, impinging jet flow can be categorised as either laminar or turbulent, depending on the Reynolds number, defined as

$$Re = \frac{U_j D}{v}$$

where U_j is the jet's bulk velocity at exit, D is the characteristic length, and v is the kinematic viscosity. For the plane jet the diamater is replaced by the slot width (B).

The different flow regimes of an impinging jet, as identified by Viskanta in 1993, are

The dissipated laminar jet ($Re < 300$)

A fully laminar jet ($300 < Re < 1000$)

A transition or semi-turbulent jet ($1000 < Re < 3000$)

A fully turbulent jet ($Re > 3000$)

An impinging jet structure can be divided into three different zones (see Figure 10.14):

Free jet zone

Impingement zone

Radial wall jet zone

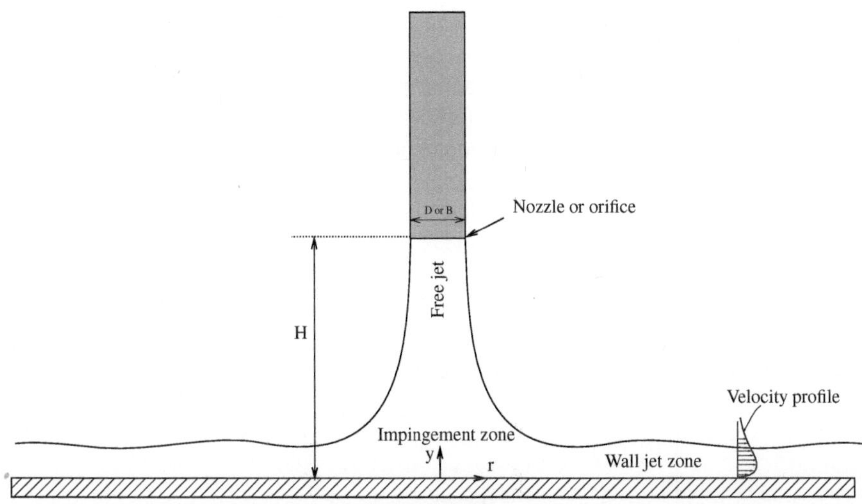

Figure 10.14   Different zones in an impinging jet.

A free jet is characterised by large eddies that are repetitive in structure and remain coherent for downstream distances significantly greater than their length scales. These eddies contribute substantially to the turbulent flow properties and are commonly referred to as coherent structures. Near the outlet, the jet's shear layer is initially governed by the Kelvin-Helmholtz (KH) instability mechanism.

The KH instabilities emerge from the interaction between the fast-moving jet flow and the slow-moving surrounding fluid. In the case of a circular jet, ring vortices are generated as a result of these Kelvin-Helmholtz instabilities. The formation of a ring vortex is linked to the most amplified frequency of a small disturbance, which grows exponentially. This complex behaviour underscores the intricate nature of impinging jet dynamics and the underlying mechanisms that govern its behaviour.

The jet is deflected and spreads radially due to the presence of a wall, forming a region known as the wall jet. This region consists of a very thin accelerating boundary layer, resulting in a complex flow characterised by strong shear. Even though the flow becomes parallel to the wall, the turbulence levels within the wall jet zone are significantly higher than those in parallel flows.

The radial distribution of the Nusselt number in an impinging jet, which is a key parameter in understanding the heat transfer characteristics of the flow, can be defined as follows:

$$Nu = \frac{q_w}{T_w - T_j} \frac{D}{k} \tag{10.12}$$

The radial distribution of the Nusselt number in an impinging jet can be expressed in terms of various parameters, including the jet inlet temperature T_j, the target wall temperature T_w, the heat flux at the wall q_w, the diameter of the jet D, and the thermal conductivity k.

The heat transfer within an impinging jet is strongly influenced by the nozzle-to-wall distance, denoted as H. When the ratio of this distance to the jet diameter, $\frac{H}{D}$, is less than 4, prominent peaks become apparent in the radial distribution of the Nusselt number.

Experimental studies have revealed that the Nusselt number at the stagnation point reaches its maximum value around $H/D \approx 8$ (Han and Goldstein 2001). Figure 10.15 illustrates the distribution of the Nusselt number at the target wall, highlighting three significant values in this distribution: the stagnation point Nusselt number, the first peak in the Nusselt number (indicated as Nu_1), and the secondary peak in the Nusselt number distribution (indicated as Nu_2).

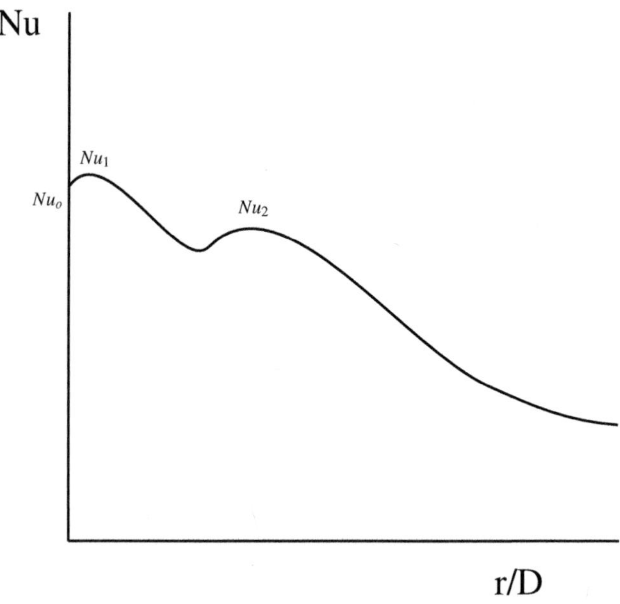

Figure 10.15   The distribution of Nusselt number on the target wall.

The appearance of secondary peaks in the Nusselt number distribution for low nozzle-to-target plate spacing in jet impingement heat transfer is a recognised phenomenon. Research has shown that the occurrence of these secondary peaks is dependent not only on the nozzle-to-plate spacing but also on the turbulence levels within the jet (Hoogendoorn, 1977).

Impinging jet heat transfer is not governed by a single parameter but a combination of factors such as nozzle-to-wall distance, jet velocity, turbulence levels, and thermal properties of the fluid. Each of these factors can significantly influence the heat transfer characteristics, including the Nusselt number distribution. The turbulence levels within the jet play a vital role in the heat transfer process. Turbulence can enhance mixing and heat transfer but also adds complexity to the flow behaviour. Predicting and controlling turbulence in multiple impinging jet can be challenging and sophistic numerical simulations are routinely conducted to estimate them.

Martin (1977) conducted an investigation into the heat and mass transfer properties of an impinging round jet. The research found that the Nusselt number at the point of impingement begins to rise as the ratio of the jet diameter to the distance between the jet and the surface (D/H) grows from 2. This increase continues until it reaches a peak at H/D = 6, after which it declines until H/D = 14. Interestingly, the relative maximum Nusselt number is seen exclusively at H/D = 2. Martin's correlation, which was formulated based on these observations, stems from the noticeable enhancement in heat transfer following the end of the accelerated flow region. This escalation is ascribed to the diminishing stabilising streamwise pressure gradient, leading to a marked increase in turbulence levels (Martin, 1977). The expression for Martin's correlation concerning the average heat transfer coefficient is:

$$\frac{\overline{Nu}}{Pr^{0.42}} = \frac{D}{r}\left[\frac{1-1.1D/r}{1+0.1(H/D-6)D/r}\right]F(Re) \tag{10.13}$$

Different Reynolds number ranges result in different functions for F(Re).

$$2000 < Re < 30,000, \qquad F(Re) = 1.36Re^{0.574} \tag{10.14}$$

$$30,000 < Re < 120,000, \qquad F(Re) = 0.54Re^{0.667} \tag{10.15}$$

$$120,000 < Re < 400,000, \qquad F(Re) = 0.151Re^{0.775} \tag{10.16}$$

Martin further proposed a function to be used with this correlation for the mean Nusselt number (Martin, 1990):

$$\overline{Nu} = Pr^{0.42}G(A_r, H/D)f(Re) \tag{10.17}$$

where, the functions $G(A_r, H/D)$ and $f(Re)$ can be calculated from

$$G(A_r, H/D) = 2\sqrt{A_r}\left[\frac{1-2.2\sqrt{A_r}}{1+0.2\sqrt{A_r}(H/D-6)}\right]$$

$$f(Re) = 2\sqrt{Re\left(1+\frac{Re^{0.55}}{200}\right)}$$

and, $A_r = 1/(4r^*)$ and $r^* = r/D$. Martin's correlation is valid for

$$2000 \leq \text{Re} \leq 400,000$$
$$2.5 \leq r/D \leq 7.5$$
$$2 \leq H/D \leq 12$$

Correlation for heat trransfer from cylinder and sphere

Heat transfer from cylinder Churchill and Bernstein (1977) correlation:

$$\overline{Nu}_D = 0.3 + \frac{0.62 Pr^{1/3} \sqrt{\text{Re}_D}}{\sqrt[4]{1 + \left(\frac{0.4}{Pr^{2/3}}\right)}} \left[1 + \left(\frac{\text{Re}_D}{282000}\right)^{5/8}\right]^{4/5} \qquad Pe \geq 0.2$$

Žukauskas and Žiugžda (1985) Correlation:

$$Nu_D = 0.29 \text{Re}_D^{0.6} \qquad 10^3 < \text{Re}_D < 10^5$$

Sarma and Sukhatme (1977) Correlation:

$$Nu_D = 0.62 \text{Re}_D^{0.505} \qquad 1200 < \text{Re}_D < 4700$$

Heat transfer from Sphere
Kramers (1946) correlation:

$$Nu_{sphere} = 1 + 1.3 Pr^{0.915} Re^{0.5} \qquad \begin{array}{l} 0.4 < Re < 2100 \\ 0.71 < Pr < 380 \end{array}$$

Yuge (1960) correlation:

$$Nu_{sphere} = 2 + 0.493 Re^{0.5} \qquad 10 < Re < 1.8E3$$
$$Nu_{sphere} = 2 + 0.3 Re^{0.57} \qquad 1.8E3 < Re < 1.5E5$$

Whitaker (1972) correlation:

$$\overline{Nu}_D = 2 + \sqrt[4]{\left(\frac{\mu}{\mu_s}\right)} \left[0.4\sqrt{\text{Re}_D} + 0.06 \sqrt[3]{\text{Re}_D^2}\right] Pr^{0.4}$$

valid for

$$0.71 \leq Pr \leq 380$$
$$3.5 \leq \text{Re}_D \leq 76000$$
$$1 \leq \left(\frac{\mu}{\mu_s}\right) \leq 3.2$$

Eastop and Smith (1972) correlation:

$$Nu = 0.42\sqrt{Re_D} + 0.0035Re_D^{0.92}$$

valid for $3000 < Re_D < 10^5$.
Ranz and Marshall (1952) correlation:

$$\overline{Nu}_D = 2 + 0.6Pr^{1/3}\sqrt{Re_D}$$

Will, Kruyt, and Venner (2017) correlation:

$$Nu_D = A \cdot \sqrt{Re_D} + B \cdot Re_D \} \qquad 7800 \leq Re \leq 2.9E5$$

where,

$$A = 0.493 \pm 0.015$$
$$B = 0.0011 \times (1 \pm 0.035)$$

PROBLEMS

10P-1 Engine oil at 1 atm and 100 °C is flowing across a 10 cm diameter cylinder at a speed of 7 m/s. The surface of the cylinder is maintained at 200 °C. What is the rate of heat transfer per meter length of the cylinder?

10P-2 Glycerin at 1 atm and 20 °C is flowing across a 1 cm diameter sphere at a speed of 1.2 m/s. The surface of the sphere is maintained at 300 °C. What is the rate of heat transfer per meter length of the sphere?

10P-3 An aluminium can of a fizzy drink has a surface temperature of 2°C. The can height is 12 cm and the diameter is 5 cm. The ambient air temperature is 20°C. Find the convective heat transfer coefficient for the air flow over the can in W/m^2 ·°C.

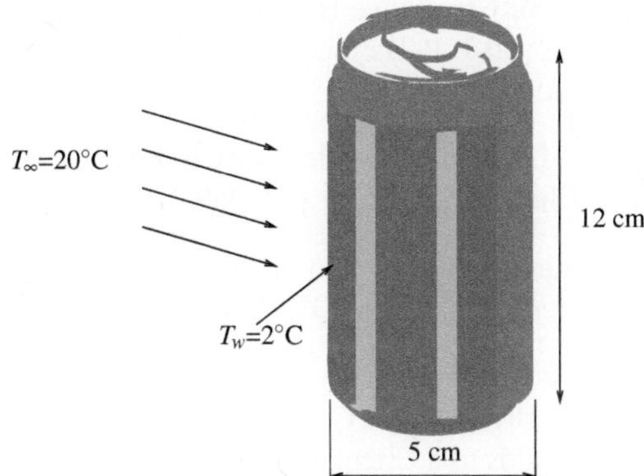

10P-4 A 1 cm in diameter red brass (85 Cu, 9 Sn, 6 Zn) spherical ball is suspended with a thread in a medium where temperature is 600K. Air is blown over ball at 5 m/s when ball temperature was 30°C. Find heat transfer.

10P-5 The tubes with 38 mm OD are arranged in an inline arrangement with $S_T=S_L=74.8$ mm. The fluid with properties $\mu=0.00082060$, k= 0.610 W/m·K, $\rho=996$ kg/m^3, $cp=4179$ J/kg·K, is moving through the tube bank and the maximum fluid velocity reaches 0.4704 m/s. Find the convective heat transfer coefficient using suitable correlation.

10P-6 Air is flowing over a heated cylinder at a velocity of 12 m/s, and heat is transmitting to air. The diameter of the cylinder is 10 mm. Find the mean Nusslt number using Martinelli et al. (1943), and Knudsen and Katz (1958) correlations.

REFERENCES

N. Uddin, Fluid Mechanics: A Problem-Solving Approach, 1st ed, 2022, Taylor and Francis Press.

E. Schmidt, K. Wenner, Wirmeabgabe ueber Umfang eines angeblasenen geheizten Zylinders. 18. Forschung, 12, 65–73 (1941). (English transl., R.A.E. Transl. No. 280).

G. P. Merker and H. Hanke, Heat transfer and pressure drop on the shell-side of tube-banks having oval-shaped tubes, International Journal of Heat and Mass Transfer 29(12), 1903–1909, 1986.

D. B. Spalding and W. M. Pun, A review of methods for predicting heat-transfer coefficients for laminar uniform-property boundary layer flows, 5(3-4), 239–249, 1962.

H. C. Perkins Jr. and G. Leppert, Local heat-transfer coefficients on a uniformly heated cylinder, 7(2), 143–158, 1964.

V. T. Morgan, The Overall Convective Heat Transfer from Smooth Circular Cylinders, in T. F. Irvine and J. P. Hartnett, (eds.). Advances in Heat Transfer, vol. 11, New York: Academic Press, 1975.

I. S. Pasternak and W. H. Gauvin, Turbulent heat and mass transfer from stationary particles. Can. J. Chem. Eng. 38, 35–42, 1960.

V. Gnielinski, Berechnung mittlerer Wärme- und Stoffübergangskoeffizienten an laminar und turbulent überströmten Einzelkörpern mithilfe einer einheitlichen Gleichung. Forsch. Ing.-Wes. 41(5), 145–153, 1975.

S. W. Churchill and M. Bernstein. A Correlating Equation for Forced Convection from Gases and Liquids to a Circular Cylinder in Cross-flow, J. Heat Transfer, vol. 99, pp. 300–306, 1977.

J. B. Will, N. P. Kruyt, and C. H. Venner, An experimental study of forced convective heat transfer from smooth, solid spheres, Int. J. Heat Mass Transf. 109, 1059–1067, 2017.

S. Whitaker, Forced convection heat transfer correlations for flow in pipes, past flat plates, single cylinders, single spheres, and for flow in packed beds and tube bundles, AIChE J. 18, 361–371, 1972.

T. D. Eastop and C. Smith, Heat transfer from a sphere to an air stream at subcritical Reynolds numbers, Trans. Inst. Chem. Eng. 50, 26–31, 1972.

W. Ranz, W. Marshall, Chem. Eng. Prog., 48, 141, 1952.

D. C. Collis, A. J. Williams, J . Fluid Mech., 6, 357, 1959.

R. C. Martinelli et al. – Trans. Am. Inst. Chem. Engrs. 38, p. 943, 1943.

J. G. Knudsen and D. L. Katz, Fluid Dynamics and Heat Transfer, McGraw-Hill, New York, 1958.

T. S. Sarma and S. P. Sukhatme, Local Heat Transfer from a Horizontal Cylinder to Air in Cross Flow: Influence of Free Convection and Free Stream Turbulence, Int. J. Heat Mass Transfer, 20, pp. 51–56, 1977.

A. Žukauskas and J. Žiugžda, Heat Transfer of a Cylinder in Cross-flow, Hemisphere, New York, 1985.

A. Žukauskas, Advances in Heat Transfer, Academic Press, New York, pp. 93–160. 1972.

W. Schönauer, Ein Differenzenverfahren zur Lösung der Grenzschichtgleichung für stationäre, laminare, inkompressible Strömung, Ing.-Arch., Vol. 33, p. 173, 1964.

K. Hiemenz, Die Grenzschicht an einem in den gleichformigen Flüssigkeitsstrom eingetauschten geraden Kreiszylinder, Dinglers Polytech J., Vol. 326, p. 326, 1911.

K. Pohlhausen, Zur Näherungsweise Integration der Differential Gleichung der Laminaren Reibungschicht, Zeitschrift für angewandte Mathematic und Mechanic, Vol. 1, pp. 252–268, 1921.

M. M. Zdravkovich, Review of flow interference between two circular cylinders in various arrangements. ASME Journal of Fluids Engineering 99:618-633, 1997.

T. Yuge, Experiments on heat transfer from spheres including combined natural and forced convection, J. Heat Transf. (Trans. ASME) 82, 214-220, 1960.

H. Kramers, Heat transfer from spheres to flowing media, Physica 12, 61-80, 1946.

J. B. Will, N. P. Kruyt, and C.H. Venner, An experimental study of forced convective heat transfer from smooth, solid spheres, Int. J. Heat Mass Transf. 109, 1059-1067, 2017.

N. Uddin, Turbulence modeling of complex flows in CFD, PhD Thesis, Universitaet Stuttgart, Germany, 2008.

H. Martin, Heat and Mass Transfer between Impinging Gas Jets and Solid Surfaces, Advances in Heat Transfer 13 (1977) 1-60.

B. Han and R. J. Goldstein, *Jet-impingement heat transfer in Gas Turbine systems*, Annals of New York Academy of Sciences, Vol. 934, 147 - 161, 2001.

C. J. Hoogendoorn, *The effect of turbulence on heat transfer at a stagnation point*, Int. J. Heat and Mass Transfer, Vol. 20, 1333 - 1338, 1977.

R. Viskanta, *Heat transfer to impinging Isothermal Gas and Flame jets*, Exp. Thermal and Fluid Sciences , 6: 111 - 134, 1993.

H. Martin, Impinging jets, in Handbook of Heat Exchanger Design, Hewitt, G.F., Ed., Hemisphere, New York, 1990.

11 Natural and Mixed Convection

In this chapter, we will discuss the heat transfer caused by the buoyancy driven flow, which is classified as natural or free convection. In engineering applications, we sometimes design equipment which have no fans and thus the flow is generated due to density changes. In some cases, both forced and free convection are present. We will develop an analytical solution for the laminar free convection boundary layer along a vertical flat plate, and we will also explore correlations to be used for variety of situations. After finishing this chapter, one will be able to:

- Calculate the natural convection related heat transfer from vertical flat surface.
- Calculate the natural convection related heat transfer in enclosures.
- Decide about the type of convective heat transfer involved using the criterion given in this chapter.
- Compute heat transfer in case of mixed convection scenarios.

Natural convection is important not only for the design of devices which involves no fans but also for the buoyancy-induced motions in the atmosphere, and in bodies of water. Natural convection has gained more importance since the use of nano-scale devices has been started at the turn of the century. The phenomenon of natural or free convection is different from forced convection. The fluid close to the heated wall becomes lighter and rises. This fluid is shed at the top of the object as a wake, which may arise as a plume. Hence, fluid from the extensive medium continually flows into the convection region near the surface to replace the rising material.

11.1 OBERBECK-BOUSSINESQ-APPROXIMATION

In case of free or natural convection, the movement of the fluid occurs exclusively due to density change as a result of temperature differences. We know that density is a function of temperature and because of its change the Buoyancy force will be the main driving force for the fluid. The Navier-Stokes equation is

$$\rho \frac{Du}{Dt} = -\nabla p + \nabla \tau_{ij} + \rho g$$

We decompose the pressure into a static and a dynamic part:

$$p = p_{static} + p_{dynamic}$$

DOI: 10.1201/9781003428404-11

where static pressure is the hydrostatic pressure,and hence the index o is added with the density of the fluid. The dynamic pressure due to the motion of the fluid:

$$\nabla p_{static} \approx p_{static} = \rho_o g$$

This gives pressure field as

$$\nabla p = \rho_o g + \nabla p_{dynamic}$$

The body force term can be written as

$$\rho \frac{Du}{Dt} = -\nabla p_{dynamic} + (\rho - \rho_o)g + \nabla \tau_{ij}$$

The dynamic pressure in modified Navier-Stokes equation is no longer thermodynamic pressure. However, since we are not taking the material properties dependence on pressure into account this is not a major issue and hence we can still approximate the pressure term as a regular pressure with negligible error. The temperture variation as depicted by energy equation would only influence one term in momentum equation. We are only considering the density variation in the body force term and this is a case of *weakly coupled flow*.

$$\beta = \frac{-1}{\rho} \left(\frac{\partial \rho}{\partial T} \right)_p$$

We use Taylor series to analyse the limits of density approximation based on temperature difference:

$$\rho(T) = \rho(T_o) + \left(\frac{\partial \rho}{\partial T} \right)_o \frac{(T - T_o)}{1!} + + \left(\frac{\partial \rho}{\partial T} \right)_o \frac{(T - T_o)^2}{2!} + \dots$$

Ignoring higher order terms

$$\frac{\rho(T) - \rho(T_o)}{\rho(T_o)} = \frac{1}{\rho(T_o)} \left(\frac{\partial \rho}{\partial T} \right)_o (T - T_o)$$

Rearranging

$$\frac{\rho(T)}{\rho(T_o)} = 1 + \frac{1}{\rho(T_o)} \left(\frac{\partial \rho}{\partial T} \right)_o (T - T_o)$$

Introducing the isobaric coefficient of thermal expansion:

$$\frac{\rho(T)}{\rho(T_o)} = 1 - \beta_o(T - T_o)$$

or simply

$$\frac{\rho}{\rho_o} = 1 - \beta_o(T - T_o)$$

The coefficient of volumetric expansion β is of the order of magnitude of 10^{-4} to 10^{-2} and thus, for gases, β can be treated as constant.

With assumption that in free convection boundary layers $\beta_o(T - T_o) \ll 1$ we can approximate density solely as function of temperature and this means that fluid can still be treated as incompressible though there is a density change.

$$\frac{Du}{Dt} = -\frac{1}{\rho}\nabla p_{dynamic} + \frac{(\rho - \rho_o)}{\rho}g + \frac{1}{\rho}\nabla \tau_{ij}$$

With the discussed approximations, the new form of Navier-Stokes equation is

$$\frac{Du}{Dt} = \frac{-1}{\rho_o}\nabla p - \beta_o g(T - T_o) + \nu_o \nabla^2 u$$

The energy equation valid for buoyancy driven flows can be obtained by simplifying the generalised energy equation 7.37:

$$\rho c_p \frac{DT}{Dt} = -\nabla.q + \tilde{q} + \beta T \frac{Dp}{Dt} + \mu \phi_{diss}$$

We neglect the internal energy generation term and also the energy dissipation due to relatively low velocity of rising fluid. Also, there is a negligible dissipated kinetic energy or frictional heating involved.

$$\rho c_p \frac{DT}{Dt} = -\nabla.q + \beta T \frac{Dp}{Dt}$$

Assuming no pressure change due to enthalpy change, we have

$$\rho_o c_p \frac{DT}{Dt} = k \left(\frac{\partial^2 T}{\partial y^2} \right)$$

$$\frac{DT}{Dt} \equiv \alpha_o \left(\frac{\partial^2 T}{\partial y^2} \right)$$

The above simplifications are also referred to as Oberbeck-Boussinesq approximation. They are based on works by Oberbeck (1879) and Boussinesq (1903) and presented by Joseph (1976). This simplification is based essentially on the three assumptions:

i the density is considered constant in all terms except in the buoyancy term of the momentum equation,

ii All other material properties are assumed to be constant; and

iii the energy dissipation is neglected;

$$\frac{\partial u}{\partial x} + \frac{\partial u}{\partial y} = 0$$

$$u\frac{\partial u}{\partial x} + v\frac{\partial u}{\partial y} = v_o\frac{\partial^2 u}{\partial y^2} + \beta_o(T - T_o)$$

$$u\frac{\partial T}{\partial x} + v\frac{\partial T}{\partial y} = \alpha_o\frac{\partial^2 T}{\partial y^2}$$

Figure 11.1 shows the empirical obtained temperature and velocity distributions inside the free convection related thermal and momentum boundary layers. We notice the differences with the forced convection boundary layers where the velocity varies from zero to freestream values, here in case of natural convection the the velocity varies from zero to maximum value somewhere inside boundary layer and then it goes to small values at the edge of the boundary layer.

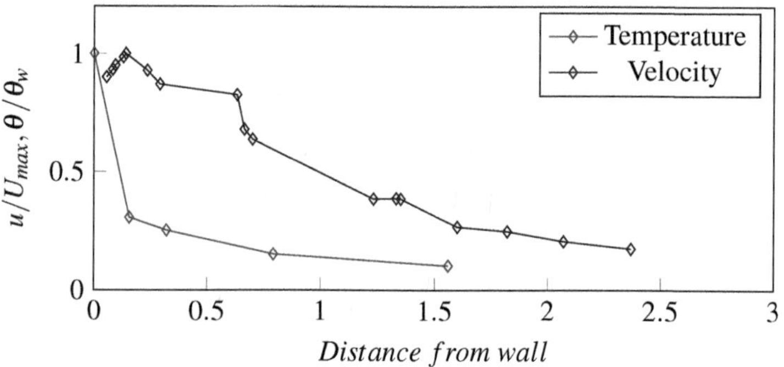

Figure 11.1   The plot of experimental distribution of velocity and temperature distribution inside free convection boundary layer.

11.2   LAMINAR FREE CONVECTION FROM VERTICAL ISOTHERMAL PLATE

The continuity, momentum, and energy equations for two-dimensional free convection boundary layers are

$$\frac{\partial u}{\partial x} + \frac{\partial v}{\partial y} = 0$$

$$u\frac{\partial u}{\partial x} + v\frac{\partial u}{\partial y} = \frac{\mu}{\rho}\frac{\partial^2 u}{\partial y^2} + g\beta(T - T_\infty)$$

$$u\frac{\partial T}{\partial x} + v\frac{\partial T}{\partial y} = \left(\frac{k}{\rho c_p}\right)\frac{\partial^2 T}{\partial y^2} + \underbrace{\frac{\mu}{\rho}\left(\frac{\partial u}{\partial y}\right)^2}_{negligible}$$

Pohlhausen has reduced the above natural convection boundary-layer equations for negligible viscous dissipation to two ordinary differential equations by introducing a stream function $\psi(x,y)$ and a similarity variable η. The similarity variable η is not the one we used for forced convection instead we need to introduce another similarity variable as it would cater for the buoyancy term in conservation equation.

We introduce the dimensionless temperature φ:

$$\varphi = \frac{(T - T_\infty)}{(T_w - T_\infty)}$$

stream function:

$$\psi(x,y) = 4vf(\eta)\sqrt[4]{\frac{Gr}{4}}$$

where, local Grashof number is defined as

$$Gr_x = \frac{g\beta x^3(T_w - T_\infty)}{v^2}$$

and similarity variable

$$\eta = \frac{y}{x}\sqrt[4]{\frac{Gr_x}{4}}$$

we can transform the partial differential equations into ordinary differential equations. The velocity components can be solved using

$$u = \frac{\partial \psi}{\partial y}, \qquad v = -\frac{\partial \psi}{\partial x}$$

hence,

$$u = \frac{\partial \psi}{\partial y}$$

$$= 4v\left(\sqrt[4]{\frac{Gr}{4}}\right)f'(\eta)$$

$$= \frac{2vf'(\eta)}{x}Gr^{1/2}$$

This would cast the conservation equation into a set of ODE:

$$f''' - 2(f')^2 + 3ff'' + \varphi = 0 \tag{11.1}$$

$$\varphi'' + 3Pr(\varphi'f) = 0 \tag{11.2}$$

Boundary conditions are

$$\begin{array}{cccc} \eta = 0 & f = 0 & f' = 0 & \varphi = 1 \\ \eta \to \infty & & f' \to 0 & \varphi \to 0 \end{array} \tag{11.3}$$

The first solution was given by Pohlhausen in 1930 for an isothermal surface and Pr = 0.733. Schuh, in 1948, solved these equations for Prandtl numbers of 10, 100, and 1000.

The numerical solution of ordinary differential equations 11.1, and 11.2 is tabulated in Table 11.1 and plotted in Figure 11.2. We can find the heat transfer as

$$q'' = -k\left(\frac{\partial T}{\partial y}\right)_{y=0} = -k\varphi'(0)\left(\frac{T_w - T_\infty}{x}\right)\left(\frac{Gr_x}{4}\right)^{1/4}$$

$$h_x = \frac{-k}{x}\varphi'(0)\left(\frac{Gr_x}{4}\right)^{1/4}$$

$$Nu_x = \frac{-1}{\sqrt{2}}\varphi'(0)(Gr_x)^{1/4}$$

$$Nu_L = \frac{4}{3}\left(\frac{-1}{\sqrt{2}}\varphi'(0)(Gr_x)^{1/4}\right)$$

$$\frac{Nu_L}{(Gr_x)^{1/4}} = \frac{-4}{3.\sqrt{2}}\varphi'(0) = \frac{-4}{3.\sqrt{2}}\varphi'(0)$$

TABLE 11.1

Numerical Solution of ODE for Pr=0.7

η	φ	φ'	f	f'	f''
0	1.0000	-0.4995	0.0000	0.0000	0.6789
1	0.5213	-0.4256	0.1970	0.2782	-0.0131
2	0.1993	-0.2157	0.4315	0.1722	-0.1336
3	0.0628	-0.0759	0.5461	0.0679	-0.0711
4	0.0183	-0.0229	0.5875	0.0223	-0.0260
5	0.0052	-0.0066	0.6005	0.0068	-0.0082
6	0.0014	-0.0018	0.6044	0.0020	-0.0025
7	0.0004	-0.0005	0.6055	0.0005	-0.0007
8	0.0001	-0.0001	0.6058	0.0001	-0.0002
9	0.0000	0.0000	0.6059	0.0000	-0.0001
10	0.0000	0.0000	0.6059	0.0000	0.0000

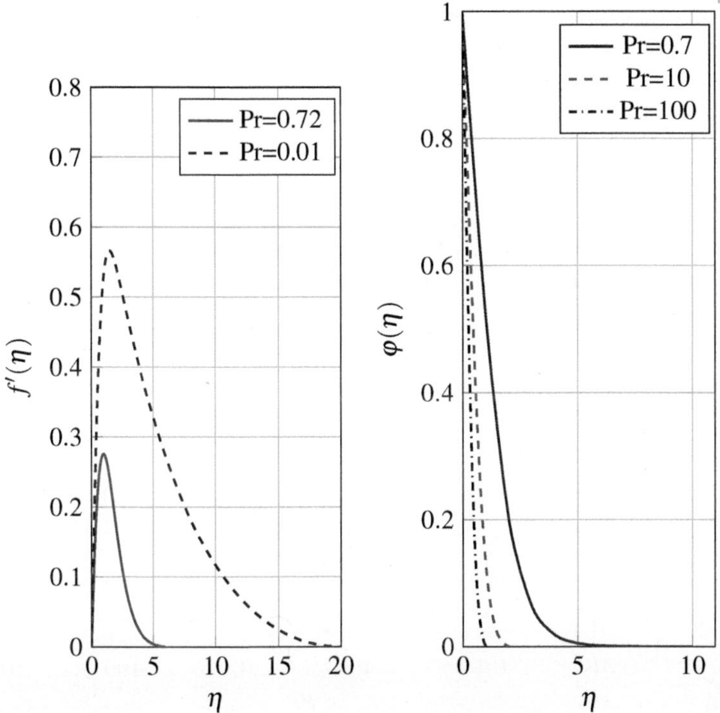

Figure 11.2   (Left) Ostrach solution (1953) for velocity distribution (Pr=0.72 and 0.01); (Right) Numerical solution of temperature distribution for Pr=0.7,10,100.

Using this we can estimate the mean Nusselt number of wall at different Prandtl numbers. For example, for Pr=100 we have $\varphi(0)$=-2.02 so Nu is

$$\frac{Nu_L}{(Gr_x)^{1/4}} = \frac{-4}{3.\sqrt{2}}\varphi'(0) = \frac{-4}{3.\sqrt{2}}(-2.02) = 1.9 \qquad Pr = 100$$

$$\frac{Nu_L}{(Gr_x)^{1/4}} = \frac{-4}{3.\sqrt{2}}\varphi'(0) = \frac{-4}{3.\sqrt{2}}(-0.499) = 0.47 \qquad Pr = 0.7$$

Ostrach and Lynn (1953) solved a similar format of equations for the large Grashof number. Their equations were

$$f''' - 2(f')^2 + 3ff'' + \varphi = 0$$

$$\varphi'' + 3Pr(\varphi'f) = 0$$

where, the parameters were defined a slight different manner.

$$\eta = \frac{y}{\sqrt[4]{4x}}, \psi = f(\eta)\sqrt[4]{(4x)^3} \quad and \quad \varphi = \overline{\theta}\beta T_\infty$$

Figure 11.3 shows the distribution of function of Nusselt number and Grashof number for range of Prandtl number.

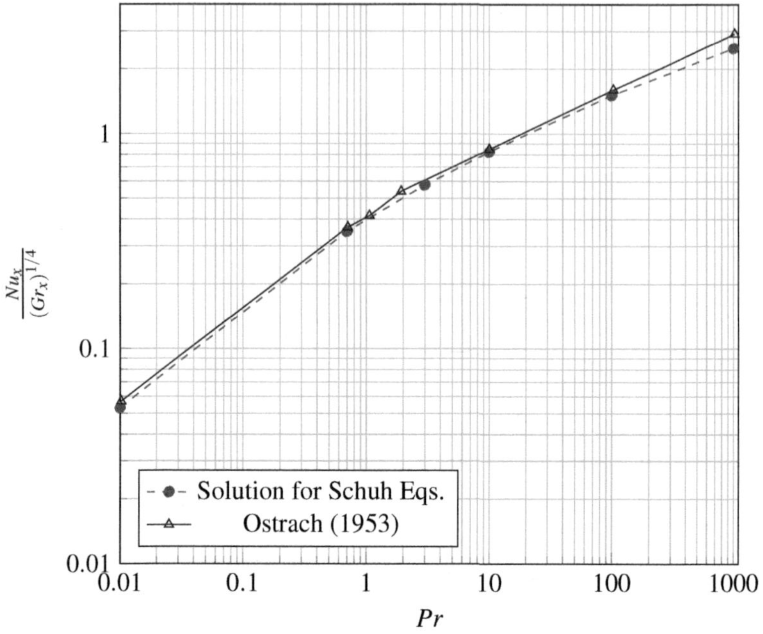

Figure 11.3 $\mathrm{Nu}_x/\mathrm{Gr}_x^{1/4}$ distribution against Prandtl number. Heat transfer due to natural convection from a vertical isothermal plate.

The solution of above equation as shown in the figure reveal that

$$Nu_x = 0.351(Gr_x)^{1/4} \quad Pr = 0.7$$

Warner and Arpaci (1968) proposed the correlation for mean Nusselt number for laminar flow:

$$Nu_{m,T} = 0.59Ra^{1/4} \qquad 10^4 < Ra < 10^9$$

The correlation is in close agreement with Ostrach's similarity solution for fluids with Prandtl number 0.7 and higher.

Figure 11.4 shows the distribution of function of Nusselt number and Grashof number for range of Prandtl number along with Warner and Arpaci correlation.

The analytical solution for mean Nusselt number can be cast into form

$$Nu_{L,T} = \frac{4}{3}\left(\frac{0.902\sqrt{Pr}}{(0.861+Pr)^{1/4}}\right)\left(\frac{Gr_L}{4}\right)^{1/4}$$

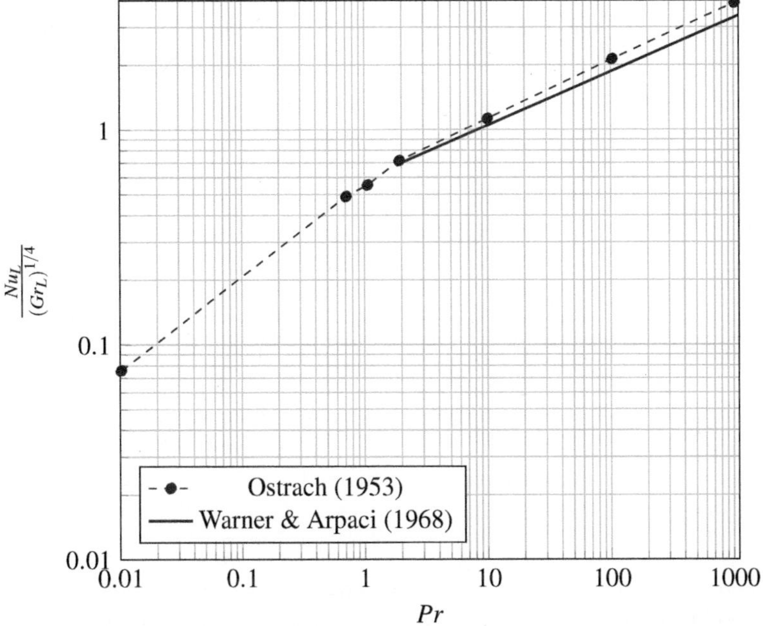

Figure 11.4 $\mathrm{Nu}_L/\mathrm{Gr}_L^{1/4}$ distribution against Prandtl number. Heat transfer due to natural convection from a vertical isothermal plate.

William H. McAdams has presented the following correlation for air in case of convection from a vertical plate.

$$Nu_m = 0.548(Ra)^{1/4} \quad Air$$

Ostrach (1953) suggested slight change to consider other fluids:

$$Nu_m = 0.555(Ra)^{1/4} \quad Oils$$
$$Nu_m = 0.33(Ra)^{1/4} \quad Liquid\ Metals$$

Free Convection Correlation for Vertical Plate

The ratio of mean Nusselt number with constant wall temperature and heat flux boundary conditions is

$$\frac{Nu_{m,T}}{Nu_{m,H}} = 1.044\left[\frac{0.8 + Pr}{0.952 + Pr}\right]^{1/4}$$

Churchill and Chu (1975) presented the correlation for vertical isothermal surface

$$Nu_{m,T} = \left[0.825 + \frac{0.387 Ra^{1/6}}{\left[1 + \left(\frac{0.492}{Pr} \right)^{9/16} \right]^{8/27}} \right]^2 \qquad 10^{-1} < Ra < 10^{12}$$

(11.4)

Example 11.1

A 20cm (height) $\times$ 5cm (width) flat surface suspended vertically in quiescent air medium that is held at 700 K. The air temperature is 300 K. Estimate the velocity boundary layer thickness at the end of the plate and find the heat transfer from the flat surface.

Solution

We use the properties at the mean temperature of 500 K:
k = 0.0389 W/m·K, v = 0.0000373 m^2/s, β=1/500, and Pr = 0.69.
The temperature difference ΔT = 700-300= 400K, we compute the Rayleigh number as

$$Ra_D = \frac{\beta \cdot \Delta T \cdot g \cdot D^3 \cdot Pr}{v^2} = 3.11 \times 10^7$$

As $Ra_D < 10^9$ the buoyancy driven boundary layer is laminar. From Figure 11.2 it is clear that velocity is approaching zero at $\eta = 6$, we can estimate the boundary layer thickness for free convection boundary layer:

$$Gr = \frac{\beta \cdot \Delta T \cdot g \cdot L^3}{v^2} \; 4.512 \times 10^7$$

$$\delta = \frac{6 \cdot L}{\left(\frac{Gr}{4} \right)^{\frac{1}{4}}} = 0.0207 \text{ m}$$

Using Churchill and Chu correlation, we have the Nusselt number:

$$Nu_{L,T} = \left(0.825 + \frac{0.387 \cdot Ra^{\frac{1}{6}}}{\left(1 + \left(\frac{0.492}{Pr} \right)^{\frac{9}{16}} \right)^{\frac{8}{27}}} \right)^2 = 43.12$$

The convective heat transfer coefficient is

$$h = 8.387 W/m^2 \cdot K$$

The area is 0.01 m^2 and heat transfer is

$$Q = h \cdot A \cdot \Delta T = 8.387 \times 0.01 \times (700 - 300) = 33.55W$$

Using Warner and Arpaci correlation, we have the Nusselt number:

$$Nu_{L,T} = 0.59 \cdot Ra^{\frac{1}{4}}$$

The convective heat transfer coefficient is

$$h = 8.572W/m^2 \cdot K$$

heat transfer is

$$Q = h \cdot A \cdot \Delta T = 34.288W$$

Using approximation for Ostrach correlation:

$$Nu_{L,T} = \frac{4}{3}\left(\frac{0.902\sqrt{Pr}}{(0.861+Pr)^{1/4}}\right)\left(\frac{Gr_L}{4}\right)^{1/4} = 51.8811$$

The convective heat transfer coefficient is

$$h = 10.09W/m^2 \cdot K$$

heat transfer is

$$Q = h \cdot A \cdot \Delta T = 40.36W$$

11.3 LAMINAR FREE CONVECTION FROM HORIZONTAL OR INCLINED SURFACE

The problem of considerable importance is that of the natural convection over a horizontal surface for a semi-horizontal surface with a single leading edge. The governing equation for the momentum transfer can be written as:

$$u\frac{\partial u}{\partial x} + v\frac{\partial u}{\partial y} = \frac{1}{\rho}\frac{\partial}{\partial y}\left(\mu\frac{\partial u}{\partial y}\right) - \frac{1}{\rho}\frac{\partial p_d}{\partial x}$$

$$\frac{1}{\rho}\frac{\partial p_d}{\partial y} = g\beta(T - T_w)$$

Pera and Gebhart (1972) have considered the flow over such a surface with a slight inclination from the horizontal axis. Experiments indicate the boundary near the leading edge of the upper side of the heated horizontal surface can be related to the power of the temperature difference as $T_w - T_\infty = N.x^m$.

The dynamic motion pressure (p_d) drives the flow from the heated surface as it heats the adjacent layers, causing them to have a low density, which leads to the rising of the plume from the surface. This results in a negative pressure gradient, which causes a boundary layer type flow formation over the surface. Pera and Gebhart (1972) have solved this problem using the similarity solution. They found the Nusselt number is proportional to the Prandtl number in the range of 0.1 to 100. For a uniform surface temperature distribution, the Nusselt number is

$$Nu_{x,T} = 0.394 \frac{Gr_x^{1/5}}{Pr^{1/4}}$$

and for the uniform heat flux distribution, the Nusselt number is

$$Nu_{x,H} = 0.5013 \frac{Gr_x^{1/5}}{Pr^{1/4}}$$

Note that for the isothermal surface, the average Nusselt number would be 5/3 times the value of the local Nusselt number at the end of the surface.

11.4 TURBULENT CONVECTION

It has been found that the transition from laminar to turbulent boundary layer flow occurs between $Gr_x = 10^9$ and 10^{10}. In case of boundary layer transition the effects like surface inclination and the heat flux rate would also influence the transition phenomenon.

Owing to the complexities of turbulent flow in 1950s, Eckert and Jackson presented an integral analysis for the estimation of heat transfer in case of turbulent free convection. The approach is based on the idea that the structure of the turbulent free convection boundary layer flow in the immediate vicinity of the wall is independent of mechanism of the of turbulence generation. The reasonable velocity profile for buoyancy driven flow is

$$\frac{u}{U_\infty} = \left(\frac{y}{\delta}\right)^{1/7}\left[1 - \left(\frac{y}{\delta}\right)\right]^4$$

In accordance with experimental data, the 1/7 power law known from the forced convection can be taken for turbulent free convection boundary layer as well

$$\theta = \frac{T - T_\infty}{T_w - T_\infty} = \left[1 - \left(\frac{y}{\delta}\right)^{1/7}\right]$$

Using the experimental data, the polynomials obtained by curve-fitting for velocity distributions in bouyancy driven flows shows the following type of profiles:

$$\frac{u_1}{U_\infty} = 1.69\left(\frac{y}{\delta}\right)^{1/7}\left[1 - \left(\frac{y}{\delta}\right)\right]^2$$

or

$$\frac{u_2}{U_\infty} = 1.87\left(\frac{y}{\delta}\right)^{1/7}\left[1 - \left(\frac{y}{\delta}\right)\right]^4$$

The two profiles are plotted in Figure 11.5. This shows that Eckert and Jackson's profile considerations have some empirical basis.

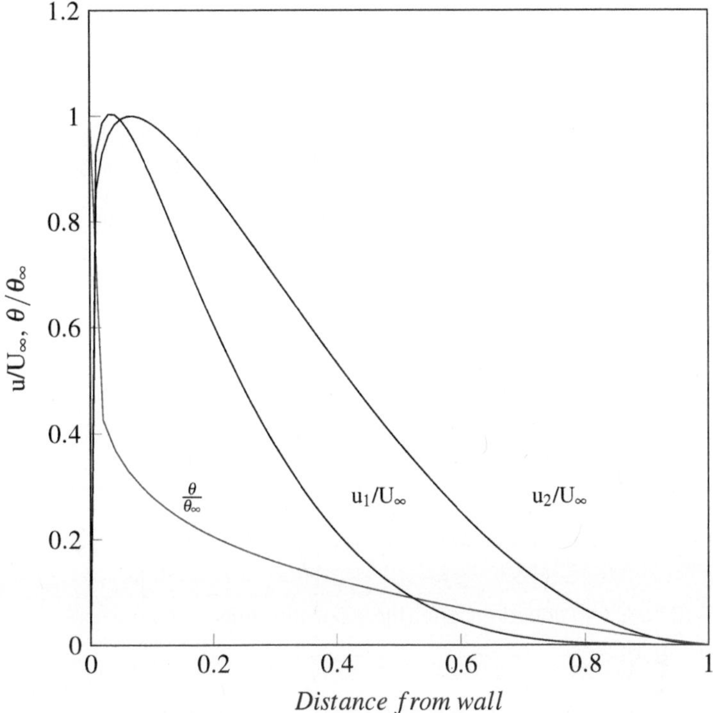

Figure 11.5   The reasonable polynomials for the velocity and temperature distributions inside turbulent free convection boundary layer.

In the case of free convection, Ostrach's exact solution (1953) shows that the laminar layer $\delta_T << \delta$. This can also the case for the turbulent flows for fluids having Prandtl number close to unity.

With the above assumption that the structure of the turbulent in free convection bouyancy driven flow in the immediate vicinity of the wall is independent of the formation mechanism, Blasius law for the wall shear stress can be used

$$\tau_w = 0.0225\rho_o U_o^2 \left(\frac{\nu}{U_o\delta}\right)^{1/4}$$

Stanton number for such case can be

$$St = \frac{q_w U_o}{\tau_w c_p(T_w - T_\infty)} = 1 \quad for \quad Pr = 1$$

also we consider

$$St \sim Pr^{-2/3}$$

this leads to approximate heat flux at wall as

$$q_w \equiv \left[\tau_w c_p \frac{(T_w - T_\infty)}{U_o} \right] Pr^{-2/3}$$

The integral energy equation is

$$\frac{1}{R} \left\{ \frac{d}{dx} \left[R \int_0^{\delta_t} u(T - T_\infty)dy \right] + \frac{dT_\infty}{dx} \left[\int_0^{\delta_t} u\,dy \right] \right\} = \left(\frac{q_w}{\rho c_p} \right)$$

For flat plate $R \neq R(x)$ and for case of $dT_\infty/dx = 0$ we have the integral energy equation as

$$\frac{d}{dx} \left[\int_0^{\delta_T} u(T - T_\infty)dy \right] = \left(\frac{q_w}{\rho c_p} \right)$$

The substituting shear stress and reasonable profiles into the momentum integral equation for buoyancy driven flows is

$$\frac{d}{dx} \left[\int_0^{\delta} u^2 dy \right] = \frac{\tau_w}{\rho} - g \left[\int_0^{\delta} \beta(T - T_\infty)dy \right]$$

We will have two equations after these substitutions

$$0.05231 \frac{d}{dx}(U_o^2 \delta) = -0.0225 U_o^2 \left(\frac{\nu}{U_o \delta} \right)^{1/4} + 0.125 g \beta \delta (T_w - T_\infty)$$

$$0.0366 \frac{d}{dx}(U_o \delta) = 0.0225 U_o \left(\frac{\nu}{U_o \delta} \right)^{1/4} Pr^{-2/3}$$

Here we need to prescribe the external flow velocity and boundary layer thicknesses. For boundary layer thickness we assume the relationship with x as $\delta(x) = C_2 x^n$ and for the velocity distribution $U_o(x) = C_1 x^m$ we compare the solution and get the equation

$$5n + m = 4$$

this gives

$$m = 1/2 \quad and \quad n = 7/10$$

$$\frac{U_o x}{\nu} = 1.185 \sqrt{Gr_x} \sqrt{\left[\frac{1}{1 + 0.494 Pr^{2/3}} \right]}$$

$$\frac{\delta}{x} = 0.565 Gr_x^{-1/10} \left[\frac{1}{1 + 0.494 Pr^{2/3}} \right]^{1/10}$$

Using these equations Eckert and Jackson (1951) arrived at the following correlation
for Nusselt number based on reasonable profiles for velocity and temperature

$$Nu_x = 0.0295(Ra_x)^{2/5}\frac{Pr^{1/15}}{\left[1+0.494Pr^{2/3}\right]^{2/5}}$$

Integrating this for turbulent boundary layer, we have $Nu_L = \frac{5}{6}Nu_x$. For Prandtl
numbers close to unity, the parenthetical term in above equation, we obtain :

$$Nu_L = 0.021(Ra_L)^{2/5}$$

Figure 11.6 and 11.7 show the mean Nusselt number plotted along the Rayleigh
number.

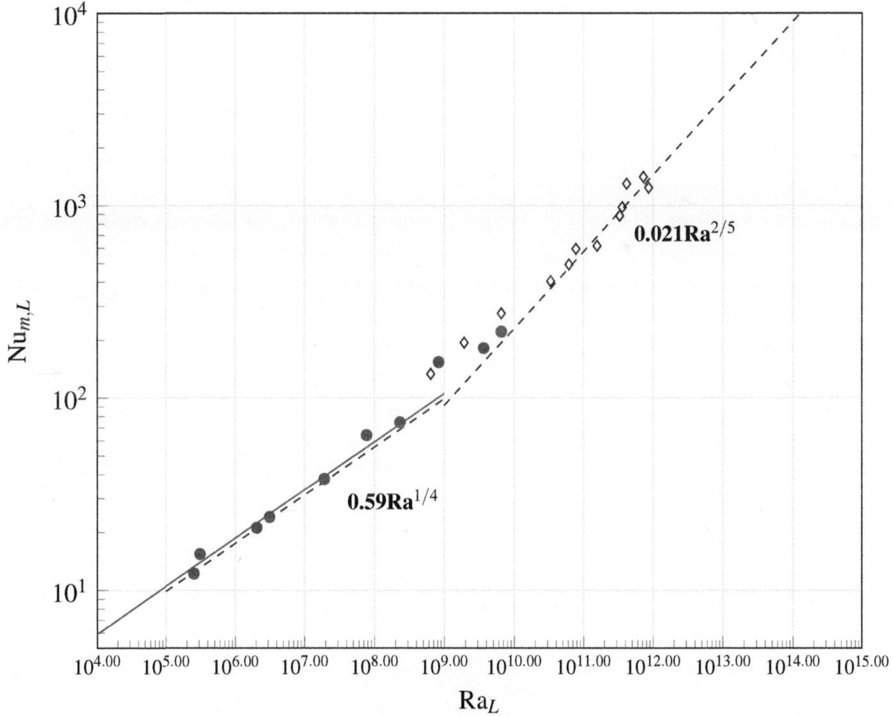

Figure 11.6 $Nu_{m,T}$ for free convection from vertical plate (Pr=0.7). Experimental data plotted along with integral solution based correlations of Eckert and Jackson (1950) (dashed line) and the empirical relation of Warner and Arpaci (1968) (solid line).

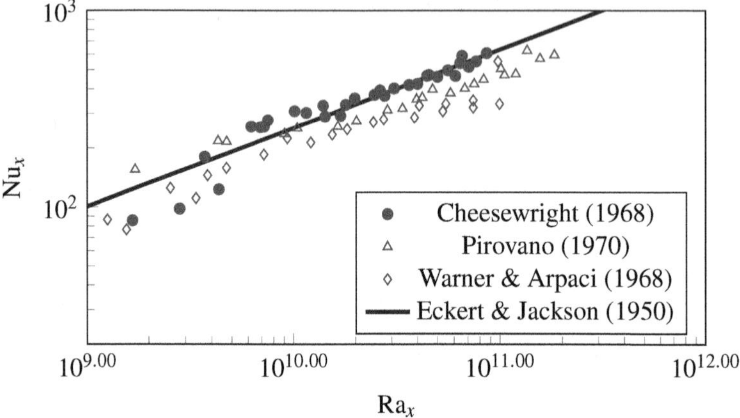

Figure 11.7   Distribution of Nusselt vs. Reynolds number on a vertical plate in case of turbulent free convection (Pr=0.7).

11.5   FREE CONVECTION FROM LONG HORIZONTAL CYLINDER

Figure 11.8 shows the phenomenon of natural convection from a horizontal cylinder. The surface temperature of the cylinder is T_w, and T_a is temperature of ambient quiescent medium. The gravity vector is acting downwards towards earth. The heat transfer is influenced by the plume development, which starts from the base of the cylinder and then it will ascend in the quiescent medium from the surface of the cylinder. Kuehn and Goldstein (1982) have found that the heat transfer is highest at the bottom of the cylinder, and it reduces gradually towards the top of the cylinder as plume rises from it. The reduction in heat transfer is attributed to the thickening of the thermal boundary layer. Researchers have proposed correlations for the mean Nusselt number the entire circumference of an isothermal cylinder:

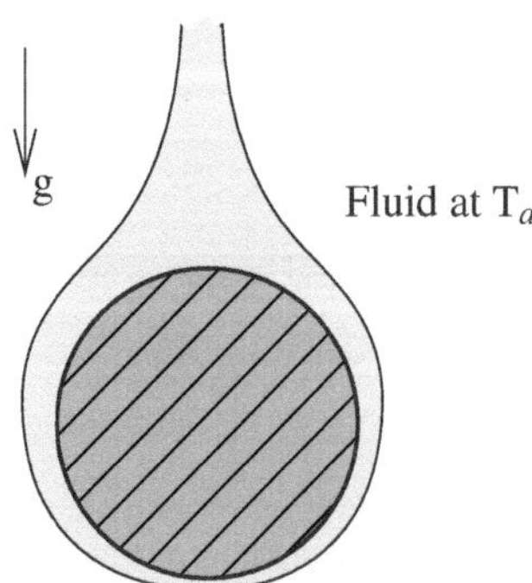

Figure 11.8   The plume rising from the horizontal cylinder. T_a is temperature of ambient quiescent medium.

McAdams (1954) proposed the correlation:

$$Nu_D = 0.53 Ra^{1/4} \quad 10^4 < Ra_D < 10^9$$

$$Nu_D = 0.13 Ra^{1/3} \quad 10^9 < Ra_D < 10^{12}$$

Churchill and Chu (1975) proposed correlation for the case of horizontal cylinder:

$$\mathrm{Nu}_D = 0.36 + \frac{0.518 Ra^{1/4}}{\left[1 + \left(\frac{0.559}{Pr}\right)^{9/16}\right]^{4/9}} \quad 10^{-6} \leq Ra_D \leq 10^9$$

The above correlation does not match well with the experimental data for very low values of $Gr_D \cdot Pr \leq 10^{-6}$. Churchill and Chu (1975) further improved this correlation as [see Boetcher (2014)]:

$$Nu_D = \left[0.6 + 0.387 \left(\frac{Ra_D}{\left[1 + \left(\frac{0.559}{Pr}\right)^{9/16}\right]^{16/9}}\right)^{1/6}\right]^2 \quad 10^{-11} \leq Ra_D \leq 10^9$$

Note that above correlation is not for turbulent flow only. The range of validity of this correlation dictate that it can be used for laminar and turbulent flows.

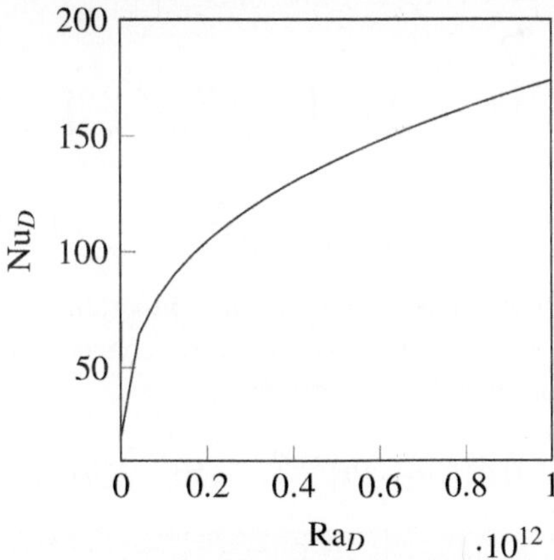

Figure 11.9 Distribution of Nusselt versus Ra number in case of free convection from cylinder (Churchill and Chu (1975) correlation).

Figure 11.9 shows the distribution of Nusselt number as predicted by Churchill and Chu correlation.

Morgan (1975) proposed the correlation for cylinder maintained at constant surface temperature as

$$\mathrm{Nu}_D = C \cdot Ra_D^m$$

where C and m are listed in Table 11.2.

TABLE 11.2

C and m to be Used with Morgan Correlation

Ra_D		C	m
10^{-10}	0.01	0.675	0.058
0.01	100	1.02	0.148
100	10^4	0.85	0.188
10^4	10^7	0.48	0.25
10^7	10^{12}	0.125	0.333

Kuehn and Goldstein (1976,1980) found that the position of the outer boundary has an influence on the values of heat transfer coefficient if the outer boundary is not set far enough from the surface of the cylinder. They recommended the correlation for any Prandtl and Rayleigh numbers:

$$Nu_D = \frac{2}{\ln\left(1 + \frac{2}{\Phi}\right)}$$

where,

$$\Phi = (f_1 + f_2)^{\frac{1}{15}}$$

$$f_1 = \left(0.518 \cdot Ra^{\frac{1}{4}} \cdot \left(1 + \left(\frac{0.559}{Pr}\right)^{\frac{3}{5}}\right)^{-\frac{5}{12}}\right)^{15}$$

$$f_2 = \left(0.1 \cdot Ra^{0.3333}\right)^{15}$$

The above correlation is based on constant temperature conditions. Dyer (1965) recommended the following correlation for natural convection from horizontal cylinder subjected to constant heat flux condition:

$$Nu_{D,H} = 0.61[Gr_D \cdot Pr]^{0.192} \qquad 10^3 \leq Gr_D \cdot Pr \leq 10^{10}$$

Example 11.2

A long horizontal cylinder having external diameter 7m and length 13m is placed in a quiescent air where ambient temperature is kept at 300 K. The cylinder wall temperature is 700 K. Find the convective heat transfer coefficient using various correlation presented in this chapter.

Solution The external diameter D = 7 m. We use the properties at the mean temperature of 500 K:

k = 0.0389 W/m·K, v = 0.0000373 m²/s, β=1/500, and Pr = 0.69.

The temperature difference $\Delta T = 700 - 300 = 400 K$, we compute the Rayleigh number as

$$Ra_D = \frac{\beta \cdot \Delta T \cdot g \cdot D^3 \cdot Pr}{\nu^2} = 1.33 \times 10^{12}$$

The plume is turbulent as $Ra > 10^9$. We use Churchill and Chu (1975) correlation for determination of the natural convection Nusselt number:

$$Nu_D = \left[0.6 + 0.387 \left(\frac{Ra_D}{\left[1 + \left(\frac{0.559}{Pr} \right)^{9/16} \right]^{16/9}} \right)^{1/6} \right]^2$$

$$Nu_D = 1172.21$$

This gives the convective heat transfer coefficient as

$$h = 6.514 \frac{W}{m^2 \cdot K}$$

Using Morgan Correlation we have

$$Nu_D = C \cdot Ra_D^m$$
$$Nu_D = 0.125 \cdot Ra^{0.333} = 1363.62$$

This gives the convective heat transfer coefficient as

$$h = 7.577 \frac{W}{m^2 \cdot K}$$

We use Kuehn and Goldstein correlation:

$$Nu_D = \frac{2}{\ln\left(1 + \frac{2}{\Phi}\right)}$$

$$\Phi = (f_1 + f_2)^{\frac{1}{15}}$$

$$f_1 = \left(0.518 \cdot Ra^{\frac{1}{4}} \cdot \left(1 + \left(\frac{0.559}{Pr} \right)^{\frac{3}{5}} \right)^{-\frac{5}{12}} \right)^{15}$$

$$f_2 = \left(0.1 \cdot Ra^{0.3333} \right)^{15}$$

This gives

$$Nu_D = 1101.07$$

and the convective heat transfer coefficient as

$$h = 6.118 \frac{W}{m^2 \cdot K}$$

Usin McAdams (1954) correlation, we have

$$Nu_D = 0.13 Ra^{1/3} \quad 10^9 < Ra_D < 10^{12}$$

$$Nu_D = 1430.10$$

and the convective heat transfer coefficient as

$$h = 7.947 \frac{W}{m^2 \cdot K}$$

Comparison of convective heat transfer coefficients

Correlation	$h \ \frac{W}{m^2 \cdot K}$
McAdams (1954)	7.947
Morgan (1975)	7.577
Churchill and Chu (1975)	6.514
Kuehn and Goldstein (1976)	6.118

11.6   FREE CONVECTION FROM A SPHERE

Yuge (1960) proposed the correlation for fluid with Prandtl number unity:

$$Nu_D = 2 + 0.43 Ra^{1/4} \quad 1 < Ra_D < 10^5$$

Churchill (1983, 2002) proposed the correlation:

$$Nu_D = 2 + \frac{0.589 Ra^{1/4}}{\left[1 + \left(\frac{0.469}{Pr} \right)^{9/16} \right]^{4/9}} \quad Ra \leq 10^{11}, Pr \geq 0.5$$

11.7   NATURAL CONVECTION IN ENCLOSURES

Enclosures are regularly encountered in engineering applications like solar collectors, double-glazed windows, computer casing, etc. The transfer of heat in such confined spaces does not remain stationary. In an enclosure, the fluid adjacent to the vertical hotter surface ascends due to buoyancy and the fluid adjacent to the cooler one descends. This generates a rotational movement within the enclosure that increases the heat transfer through the enclosure. The flow patterns in a rectangular enclosure is shown in Figure 11.10.

The density of cold fluid is higher than the hot fluid and this would create the buoyancy driven flow. Rayleigh–Bénard convection is a type of free convection, which happens in a planar horizontal layer of fluid heated from below, in which the fluid creates a pattern of convection cells known as Bénard cells. Initially the flow will be laminar, and the convective Bénard cells would tend to appear like hexagonal prisms. However, when the temperature of the bottom plane is increased, the flow structure would become more complex and the flow would become chaotic and turbulent.

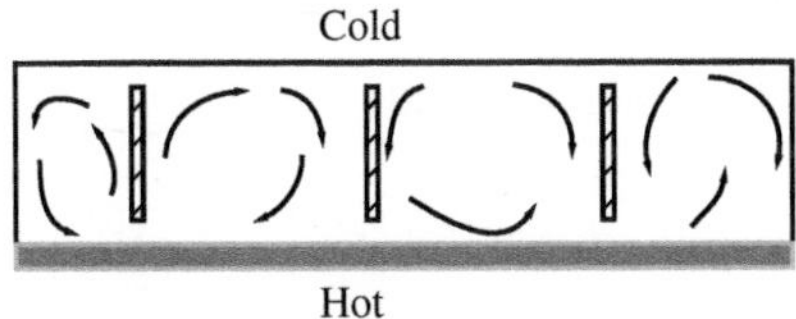

Figure 11.10   Convective currents in a horizontal enclosure with top cold plate.

Rayleigh numbers less than a critical value of $Ra_{L,c} < 1708$, the buoyancy forces cannot overcome the viscous forces and there is no advection within the cavity. When Rayleigh numbers is greater than 1708, the bouyant flow becomes thermally unstable and there is advection within the enclosure.

$$Ra_L = \frac{\beta \cdot g \cdot L^3 \cdot (T_h - T_c)}{\alpha \cdot v} > 1708$$

Jakob (1949) recommended the following correlations for horizontal enclosures $(0.5 < Pr < 2)$ when the hot plate is at the base:

$$Nu_L = 0.195 \cdot Ra_L^{\frac{1}{4}} \quad 10^4 \leq Ra_L \leq 4 \times 10^5$$

$$Nu_L = 0.068 \cdot Ra_L^{\frac{1}{3}} \quad 4 \times 10^5 \leq Ra_L \leq \times 10^7$$

Globe and Dropkin (1959) proposed the following correlation for horizontal enclosures heated from below

$$Nu_L = 0.069 \cdot Ra_L^{\frac{1}{3}} Pr^{0.074} \quad 3 \times 10^5 \leq Ra_L \leq 7 \times 10^9$$

The properties are evaluated at the average temperature and the correlation is valid for water, silicone oil, and mercury.

Example 11.3

A horizontal enclosure comprising of two parallel plates is separated by a gap of 8 mm. The lower plate is kept at a uniform temperature of 600 K, and the upper plate is kept at a uniform temperature of 300 K. A fluid has been filled in between this enclosure, with properties:

$$v = 0.658 \times 10^{-6} m^2/s, Pr = 5.83, \alpha = 0.1512 \times 10^{-6} m^2/s, k = 0.61 W/m \cdot K$$

Find the convective heat transfer coefficient if the side walls are insulated.

Solution

The β at mean temperature is $1/300 \text{ K}^{-1}$. As L = 8 mm, The Rayleigh number is

$$Ra = \frac{\beta \cdot 9.81 \cdot L^3 \cdot (600 - 300)}{\alpha \cdot \nu} = Ra := 5.048487482 \times 10^7$$

Using Globe and Dropkin (1959) correlation, the Nusselt number is

$$\text{Nu} = 0.069 \cdot \sqrt[3]{Ra} \cdot Pr^{0.074} = 29.05$$

The convective heat transfer coefficient is

$$h = 2215.485 W/m^2 \cdot K$$

11.7.1 VERTICAL ENCLOSURES (H/L ≫ 1)

The heat transmission for the case of double glazed windows or the heat absorption in the solar collectors, are the cases of heat transfer in an enclosure. Heat transfer in vertical and inclined enclosures with plane parallel side walls has been investigated very intensively. Figure 11.11 shows a slender vertical enclosure with high aspect ratio (H ≫ width L). The heat transport in the gap can be regarded as a two-dimensional problem. The left and right walls are kept at constant but different temperatures and, the lower and upper walls are considered as adiabatic.

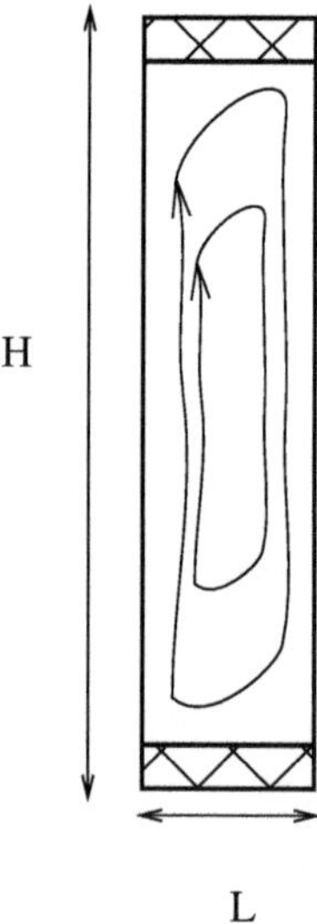

Figure 11.11   An enclosure with high aspect ratio H/L ≫ 1.

Catton (1978) proposed the following relation:

$$Nu_L = 0.18 \left[\frac{Pr \cdot Ra_L}{Pr + 0.2} \right]^{0.29} \left. \right\} \quad \begin{matrix} 1 \leq H/L \leq 2, \\ 0.001 \leq Ra_L \leq 10^5 \\ \left(\frac{Pr \cdot Ra_L}{Pr+0.2} \right) \geq 10^3 \end{matrix} \tag{11.5}$$

$$Nu_L = 0.22 \left(\frac{L}{H} \right)^{1/4} \left[\frac{Pr \cdot Ra_L}{Pr + 0.2} \right]^{0.28} \left. \right\} \quad \begin{matrix} 2 \leq H/L \leq 10 \\ 1000 \leq Ra_L \leq 10^{10} \\ Pr \leq 10^5 \end{matrix} \tag{11.6}$$

The case of enclosure with constant wall flux was experimentally investigated by MacGregor and Emery. For the laminar boundary layer flow, MacGregor and Emery (1969) recommended the following correlation for the vertical enclosures with larger aspect ratios

$$Nu_L = 0.42 \cdot Ra_L^{1/4} \cdot Pr^{0.012} \left(\frac{L}{H} \right)^{0.3} \left. \right\} \quad \begin{matrix} 10 \leq H/L \leq 40, \\ 10^4 \leq Ra_L \leq 10^7, \\ 1 < Pr < 20,000 \end{matrix} \tag{11.7}$$

For turbulent regime the correlation is

$$Nu_L = 0.046 \cdot Ra_L^{1/3} \left. \right\} \quad \begin{matrix} 1 \leq H/L \leq 40, \\ 10^6 \leq Ra \leq 10^9, \\ 1 < Pr < 20 \end{matrix} \tag{11.8}$$

Figure 11.12 shows the Nusselt number plotted along the Rayliegh number.

11.7.2 INCLINED RECTANGULAR ENCLOSURES

An inclined rectangular enclosure with isothermal surfaces is shown in Figure 11.13. Hollands et al. (1976) proposed the following relation:

$$Nu_L = 1 + \left\langle \frac{\sqrt[3]{\chi}}{18} - 1 \right\rangle + 1.44 \left\langle 1 - \frac{1708}{\chi} \right\rangle \cdot \left(1 - \frac{1708 \cdot (\sin 1.8\theta)^{1.6}}{\chi} \right) \quad \begin{matrix} Ra_L < 10^5 \\ 0 < \theta < 70° \\ H/L \geq 12 \end{matrix}$$

where, $\chi = Ra_L \cos \theta$, and if $\langle . \rangle$ terms are negative, then they are set to zero.

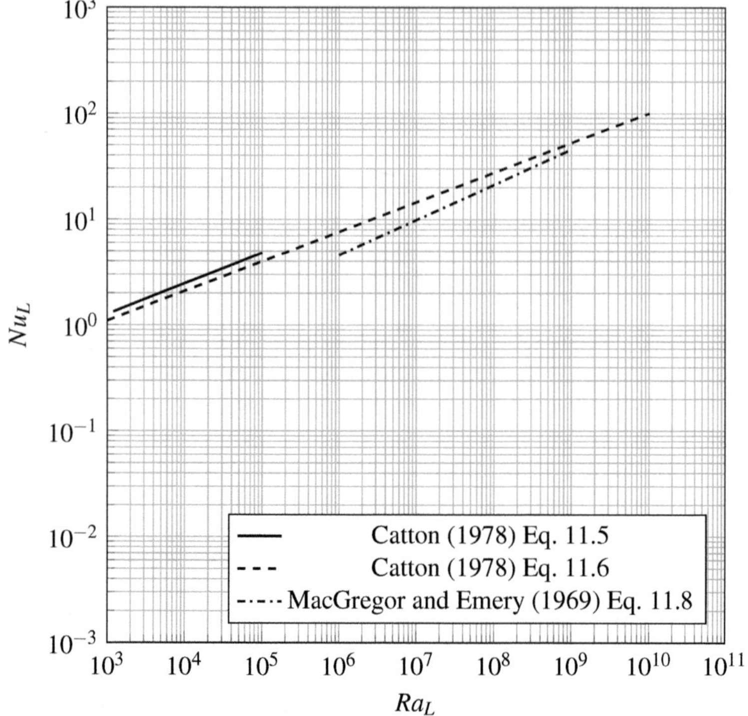

Figure 11.12   Qualitative trend of Nu_L as a function of the Rayleigh number for a slim vertical enclosure ($H/L \approx 3$) and $Pr \approx 1$.

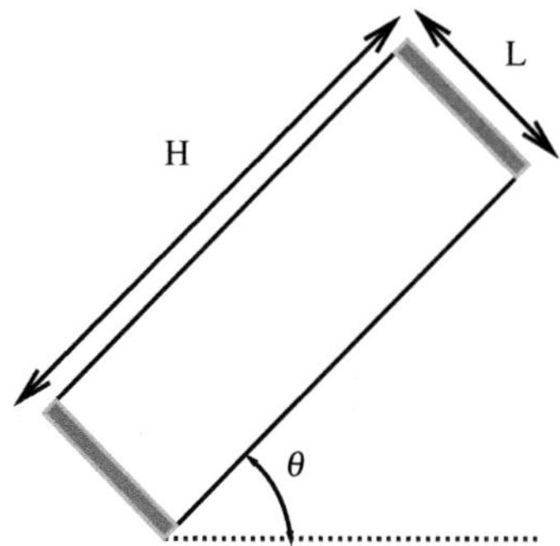

Figure 11.13   An inclined rectangular enclosure with isothermal surfaces.

Example 11.4

A solar collector is 4m (height) $\times$ 3cm in cross section has length of 5m. The collector is placed with angle of $50°$ facing sun. Find the heat flux coming on the collector. The temperature of the cover and absorber are $25°$, and $80°$, respectively. The fluid properties at the mean temperature are k= 0.028 W/m $\cdot$ K, $v = 0.0000184$ m^2/s, Pr = 0.708, and $\alpha = 0.0000259$ m^2/s.

Solution

The mean temperature is $T_m = 52.5°$, and the thermal expansion coefficient is $\beta = 1/T_m(K) = 0.003072196$. The Rayleigh number is

$$Ra_L = \frac{\beta \cdot 9.81 \cdot L^3 \cdot (T_1 - T_2)}{v \cdot \alpha} = 25{,}018$$

The $\theta = 50° \cdot \pi/180 = 0.87266$ radians, which gives $\chi = Ra_L \cos\theta = 16081$

$$Nu_L = 1 + \left\langle \frac{\sqrt[3]{\chi}}{18} - 1 \right\rangle + 1.44 \left\langle 1 - \frac{1708}{\chi} \right\rangle \cdot \left(1 - \frac{1708 \cdot (\sin 1.8\theta)^{1.6}}{\chi} \right)$$

None of the $\langle . \rangle$ terms are negative, so

$$Nu_L = 2.5526$$

The convective heat transfer coefficient is

$$h = 3.57369 \, W/m^2 \cdot K$$

The heat flux is

$$q'' = h \cdot (T_1 - T_2) = 2144.21 W/m^2$$

11.7.3   SQUARE ENCLOSURES (H=L)

The square enclosures whose side walls are maintained at constant but different temperatures and whose lower and upper walls are adiabatic have been investigated by Berkovsky and Polevikov (1976), and they proposed the correlation:

$$Nu_L = 0.18 \left(\frac{Pr \cdot Ra_L}{0.2 + Pr} \right)^{0.29} \left. \right\} \quad \begin{array}{l} 0.001 < Pr < 10^5 \\ \frac{Pr \cdot Ra_L}{0.2 + pr} > 10^3, 1 < H/L < 2 \end{array}$$

11.8 FREE CONVECTION HEAT TRANSFER BETWEEN CONCENTRIC SPHERES

Raithby and Hollands (1975) proposed the following relations to estimate the natural convection heat transfer for an enclosed space between two concentric spheres.

The Rayleigh number is based on

$$\ell_{sphere} = \frac{\sqrt[3]{\left(\frac{1}{r_i} - \frac{1}{r_o}\right)^4}}{1.26 \cdot \sqrt[3]{\left(\frac{1}{r_i^{7/5}} + \frac{1}{r_o^{7/5}}\right)^5}}$$

and the heat transfer per unit gap space can be calculated as

$$Q = \frac{4\pi k_{eff}(T_i - T_o)}{(1/r_i) - (1/r_o)} \quad (W/m)$$

where, the effective thermal conductivity can be obtained from a relation:

$$\frac{k_{eff}}{k} = 0.74 \cdot \sqrt[4]{\left(\frac{Ra_{sphere} \cdot Pr}{Pr + 0.861}\right)}$$

where, k is thermal conductivity of fluid in the gap. Note that if $k_{eff} < k$ then take $k_{eff} = k$.

Example 11.5

The gap space between the two concentric spheres is filled with a fluid. The inner sphere radius is 0.65 m and the outer sphere radius is 0.7 m. The inner and outer spheres temperature are 323 K and 293 K, respectively. Find the heat transfer per unit gap space between the sphere, if the fluid properties are k=0.027 W/m·K, ν=3.3E-5, Pr = 0.72, and β=0.003246.

Solution

We first calculate the characteristic length for the sphere as

$$\ell_{sphere} = \frac{\sqrt[3]{\left(\frac{1}{r_i} - \frac{1}{r_o}\right)^4}}{1.26 \cdot \sqrt[3]{\left(\frac{1}{r_i^{7/5}} + \frac{1}{r_o^{7/5}}\right)^5}} = 0.005239$$

The Rayleigh number is

$$Ra_{sphere} = \frac{\beta \cdot g \cdot \ell_{sphere}^3 \cdot (T_h - T_c)}{\alpha \cdot \nu} = 90.84$$

the effective thermal conductivity ratio is

$$\frac{k_{eff}}{k} = 0.74 \cdot \sqrt[4]{\left(\frac{Ra_{sphere} \cdot Pr}{Pr + 0.861}\right)} = 1.876 > 1$$

This gives

$$k_{eff} = 0.05086 W/m \cdot K$$

$$Q_{spheres} = \frac{4\pi k_{eff}(T_i - T_o)}{(1/r_i) - (1/r_o)} = 174.394 W/m$$

11.9 FREE CONVECTION HEAT TRANSFER BETWEEN CONCENTRIC CYLINDERS

For the case of natural convection between concentric cylinders, Raithby and Hollands (1975) had recommended the following set of equations in order to compute the characteristic length, the effective conductivity and heat transfer:

$$\ell_{cylinder} = \frac{2\ln\left(\sqrt[3]{\left(\frac{r_o}{r_i}\right)^4}\right)}{\sqrt[3]{\left(\frac{1}{r_i^{3/5}} + \frac{1}{r_o^{3/5}}\right)^5}}$$

$$\frac{k_{eff}}{k} = 0.386 \cdot \sqrt[4]{\left(\frac{Pr \cdot Ra_{cylinder}}{Pr + 0.861}\right)}$$

If $k_{eff} < k$ then take $k_{eff} = k$.

$$Q_{cylinders} = \frac{2\pi k_{eff}(T_i - T_o)}{\ln(r_o/r_i)} \quad (W/m)$$

The correlation is valid as long as $0.7 \le Pr \le 6000$, and $Ra_{cylinder} \le 10^7$.

11.10 CRITERION FOR FREE, FORCED, AND MIXED CONVECTION FLOWS

Sparrow et al. (1959) have suggested that a flow can be considered as effectively pure (either forced or free) if the heat transfer or the friction coefficient deviates by no more than 5 percent from the value associated with the completely pure flow.

$$\left.\begin{array}{ll} 0 < \left(\frac{Gr}{Re^2}\right) < 0.3 & \textit{Forced Convection} \\[2mm] 0.3 < \left(\frac{Gr}{Re^2}\right) < 16 & \textit{Mixed Convection} \\[2mm] \left(\frac{Gr}{Re^2}\right) > 16 & \textit{Free Convection} \end{array}\right\} \tag{11.9}$$

The mixed convection regime is the flow for which the Nusselt number deviates more than 10% from the pure forced or free convection value. The mixed convection mechanism in duct flows is a bit more complicated than that in external flows. Buoyancy force increases heat transfer in laminar flows in vertical ducts; however, it decreases heat transfer in turbulent flows. Heating a fluid flowing in a horizontal pipe produces secondary currents which increase heat transfer. Cooling on other hand reverses the direction of secondary currents but it also enhances the heat transfer.

The fluid near the tube or pipe wall will have high temperature and low density. This causes the formation of plume and circulation in upward direction. Whereas, the fluid near the core region of the pipe, will have low temperature and a high density, and circulates downward. Such counter-rotating transverse vortices are superimposed on the main streamwise flow and they can significantly increase the forced convection heat transfer.

Figures 11.14 and 11.15 show Metais and Eckert (1964) plot for the different convective flow regimes in a pipe with constant wall temperature condition. For horizontal flow, the boundary between mixed and free convection is not presented as it is not yet established.

Kem and Othmer (1943), Metais (1963) have found via experiments that the transition Reynolds number range between 600 and 800 for bouyancy driven flows.

Laminar Forced convection: Sieder-Tate proposed the correlation for laminar forced convection flows as

$$Nu_D = 1.86 Pe^{1/3}\left(\frac{\mu}{\mu_w}\right)^{0.14}\left(\frac{D}{L}\right)^{1/3} \tag{11.10}$$

Turbulent Forced convection:

Hausen proposed the correlation for turbulent forced convection flows as

$$Nu_D = 0.116\left[1+\left(\frac{D}{x}\right)^{2/3}\right](Re^{2/3}-125)Pr^{1/3}\left(\frac{\mu}{\mu_w}\right)^{0.14} \tag{11.11}$$

Mixed convection

Oliver proposed the correlation for laminar flows as

$$Nu_D = 1.75\left(\frac{\mu}{\mu_w}\right)^{0.14}\left[Gz+0.0083(Gr\cdot Pr)^{0.75}\right]^{1/3} \tag{11.12}$$

Metais proposed the correlation for turbulent flows as

$$Nu_D = 4.69 Re^{0.27}Pr^{0.21}Gr^{0.07}\left(\frac{D}{x}\right)^{0.36} \tag{11.13}$$

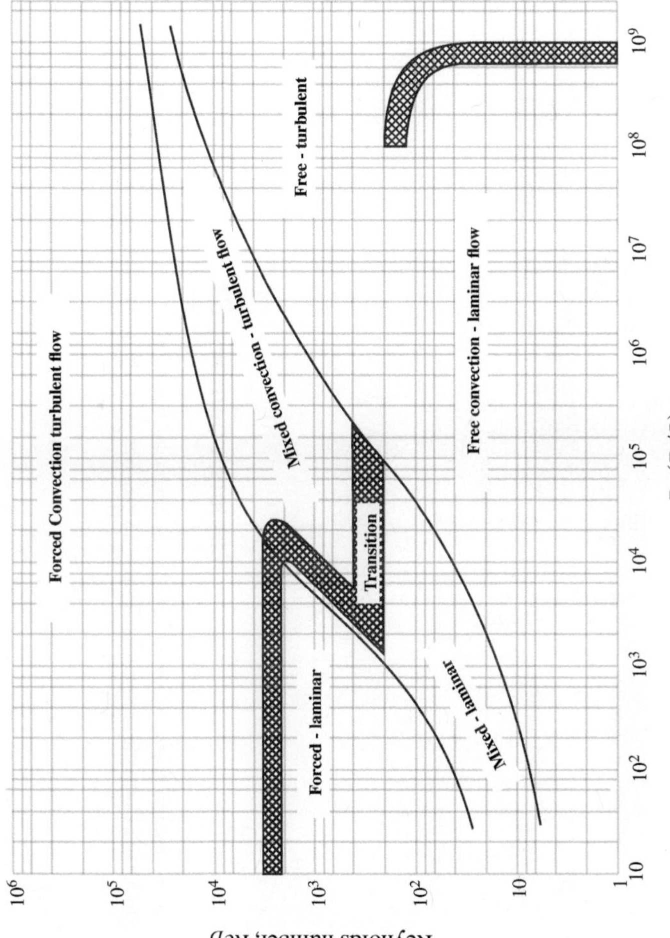

Figure 11.14 Mixed forced and natural convection flow regimes in vertical tube ($10^{-2} < $ Pr $\cdot$ D/L $ < 1$). (Source: Courtesy of B. Metais and E. R. G. Eckert, Forced, Free, and Mixed Convection Regimes, Trans. ASME. Ser. C. J. Heat Transfer, Vol. 86, pp. 295-298, 1964.)

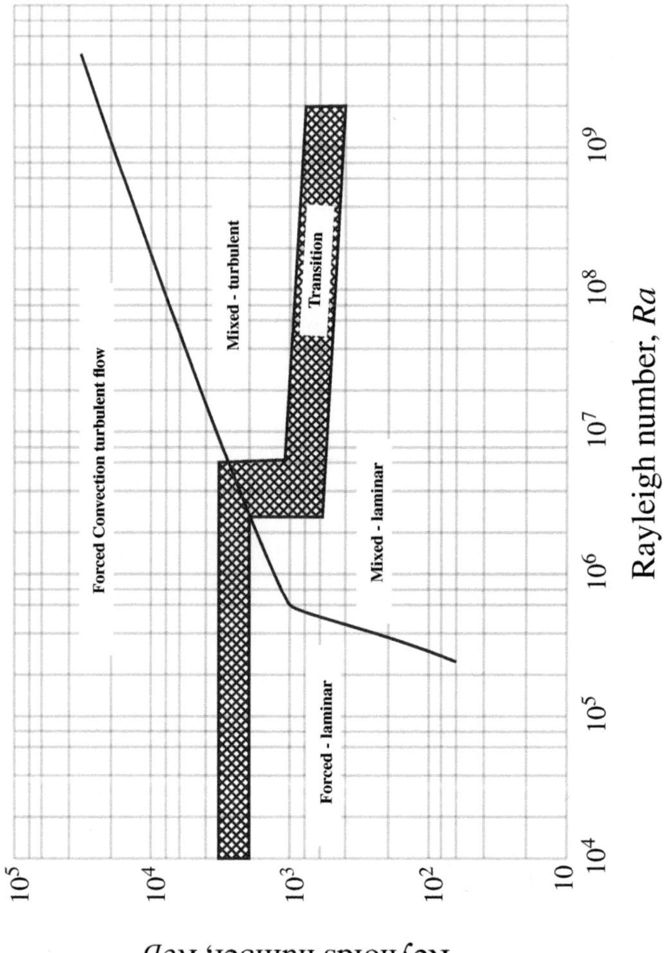

Figure 11.15   Mixed forced and natural convection flow regimes in horizontal tubes ($10^{-2} < Pr \cdot D/L < 1$. (Source: Courtesy of B. Metais and E. R. G. Eckert, Forced, Free, and Mixed Convection Regimes, Trans. ASME. Ser. C. J. Heat Transfer, Vol. 86, pp. 295-298, 1964.)

Example 11.6

Air at 7 °C is flowing through a vertical pipe, which is maintained at constant temperature of 153 °C. The internal diameter of pipe is 30 mm, and the speed of air flow is 0.2 m/s. The length of pipe is 0.3048 m/s. Find the convective heat transfer coefficient and heat transfer if flow is in (i) mixed convection regime (ii) in forced convection regime.

Solution From property data for air, we have the properties at mean temperature of 80 °C.

$$\mu_w = 0.0000235 \; kg/m.s, \; \nu = 0.00002094 \; m^2/s, \; \mu = 0.0000210 \; kg/m.s$$
$$k = 0.03 W/m.K, \; \beta = 0.00283 \; K^{-1}, \; Pr = 0.708$$

The Reynolds number of the flow is

$$\text{Re} = \frac{u \cdot D}{\nu} = 286.532$$

The flow is laminar.
The Grashof number of the flow is

$$Gr = \frac{g \cdot \beta \cdot \Delta T \cdot D^3}{\nu^2} = \frac{9.81 \times 0.00283 \times 20 \times 0.03^3}{0.00002094^2} = 249585.13$$

We compute the ratio Gr/Re^2:

$$Gr/Re^2 = 3.039$$

As ratio falls in range
$$0.3 < Gr/Re^2 < 16$$

the flow is having the mixed convection condition. Also from Figure 11.14 we can see that the flow condition is in laminar mixed convection regime. The corresponding Rayleigh number is

$$Ra = Gr \cdot Pr = 176706.272$$

and this gives $Ra \cdot (D/L) = 17392.349$. Ignoring the transition, we take this flow regime as mixed convection laminar flow and the applicable correlation is Oliver correlation

$$Nu_D = 1.75 \left(\frac{\mu}{\mu_w}\right)^{0.14} \left[Gz + 0.0083(Gr \cdot Pr)^{0.75}\right]^{1/3}$$

$$Nu_D = 7.762$$

The convective heat transfer coefficient is

$$h = 7.736 \frac{W}{m^2 \cdot K}$$

$$Q = h \cdot \pi \cdot D \cdot L \cdot \Delta T = 32.448 W$$

(ii) In case of laminar forced convection, we can use the Sieder and Tate correlation

$$Nu_D = 1.86 Pe^{1/3} \left(\frac{\mu}{\mu_w} \right)^{0.14} \left(\frac{D}{x} \right)^{1/3}$$

$$h = 4.967 \frac{W}{m^2 \cdot K}$$

$$Q = h \cdot \pi \cdot D \cdot L \cdot \Delta T = 20.763 W$$

Example 11.7

Air at 7 °C is flowing through a vertical pipe, which is maintained at constant temperature of 153 °C. The internal diameter of pipe is 0.6 m and the speed of air flow is 1.5 m/s. The length of pipe is 0.5048 m. Find the convective heat transfer coefficient and corresponding heat transfer.

Solution
From property data for air we have the properties at mean temperature of 80°C.

$$\mu_w = 0.0000235 \ kg/m.s, \quad v = 0.00002094 \ m^2/s, \quad \mu = 0.0000210 \ kg/m.s$$
$$k = 0.03 W/m.K, \quad \beta = 0.00283 \ K^{-1}, \quad Pr = 0.708$$

The Reynolds number of the flow is

$$Re = \frac{u \cdot D}{v} = 42979.94$$

The Grashof number of the flow is

$$Gr = \frac{g \cdot \beta \cdot \Delta T \cdot D^3}{v^2} = 1.99668 \times 10^9$$

We compute the ratio Gr/Re^2:

$$Gr/Re^2 = 1.080$$

As ratio falls in range

$$0.3 < Gr/Re^2 < 16$$

the flow is having the mixed convection condition. The corresponding Rayleigh number is

$$Ra = Gr \cdot Pr = 176706.272$$

and

$$Ra \cdot D/L = 1.6802 \times 10^9$$

From Figure 11.14 we can see that the flow condition is close to turbulent mixed convection regime. Note that Pr $\cdot$ D/L = 0.8415.

We apply the Metais correlation to estimate the convective heat transfer coefficient.

$$Nu_D = 4.69 \text{Re}^{0.27} Pr^{0.21} Gr^{0.07} \left(\frac{D}{x}\right)^{0.36} = 370.4202289$$

This gives

$$h_{mixed} = 18.459 \, W/m^2 \cdot K$$

This gives heat transfer as

$$Q = 2564.41W$$

In case the flow is purely forced convection, then we may use the Hausen correlation:

$$Nu_D = 0.116 \left[1 + \left(\frac{D}{x}\right)^{2/3}\right] (\text{Re}^{2/3} - 125)Pr^{1/3} \left(\frac{\mu}{\mu_w}\right)^{0.14} = 237.99$$

This gives

$$h_{forced} = 11.860 \, W/m^2 \cdot K$$

$$Q = 1647.649 \, W$$

In the mixed convection problem, we also consider the direction of movement of air and plume. If the buoyancy-driven flow is in the same direction as the forced convection flow, then such a flow is classified as aiding flow else the the buoyant motion acts against the forced motion, and then opposing flow would be present. For aiding flow condition in case of flow rising along a vertical flat plate, often a rule of thumb is proposed for the estimation of overall Nusselt number:

$$Nu_{mixed} = \sqrt[3]{\left[Nu_{free}^3 + Nu_{forced}^3\right]} \tag{11.14}$$

Free Convection Correlation for Iso-thermal Flat Surfaces

Geometry	Correlation $Nu_{m,T}$	Range
Vertical and inclined Flat Plate ($\theta < 60°$)		
Churchill & Chu (1975)	$\left[0.825 + \dfrac{0.387 Ra^{1/6}}{\left[1+(0.492/Pr)^{9/16}\right]^{8/27}}\right]^2$ inclination: $g \to g.\cos(\theta)$	$10^{-1} < Ra < 10^{12}$
Warner and Arpaci (1968)	$0.59 Ra^{1/4}$	$10^4 < Ra < 10^9$
	$0.1 Ra^{1/3}$	$10^9 < Ra < 10^{13}$
Eckert and Jackson (1950)	$0.555 Ra^{1/4}$	$10^5 < Ra < 10^9$
	$0.021 Ra^{2/5}$	$10^9 < Ra < 10^{15}$
Horizontal Flat Plate		
Fishenden and Saunders (1950)	$0.54 Ra^{1/4}$	$10^5 < Ra < 2.10^7$
	$0.14 Ra^{1/3}$	$2.10^7 < Ra < 3.10^{10}$

Free Convection Correlation for Iso-thermal Curved Surfaces

Geometry	Correlation $Nu_{m,T}$	Range
Horizontal Cylinder		
Churchill & Chu (1975)	$\left[0.6 + 0.387 \left(\dfrac{Ra_D}{\left[1+(0.559/Pr)^{9/16}\right]^{16/9}}\right)^{1/6}\right]^2$	$10^{-11} < Ra < 10^9$
Sphere		
Yuge (1960)	$2 + 0.43 Ra^{1/4}$	$1 < Ra_D < 10^5,$ $Pr = 1$
Churchill (2002)	$\left[2 + \dfrac{0.589 Ra^{1/4}}{\left[1+(0.469/Pr)^{9/16}\right]^{4/9}}\right]$	$Ra \leq 10^{11}, Pr \geq 1$

PROBLEMS

11P-1 A plate is hanging vertically in water which is at 298 K. The plate is 10 cm wide and 23 cm high, and its is maintained at a constant temperature of 350 K. Find the rate of heat transfer from the surface?

11P-2 A thin plate (1.5 m high $\times$ 0.5 m wide) hanging vertically. One face of the plate is insulated while the another face is exposed to constant heat flux of 220 W/m^2. Find the free convection heat transfer from the plate to the ambient 20 °C air.

11P-3 A 3 m high and 2.5 m wide flat plate solar collector consists of a glass cover and an absorber plate. The collector is tilted at an angle of 63 degrees from the horizontal. The gap between the cover and the absorber plate is 2.5 cm filled with air. Considering the absorber plate temperature as 350 K and the cover temperature as 305 K, find the heat transfer rate by free convection from the absorber plate?

11P-4 A long horizontal cylinder having external diameter 10 m and length 15 m is placed in a quiescent air where ambient temperature is kept at 400 K. The cylinder wall temperature is 800 K. Find the convective heat transfer.

11P-5 A sphere having diameter 0.7 m is placed in a quiescent air where ambient temperature is kept at 500 K. The sphere wall temperature is 300 K. Find the heat transfer.

11P-6 A horizontal enclosure comprising of two parallel plates is separated by a gap of 8 mm. The lower plate is kept at a uniform temperature of 600 K and the upper plate is kept at a uniform temperature of 300 K. A nanofluid has been filled in between this enclosure with water-Al$_2$O$_3$ with φ=0.2%. Find the convective heat transfer coefficient if the side walls are insulated.

11P-7 The gap space between the two concentric spheres is filled with ait. The inner sphere radius is 0.6 m and the outer sphere radius is 0.75 m. The inner and outer spheres temperature are 350 K and 800 K, respectively. Find the heat transfer per unit gap space between the sphere,

11P-8 Consider the natural convection between concentric cylinders of inner and outer radius 5 and 6 mm, respectively. The fluid Raylieigh is 10^6, and Prandtl number is 240. Find the Nussle number.

11P-9 Air at 10 °C is flowing through a vertical pipe, which is maintained at constant temperature of 800 K. The internal diameter of pipe is 0.55 m and the speed of air flow is 0.7 m/s. The length of pipe is 0.5048 m. Find the convective heat transfer coefficient and corresponding heat transfer.

11P-10 Estimate the convective heat transfer coefficient for a vertical double wall of 4 m height. The air gap is 10 cm. The wall temperatures are maintained at 304 K and 298 K, at pressure of 1 atm.

11P-11 Natural convection over a room wall is planned to be modeled in a laboratory. The prototype room wall is 3 m tall and the air temperatures on wall sides are 25 and 10 °C. The Rayleigh number of natural convection boundary layer over the room wall is 43.33×10^9. If water is selected as a working fluid for the model in laboratory, what shall be the vertical plate height? Take necessary assumptions.

11P-12 A large body of water at 24 °C is in contact with a infinite surface kept at 40 °C. The water has no appreciable movement and thermal plume is forming near wall and rising in the pool of water. The Rayleigh number for the rising plume based on the conduction layer thickness δ is 1000. Find the time interval between the rise of two thermal plumes if $\delta \approx \sqrt{(\alpha \cdot \tau)}$, where τ is the time scale, and α is thermal diffusivity of water.

REFERENCES

A. Oberbeck, Uber die Waerrmeleitung der Fluessigkeiten bei Beruecksichtigung der Stroemung infolge von Temperaturdifferenzen. Ann. Phys. Chern. 7, 271 - 292, 1879.

M. J. Boussinesq, Theorie Analytique de la chaleur. Vol. 2, Paris: Gauthier-Villars 1903 Carey, Van P.; Mollendorf, J.e.: Natural convection in liquids with temperature dependent viscosity. In: Heat Transfer 1978. Vol. 2, pp 211-216 Washington: Hemisphere 1978.

D. D. Joseph, Stability of fluid motions. Vol. 1 and 2. Berlin: Springer 1976.

B. R. Rich, Trans. ASME, vol. 75, p. 489, 1953.

O. A. Saunders, Proc. Roy. Soc. (London), ser. A, vol. 157, p. 278, 1936.

H. Schlichting, Boundary Layer Theory, McGraw-Hill Book Company, Inc., New York, 1955.

E. Schmidt, and W. Beckmann: Tech. Mech. u. Thermodynam., vol. 1, pp. 341, 391, 1930.

E. M. Sparrow, R. Eichhorn, and J. L. Gregg: Phys. Fluids, vol. 2, p. 319, 1959.

E. M. Sparrow and J. L. Gregg, Laminar-Free-Convection Heat Transfer From the Outer Surface of a Vertical Circular Cylinder, Trans. ASME , 78 , pp. 1823–1829, 1956.

T. Fuji and M. Fuji, The Dependence of Local Nusselt Number on Prandtl Number in the Case of Free Convection Along a Vertical Surface with Uniform Heat Flux, Int. J. Heat Mass Transfer, Vol. 19, 121-122, 1976.

Kwang-Tzu Yang, Possible Similarity Solutions for Laminar Free Convection on Vertical Plates and Cylinders, ASME J. Appl. Mech., 27, pp. 230 -236, 1960.

S. W. Churchill and H.H. Chu, Correlating equations for the laminar and turbulent for convection from vertical plate, Int. J. Heat mass transfer, vol 18, 1323-1329, 1975.

T. Yuge, Experiments on the heat transfer from the spheres including combined natural and forced convection, J. Heat transfer, vol 82, 214-220, 1960.

C. Y. Warner and V. S. Arpaci, An experimental investigation of turbulent natural convection in air at low pressure along a vertical heated flat plate, Int. J. Heat Mass Transfer, vol 11, pp 397 - 406, 1968.

S. W. Churchill, Free Convection Around Immersed Bodies, 2.5.7–24, in G. F. Hewitt (ed.), Heat Exchanger Design Handbook, Washington, D.C.: Hemisphere Publishing Corp., 1983.

S. W. Churchill, Free Convection Around Immersed Bodies, in G. F. Hewitt, Exec. Ed., Heat Exchanger Design Handbook, Section 2.5.7, Begell House, New York, 2002.

M. Fishenden and O. A. Saunders, An Introduction to Heat Transfer, Oxford University Press, London, 1950.

E. R. G. Eckert and T. W. Jackson, Free convection boundary layer on flat plate. NACA-TN 2207, 1950.

H. Schuh, Boundary Layers of Temperatum. Reps. and Trans. 1007, AVA Monographs, British M. A. P., April 15, 1948.

S. Ostrach, An Analysis of Laminar Free-Convection Flow and Heat Transfer about a Flat Plate Parallel to the Direction of the Generating Body Force. NACA Report, 1111, 1953

S. Ostrach, Natural Convection in Enclosures, Advances in Heat Transfer (ed. T. F. Irvine, Jr., and J. P. Hartnett), Vol. 8, 171-227, Academic Press, New York.

W. H. McAdams, Heat Transmission, Second cd., MoGraw-Hill Book Co., 1942.

H. Schlichting, Boundary layer theory (trans: Kestin J), 4th edn. McGraw-Hill, New York, USA, 1960.

S. Ostrach, An Analysis of Laminar-Free-Convection Flow and Heat Transfer About a Flat Plate Parallel to the Direction of the Generating Body Force, NACA Tech Rep I 1 1I , 1953.

C. Y. Warner and V. S. Arpaci, An Experimental Investigation of Turbulent Natural Convection in Air at Low Pressure Along a Vertical Heated Flat Plate, Int J Heat and Mass Transfer, 1968.

F. M. Sparrow and J. L. Gregg, Laminar Free Convection from a Vertical Flat Plate, Trans ASME, v. 78, 1956, p. 435, 1956.

B. Metais, E. R. G. Eckert, Forced, mixed, and free convection regimes, J. Heat Transfer, 86, 295-296, 1964.

D. Q. Kem, D. F. Othmer, Effect of Free Convection on Viscous Heat Transfer in Horizontal Tubes, AIChE J. 39, 517-555, 1943.

B. Metais, Criteria for Mixed Convection, HTL Tech. Rep. No. 51, Heat Transfer Laboratory, Univ. Minnesota, Minneapolis, MN, 1963.

V. T. Morgan, The Overall Convective Heat Transfer from Smooth Circular Cylinders, in T. F. Irvine and J. P. Hartnett, Eds., Advances in Heat Transfer, Vol. 11, Academic Press, New York, 199 - 264, 1975.

T. H. Kuehn and R. J. Goldstein, Correlating equations for natural convection heat transfer between horizontal circular cylinders, Int. J. Heat Mass Transf. 1976, 19, 1127- 1134, 1976.

T. H. Kuehn and R. J. Goldstein, International Journal of Heat and Mass Transfer, 23, 971, 1982.

J. Dyer, Laminar natural convection from a horizontal cylinder with a uniform convective heat flux. Trans Inst Eng Aust MC1:125-128, 1965.

S. K. S. Boetcher, Natural Convection from Circular Cylinders, Springer Briefsin Thermal Engineering and Applied Science, 2014.

I. Catton, Natural Convection in Enclosures, Proc. 6th Int. Heat Transfer Conf., Toronto, Canada, Vol. 6, pp. 13–31, 1978.

K. G. T. Hollands, G. D. Raithby, and L. Konicek, Int. J. Heat Mass Transfer, 18, 879, 1975.

G. D. Raithby and K. G. T. Hollands, A General Method of Obtaining Approximate Solutions to Laminar and Turbulent Free Convection Problems, in T. F. Irvine and J. P. Hartnett, Eds., Advances in Heat Transfer, Vol. 11, Academic Press, New York, 1975, USA.

12 Heat Transfer along Phase Change

For the designing of heat transfer devices in which fluid is going through phase change, the knowledge of the condensation and boiling heat transfer phenomenon is necessary. When vapour comes into contact with a surface maintained at a temperature less than the saturation temperature then the condensation will occur. The flow of the film is initially laminar, and then with increasing film thickness, it will be classified as wavy or turbulent film. When liquid is heated in the pool a point will come when liquid started to volumetrically converted into vapors, the phenomenon we know as boiling. After finishing this chapter, one will be able to:

- Drive the convective heat transfer in case of laminar film condensation using Nusselt theory.
- Delineate between laminar, wavy and turbulent condensations.
- Understand differences between pool and film boiling phenomenon.
- Calculate heat transfer due to condensation and boiling.

In this chapter, we focus on the latent heat transfer associated with two phase convection problems. Here we will mathematically model the heat transfer in case of condensation and boiling, which are significantly complex phenomenon compared with single-phase heat transfer phenomenon. The convective heat transfer coefficient associated with phase change phenomenon are much higher than the convective heat transfer coefficient associated with single phase heat transfer.

We will model the film condensation heat transfer on a flat vertical plate and formulate a relation for the determination of the heat transfer coefficients. We will characterise the pool boiling behaviour as it is a simple setup to learn about the complex boiling phenomenon and its different regimes. We will learn about the critical heat flux at which burnout of metallic wire occurs and is very important in nuclear reactor boilers. We will also discuss different flow regimes of forced convection boiling.

Some representative values of convective heat transfer coefficients for phase change heat transfer are listed in Table 12.1.

Condensation happens when fluid in a gaseous state (like steam or vapors) comes into physical contact with a cooled surface, maintained at a temperature lower than the fluids saturation temperature at that pressure. There are two kinds of condensations: dropwise condensation, and film condensation. In dropwise condensation, the

DOI: 10.1201/9781003428404-12

TABLE 12.1

Some Representatives Convective Heat Transfer Coefficient Values for Phase Change Heat Transfer

$$h\left(\frac{W}{m^2 \cdot K}\right)$$

Water	Nucleate boiling	5×10^3
Liquid Helium	Nucleate boiling	8×10^3
Steam	Film condensation	10^4
Steam	Dropwise condensation	5×10^4
Water	Forced convection boiling	10^5

condensate liquid does not completely wet the surface, whereas with film condensation, the liquid will form a thin layer over the surface and this layer moves down under the action of gravity. Hence, the film condensate will have some speed.

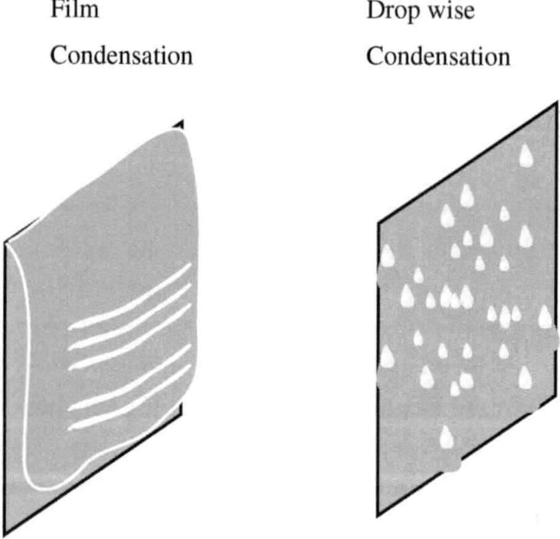

Figure 12.1　Film condensation and dropwise condensation on a vertical surface.

Figure 12.1 shows the two distinct types of condensation on a vertical surface: film condensation and dropwise condensation. In the case of film condensation, the wall is covered by a film of liquid which gradually increases in thickness. This liquid condensate layer acts as a resistance to heat transfer. The heat of vaporisation h_{fg} released as a result of phase change must overcome this resistance before it reaches the wall or surface. On the other hand, in case of the dropwise condensation, the liquid droplets will form over the surface and they fall or slide down due to weight, and hence the surface is not completely blanketed by the condensate. The rate of heat transfer in the case of dropwise condensation is around ten times more than those associated with film condensation. In engineering practices, often dropwise condensation is preferred because of this increased heat transfer.

Dropwise condensation is often observed on hydrophobic surfaces, like some plant leaves (like lotus leaf) we observe dropwise condensation. The leaves surfaces have the naturally forming microscopic wax crystals that create a rough texture and reduce the surface area in contact with water. Also, the wax crystals often contain hydrophobic molecules, such as long-chain fatty acids and alkanes, which further enhance the water-repellent properties of the surface. Also, ferns, have hydrophobic fronds that help to channel rainwater towards the roots of the plant, and pine needles. Dropwise condensation is also desirable in clothing industry. Polyester or nylon inherently have hydrophobic properties, due to their non-polar chemical structure and can be used to make fabrics that are highly water-repellent.

12.1 NUSSELT MODEL OF LAMINAR FILM CONDENSATION

In 1916, Nusselt presented the derivation for the heat transfer in laminar film condensation. The equations he developed can be used both for vertical/inclined walls and pipes having large diameter. We will now present the Nusselt's film condensation theory. Figure 12.2 shows the schematic diagram of the condensation phenomenon on a a vertical flat plate of width b in 2D Cartesian coordinates. The plate is maintained at a constant temperature T_w and the vapor in contact is condensing like a film on the plate. We will now formulate a mathematical description of film condensation phenomenon with assumptions:

(i) It is assumed that the film is two dimensional and has same cross-section in depth.

(ii) The condensate is flowing under action of gravity on a vertical plate with steady-state condition.

(iii) The vapour is releasing the latent heat, and there is no subcooling of the condensate, i.e when the condensate forms its temperature remains constant.

(iv) Inside condensate layer we consider conduction only, and ignore any convection in the laminar flow.

The thickness of condensation film is growing in z direction so δ is function of z. We consider a small control volume of volume size $(\delta\text{-y})b$, where, b is the width of the control volume. The balance of the forces on the volume surface, we consider the following forces:

(i) The force acting on the CV due to gravity (acting in positive z direction),

(ii) The buoyancy force (in the negative z direction), and

(iii) The viscous shear force that act in flow opposing direction.

The force balance on a non-accelerating fluid element is

$$\rho_\ell g\,(\delta - y)\,bdz = \rho_v g\,(\delta - y)\,bdz + \tau_{yz}(bdz)$$

where, W is velocity of the condensing film and shear stress can be estimated from Newton's law of viscosity.

$$\tau_{yz} = \mu \frac{dW}{dy}$$

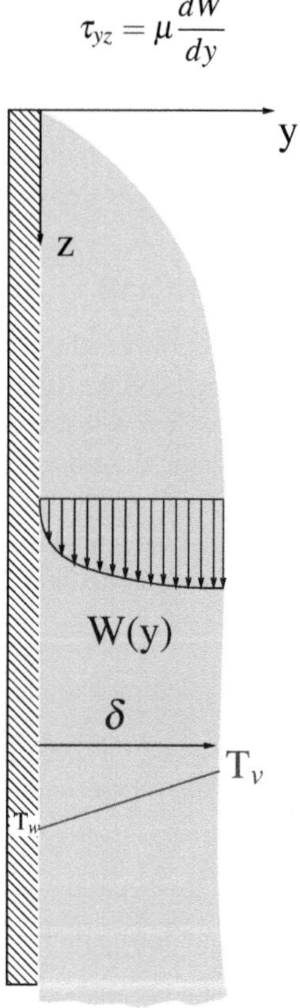

Figure 12.2   Schematic diagram of condensation phenomenon. T_w and T_v are wall and vapor temperature. For condensation to occur $T_v < T_{sat}$.

where ρ_ℓ is the density of the condensing liquid and ρ_v is the density of the vapour in contact with cold surface. For the condensation to happen, the ambient vapour temperature T_v is less than the saturation temperature T_{sat} of the vapour.

Rearranging we get

$$\frac{dW}{dy} = \frac{\rho_\ell g\,(\delta - y)\,bdz - \rho_v g\,(\delta - y)\,bdz}{\mu.bdz}$$

$$\frac{dW}{dy} = \frac{(\delta - y)\,g(\rho_\ell - \rho_v)}{\mu}$$

Integrating this we have

$$W(y) = \frac{g(\rho_\ell - \rho_v)}{\mu}\left(\delta y - \frac{y^2}{2}\right) + C_1$$

at y=0, W=0 so

$$W(y) = \frac{\delta^2 g\left[1 - (\rho_v/\rho_\ell)\right]}{\mu/\rho_\ell}\left(\frac{y}{\delta} - \frac{y^2}{2\delta^2}\right)$$

$$W(y) = \frac{\delta^2 g\left[1 - (\rho_v/\rho_\ell)\right]}{v_\ell}\left(\frac{y}{\delta} - \frac{y^2}{2\delta^2}\right)$$

We can rearrange this and formulate an equation in dimensionless velocity (ξ) form

$$\xi(y) = \frac{W(y)}{\delta^2 g\left[1 - (\rho_v/\rho_\ell)\right]/v_\ell} = \left(\frac{y}{\delta} - \frac{y^2}{2\delta^2}\right)$$

The amount of vapor condensed in displacement dz can be estimated using Taylor's series:

$$\dot{m}_{z+dz} = \dot{m}_z + \frac{d}{dz}(\dot{m}_z)dz$$

$$\dot{m}_{z+dz} = \dot{m}_z + \frac{d(\dot{m}_z)}{d\delta}\frac{d\delta}{dz}dz$$

The net condensate rate is

$$\Delta\dot{m} = \left(\frac{d(\dot{m}_z)}{d\delta}\right)d\delta$$

where, the mass condensate rate is

$$\dot{m}_z = \rho_\ell \cdot Area \cdot W$$

$$\dot{m}_z = \int_0^b \int_0^\delta \rho_\ell \cdot \left\{\frac{\delta^2 g\left[1 - (\rho_v/\rho_\ell)\right]}{v_\ell}\left(\frac{y}{\delta} - \frac{y^2}{2\delta^2}\right)\right\} dy \cdot dx$$

which gives

$$\dot{m}_z = \frac{\rho_\ell \cdot b \cdot g \cdot \delta^3 \left(1 - \frac{\rho_v}{\rho_\ell}\right)}{3 \cdot v_\ell}$$

$$\Delta\dot{m} = \left(\frac{d(\dot{m}_z)}{d\delta}\right)d\delta = \frac{\rho_\ell \cdot b \cdot g \cdot \delta^2 \left(1 - \frac{\rho_v}{\rho_\ell}\right)}{v_\ell}d\delta$$

The heat transferred to the condensing film within length dz shall be equal the to the condensate mass flow times the specific enthalpy of condensation of the vapour

$$Q = \Delta \dot{m} \cdot h_{fg} = k_\ell b \cdot dz \cdot \left(\frac{T_{sat} - T_w}{\delta} \right)$$

$$h_{fg} \left\{ \frac{\rho_\ell \cdot b \cdot g \cdot \delta^2 \left(1 - \frac{\rho_v}{\rho_\ell} \right)}{\nu_\ell} \right\} d\delta = k_\ell b \cdot dz \cdot \left(\frac{T_{sat} - T_w}{\delta} \right)$$

Rearranging we have

$$\delta^3 d\delta = \frac{k_\ell \cdot \nu_\ell \left(T_{sat} - T_w \right)}{h_{fg} \left\{ \rho_\ell \cdot g \cdot \left(1 - \frac{\rho_v}{\rho_\ell} \right) \right\}} dz$$

Integrating we get

$$\delta^4 = \frac{4 k_\ell \cdot \nu_\ell \left(T_{sat} - T_w \right)}{h_{fg} \left\{ \rho_\ell \cdot g \cdot \left(1 - \frac{\rho_v}{\rho_\ell} \right) \right\}} z + C_2$$

We assume that at z=0 , δ =0 this gives C_2=0 and we have condensate thickness as

$$\delta = \left[\frac{4 k_\ell \mu_\ell \left(T_{sat} - T_w \right)}{h_{fg} \rho_\ell g \left(\rho_\ell - \rho_v \right)} z \right]^{1/4}$$

We can now formulate an expression for convective heat transfer by balancing heat conduction into wall and heat convection over the surface

$$q_{conv} = q_{cond}$$

$$h \cdot \left(T_{sat} - T_w \right) = k_\ell \cdot \left(\frac{T_{sat} - T_w}{\delta} \right)$$

$$h = \frac{k_\ell}{\delta} = \frac{k_\ell}{\left[\frac{4 k_\ell \mu_\ell \left(T_{sat} - T_w \right)}{h_{fg} \rho_\ell g \left(\rho_\ell - \rho_v \right)} z \right]^{1/4}}$$

Since h is varying in z direction, we introduce subscript z with h:

$$h_z = \frac{k_\ell}{\delta} = \frac{k_\ell}{\left[\frac{4 k_\ell \mu_\ell \left(T_{sat} - T_w \right)}{h_{fg} \rho_\ell g \left(\rho_\ell - \rho_v \right)} z \right]^{1/4}} = \left[\frac{h_{fg} \rho_\ell g \left(\rho_\ell - \rho_v \right)}{4 \mu_\ell \left(T_{sat} - T_w \right) z} \right]^{1/4} k_\ell^{1 - 1/4}$$

$$h_z = \left[\frac{h_{fg} \rho_\ell g \left(\rho_\ell - \rho_v \right) k_\ell^3}{4 \mu_\ell \left(T_{sat} - T_w \right) z} \right]^{1/4}$$

The mean convective heat transfer coefficient is

$$h_m = \frac{-1}{b \cdot L} \int_0^b \int_0^L h_z \cdot dx \cdot dz = \frac{4}{3} h_{z=L}$$

$$h_m = \left(\frac{4}{3 \cdot \sqrt[4]{4}}\right) \left[\frac{h_{fg}\rho_\ell g (\rho_\ell - \rho_v) k_\ell^3}{\mu_\ell (T_{sat} - T_w) L}\right]^{1/4}$$

$$h_m = 0.943 \left[\frac{h_{fg}\rho_\ell g (\rho_\ell - \rho_v) k_\ell^3}{\mu_\ell (T_{sat} - T_w) L}\right]^{1/4}$$

We define Nusselt number as $\mathrm{Nu}_{m,Lam} = h_m \cdot L / k_\ell$.

$$\boxed{Nu_{m,Lam} = 0.943 \left[\frac{h_{fg} \cdot \rho_\ell \cdot g (\rho_\ell - \rho_v) \cdot L^3}{\mu_\ell \cdot k_\ell (T_{sat} - T_w)}\right]^{1/4}} \tag{12.1}$$

12.2   LAMINAR CONDENSATION CORRELATIONS FOR SOME CASES

1 **Condensation over an inclined wall:** If the condensation occurs on inclined walls, the influence of gravity would decrease according to the angle of inclination (φ), which is measure from the horizontal line (see Figure 12.3).

$$h_{m,inclined} = h_{m,vertical} \cdot (\cos \varphi)^{0.25} \tag{12.2}$$

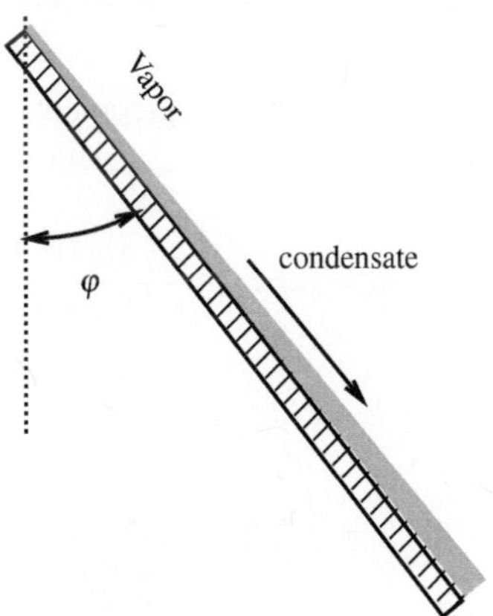

Figure 12.3   Condensate is flowing down at angle φ.

2 **Nonlinear temperature distribution in condensate and effect of subcooling:**
The account for the nonlinear temperature profile in the condensate film, we use modified latent heat of vaporisation h'_{fg}, defined as

$$h'_{fg} = h_{fg} + 0.375 c_{p,\ell} (T_{sat} - T_w) \tag{12.3}$$

where, T_{sat} is the saturation temperature. This correction is mostly invoked in case of condensations over curved surfaces specially inside tubes.

Many refrigerants have relatively low enthalpies of phase change, and thus we should make a correction for the subcooling effect. To account for the effects of low enthalpies of phase change, like in case of refrigerants, we should make a correction for the subcooling term. Rohsenow (1956) suggested that the cooling of the liquid below the saturation temperature can be corrected by replacing h_{fg} to h'_{fg}.

$$h'_{fg} = h_{fg} + 0.68 c_{p,\ell} (T_{sat} - T_w) \tag{12.4}$$

Sadasivan and Lienhard (1987) suggested to adjust the latent heat as

$$h'_{fg} = h_{fg} \left[1 + \left(0.683 - \frac{0.228}{Pr} \right) Ja \right]$$

where,

$$Ja = \frac{c_p (T_{sat} - T_w)}{h_{fg}}$$

3 **Condensation over a horizontal tube:** Figure 12.4 shows film formation over a tube. For laminar film condensation on horizontal tubes, Nusselt obtained the relation

$$h_m = 0.728 \left[\frac{k_\ell^3 h_{fg} g \rho_\ell (\rho_\ell - \rho_v)}{\mu_\ell (T_{sat} - T_w) D} \right]^{1/4} \tag{12.5}$$

Selin (1961) found that improved agreement with film condensation experimental data can be obtained, if the constant value of 0.61 is used instead of 0.728.

$$h_m = 0.61 \left[\frac{k_\ell^3 h_{fg} g \rho_\ell (\rho_\ell - \rho_v)}{\mu_\ell (T_{sat} - T_w) D} \right]^{1/4} \tag{12.6}$$

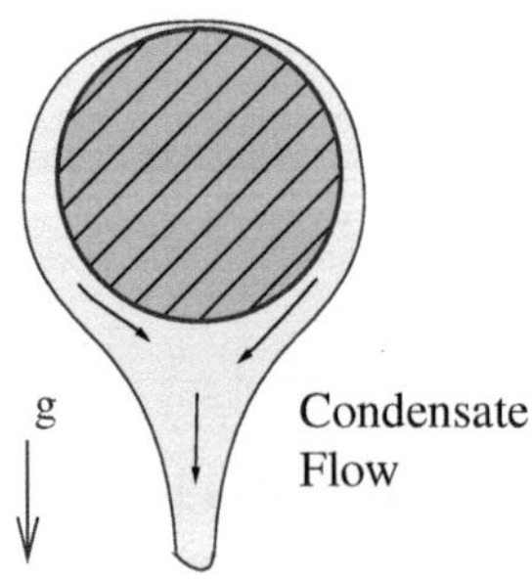

Figure 12.4   The condensate film over the surface of a tube.

4 Condensation inside a tube: For condensation of refrigerants at small vapour velocities inside horizontal tubes, Chato suggested the correlation:

$$h_m = 0.555 \left[\frac{k_\ell^3 h'_{fg} g \rho_\ell (\rho_\ell - \rho_v)}{\mu_\ell (T_{sat} - T_w) D} \right]^{1/4} \tag{12.7}$$

This correlation is valid as long as

$$\text{Re}_v = \left(\frac{\rho_v \cdot D \cdot u_v}{\mu_v} \right) < 35,000$$

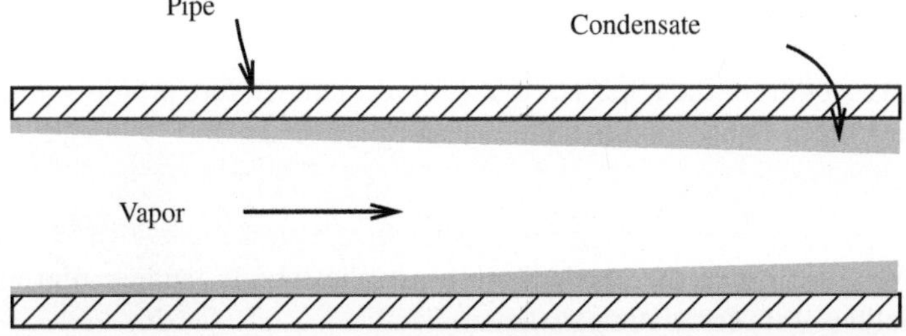

Figure 12.5	Condensation at the horizontal pipe walls and the accelerating vapor flow in the core of the pipe.

5 Condensation over sphere: For condensation over sphere, the following correlation can be used

$$h_m = 0.828 \left[\frac{k_\ell^3 h'_{fg} g \rho_\ell (\rho_\ell - \rho_v)}{\mu_\ell (T_{sat} - T_w) D} \right]^{1/4} \tag{12.8}$$

6 Condensation over several tubes: For condensation over n number of tubes

$$Nu_{m,D,n\ tubes} = \frac{Nu_{m,D,\ one\ tube}}{n^{1/4}} \tag{12.9}$$

Kern (1950) suggested for n-tubes in a vertical column arrangement for the average heat transfer coefficient:

$$Nu_{m,D,n\ tubes} = \frac{Nu_{m,D,\ one\ tube}}{n^{1/6}} \tag{12.10}$$

Example 12.1

Steam at 45°C is condensing on the outer surface of a 2 cm pipe. The outer surfaces of the pipe is maintained at 25°C. Find (a) the rate of heat transfer to the fluid in the pipe, and (b) the rate of condensation of steam per unit length of a horizontal pipe.

Solution We use the properties at mean temperature of 35°C.

$$\rho_\ell = 994 \ \text{kg/m}^3, k_\ell = 0.623 \text{W/m} \cdot \text{K}, c_{p,\ell} = 4178 \text{J/kg} \cdot \text{K},$$
$$\mu_\ell = 0.00072 \text{Pa} \cdot \text{s}, h_{\text{fg}} = 0.2407 \times 10^7 \text{J/kg}$$

As per problem data

$$T_w = 25°C, T_{sat} = 45°C, D = 0.02m, L = 1m, A = \pi D \cdot L = 0.06283 m^2$$

To account for the effects of the nonlinear temperature profile in the condensate film, we use modified latent heat of vaporisation:

$$h_{fg}^* = h_{fg} + 0.68 \cdot c_{p,\ell} \cdot (T_{sat} - T_w) = 2.46 \times 10^6 J/kg$$

We assume that the condensation over the tube is laminar film and using Selin (1961) correlation we have

$$h_m = 0.61 \left[\frac{k_\ell^3 h_{fg}^* g \rho_\ell (\rho_\ell - \rho_v)}{\mu_\ell (T_{sat} - T_w) D} \right]^{1/4}$$

As $\rho_v \ll \rho_\ell$, we approximate
This gives

$$h_m = 0.61 \left[\frac{k_\ell^3 h_{fg}^* g \rho_\ell^2}{\mu_\ell (T_{sat} - T_w) D} \right]^{1/4}$$

$$h_m = 8674.77 \frac{W}{m^2 \cdot K}$$

The heat transfer is

$$Q = h_m \cdot A \cdot (T_{sat} - T_w) = 10901.04 W$$

The rate of condensation of steam is

$$\dot{m}_{condensate} = \frac{Q}{h_{fg}^*} = 0.00442 \text{kg/s}$$

Example 12.2

Using data of Example 12.1, consider condensation on horizontal tubes with 9 horizontal tubes arranged in a column style such that the condensate from upper tubes is dripping over the lower tubes. Find the condensate rate from all nine tubes.

Solution

$$Nu_{m,D,n\ tubes} = \frac{Nu_{m,D,\ one\ tube}}{n^{1/4}} = \frac{8674.7 \times 0.02/0.623}{9^{1/4}} = 134.53$$

Using, Kern (1950) proposal for the average heat transfer coefficient:

$$Nu_{m,D,n\ tubes} = \frac{Nu_{m,D,\ one\ tube}}{n^{1/6}} = \frac{8674.7 \times 0.02/0.623}{9^{1/6}} = 161.57$$

We use this value and corresponding convective heat transfer coefficient over nine tubes is

$$h_n = \frac{h \cdot k_\ell}{D} = \frac{161.57 \times 0.623}{0.02} = 5032.920\,W/m^2 \cdot K$$

The area of nine tube walls is $A_n = n \cdot \pi \cdot D \cdot (1) = 0.5652\ m^2$. The heat transfer is

$$Q_n = h_n \cdot A_n \cdot (T_{sat} - T_w) = 5032.920 \times 0.5652 \times (45 - 25) = 9121.58W$$

The condensate rate from all 9 tubes is

$$\dot{m}_{condenstae,n} = Q_n/h_{fg}^* = 0.0230 kg/s$$

12.3 WAVY AND TURBULENT CONDENSATION

in the previous analysis, we assumed that the condensate flow is laminar. In many cases, condensate flow can have a wavy pattern and we can no longer use the Nusselt model for such problems. Figure 12.6 schematically depicts different flow regimes encountered in film condensation. Kutateladze (1963) proposed the theory to tackle the wavy condensate flow based on Reynolds number of the condensate.

For condensation problems, usually two definitions of Reynolds number are used in literature. The Reynolds number of condensate based on mass flow rate is

$$Re_\ell = \frac{4\dot{m}}{\mu_\ell P}$$

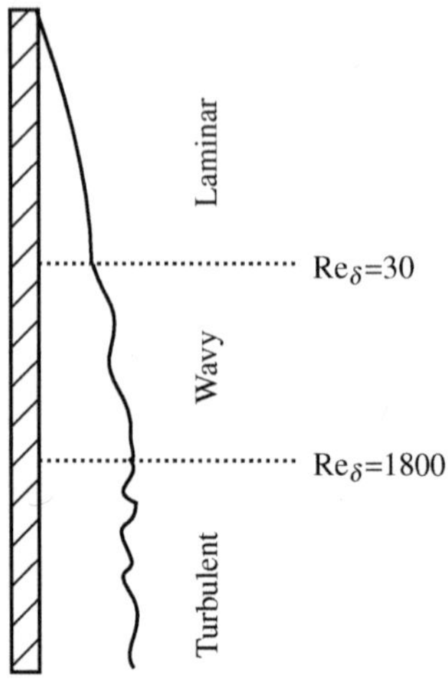

Figure 12.6   Flow regimes for film condensate.

where, P is perimeter and mass flow rate can be arrived at from balance of energy:

$$\dot{m} = \frac{h_m \cdot A \cdot (T_{sat} - T_w)}{h_{fg}}$$

$$\mathrm{Re}_\ell = \frac{4\dot{m}}{\mu_\ell P} = \frac{4}{\mu_\ell P}\left[\frac{h_m \cdot P \cdot L \cdot (T_{sat} - T_w)}{h_{fg}}\right]$$

$$\mathrm{Re}_\ell = \frac{4}{\mu_\ell}\left[\frac{h_m \cdot L \cdot (T_{sat} - T_w)}{h_{fg}}\right]$$

Note that although these definitions of Reynolds number look different, however they give same results when calculated.

Different flow regimes can be identified by Reynolds number based on condensate layer thickness, defined as

$$\mathrm{Re}_\delta = \frac{4\rho_\ell(\rho_\ell - \rho_v)\delta^3 g}{3\mu_\ell^2} \qquad (12.11)$$

The laminar condensation equations derived here match well with experimental predictions as long as the film condensation is laminar. It is found that ripples may form in the condensate film for Reynolds numbers as low as 30 or 40. The equation is also approximated as

$$\boxed{Nu_{m,Lam,film} = 1.47Re_\delta^{-1/3} \qquad Re_\delta \leq 30} \qquad (12.12)$$

Kutateladze (1963) proposed the correlation for laminar way regime:

$$\boxed{Nu_{m,Lam,wavy} = \frac{Re_\delta}{1.08(Re_\delta^{1.22}) - 5.2} \qquad 30 \leq Re_\delta \leq 1800} \qquad (12.13)$$

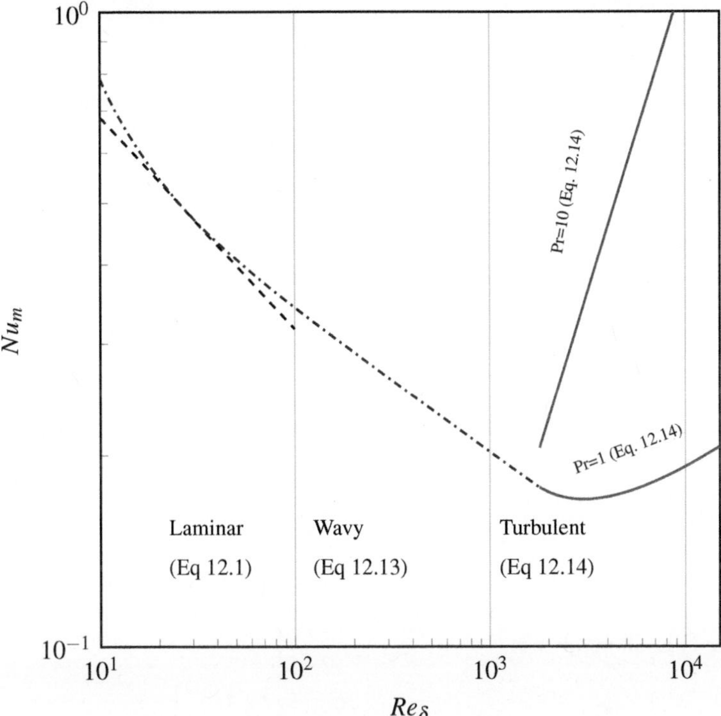

Figure 12.7 Plot of Nusselt (dashed), Kutateladze (dash-dotted) and Labuntsov correlations for Nusselt number on a vertical plate for condensation in laminar, wavy and turbulent regimes.

Labuntsov (1957) recommended the correlation for the turbulent regime as

$$Nu_{Turb} = \frac{Re_\delta}{8750 + 58Pr_\ell^{-0.5}(Re_\delta^{0.75} - 253)} \qquad Re_\delta \geq 1800, Pr \geq 1 \qquad (12.14)$$

and based on Re_ℓ Nusselt number is defined as

$$Nu_{m,Turb} = \frac{0.020 \cdot Re_\ell^{7/24} Pr_\ell^{1/3}}{1 + 20.52 Re_\ell^{-3/8} Pr_\ell^{-1/6}}$$

Figure 12.7 shows the plots of Nusselt correlations for different condensate flow regimes. The combined Nusselt number for the complete plate is

$$Nu_{m,L} = \left(\frac{\mu_\ell}{\mu_w}\right)^{0.25} \left[Nu_{Lam}^{1.2} + Nu_{Turb}^{1.2}\right]^{1/1.2}$$

Example 12.3

A vertical rectangular plate with 20 cm height and 10 cm width is exposed to steam at atmospheric pressure. The plate is maintained at temperature of 98 °C. Calculate the thickness of condensate layer and Reynolds number at the end of the plate. Also, find the heat transfer and the mass of steam condensed per second.

Solution

We use the properties at the film temperature of 99°C:

$$\rho_f = 960 \ \text{kg/m}^3, \text{h}_{\text{fg}} = 0.2257 \times 10^7 \ \text{J/kg} \cdot \text{K}, \mu_{\text{f}} = 0.000282 \text{Pa} \cdot \text{s}, \text{k}_{\text{f}}$$
$$= 0.68 \text{W/m} \cdot \text{K}$$

As $\rho_\ell \gg \rho_v$, we can approximate the density terms as $\rho_\ell(\rho_\ell - \rho_v) \approx \rho_\ell^2$. We assume that condensate is flowing down as a laminar flow. The thickness of condensate is

$$\delta = \left(\frac{4 \cdot k_\ell \cdot \mu_\ell \cdot (T_{sat} - T_w) \cdot L}{h_{fg} \cdot \rho_\ell^2 \cdot g} \right)^{\frac{1}{4}} = 6.2\text{E} - 5 \ \text{m}$$

The Nusselt number is

$$\text{Nu} = 0.943 \cdot \left(\frac{h_{fg} \cdot \rho_\ell^2 \cdot 9.81 \cdot L^3}{\mu_\ell \cdot k_\ell \cdot (T_{sat} - T_w)} \right)^{\frac{1}{4}} = 4283.24$$

This give convective heat transfer coefficient as

$$h = 14563.03 W/m^2 \cdot K$$

The Reynolds number of flow is

$$Re_\ell = \frac{4}{\mu_f} \cdot \left(\frac{h \cdot L \cdot (T_{sat} - T_w)}{h_{fg}} \right) = 36.6$$

Also $Re_\delta = 36.6$. This shows that condensate flow is laminar and estimate of thickness done above is reliable. The heat transfer is

$$Q = h \cdot (w \cdot L) \cdot (T_{sat} - T_w) = 582.52 W$$

The condensate rate is

$$\dot{m}_{condensate} = \frac{Q}{h_{fg}} = 0.000258 \text{kg/s}$$

Example 12.4

An organic fluid in vapour form and is being condensed over a copper tube. The properties of organic fluid are

$$k_\ell = 0.090\,W/m \cdot K, \rho_\ell = 560\ kg/m^3, \rho_v = 13.1\ kg/m^3,$$
$$h_{fg} = 282000\,J/kg, \mu_\ell = 0.000127\ Pa \cdot s$$

Inside the copper tube water is flowing as a coolant and its temperature is rising. The velocity of water is 1.3 m/s and the inner and outer diameter of cooper tube is 3.5 cm and 2.4 cm, respectively. The properties of water are

$$\rho_w = 997 kg/m^3, c_{pw} = 4184\,J/kg \cdot K, \mu_w = 0.001003\,Pa \cdot s, k_w$$
$$= 0.6\,W/m \cdot K, Pr_w = 6.99$$

Find the heat transferred to the coolant water and the rate of condensate formed.

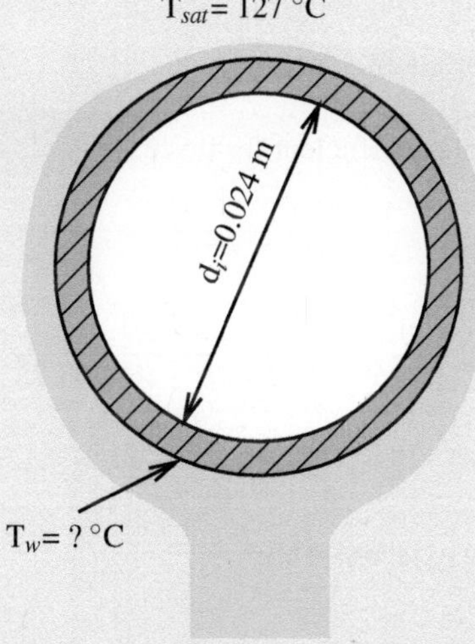

Figure 12.8 Laminar film condensation over a copper tube.

Solution
Water flow inside copper tube:
We first find the Reynolds number and Nusselt number for the water flow inside a copper tube:

$$Re = \frac{d_i \cdot \rho_w \cdot u}{\mu_w} = 31{,}013$$

The flow inside tube is turbulent. The Nusselt number according to Dittus-Boelter relation:

$$Nu_w = 0.023 \cdot Re^{0.8} \cdot Pr_w^{\frac{1}{4}} = 146.60$$

The convective heat transfer coefficient is

$$h_w = \frac{Nu^w \cdot k_w}{d_i} = 3665.15 W/m^2 \cdot K$$

Condensation over copper tube:

The saturation temperature, $T_{sat} = 400$ K and, as a first approximation, we assume the wall temperature as $T_w = 290$ K. We use the Selin equation:

$$h_{cond} = 0.61 \cdot \left(\frac{(k_\ell^3 \cdot \rho\ell \cdot (\rho\ell - \rho v) \cdot 9.81 \cdot h_{fg})}{\mu\ell \cdot d_{out} \cdot (T_{sat} - T_w)} \right)^{\frac{1}{4}} = 646.69 W/m^2 \cdot K$$

where, h_{cond} is condensation related heat transfer coefficient at the outer wall of pipe.

Overall heat transfer coefficient:

The overall heat transfer coefficient for this case is

$$\frac{1}{U} - \frac{1}{h_{cond}} - \frac{d_{out}}{d_i \cdot h_w} = 0$$

$$\frac{1}{U} - \frac{1}{646.69} - \frac{3.5 \times 10^{-2}}{2.4 \times 10^{-2} \times 3665.15} = 0$$

Solving for U, we have

$$U = 514.34 W/m^2 \cdot K$$

Heat flux at the pipe external wall:

$$q_w'' = U \cdot (T_{sat} - T_w) = 514.34(400 - 293) = 55035 W/m^2$$

The new estimate for the wall temperature is computed as

$$T_w := T_{sat} - \frac{q_w''}{h_{cond}} = 314.89 K$$

We repeat this calculation procedure and finally the $T_w = 316.1$ K.

Condensate rate:

$$mass_{cond} = \frac{q_w'' \cdot \pi \cdot d_{out}}{h_{fg}} = 0.02263 kg/s$$

12.4 BOILING

Boiling is the phenomenon in which the fluid state changes from the saturated liquid state to the saturated gas state. In this case, the evaporation occurs at a solid-liquid interface, where the surface temperature exceeds the saturation temperature corresponding to the liquid pressure. The boiling phenomenon is comprised of the formation of cavities that grow and detach from the surface.

12.5 POOL BOILING

Japanese scientist Shiro Nukiyama in 1934, identified different boiling regimes in case of pool boiling of water. The schematic diagram of his simple apparatus is shown in Figure 12.9. If ΔT is the temperature difference between the surface and the water, then the heat transmitted from the metal surface per unit area per unit time to the water, q are related directly to each other as

$$\mathbf{q}_s'' = h\Delta T$$

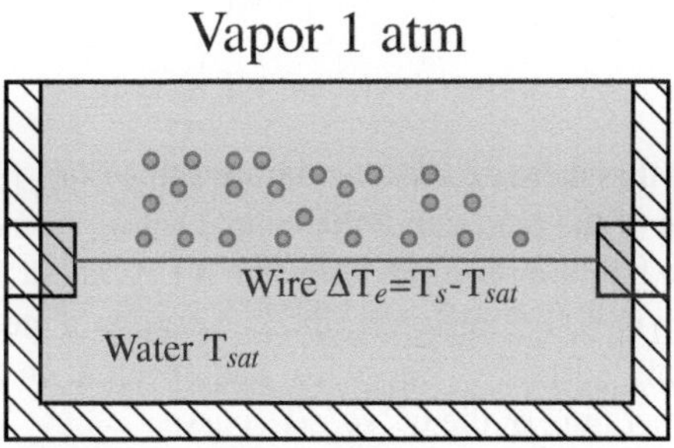

Vapor 1 atm

Wire $\Delta T_e = T_s - T_{sat}$

Water T_{sat}

Figure 12.9 Schematic diagram of Shiro Nukiyama's power-controlled heating apparatus used for investigation of the boiling phenomenon.

or

$$\mathbf{q}_s'' = h\Delta T_e$$

where ΔT_e is also called the excess temperature and h is convective heat transfer coefficient in units $\text{J/(m}^2.\text{s.K)}$.

A thin metal wire is placed in the water pool and was supplied with the necessary heat by running electric current through the wire. Nukiyama ran experiments with different metals of various diameters. He obtained the minimum value of heat transfer with a thin platinum wire.

where d is the diameter of the metal wire in cm, L is the length in cm, the reading of the ammeter and the voltmeter are symolised as I and V, respectively, and the value of heat flux is then calculated by the equation:

$$\mathbf{q}_s'' = \left(\frac{1}{\pi d\ell}\right)\left(\frac{I.V}{4.184}\right) \quad \frac{cal}{s.cm^2}$$

It is made sure that the heat generation from the metal wire due to the current is uniform throughout the wire.

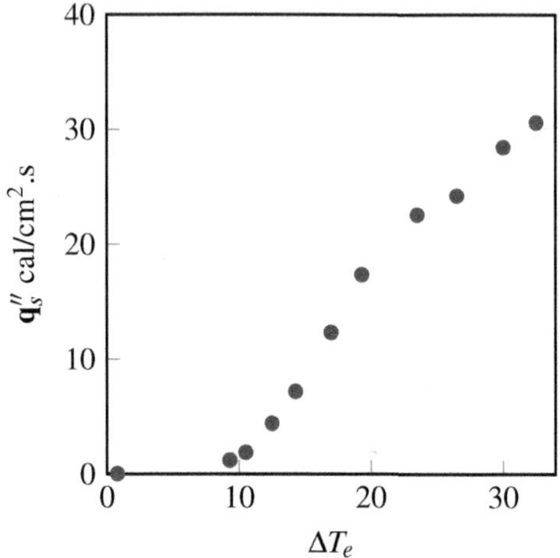

Figure 12.10   Heat flux vs. excess temperature with nickel wire.

Nukiyama found that there exists a maximum value of $\mathbf{q}_s''$ for the $\mathbf{q}$ vs. ΔT_e curve, which is independent of the material of the metal heating surface. He also found that when $\mathbf{q}_s''$ is small, the bubbling is mild, but when ΔT_e is about $12°$ C, no matter if the temperature of water is $100°$C or lower, the generation of steam bubbles is violent and the whole heating surface of the wire was covered. In this case $\mathbf{q}_s''$ abruptly increases at this temperature in the $\mathbf{q}_s''$ vs. ΔT_e curve.

The pool boiling process goes through several stages and regimes.

 i. Free convection boiling

 ii. Nucleate boiling

 iii. Transition boiling

 iv. Film boiling

Free convection boiling: This occurs at the onset of the boiling process when we start heating the liquid. The liquid in this regime is not agitating and the fluid currents near the surface are mainly due to the density change. Thus, this phenomenon is called free convection boiling. Free convection boiling is considered to be present when the excess temperature is less than or equal to $5°$C.

Nucleate boiling: When the excess temperature is increased beyond $5°$C, the boiling mode change to nucleate boiling. The *Onset of Nucleate Boiling* is marked as ONB in Figure 12.11. The vapor bubbles are produced at certain favorable spots referred to as a nucleation or active site, which are present at the surface cervices and the solid impurities like the dust particle in the body of liquid.The liquid will be in the nucleate boiling regime from 5° to 30°C. Here one can witness the isolated bubble formation from the nucleation sites on the surface and the detachment of these bubbles from the surface. This would result in a continuous increase of heat flux and the convective heat transfer coefficient. This is mostly the case in the boiling zone AB. When the temperature is increased beyond 10°C, one can witness the agglomeration of bubbles and the rise of the bubbles like the columns from the surface. This will be the boiling mode from 10° to 30°C costly and indicated in Figure 12.11 as BC. At point C, the heat flux is maximum and this is called the *critical heat flux* in the boiling heat transfer literature. At point C there would be a considerable vapor formation which makes it difficult for a liquid to continuously wet the heated surface.

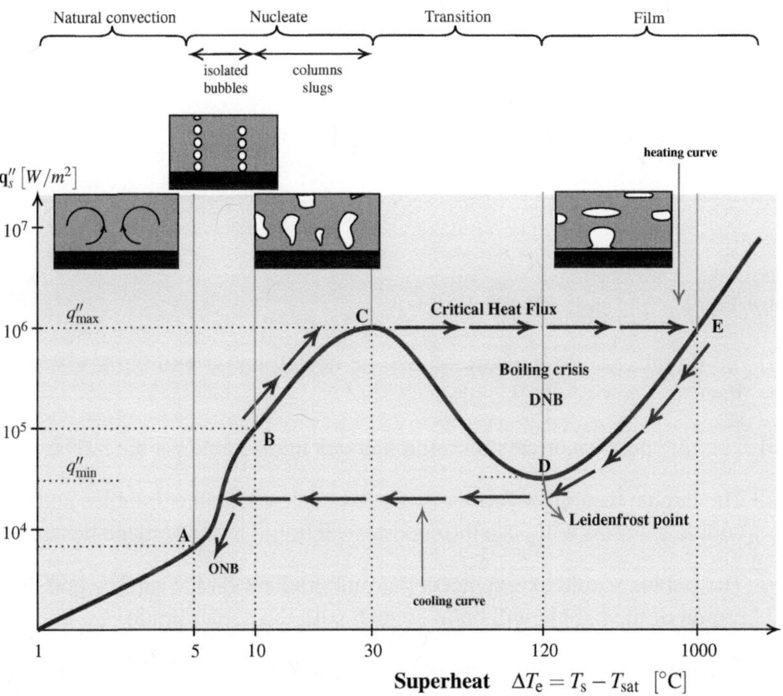

Figure 12.11 q_s'' vs. excess temperature.

Transition boiling: This regime will occur when the excess temperature is between 30° and 120°C. The zone is also called *unstable film boiling* or *partial film boiling* regime.

Film boiling: The boiling zone beyond 120°C is referred as the film boiling regime and it is indicated by the region DE in the boiling curve. Beyond point C, one can witness a sudden drop in the heat flux. The condition is marked as point the D in Figure 12.11. The point D is called *Leidenfrost* point, which is named after German doctor Johann Gottlob Leidenfrost, who had first discovered this phenomenon in 1756 long before Nukiyamas experiment. The heat flux is minimal at this point, and the surface is completely covered by the vapours. The liquid in this condition has no chance to wet the surface. In the gaseous fluid zone, the dominant modes of heat transfer are conduction and radiation.

Some important observations in the boling phenomenon are depicted in Figure 12.12.

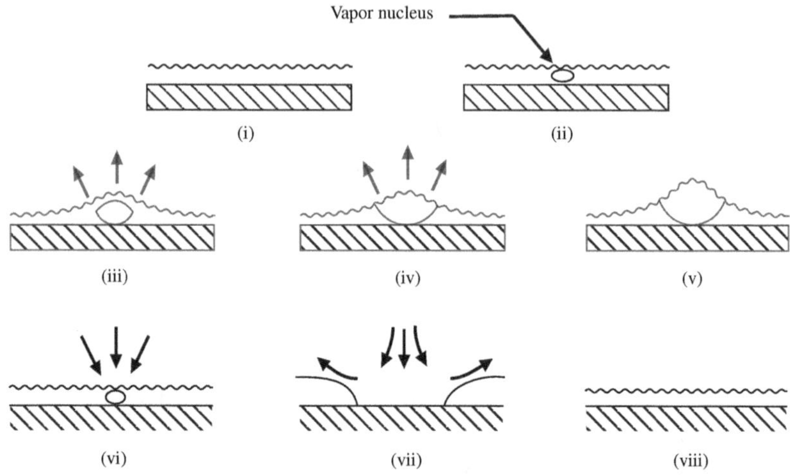

Figure 12.12   The onset of boiling.

[i]   the liquid close to the surface will be heated and it will be in a superheated state.

[ii] The tiny bubbles or cavities agglomerates into sizable visible bubbles.

[iii] The bubble then grow further in size by coalescing other bubbles and the large bubble columns will push the superheated liquid away from the heated surface.

[iv] The bubble would experience a thermal gradient on the surface and the upper region of the bubble will be in contact with the cooler liquid.

[v]   The inertia of the liquid and the bubble will cause the bubble to grow in size and rise further. During this process, bubbles will transfer heat to the cooler liquid.

[vi] The bubble starts to collapse as the cooler liquid quickly fill in the bubble volume.

[vii] The bubble experiences a total collapse and the inertia of the cooler liquid brings it into contact with the heating surface.

[viii] Finally, the cooler liquid gets heated above the saturation temperature.

12.6 MODELING AND EXPERIMENTS OF NUCLEATE BOILING

The nucleate boiling is all about bubble formation and the balance of force at the inception stage can be simply modeled as balnce of surface tension force and the buoyancy force:

$$F_b \sim F_\sigma$$

where, F_b is buoyancy force and F_σ is surface tension force. The buoyant force acting on bubble is $(\gamma_f - \gamma_g)\forall$ this leads to balance:

$$(\gamma_f - \gamma_g)\forall = \sigma.(2\pi R_b)$$

where, R_b is the bubble radius.

$$(\rho_f - \rho_g)g\forall = \sigma.(2\pi R_b)$$

$$(\rho_f - \rho_g)g \left(\frac{4}{3}\pi R_b^3\right) = \sigma.(2\pi R_b)$$

$$R_b = \sqrt{\frac{3\sigma}{2(\rho_f - \rho_g)g}}$$

The characteristic bubble length is usually taken without number as

$$L_b \approx \sqrt{\frac{\sigma}{(\rho_f - \rho_g)g}} \tag{12.15}$$

The Nusselt number for this boiling regime can be defined as

$$Nu = \frac{hL_b}{k_f}$$

where, k_f is liquid's thermal conductivity. For nucleate number, the Jakob number is defined as:

$$Ja = \left[\frac{c_{p,f}}{h_{fg}}(T_s - T_{sat})\right]$$

Jakob number is defined as the ratio of the maximum sensible heat absorbed by the liquid to the latent heat of the liquid.

Max Jakob

Max Jakob (July 20, 1879 – January 4, 1955) was a German physicist who did pioneering work on phase change phenomena. Jakob studied engineering at Technical University Munich, from which he graduated in 1903. In 1961, after Jakob's death, in honor of Jakob, the American Society of Mechanical Engineers (ASME) Heat Transfer Division started the Max Jakob Memorial Award, which is now considered to be one of the highest honor in the field of heat transfer.

Rohsenow formulated a correlation for the nucleate boiling:

$$\frac{Q}{A} \sim \mu \cdot h_{fg} \left(\sqrt{\frac{\sigma}{(\rho_f - \rho_g)g}} \right)^{-1} \left(\frac{Ja}{Pr} \right)^3$$

From experimental data, the final form is

$$\frac{Q}{A} = \mu_f \cdot h_{fg} \left(\sqrt{\frac{\sigma}{(\rho_f - \rho_g)g}} \right)^{-1} \left(\frac{1}{Pr_f{}^m C_{nb}} \right)^3 \left[\frac{c_{p,f}}{h_{fg}} (T_s - T_{sat}) \right]^3$$

where f and g refers to liquid and gaseous states as used in thermodynamics. The values of the constant C_{nb} and exponent m are tabulated in Table 12.2. In terms of dimensionless parameter, the equation takes the form

$$q_{nucleate} = \left(\frac{Q}{A} \right)_{nucleate} = \mu_f \cdot h_{fg} \left(\frac{1}{L_b} \right) \left(\frac{1}{Pr_f^m C_{nb}} \right)^3 Ja^3 \qquad (12.16)$$

TABLE 12.2

The Constant C_{nb} and Exponent m for the Nucleate Boiling Correlation (Eq. 12.16)

Liquid	Metal	C_{nb}	m
Water	Brass	0.006	2
Water	Nickel	0.006	2
Water	Copper scored	0.0068	2
Water	Stainless steel (ground and polished)	0.008	2
Water	Copper polished	0.013	2
Water	Stainless steel chemically etched	0.013	2
Water	Stainless steel mechanically polished	0.013	2
Water	Platinum	0.013	2
Carbon tetrachloride	Copper	0.013	4.1
Benzene	Chromium	0.01	4.1
Ethanol	Chromium	0.0027	4.1
Isopropyl alcohol	Copper	0.0023	4.1
n-butyl alcohol	Copper	0.003	4.1

Example 12.5

Water is being boiled in an open atmospheric condition in a grounded and polished stainless steel pot. As water is boiling in 1 atm, the saturation temperature is 100°C. The inner surface of the bottom of the pot is at 110°C.

The area of bottom of the pot is 0.08 m^2. Find the rate of heat transfer for nucleate boiling and the rate of evaporation of water.

Solution

The properties of water at the saturation temperature are

$\rho_f = 957.9 \ kg/m^3; \rho_g = 0.6 \ kg/m^3; h_{fg} = 2257 \times 10^3 \ J/kg,$
$Pr_f = 1.75; c_{pf} = 4217 J/kg \cdot K; \sigma = 0.0589; \mu_f = 0.282 \times 10^{-3} \ Pa \cdot s$

The length scale associated with nucleate boiling is

$$L_b \approx \sqrt{\frac{\sigma}{(\rho_f - \rho_g)g}} = 0.002504 \text{m}$$

The Jakob number is

$$Ja = \left[\frac{c_{p,\ell}}{h_{fg}}(T_s - T_{sat})\right] = 0.01868$$

The constant C_{nb} and exponent m for stainless steel (ground and polished) are 0.008 and 2. The heat flux required for the nucleate boiling as per Rohsenow correlation is

$$q_{nucleate} = \left(\frac{Q}{A}\right)_{nucleate} = \mu_f \cdot h_{fg}\left(\frac{1}{L_b}\right)\left(\frac{1}{Pr_f^m C_{nb}}\right)^3 Ja^3 = 112719.59 \, W/m^2$$

The total heat transfer is

$$Q = q_{nucleate} \cdot A = 9017.56 W$$

The rate of evaporation of water will be

$$\dot{m}_{evap} = \frac{Q}{h_{fg}} = \frac{9017.56}{2257 \times 10^3} = 0.00399 \text{kg/s}$$

12.7 MAXIMUM OR CRITICAL HEAT-FLUX

S. Kutateladze hypothesised that we can find the maximum heat flux by equating the kinetic energy of the vapour to work done by buoyant forces on the characteristic length of the bubble L_b. Also, we can balance the inertia and buoyancy forces on the bubble. The inertia for experienced by bubble is:

$$F_{inertia} = m \cdot a = \rho \cdot L_b^3 \times u_b/t \approx \rho \cdot u_b \cdot \left(\frac{L_b}{t}\right) \cdot L_b^2$$

$$F_{inertia} \equiv \rho \cdot u_b \cdot u_b \cdot L_b^2 \equiv \rho_g \cdot u_b^2 \cdot L_b^2$$

The the buoyancy force is

$$F_{buoyancy} \equiv \left[(\rho_f - \rho_g)g\forall\right] \equiv (\rho_f - \rho_g)gL_b^3$$

Balancing the two forces

$$F_{inertia} \sim F_{bouyancy}$$

we have

$$(\rho_f - \rho_g)gL_b \sim \rho_g \cdot u_b^2$$

Inserting the definition of bubble characteristic length scale

$$L_b = \sqrt{\frac{\sigma}{(\rho_f - \rho_g)g}}$$

we have

$$\rho_g \cdot u_b^2 \sim (\rho_f - \rho_g)g \cdot \sqrt{\frac{\sigma}{(\rho_f - \rho_g)g}}$$

Rearranging we have

$$u_b^2 \sim \sqrt{\left(\frac{(\rho_f - \rho_g)g}{\rho_g^2}\right)\sigma}$$

Note that u_b is the bubble velocity associated with maximum heat flux, in heat transfer literature it is also written as u_{max}, and we arrive at relation:

$$u_{max} \equiv \sqrt[4]{\frac{(\rho_f - \rho_g)g\sigma}{\rho_g^2}} \tag{12.17}$$

Kutateladze proposed that maximum heat flux is related to maximum bubble velocity as

$$q''_{s,max} \sim \rho_g u_{max} h_{fg}$$

$$q''_{s,max} \approx C_{max}\rho_g u_{max} h_{fg}$$

where C_{max} is an empirical constant. The critical (or maximum) heat flux in nucleate pool boiling was theoretically modelled by Russian scientist S. S. Kutateladze in 1948. Later, American scientist N. Zuber in 1958 expressed the formula as

$$\boxed{q''_{s,max} = C_{max} h_{fg}\left(\rho_g^2(\rho_f - \rho_g)g\sigma\right)^{1/4}} \tag{12.18}$$

Lienhard and Dhir (1973) proposed expressions and values to estimate the constant C_{max} which are tabulated in Table 12.3.

TABLE 12.3

Values and Formula for $C_{\max}$ to be Used in Equation 12.18 (Note: $L^* = L/L_b$)

Geometry	C_{max}	Characteristic Dimension	L^*
Finite Length object	0.12	-	-
Small flat heater	$0.15(12\,\pi L_b^2/A)$	Diameter/width	$9 < L^* < 20$
Infinite flat heater	0.15	Diameter/width	$L^* > 27$
Small horizontal cylinder	$0.12(\sqrt[4]{L^*})$	Cylinder radius	$0.15 < L^* < 1.2$
Large horizontal cylinder	0.12	Cylinder radius	$L^* > 1.2$
Small sphere	$0.27(\sqrt[2]{L^*})$	Sphere radius	$0.15 < L^* < 4.26$
Large sphere	0.11	Sphere radius	$L^* > 4.26$

The formation of too many bubbles near the surface is not desirable as bubbles might coalesce and create a vapour film near the surface which may act as a resistance to heat transfer process. One possible solution is to create optimally spaced micro-scale cavities on the surface which helps in preventing the bubbles to coalesce. This helps in better control over the critical heat flux (CHF) in boiling heat transfer.

Example 12.6

Water is being boiled in an open atmosphere by a heating element equipped with electrical resistance. The element has 2 mm diameter. Estimate the maximum heat flux which the nucleate boiling regime will have at the heating element.

Solution

The properties of water at the saturation temperature of 100°C are
$$\rho_f = 957.9 \ kg/m^3; \rho_g = 0.6 \ kg/m^3; h_{fg} = 2257 \times 10^3 \ J/kg,$$
$$Pr_f = 1.75; \ c_{pf} = 4217 J/kg \cdot K; \sigma = 0.0589; \mu_f = 0.282 \times 10^{-3} \ Pa \cdot s$$
The characteristic dimension for a small cylinder is its radius.

$$L = r = 0.002 m$$

The characteristic length for boiling is

$$L_b \approx \sqrt{\frac{\sigma}{(\rho_f - \rho_g)g}} = 0.002504 \text{m}$$

The dimensionless characteristic length ratio is $L^* = L/L_b = 0.7986$. The maximum or critical heat flux according to Zuber-Kutateladze correlation is

$$q''_{s,\max} = C_{\max} h_{fg} \left(\rho_g^2 (\rho_f - \rho_g) g \sigma \right)^{1/4}$$

where, the constant C_{max} can be taken from Table 12.3. This gives C_{max} for a short cylinder:

$$C_{max} = 0.12 \cdot \sqrt[4]{L^*} = 0.1134$$

$$q''_{s,max} = 0.961 MW/m^2$$

Example 12.7

Find the maximum heat flux for pool boiling of water over an infinite flat heating surface.

Solution

The properties of water at the saturation temperature of 100°C are
$\rho_f = 957.9 \ kg/m^3; \rho_g = 0.6 \ kg/m^3; h_{fg} = 2257 \times 10^3 \ J/kg,$
$Pr_f = 1.75; \ c_{pf} = 4217 J/kg \cdot K; \sigma = 0.0589; \mu_f = 0.282 \times 10^{-3} \ Pa \cdot s$
The constant C_{max} can be taken from Table 12.3 for Infinite flat heater as 0.15. This gives

$$q''_{s,max} = C_{max} h_{fg} \left(\rho_g^2 (\rho_f - \rho_g) g \sigma \right)^{1/4}$$

$$q''_{s,max} = 0.15 h_{fg} \left(\rho_g^2 (\rho_f - \rho_g) g \sigma \right)^{1/4} = 1.27 \times 10^6 W$$

12.8　MINIMUM HEAT FLUX AT LEIDENFROST POINT

At Leidenfrost point, a bubble hovers over the surface, and thus the physical contact with the surface is gone. Between bubble and surface there are vapors present and hence in modelling the heat transfer from the surface both phases should be taken into an account. The Leidenfrost point is important as it signifies the onset of stable film boiling, and at this point the heat flux is at its minimum as the heating surface is completely layered by the blanket or layer of vapors. Figure 12.13 shows the schematic presentation of a droplet hovering over the surface.

On the basis of scale analysis, we can formulate the same type of relationship for minimum heat flux as we did for the development of maximum heat flux relation, i.e.:

$$q''_{s,min} \sim \rho_g u_b h_{fg}$$

where bubble velocity for this point can be formulated as we have developed for maximum heat flux condition:

$$u_b \equiv \sqrt[4]{\frac{(\rho_f - \rho_g) g \sigma}{\rho_g^2}}$$

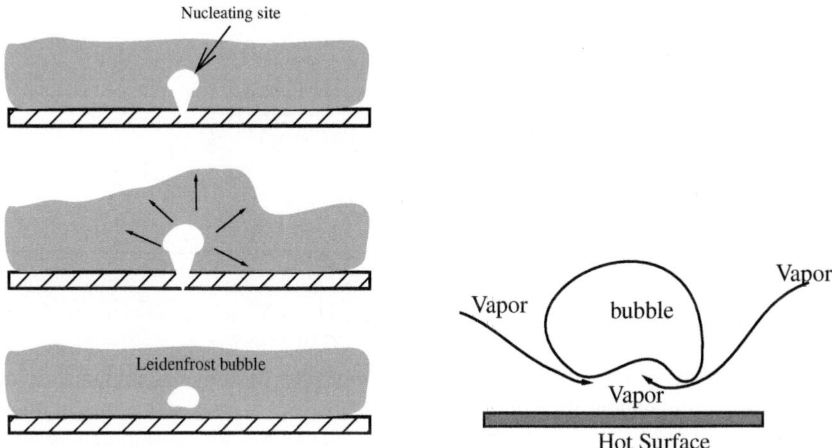

Figure 12.13   Leidenfrost bubble hovering over the hot surface.

This gives the expression

$$q''_{s,min} \sim \rho_g \cdot \sqrt[4]{\frac{(\rho_f - \rho_g)g\sigma}{\rho_g^2}} \cdot h_{fg}$$

Zuber derived the an expression for the minimum heat flux at for the case of boiling over a large horizontal plate using the stability theory.

$$q''_{s,min} = C_{LF} \cdot \rho_g \cdot h_{fg} \left[\frac{\sigma \cdot g \cdot (\rho_f - \rho_g)}{(\rho_f + \rho_g)^2} \right]^{1/4} \tag{12.19}$$

Originally, Zuber suggested the constant value as $C_{LF} = \pi/24$. However, Berenson in 1961, suggested the value $C_{LF} = 0.09$, to be used in the expression 12.19 for better predictions. A note of caution here is that the equation 12.19 is accurate enough up till 50% for most of the fluids as long as the pressure is moderate. However, if the pressure is too high the results are not reliable.

Example 12.8

Calculate the minimum heat flux for the case water boiled over an infinite surface.

Solution
The properties of water at the saturation temperature of 100°C are
$\rho_f = 957.9 \ kg/m^3; \rho_g = 0.6 \ kg/m^3; h_{fg} = 2257 \times 10^3 \ J/kg,$
$Pr_f = 1.75; c_{pf} = 4217 J/kg \cdot K; \sigma = 0.0589; \mu_f = 0.282 \times 10^{-3} \ Pa \cdot s$

We use equation 12.19

$$q''_{s,min} = 0.09 \cdot \rho_g \cdot h_{fg} \left[\frac{\sigma \cdot g \cdot (\rho_f - \rho_g)}{(\rho_f + \rho_g)^2} \right]^{1/4}$$

$$q''_{s,min} = 19.091 kW/m^2$$

12.9   FILM POOL BOILING

It is found that the conditions in the stable vapor film strongly resemblance to those of laminar film condensation on a cylinder or sphere. It is proposed to model the film boiling by taking an analogy between film boiling and film condensation over curved surfaces, leading to a form

$$Nu = \frac{\overline{h}_{conv} \cdot D}{k} = C \left[\frac{g(\rho_f - \rho_g)h^*_{fg}D^3}{v_g k_g (T_w - T_{sat})} \right]^{1/4}$$

Properties in above relation must be cmputed at mean film temperature, $T_f=0.5(T_w+T_{sat})$. Bromley (1950) suggested the value of C=0.62 for horizontal cylinders and Lienhard (1987) suggested C=0.67 for spheres. D is the diameter of cylinder or sphere, and latent heat should be adjusted as

$$h^*_{fg} = h_{fg} + 0.8c_{p,g}(T_w - T_{sat})$$

The convective heat transfer in film pool regime depends strongly on the radiation and in the modelling it is necessary to include the radiative heat transfer coefficient defined after balancing the energy:

$$\sqrt[3]{(\overline{h}_{film})^4} = \sqrt[3]{(\overline{h}_{conv})^4} + \overline{h}_{rad} \cdot \sqrt[3]{(\overline{h}_{conv})}$$

$$\overline{h}_{rad} = \frac{\varepsilon \cdot \sigma(T_w^4 - T_{sat}^4)}{(T_w - T_{sat})}$$

where, ε is the surface emissivity and σ is the Stefan-Boltzmann constant. Note that the analogy may not be valid for surfaces with high curvature.

12.10   FLOW BOILING

In the previous section, we discussed the pool boiling in which vapor bubbles were rising to the top as a result of buoyancy effects in a pool of a static liquid. The different flow regimes through which the flow will go through are depicted in Figure 12.14. As the liquid enters a subcooled state, the heat will be transferred by forced convection. The bubbles formation ensued in the region close to the heated wall and these bubbles would be detached from the surface and enter the core of the tube. This gives rise to the bubbly fluid flow. Upon further heating, the size of the bubbles grows and they coalesce and amalgamate further to form slugs of vapor. In the slug flow regime, almost half of the volume is engulfed by the vapors. With further heating, the liquid is converted into vapor and the core of the flow eventually is made up of vapor only, with the liquid present as an annular space between the vapor core and the tube wall. This type of flow regime is called the annular-flow regime. The flow here in this regime experiences a very high heat transfer coefficient. Upon further heating, the annular liquid layer becomes thinner and dry spots will appear on the tube wall. The formation of dry spots caused a sharp drop in the heat transfer coefficient. This is called the transition the regime and fluid quality started approaching the value of unity i.e. more and more liquid is vaporising. With further heating, the misty flow will form, and all the liquid droplets will be vaporised. The variation of the heat transfer coefficient along the tube length is also shown in Figure 12.14.

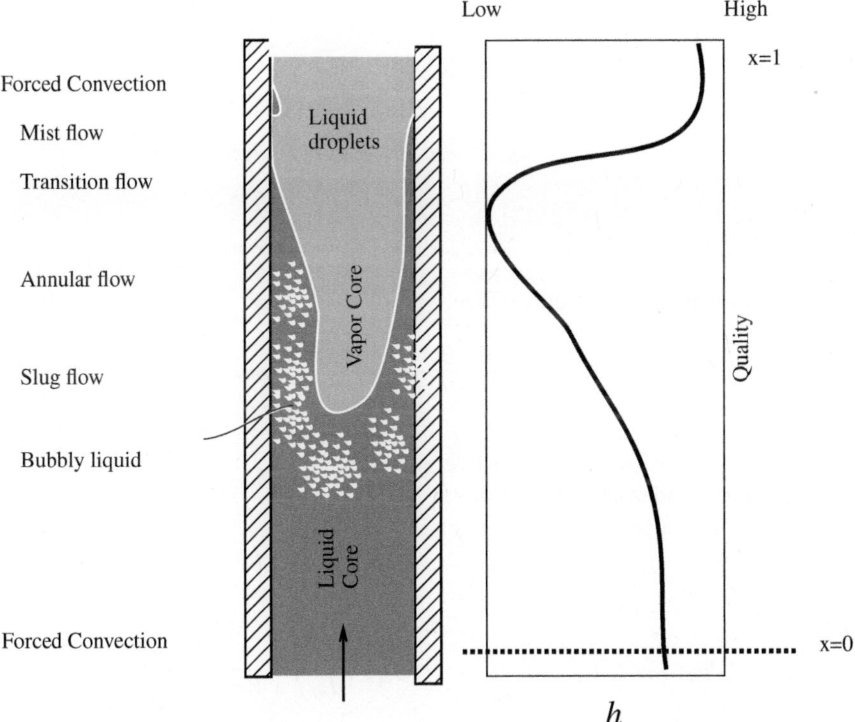

Figure 12.14 Flow boiling regimes as liquid moves through the heated tube and the variation of convective heat transfer coefficient vs. the quality.

Correlation for condensation

Correction for sub-cooling: $h'_{fg} = h_{fg} + 0.375 c_{p,\ell}(T_v - T_w)$

Condensation on flat Flat Surface:

$$Nu_{m,Lam} = 0.943 \left[\frac{h'_{fg} \cdot \rho_\ell \cdot g\,(\rho_\ell - \rho_v) \cdot L^3}{\mu_\ell \cdot k_\ell\,(T_{sat} - T_w)} \right]^{1/4}$$

$$h_{m,inclined} = h_{m,vertical} \cdot (\cos\varphi)^{0.25} \quad (condensation\,on\,inclined\,surface)$$

$$Nu_{m,Lam,film} = 1.47 Re_\delta^{-1/3} \qquad Re_\delta \leq 30 \quad (Laminar\,Regime)$$

$$Nu_{m,Lam,wavy} = \frac{Re_\delta}{1.08(Re_\delta^{1.22}) - 5.2} \qquad 30 \leq Re_\delta \leq 1800 \quad (Wavy\,Regime)$$

$$Nu_{Turb} = \frac{Re_\delta}{8750 + 58 Pr_\ell^{-0.5}(Re_\delta^{0.75} - 253)}$$

$$Re_\delta \geq 1800, Pr \geq 1 \quad (Turbulent\,Regime)$$

$$Nu_{m,L} = \left(\frac{\mu_\ell}{\mu_w} \right)^{0.25} \left[Nu_{Lam}^{1.2} + Nu_{Turb}^{1.2} \right]^{1/1.2} \quad (Combined\,Flow)$$

Condensation on curved surface

$$h_m = C \left[\frac{k_\ell^3 h_{fg} g \rho_\ell\,(\rho_\ell - \rho_v)}{\mu_\ell\,(T_v - T_w)\,D} \right]^{1/4} \quad (pipe/tube)$$

where,
C=0.728 or 0.61 (condensation over pipe/tube)
C=0.828 (condensation over sphere)
C=0.555 (condensation inside pipe)

Boiling:

$$q_{nucleate} = \left(\frac{Q}{A} \right)_{nucleate} = \mu_f \cdot h_{fg} \left(\frac{1}{L_b} \right) \left(\frac{1}{Pr_f^m C_{nb}} \right)^3 Ja^3$$

$$q''_{s,max} = C_{max} h_{fg} \left(\rho_g^2 (\rho_f - \rho_g) g \sigma \right)^{1/4}$$

$$q''_{s,min} = C_{LF} \cdot \rho_g \cdot h_{fg} \left[\frac{\sigma \cdot g \cdot (\rho_f - \rho_g)}{(\rho_f + \rho_g)^2} \right]^{1/4}$$

PROBLEMS

12P-1 In a condenser, the water vapour is condensing at pressure of 0.004 MPa with the mass flow rate of 5 kg/s. The vapour saturation temperature is T_{sat} = 28.98 °C. Cooling is provided by liquid water, whose temperature is maintained at 7 °C at inlet and 23 °C at exit. Water circulates in brass tubes at a speed of 2 m/s. The external diameter of the tube is 20 mm and the internal diameter is 17 mm. The tubes are aligned vertically in a column of 20 tubes. The thermal conductivity of brass is k = 110 W/m·K.

The properties of the water at a temperature of 15 °C, are: $\nu = 1.15 \times 10^{-6}$ m²/s, k = 0.587 W/m·K, and Pr = 8.3.

The properties of condensate and steam are determined at the saturation temperature: $\nu_\ell = 0.825 \times 10^{-6}$ m²/s, $k_\ell = 0.616$ W/m·K, $\rho_\ell = 996$ kg/m³, $\rho_v = 0.0287$ kg/m³, $h_{fg} = 2{,}433$ kJ/kg. What shall be the total length of the tubes?

12P-2 Steam at 31.2 kPa condenses over tube bank where 70 tubes are exposed to steam. The array of tubes is 10 tubes by 7 tubes, thus 7 tubes are arranged in a column. The outer surface temperature of tube walls is 10 °C. If tube length and diameter are d=5cm and L=20 cm, respectively, find rate of heat transfer and condensate formation.

[Ans $\dot{m}_{cond}$=0.17719 kg/s]

12P-3 Find peak heat flux for mercury pool boiling over an infinite surface at saturation temperature of 200 °C.

12P-4 Using the units of quantities, prove that equation 12.15 holds the dimensions of a length scale.

12P-5 Using the units of quantities, prove that equation 12.17 holds the dimensions of a speed.

12P-6 Water is boiled at 150 °C inside chemically etched stainless steel tubes using steam. The tube diameter and length are 6 cm and 55 m, respectively. The outer tube temperature has reached 162 °C. Find (i) the characteristic bubble size in nucleate boiling regime, (ii) Jakob number, (iii) the evaporation rate, (iv) critical heat-flux, and tube surface temperature corresponding to critical heat-flux.

[Ans $T_{burnout}$ = 169.1 °C]

12P-7 A nickel-platted heating element is used to pool boil water at saturation temperature of 250 °F. The heating element, 0.4-in diameter and 3 ft long has its surface temperature reached up to 290 °F. Find the electric power consumed and the convective heat transfer coefficient.

[Ans: h=70425.6 Btu/hr·ft^2·°F, Power=0.885 M Btu/hr]

12P-8 A square rod having side as 100-mm at 800 K is suddenly submerged in a large tank containing saturated water at 1 atm for quenching purpose. Estimate the maximum and minimum heat removal rate experienced by the rod if the length of the rod is 1.5 m. Ignore initial transient effects and conduction through the rod.

[Ans: Q_{max}=0.756 MW, Q_{min}= 11.45 kW]

12P-9 A 5-mm diameter horizontal chromium wire is placed in water at 1 atm. The emissivity of chromium is 0.058, and the surface temperature of chromium is 800 K. Find the convective heat transfer coefficient for film pool boiling and associated heat transfer.

$$[\text{Ans: } h_{film} = 342.536 W/m^2 \cdot K, \ q_w = 146.26 \ kW/m^2]$$

REFERENCES

D. Q. Kern, Process heat transfer. McGraw-Hill, New York, 1950.

S. S. Kutateladze, Fundamentals of Heat Transfer, Academic Press, New York, 1963.

D. A. Labuntsov, Teploenergetika, 4, 72, 1957.

W. M. Rohsenow, Film Condensation. In Handbook of Heat Transfer, ed. W. M. Rohsenow and J. P. Hartnett, Ch. 12 A. New York: McGraw-Hill, 1973.

S. Nukiyama, J. Japan Soc. Mech. Eng., 37, 367, 1934 (Translated in Int. J. Heat Mass Transfer, 9, 1419, 1966)

N. Zuber, On the Stability of Boiling Heat Transfer. ASME Transactions 80, 711 - 720, 1958.

J. H. Lienhard and V. K. Dhir, Hydrodynamic prediction of peak pool-boiling heat fluxes from finite bodies, J. Heat Transfer, 95, 152-158, 1973.

P. Berenson, Film-boiling heat transfer from a horizontal surface, J. Heat Transfer, 83, 351-358, 1961.

L. A. Bromley, Chem. Eng. Prog., 46, 221, 1950.

J. H. Lienhard, A Heat Transfer Textbook, 2nd ed., Prentice-Hall, Englewood Cliffs, NJ, 1987.

13 Fundamentals of Thermal Radiation

Radiation is a surface phenomenon in the case of solid and liquid bodies, and it happens volumetrically in gases comprising of molecules having more than two atoms. Radiations occur due to electromagnetic waves, and it is the most dominant mode at high temperatures. Radiation is the only mode of heat transfer outside the earth's atmosphere and thus plays important role in space technology designs. After finishing this chapter, one will be able to:

- State Stefan-Boltzmann law, Planck's law, Wien's displacement law, and Kirchhoff law.
- Understand the emissivity, absorptivity, reflectivity, and transmissivity of the surfaces.
- Compute black and gray body emissions.
- Understand the terminology related to extraterrestrial radiation.

Heat transfer between solid or liquid surfaces, and gases because of electromagnetic waves is thermal radiation. All matter forms constantly emit thermal radiation, which is mainly due to the vibrational, rotational motions of atoms and molecules of a substance.

The electromagnetic waves are classified in Table 13.1

In order to explain the behaviour of electromagnetic thermal radiation, scientist have created the concept of a *blackbody*. The *blackbody* is a *ideal* body which perfectly emit and absorb thermal radiation at a given absolute temperature. In laboratory, blackbody can be approximated by forming a small cavity in a copper block. The radiation would enter this cavity through a narrow aperture. By heating the cavity surface with brief pulses and preventing heat loss using the insulation, we can create different temperature levels inside the cavity. If the cavity is isothermal at each recording and well insulated, then this type of arrangement is good for approximating the cavity as a black body in a laboratory. The radiation incident through the small opening is reflected several times inside the cavity and also absorbed by the metal. Though a minuscule portion of incident radiation will exit through the aperture, most of the radiation will be absorbed.

DOI: 10.1201/9781003428404-13

TABLE 13.1

Classification of Electromagnetic Waves

Wavelength	Type
10^{-14} m - 10^{-12} m	Cosmic radiation
10^{-12} m - 10^{-10} m	Gamma radiation
10^{-10} m - 10^{-9} m	X-rays
10 nm - 400 nm	Ultraviolet radiation
380 nm - 760 nm	Visible radiation
0.1 μm - 100 μm	Thermal radiation

13.1   PLANK'S LAW

Max Planck in 1900 proposed the law relating the spectral distribution of the monochromatic emissive power $E_{b\lambda}$ with the wavelength as

$$E_{b\lambda} = \frac{c_1 \cdot \lambda^{-5}}{\exp\left(\frac{c_2}{\lambda \cdot T}\right) - 1} \tag{13.1}$$

where,

$$c_1 = 3.743 \times 10^{-16} \ J/m^2 \cdot s, \quad c_2 = 0.014438265235 \ K \cdot m$$

Pioneers of Heat Transfer

Max Planck born 23 April 1858 and died 4 October 1947. Plank obtained his baccalaureate degree at the age of 17. He studied physics in Munich and then in Berlin. Plank was interested in thermodynamics, and in 1879, in his doctoral thesis he presented his law on heat theory. He became a lecturer at the University of Munich and obtained the chair of physics at the University of Kiel in 1885. He become the professor of physics at the University of Berlin and become one of the most famous Physicists. In 1918, Planck received the Nobel Prize.

Figure 13.1 shows the spectral distribution of emissive power for a blackbody at absolute temperature values of 300K, 500K, 1000K, 3000K, and 6000K. The plot is semi-logarithmic plot as the emissive power peaks and drops sharply. Here in plot we can notice that all the emissive power curves reaches a maxima and then drops. At λ=0, the emissive power is zero. Also, most of the body body energy is emitted in wavelengths till 20μm. The Sun (surface T=6000K) emits 98% of its radiation in the wavelengths range of 0.1-3 μm.

In case in certain medium, if the speed of light is not close to $c_o = 2.998 \times 10^8 \ m/s$ then in such cases the emissive power relation with wavelength and temperature must include refractive index of the medium.

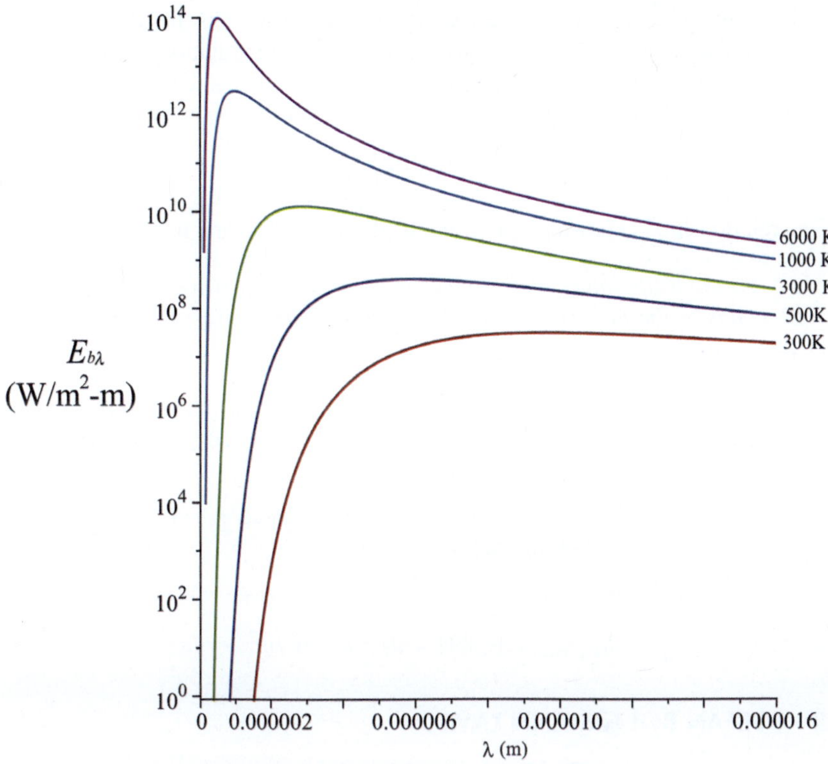

Figure 13.1   Plot of Plank's law showing monochromatic emissive power of a blackbody against wavelength at different temperatures.

13.2   WEIN'S DISPLACEMENT LAW

In order to find the relationship between the temperature of a blackbody and the wavelength at which it emits the most of its radiation, we differentiate the $E_{b\lambda}$ from Plank's law 13.1 with wavelength:

$$\frac{dE_{b\lambda}}{d\lambda} = -\frac{5c_1}{\lambda^6 \left(e^{\frac{c_2}{\lambda T}} - 1 \right)} + \frac{c_1 c_2 e^{\frac{c_2}{\lambda T}}}{\lambda^7 \left(e^{\frac{c_2}{\lambda T}} - 1 \right)^2 T} = 0$$

which on further simplification would lead to form

$$(5\lambda T - c_2) \, e^{\frac{c_2}{\lambda T}} - 5\lambda T = 0$$

Substituting the values of constants and solving them we have

$$5.\lambda T e^{\frac{0.01438265235}{\lambda T}} - 0.01438265235 e^{\frac{0.01438265235}{\lambda T}} - 5\lambda T = 0$$

Solving this equation with MAPLE gives 0.002896741480 m $\cdot$ K. Since, the wavelength here is the wavelength at which the maximum radiation is emitted, it is customary in heat transfer literature to add a subscript with this wavelength as λ_{max}. This gives a formulation:

$$\boxed{\lambda_{max} \cdot T = 2896.74 \ \mu m \cdot K}$$
(13.2)

The relationship is known as the Wien displacement law, and it can be stated as

As the temperature of the blackbody increases, the wavelength corresponds to the maximum emitted radiation shifts towards the shorter wavelengths.

Substituting the wavelength from Wien displacement law into blackbody emissive power, we have the maximum spectral monochromatic emissive power:

$$E_{b\lambda,max} = \frac{c_1 \cdot \left(\frac{0.002896741480}{T}\right)^{-5}}{\exp\left(\frac{c_2}{0.002896741480}\right) - 1} = 0.0000012894 T^5$$

This can be written as

$$E_{b\lambda,max} = 1.2894 \times 10^{-5} T^5 \ W/m^2$$

13.3 STEFAN BOLTZMANN LAW

The total emissive power of a blackbody (E_b) is defined as the total radiant energy emitted by a blackbody over the wavelength range ($\lambda = 0$ - ∞) per unit area per unit time. We will now prove that the emissive power of a blackbody is proportional to the fourth power of its absolute temperature value. We will integrate $E_{b\lambda}$ relation over the wavelengths:

$$E_b = \int_0^\infty E_{b\lambda} d\lambda$$

Substituting $\zeta = \frac{c_2}{\lambda T}$ in blackbody emissive power relation with

$$d\lambda = \frac{-c_2}{\zeta^2 T} d\zeta$$

and rearranging we have

$$E_b = \int_0^\infty E_{b\lambda} d\lambda = \int_0^\infty \frac{c_1 \cdot \lambda^{-5}}{\left[\exp\left(\frac{c_2}{\lambda \cdot T}\right) - 1\right]} d\lambda$$

$$E_b = \frac{c_1 T^4}{c_2^4} \int_0^\infty \frac{\zeta^3}{\left[\exp(\zeta) - 1\right]} d\zeta$$

Expanding the terms we have

$$\frac{\zeta^3}{[\exp(\zeta)-1]} = \zeta^3\left[e^{-\zeta} + e^{-2\zeta} + e^{-3\zeta} + \cdots\right]$$

Integrating with up till 6th term we have

$$\int_0^\infty \frac{\zeta^3}{[\exp(\zeta)-1]}d\zeta = 6.486741204$$

and the total emissive power of a blackbody is

$$E_b = \frac{c_1 T^4}{c_2^4}(6.486741204) = 5.67410 \times 10^{-8}T^4$$

$$E_b = \sigma T^4$$

The constant $\sigma = 5.67410 \times 10^{-8}\ W/m^2 K^4$ is called Stefan-Boltzmann constant. In US customary units, the value of this constant is 1.71344×10^{-9} Btu $\cdot$ hr$^{-1}\cdot$ft$^{-2}\cdot$ $^\circ$R^{-4}. The total intensity of the emitted radiation from blackbody is

$$I_b = E_b/\pi$$

> **Pioneers of Heat Transfer**
>
> Josef Stefan (1835–1893) was professor at the University of Vienna and in 1879, he inferred from some of his experiments that blackbody emission is proportional to temperature to the fourth power. Previously, Ludwig Erhard Boltzmann (1844–1906) in 1889 derived the fourth-power law purely from thermodynamic relations.

Close to our planet the biggest source of thermal radiation is our star - the sun. Solar energy coming to Earth varies over the year as the Earth orbait around sun is elliptical. The maximum irradiation received from sun is about 1412 W/m^2 in early January, and a minimum of about 1321 W/m^2 happens in early July.

> **Example 13.1**
>
> Solar energy received by the earth's atmosphere, also called the solar constant is measured to be around 1367 W/m^2. Estimate the solar constant on Jupiter? The distance from earth to sun is 147.48 million km, and distance from sun to Jupiter is 740.85 million km.

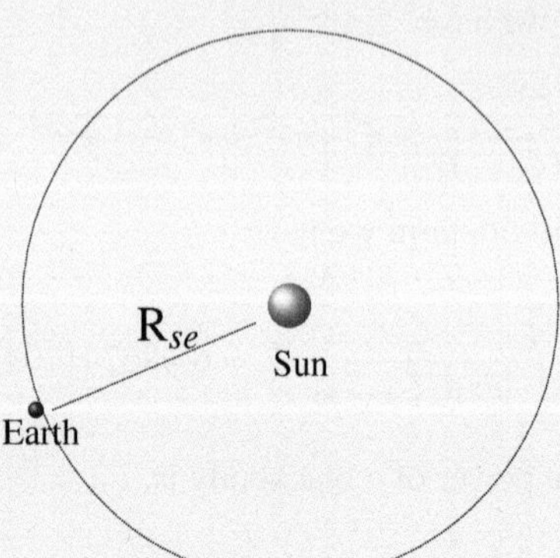

Spherical distribution of sun light

Solution

The radiation from sun surface spreading spherically and reaching Earth is

$$q_{sun-earth} = \frac{Q}{A_s} = \frac{\sigma T_{sun}^4}{4\pi R_{se}^2}$$

where R_{se} is the distance from sun to earth. The radiation from sun surface spreading spherically and reaching Jupiter is

$$q_{sun-Jupiter} = \frac{Q}{A_s} = \frac{\sigma T_{sun}^4}{4\pi R_{sj}^2}$$

where R_{sj} is the distance from sun to Jupiter.
Taking the ratio of two equations we have

$$\frac{q_{sun-Jupiter}}{q_{sun-earth}} = \frac{R_{se}^2}{R_{sj}^2} = \left(\frac{R_{se}}{R_{sj}}\right)^2$$

According to give data we have

$$q_{sun-earth} = 1367\,W/m^2; R_{se} = 147.48M\ km; R_{sj} = 740.85M\ km;$$

$$q_{sun-Jupiter} = q_{sun-earth}\left(\frac{R_{se}}{R_{sj}}\right)^2 = 1367\left(\frac{147.48}{740.84}\right) = 54.17W/m^2$$

Example 13.2

The sun is illuminating the earth with solar flux = 1367 W/m^2. Assuming that all solar energy is absorbed by earth and then emitted by earth surface spherically in all directions, find the temperature of the earth's surface.

Solution
The heat absorbed by earth is

$$Q_{abs} = q_{sol} \cdot \pi \cdot R_{earth}^2 = 1367 \times \pi \times (6371 \times 10^3)^2 = 1.7431 \times 10^{17} W$$

The heat emitted by earth is

$$Q_{emt} = \sigma \cdot T_e^4 \cdot \left(4 \cdot \pi \cdot R_{earth}^2\right) = 2.8920 \times 10^7 T_e^4$$

Comparing absorbed and emitted heat from earth

$$Q_{abs} = Q_{emt}$$

we have
$$T_e = 278.63\text{K} = 5.67^\circ C$$

13.4 BAND EMISSION

In most engineering applications, we need to determine the blackbody emissive power in a range of wavelengths. The schematic representation of band emission is shown in Figure 13.2

The fraction of energy in the hatched region in Figure 13.2 can be obtained by taking the ratio of area under the hatched region to the total area under the curve of emissive power:

$$F_{0 \to \lambda} = \frac{\int\limits_0^\lambda E_{b,\lambda} d\lambda}{\int\limits_0^\infty E_{b,\lambda} d\lambda} = \frac{\int\limits_0^\lambda E_{b,\lambda} d\lambda}{\sigma T^4}$$

$$F_{0 \to \lambda} = \frac{\int\limits_0^{\lambda T} E_{b,\lambda} d\lambda T}{\sigma T^5}$$

$$F_{\lambda_1 \to \lambda_2} = \frac{\int\limits_0^{\lambda_2 T} E_{b,\lambda} d\lambda T - \int\limits_0^{\lambda_1 T} E_{b,\lambda} d\lambda T}{\sigma T^5} = F_{0 \to \lambda_2} - F_{0 \to \lambda_1}$$

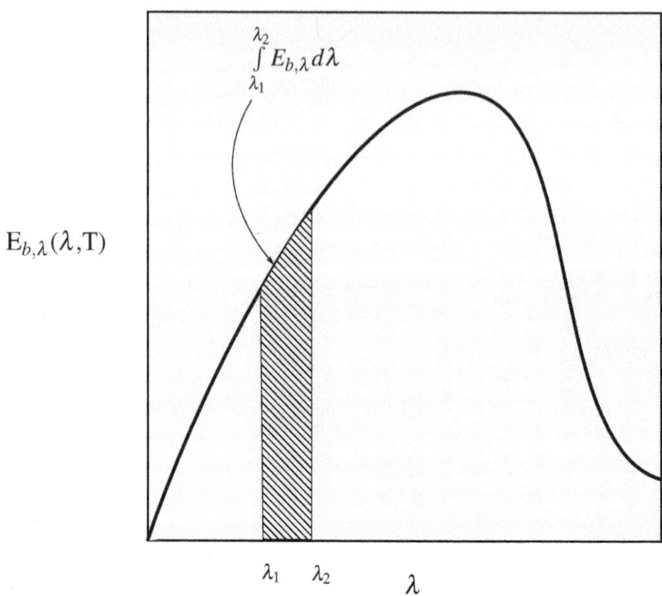

Figure 13.2   Band radiation emission from a blackbody in the spectral band λ_1 to λ_2.

The values of the function important for blackbody radiation band emission and Intensity calculations are listed in Table 13.2. The last column is used to obtain an estimate of the ratio of the spectral intensity at any wavelength to spectral intensity at wavelength where $E_{b,\lambda}$ has a peak value.

Example 13.3

An object emitting radiation at 2760 K. Estimate the percentage of total emitted energy that is emitted in the visible range. Assume that object is behaving like a blackbody.

Solution

The wavelength of visible range is from 0.4 μm to 7 μm. We assume that the spectral emission from object is following Planck's Law of radiation.
We take the product of λ with absolute temperature:

$$\lambda_1 T = 0.4\ \mu m \times 2760K = 1104\ \mu m.K$$

$$\lambda_2 T = 0.7\ \mu m \times 2760K = 1932\ \mu m.K$$

TABLE 13.2

Blackbody Radiation Band Emission Function

λT	$F_{0\to\lambda}$	λT	$F_{0\to\lambda}$
200	0	5600	0.701046
400	0	5800	0.720158
600	0	6000	0.737818
800	0.000016	6200	0.75414
1000	0.000321	6400	0.769234
1200	0.002134	6800	0.796129
1400	0.00779	7000	0.808109
1600	0.019718	7400	0.829527
1800	0.039341	7800	0.848005
2000	0.066728	8000	0.856288
2200	0.100888	9000	0.890029
2400	0.140256	10000	0.914199
2600	0.18312	11000	0.93189
2800	0.227897	12000	0.945098
2898	0.250108	13000	0.955139
3000	0.273232	14000	0.962898
3200	0.318102	15000	0.969981
3400	0.361735	16000	0.973814
3600	0.403607	18000	0.98086
3800	0.443382	20000	0.985602
4000	0.480877	25000	0.992215
4200	0.516014	30000	0.99534
4400	0.548796	40000	0.997967
4600	0.57928	50000	0.998953
4800	0.607559	75000	0.999713
5000	0.633747	100000	0.999905
5200	0.65897		
5400	0.68036		

The data from table are

λT (μm · K)	$F_{0\to\lambda}$
1000	0.000321
1200	0.002134
1800	0.039341
2000	0.066728

Interpolating for $F_{0 \to \lambda_1}$ we have

$$F_{0 \to \lambda_1} = 0.001263$$

Interpolating for $F_{0 \to \lambda_2}$

$$F_{0 \to \lambda_1} = 0.057416$$

Percentage of energy in visible band is

$$\left(F_{0 \to \lambda,2} - F_{0 \to \lambda,1}\right) \cdot 100 = 5.6152\%$$

13.5   SURFACE AND RADIATION

When radiant energy falls on a material surface, part of this radiation will be reflected, a part of it will be absorbed, and part of it will be transmitted, as depicted schematically in Figure 13.3.

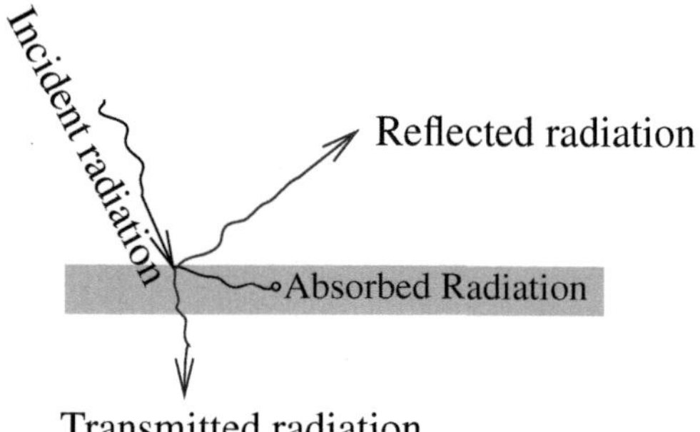

Figure 13.3   Schematic representation of incident radiation and its breakdown into transmitted, absorbed, and reflected radiation.

Radiative Heat Transfer

$$\text{Absorptivity:} \quad \alpha = \frac{Absorbed\ part\ of\ incident\ radiation}{Total\ incident\ radiation} \quad 0 \leq \alpha \leq 1$$

$$\text{Reflectivity:} \quad \rho = \frac{Reflected\ part\ of\ incident\ radiation}{Total\ incident\ radiation} \quad 0 \leq \rho \leq 1$$

$$\text{Transmissivity:} \quad \tau = \frac{Transmitted\ part\ of\ incident\ radiation}{Total\ incident\ radiation} \quad 0 \leq \tau \leq 1$$

Energy Balance: $\alpha + \rho + \tau = 1$
where, α: Absorptivity; ρ: Reflectivity; τ: Transmissivity.

Blackbody	$\alpha=1$	$\rho=0$	$\tau=0$
Opaque bodies	$\alpha+\rho=1$		$\tau=0$
Gas	$\alpha+\tau=1$	$\rho=0$	
Diatherem gas	$\alpha=0$	$\rho=0$	$\tau=1$
White body	$\alpha=0$	$\rho=1$	$\tau=0$

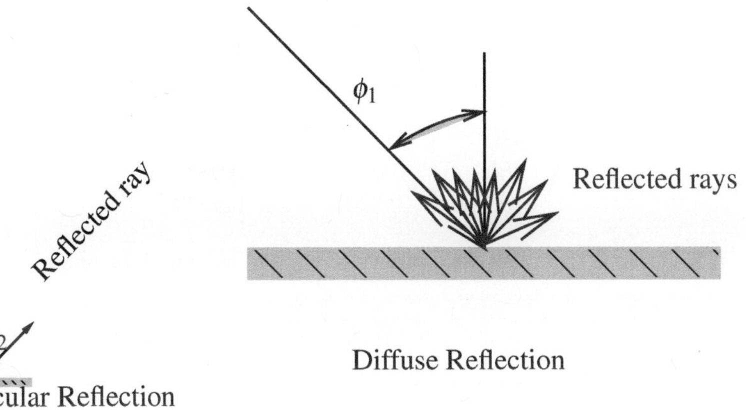

Figure 13.4 Schematic representation of specular reflection.

If the angle of incidence of radiation is equal to the angle of reflection, the reflection is called specular reflection. In case, the incident radiation is diffused in reflection, i.e. reflected in several directions then such a reflection is called diffuse reflection. These two types of reflection are shown in Figure 13.4.

Example 13.4

A glass plate of 50cm is installed as furnace window. The furnace inside temperature is 2500K. Find the heat incident, transmitted and absorbed in the visible band of radiation, if glass transmissivity is 0.5 and the emissivity is 0.3. Assume furnace as blackbody.

Solution It is given:

T=2500 K, A=0.25 m^2, τ=0.5 and ε=0.3. We know that the wavelength of radiation bands are

$$
\begin{aligned}
Ultraviolet: \quad & \lambda = 10\text{nm} - 400nm \\
Visible: \quad & \lambda = 380nm - 760nm \\
Infrared: \quad & \lambda = 780nm - 1E6nm \\
Far\,Infrared: \quad & \lambda > 1mm
\end{aligned}
$$

We calculate the energy fraction as

$$
F_{0\to\lambda} = \frac{\int_0^\lambda E_b \, d\lambda}{\sigma \cdot T^4} = \frac{\int_0^\lambda \left[\dfrac{c1 \cdot \lambda^{-5}}{\exp\left(\frac{c2}{\lambda \cdot T}\right) - 1} \right] d\lambda}{\sigma \cdot T^4}
$$

Alternatively, we can interpolate in Table 13.2 and this leads to

$\lambda\,(nm)$	$\lambda T\,(\mu m \cdot K)$	$F_{0\to\lambda}$
0	0	0
10	25	0
380	950	0.000174
760	1900	0.05229
10,000	25000	0.99397

If we assume that glass absorptivity is equal to its emissivity, then

$$
E_{incident} = (0.05229 - 0.000174) \cdot \frac{\sigma \cdot T^4 \cdot A}{1000} = 28.86\text{kW}
$$

$$
E_{transmitted} = \tau E_{incident} = 14.43 kW
$$

$$
E_{absorbed} = \alpha E_{incident} = 8.65 kW
$$

13.6   REAL AND GRAY BODY EMISSIONS

The spectral radiation emitted by a real surface is quite different from the blackbody radiation (see Figure 13.5). First of all it is not following the Planck distribution and secondly, the directional distribution is also not like a diffused distribution. Therefore, the emissivity changes with the wavelength over the surface.

The total, hemispherical emissivity, used to define the real surface thermal radiative emissions is defined as

$$\varepsilon(T) = \frac{total\ emissive\ power\ of\ a\ real\ surface\ (E(T))}{total\ emissive\ power\ of\ a\ blackbody\ surface\ (E_b(T))}$$

$$Emissivity: \quad \varepsilon = \frac{Energy\ emitted\ from\ a\ real\ surface}{Energy\ emitted\ from\ a\ blackbody} \quad 0 \leq \varepsilon \leq 1$$

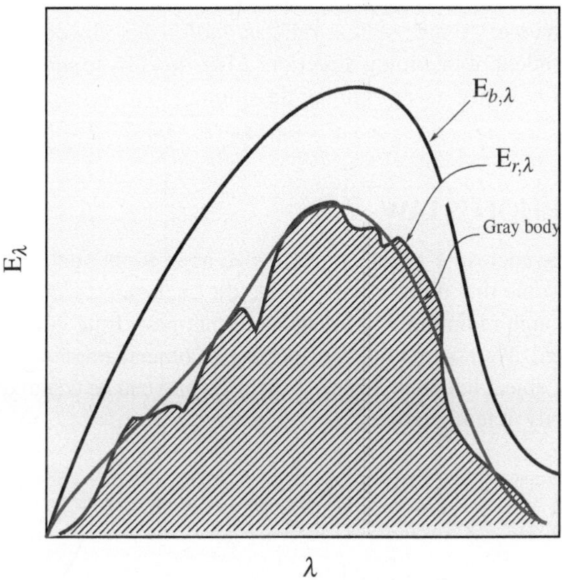

Figure 13.5   The spectral emissive powers for black, gray, and real surfaces.

A gray body is a body in which the monochromatic emissivity ε_λ of the body is independent of wavelength (λ).

For a gray surface we can write

$$E(\lambda, T) = \varepsilon(T)E_b(\lambda, T) = \varepsilon(T)\sigma T^4$$

Radiative bodies and surface types

Blackbody: A blackbody is defined as a perfect absorber and emitter of radiation, with an emissivity of unity over all wavelengths.

White body: A white body is an idealised object that reflects all incident radiation and absorbs none. Therefore, by definition, the monochromatic emissivity of a white body is zero at all wavelengths.

Gray body: A gray body is a body or medium whose radiative properties like absorptivity, emissivity, reflectivity, etc, are independent of the wavelength.

Opaque body: A theoretical object that does not allow any radiation to pass through it is called a perfect or ideal opaque body.

Specular surface: A specular surface is a surface that reflects light in a single direction according to the law of reflection.

Diffuse surface: A surface that reflects and/or that absorbs/reflects radiation independent of incoming direction. Also, the diffuse surface emits equal amounts of radiative energy into all directions.

13.7   KIRCHHOFF'S LAW

Consider a large enclosure which is maintained at an isothermal surface temperature T_w. There are some tiny objects lying inside the enclosure as shown in Figure 13.6. The objects though radiating within enclosure but have little influence on the overall radiative field. We assume that irrespective of objects orientation, the irradiation coming on any object in the enclosure is diffused and can be equated to the emission from a blackbody at temperature T_w.

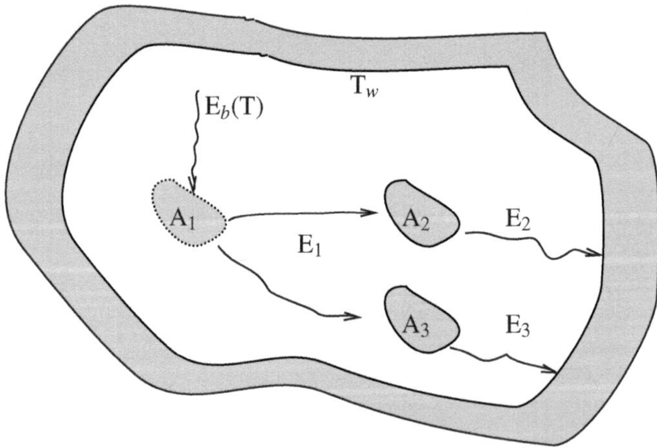

Figure 13.6   Radiative exchange in an isothermal enclosure.

For the steady-state condition condition, the thermal equilibrium must exist between the objects and the enclosure wall, which leads to condition that

$$T_1 = T_2 = \cdots = T_w$$

Doing an energy balance for object 1, we have

$$\alpha E_b(T)A_1 - E_1(T_w)A_1 = 0$$

and

$$\frac{E_1(T_w)}{\alpha} = E_b(T)$$

$$\left(\frac{Emissive\ Power}{Absorptivity}\right) = \left(\frac{E}{\alpha}\right)_1 = \left(\frac{E}{\alpha}\right)_2 = \cdots = E_b = f(T)$$

For monochromatic radiation:

$$\left(\frac{Emissive\ Power\ at\ wavelength}{Absorptivity\ at\ wavelength}\right) = \left(\frac{E_\lambda}{\alpha_\lambda}\right)_1 = \left(\frac{E_\lambda}{\alpha_\lambda}\right)_2 = \cdots = E_b = f(\lambda,T)$$

> **Gustav Robert Kirchhoff**
>
> Gustav Robert Kirchhoff was a German physicist, who by studying the phenomenon of absorption, made spectroscopy (the study of the interaction between matter and electromagnetic radiation) a means of systematic chemical analysis. In addition, he has determined the rules of electricity networks and made significant contributions to electricity, optics and elasticity as well as to the fields of hydrodynamics, and thermodynamics. In 1847, Kirchhoff and Robert Bunsen started working collectively on spectroscopy. They discovered that each chemical element emits and absorbs radiation at specific wavelengths, creating a unique spectral signature. This discovery led to the modern development of devices like spectrometers, spectrophotometers, spectrographs, or spectral analysers.

Gustav Robert Kirchhoff (1824–1887) formulated and published this law in 1860. For real surfaces in an enclosure , the spectral directional absorptivity is equal to the spectral directional emissivity.

$$\boxed{\alpha_\lambda(\lambda,T,\theta,\varphi) = \varepsilon_\lambda(\lambda,T,\theta,\varphi)} \tag{13.3}$$

Gustav Kirchhoff in 1860, proposed that for *any surface in an enclosure, the emissivity from the gray surface is equals to absorptivity coming from a blackbody at the same temperature.* For an object whose temperature is not changing, an object that absorbs radiation well at a particular wavelength will also emit radiation well at that wavelength. For spectral surfaces, we can write:

$$\boxed{\alpha(T) = \varepsilon(T)} \tag{13.4}$$

Kirchhoff Law is valid as long as the body is gray and surroundings are black, and it is very helpful in radiation analysis as surface properties of an opaque surface can be obtained from a knowledge of only one property.

The emittance from a diffuse gray surface is independent of direction. However, the emittance from a real surface vary with zenith angle β as shown in Figure 13.7.

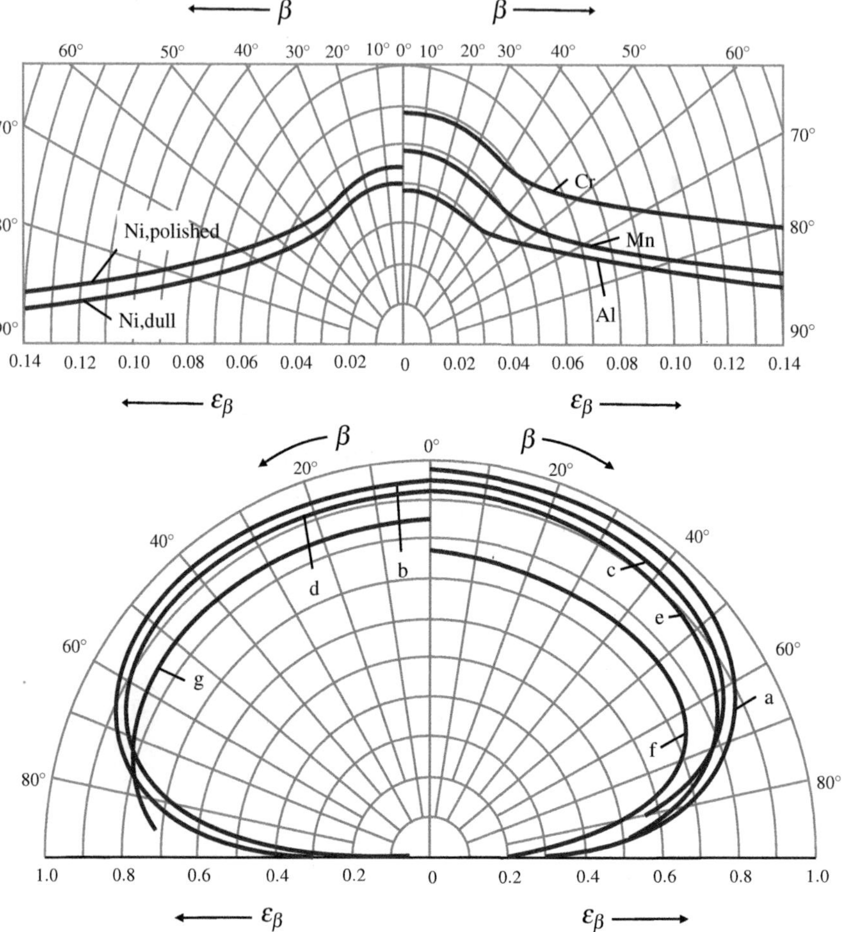

Figure 13.7   Variation of directional emittance with zenith angle β in case of metals (top figure) and non-metals (bottom figure). (a) Wet ice, (b) wood, (c) glass, (d) paper, (e) clay, (f) copper oxide, (g) aluminium oxide

If the surrounding are blackbody or gray surfaces, $\alpha = \varepsilon(\widetilde{T})$, where,

$$\widetilde{T} = \left[T_{surrounding} \cdot T_{surface} \right]^{1/2}$$

Blackbody: the radiation is completely absorbed in this body ($\alpha = \varepsilon = 1$)
White body: the radiation is completely reflected by the surface of this body ($\rho = 1$)
Gray body: the surface absorptivity of this body is same at all wavelengths ($\varepsilon < 1$)

The emission and absorption of radiation by a solid body are surface effects. The monochromatic hemispherical emissivity is defined by the relation

$$\varepsilon_\lambda = 2 \int_0^{\pi/2} \varepsilon_\lambda(\theta)\sin(\theta)\cos(\theta)d\theta$$

13.8 RADIOSITY

The total radiative energy leaving a surface per unit time and per unit area is called the radiosity and is indicated by letter J, in this book. The units of J are W/m^2. Figure 13.8 shows schematically the concept of radiosity. Here G represent the incident radiation, ρG represent the reflected radiation and εE_b represent the emitted radiation.

$$J = \varepsilon E_b + \rho G$$

for a gray and opaque surface we can write

$$J = \varepsilon E_b + (1 - \rho)G \quad (gray\ and\ opaque)$$

where E_b is the blackbody emissive power of surface. Note that a blackbody does not reflect any radiation and radiosity of a blackbody is equal to its emissive power.

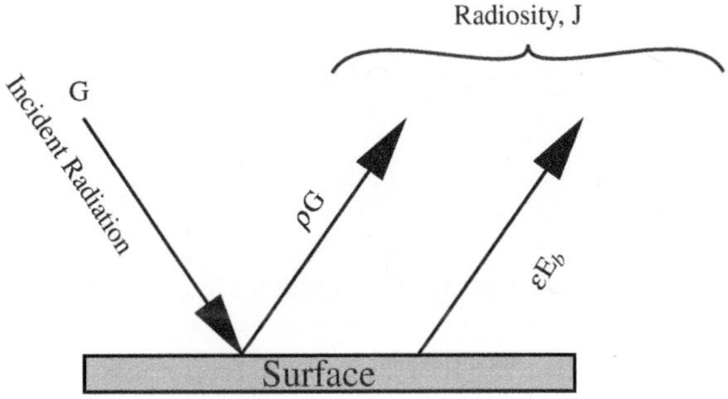

Figure 13.8   Radiosity represents the combined emitted and reflected radiative energies from a gray surface.

The heat flux through the surface can be written as

$$q = J - G$$

Introducing J relation into above equation and invoking Kirchhoff law ($\varepsilon = \alpha$) we have

$$q = \varepsilon E_b - \alpha G$$

This shows that the radiative heat transfer from a gray surface can be expressed either as radiosity minus irradiation, or as emission minus absorption.

Substituting G=J-q into above equation we have

$$q = \varepsilon E_b - \alpha(J - q)$$

$$q = \varepsilon E_b - \alpha J + \alpha q$$

$$(1 - \alpha)q = \varepsilon E_b - \alpha J$$

Since $1 - \rho = \alpha = \varepsilon$ we can write

$$q = \frac{\varepsilon}{(1 - \varepsilon)}(E_b - J)$$

The rate of heat transfer is

$$Q = \frac{A\varepsilon}{(1 - \varepsilon)}(E_b - J) \tag{13.5}$$

This gives us the rate of the net radiant energy leaving a gray surface in terms of its blackbody emissive power and radiosity. The representative electrical circuit for above equation is shown in Figure 13.9.

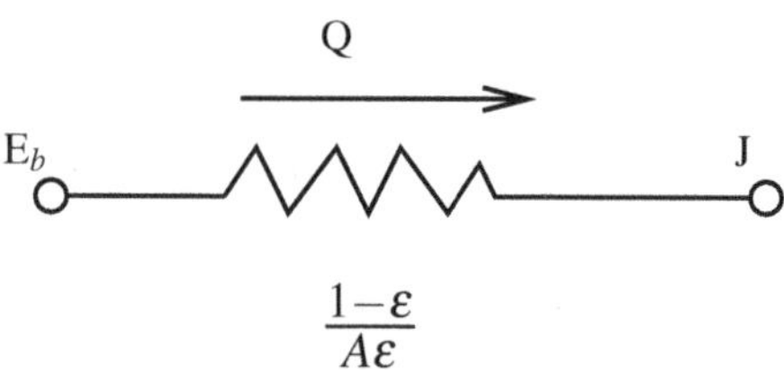

Figure 13.9   Thermal network representing Equation 13.5.

The surface resistance to radiation is

$$R = \frac{1 - \varepsilon}{A\varepsilon}$$

The quantity E_b-J represents a potential difference, and the net rate of radiation heat transfer corresponds to current in the electrical analogy.

If radiative exchange is happening between surfaces, then the network would look like as shown in Figure 13.10. The network method is introduced by A. K. Oppenheim in 1956.

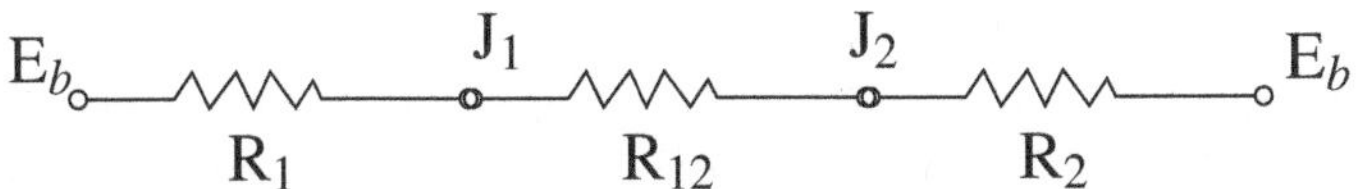

Figure 13.10   Thermal network between two gray surfaces.

The net energy transferred to any surface can be found through the energy balance as

$$q_{net} = \alpha \cdot G_{solar} - q_{loss}$$

where q represents the heat flux, G represents the incident radiation, and q_{loss} represents the energy loss. Different radiative fluxes used in radiation heat transfer are tabulated in Table 13.3

TABLE 13.3

Different Radiative Fluxes Used in Radiation Heat Transfer

Flux (W/m^2)	Description	Comment
Emissive power (E)	Radiative Emittance or flux from surface	$E = \varepsilon \sigma T^4$
Incident radiation (G)	Irradiation or radiative flux coming to surface	
Radiosity (J)	Radiative flux goes out of surface	$J = E + \rho G$
Net radiative flux (q_{rad})	Net radiative flux goes out of surface	$q = E - \alpha G$

13.9   EXTRATERRESTRIAL RADIATION

Figure 13.11 shows the spectral distribution of solar radiation. Earth's atmosphere is comprised of billions of suspended particles like sand, dust, smog, aerosol, and suspended water droplets. Solar radiation is scattered and diffused as it passes through these particles. Most of this scattering occurs in the blue colour band of visible spectrum and due to this we observe a blue sky in day light.

The drops in the spectral distribution of solar radiation coming to the earth are mainly due to the absorption by the gases like oxygen, ozone, carbon dioxide, and atmospheric moisture. The oxygen absorbs solar radiation in wavelengths close 0.76 μm, whereas the ozone absorbs most of the radiation lying in wavelengths 0.3 to 0.4 μm. That includes the absorption of ultraviolet radiation as well. The Ozone hole was taken seriously by scientific community and by the groups which believe that planet earth is going through a climatic change. Ozone hole is actually a region of rarefied ozone in the stratosphere over the continent Antarctica which appears at the beginning of southern hemisphere spring. It is also believed that chlorofluorocarbons (CFCs) might have caused the depletion in ozone layer. In the stratosphere, ultraviolet light can break the CFCs bonding and release the chlorine atoms (Cl) from the rest of the CFC molecule. A free chlorine atoms may participate in a series of chemical reactions which would affect ozone layer. On 16 September 1987, the international community signed a protocol to phased out the Hydrochlorofluorocarbons (HCFCs) – the gases used worldwide in refrigeration, air-conditioning, and foam applications.

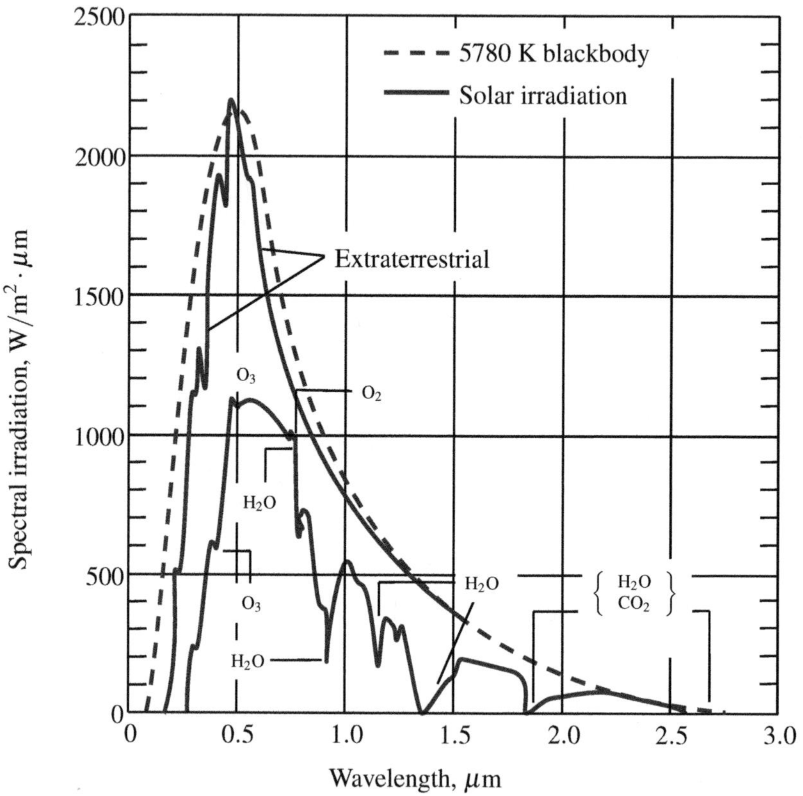

Figure 13.11   Spectral irradiation of solar radiation outside of earth's atmosphere.

The glass allows short wavelength solar radiation in the spectral region ($\lambda < 3$ μm) to enter the greenhouse, but does not permit long wavelength (medium to far infrared) radiation to leave. This is known as *Greenhouse effect*.

In radiative heat transfer, the term *Insolation* is ued to describe the intensity of direct solar radiation incident on a horizontal surface per unit area and per unit time, designated with the symbol I. Insolation is irradiation which is incident radiation on the Earth surface. The highest levels of insolation on Earth's surface are found to be in the regions known as the *doldrums* which lies along the equator.

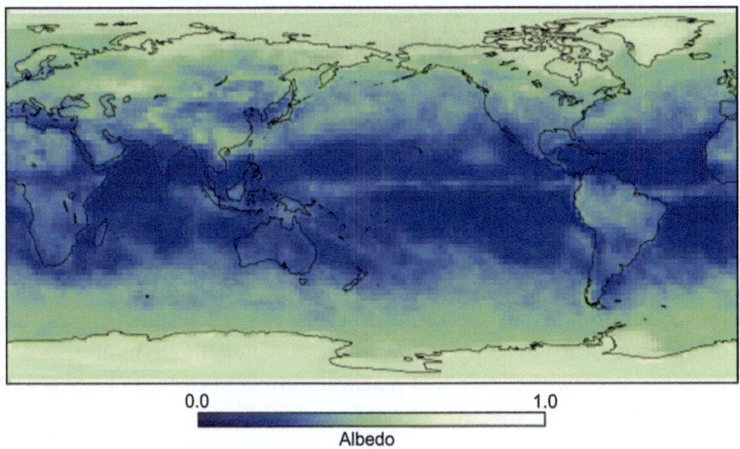

Figure 13.12 Earth's average albedo for March 2005, measured by the Clouds and Earth's Radiant Energy System (CERES) instrument aboard NASAs Terra satellite. (Image courtesy of NASA: The Earth Observatory.)

The term *Albedo* is used to represent the reflective properties of surfaces. Albedo is the fraction of Insolation or solar radiation that is reflected back into space by the Earth's surface. The term albedo is typically expressed as a percentage or in a decimal value (typically its varies from 0 when the surface absorbs all incoming radiation to a value of 1 when the surface reflects all incoming radiation). The global average albedo of the Earth is approximately 0.3, meaning that about 30% of incoming solar radiation is reflected back into space. The snowy or ice-covered regions near the poles have a much larger albedo value than the regions covered by the forest located near the equator or elsewhere. The average albedo of the entire continent of Antarctica during the austral summer (December-February) is approximately 0.8. The average albedo value for the of Borneo's rainforest during the month of June till August is around 0.13, indicating that about 13% of incoming solar radiation is reflected back into space. On average, the albedo of the open ocean is typically ranges from 0.06 to 0.08 (see Figure 13.12).

Note that the sky appears blue because of a phenomenon known as Rayleigh scattering, which happens when sunlight enters the Earth's atmosphere and interacts with gas molecules, such as oxygen and nitrogen. *Rayleigh (or molecular) scattering* is mainly caused by very small gas molecules. As sunlight enters the atmosphere, blue light is scattered in all directions by the gas molecules, giving the sky a blue appearance. The colour of the sky depends on the angle of the incoming solar radiation, the amount of atmosphere that the light must travel through, and the concentration of gas molecules in the air. *Mie scattering* happens when solar radiation strikes the dust or soot particles near atmosphere. During sunrise and sunset, the light has to pass through more atmosphere to reach us, causing more of the light to be scattered away, and giving the sky a reddish or orange hue.

In radiation heat transfer calculations, we consider the atmosphere as a blackbody with temperature that emits radiation energy equal to the emission by the CO_2 and H_2O molecules. This imaginary temperature is called the *effective sky temperature* (T_{sky}). The radiation emission from the atmosphere to the earth's surface is expressed as

$$G_{sky} = \sigma \cdot T_{sky}^4$$

The effective sky temperature value can lie in between 230 K and 285 K.

Although Mercury is the closest planet to the Sun in our solar system, its *insolation* is musch lower than that of Venus. Mercury orbit around sun is highly elliptical, which means that its distance from the Sun varies widely throughout its year. Apparently, much of the solar radiation that reaches Mercury's surface is quickly reflected back into space due to its thin atmosphere. Among the planets in our solar system, the one with the highest insolation is Venus. Venus is the second planet from the Sun, and it has a thick atmosphere that traps much of the incoming solar radiation, resulting in extremely high temperatures on its surface.

Example 13.5

A wall with a diffuse surface is maintained at 400 K. The spectral emissivity of the surface indicates the following emissivity when wall is exposed to radiative environment where temperature is maintained at 2000K.

$$\varepsilon_\lambda = \begin{cases} 0.3 & for\ 0\,\mu m < \lambda < 3\mu m \\ 0.7 & for\ \lambda > 3\mu m \end{cases}$$

(a) Find the emissive power $E_{b,\lambda}$ from the surface if it is assumed as a blackbody. (b) Find the net heat flux to the wall ?

Solution

The surface/wall temperature is T_w=400 K, and the surrounding temperature is T_{surr}=2000 K. From Wien Displacement law,

$$\lambda_{max} \cdot T = 2896.74\ \mu m \cdot K$$

the wavelength associated with maximum emission is

$$\lambda_{max} = \frac{2896.74}{400} = 7.241\mu m$$

The incident radiation (G) from surroundings on to the wall shall be absorbed as $\alpha \cdot G$, where $G = \sigma T_{surr}^4$. The emitted radiation from the wall shall be $\varepsilon \cdot E_{b,w}$, where $E_{b,w} = \sigma T_w^4$, and T_w is the wall temperature. This lead to net radiative balance as

$$q_{rad} = \alpha \cdot G - \varepsilon \cdot E_{b,wall}$$

This is the net heat flux to the wall.

We estimate the absorptivity at $T_{surr} = 2000$ K as follows with $F(3\mu m \cdot 2000) = 0.002134$:

$$\alpha = 0.3 \cdot (F(\lambda 1 \cdot T_{surr}) - F(0)) + 0.7 \cdot (1 - F(\lambda 1 \cdot T_{surr})) = 0.4048$$

and emissivity at $T_w = 400K$ as follows with $F(3\mu m \cdot 400) = 0.737818$:

$$\varepsilon = 0.3 \cdot (F(\lambda 1 \cdot T_w) - F(0)) + 0.7 \cdot (1 - F(\lambda 1 \cdot T_w)) = 0.699$$

This gives

$$q_{rad} = \alpha \cdot G_{b,surr} - \varepsilon \cdot E_{b,wall} = 3.66285 \times 10^5 W/m^2$$

PROBLEMS

13P-1 A rocket in space is flying between earth and sun at a distance of 160 million km from sun center. The sun radius is 696,000 km with temperature 6000 K. The rocket can be approximated as a cylinder with 20 m lendth and 2.2 m in diameter. Assuming both sun and rocket as black bodies find the rocket surface temperature. The applicale relation is

$$q_{sun-rocket} = q_{sun} \left(\frac{Rs}{R_{sr}} \right)^2$$

where, q is radiative flux, Rs is the radius of sun and Rsr is distance from sun core to the rocket surface.

[Ans: $T = 293.288$ K]

13P-2 A 500 W bulb is assumed to be emitting radiative flux like a sphere around it. The light flux in the visible band striking the floor 2 m directly below the bulb is 342×10^{-3} W/m^2. Assuming the bulb as a blackbody, find the light bulb temperature.

[Ans: $T = 2511K$]

13P-3 The solar flux coming to a glass window is 620.4 W/m^2. The reflectivity of the glass is 0.08 for all wave lengths. The transmissivity of the window is

$$\tau_\lambda = \begin{cases} 0.9 & for \ 0.35 \, \mu m < \lambda < 3\mu m \\ 0 & for \, other \, wavelengths \end{cases}$$

Find the component of radiation absorbed by the glass. Take the sun temperature as 5762 K.

[Ans: $q_{abs} = 63.1791$ W/m^2]

13P-4 Repeat problem **13P-3** with tinted glass having the transmissivity as

$$\tau_\lambda = \begin{cases} 0.9 & for\ 0.5\ \mu m < \lambda < 1.5\mu m \\ 0 & for\ other\ wavelengths \end{cases}$$

Find the component of absorbed radiation.

[Ans: q_{abs}=217.30 W/m^2]

13P-5 The spectral emissivity for an opaque surface at 1200 K is behaving as follows

$$\varepsilon_\lambda = \begin{cases} 0.4 & for\ 0\ \mu m < \lambda < 1.2\mu m \\ 0.6 & for\ 1.2\ \mu m < \lambda < 5\mu m \\ 0.2 & for\ 5\ \mu m < \lambda < \infty\mu m \end{cases}$$

Find the emissivity for all above ranges and the rate of radiation emission from the surface assuming opaque surface as a blackbody.

[Ans: $\varepsilon_{\lambda,T}$=0.493, 57.98 W/m^2]

13P-6 A solar collector is at T_c=450 K, and emissivity is 0.21. The absorptivity of collector is $\alpha = 0.823 \cdot \cos(\theta) + 0.011$, where, θ is the collector angle. With solar insolation of 1200 W/m^2, find the maximum ratiative heat transfer to the absorber. (ii) Find angle θ at which the radiative flux absorbed by collector is half of the maximum radiative flux.

[Ans: 512.53 W/m^2, θ=30.5°]

13P-7 What percentage of total radiative emission from the sun falls in the visible band region 380 nm $\leq \lambda \leq$ 760 nm? Take sun temperature as 5762 K.

[Ans: 44.65%]

13P-8 The emissivity of a surface coated with some metallic oxide is

$$\varepsilon_\lambda = \begin{cases} 0.23 & for\ 0\ \mu m < \lambda < 7\mu m \\ 0.85 & for\ 7\ \mu m < \lambda < \infty\mu m \end{cases}$$

Find the average emissivity of this surface at (a) 5000 K and (b) 500 K.

[Ans: (a) ε_{mean}=0.231 (b) ε_{mean} =0.6125]

13P-9 The spectral transmissivity of a glass used as a cover for a solar collector is

$$\varepsilon_\lambda = \begin{cases} 0.1 & for\ 0\ \mu m < \lambda < 0.3\mu m \\ 0.9 & for\ 0.3\ \mu m < \lambda < 7\mu m \\ 0.3 & for\ 7\ \mu m < \lambda < \infty\mu m \end{cases}$$

(a) Determine the average transmissivity of the absorber surface, (b) If the solar insolation incident on collector is 850 W/m^2, find the solar flux reaching to the absorber plate. Assume absorber plate as a blackbody and the sun temperature as 5800 K.

[Ans: (a) ε_λ=0.87273(b)741.824 W/m^2]

13P-10 The absorber plate of a solar collector is at temperature of 350 K and is coated with a layer having α_{ap}=0.85 and ε_{ap} = 0.095. The incident solar radiation is G=800 W/m^2. The sky temperature is 285 K and the ambient conditions are T$_\infty$=298 K,h$_\infty$=8 W/m^2 · K. Estimate the heat flux coming to the solar collector, the radiative heat loss, the convective heat loss and the net heat flux transferred to the absorber plate.

$$[\text{Ans: } q_{gain}=680 \text{ W/m}^2, q_{loss}= 461.3 \text{ W/m}^2]$$

13P-11 The gas turbine blade surface is expected to reach temperature of 1200 K while in operation. There are two ceramic coating in consideration: coating A and coating B having the following emissivities.

$$(\varepsilon_\lambda)_A = \begin{cases} 0.6 & 0\,\mu m \leq \lambda \leq 10\,\mu m \\ 0.2 & 10\,\mu m \leq \lambda \leq \infty\,\mu m \end{cases}$$
$$(\varepsilon_\lambda)_B = \begin{cases} 0.2 & 0\,\mu m \leq \lambda \leq 10\,\mu m \\ 0.6 & 10\,\mu m \leq \lambda \leq \infty\,\mu m \end{cases}$$

What coating will you recommend?

13P-12 An object is visible to eye in blue colour as it diffuse most of the radiation in wavelengths range associated with the blue colour. If the object temperature is approximated as 15 °C, how much is the percentage emittance from an object in blue colour band?

The wavelengths λ (nm) associated with different colours

Violet	380	424
Blue	424	491
Green	491	575
Yellow	575	585
Orange	585	647
Red	647	760

13P-13 Steel bar is going through annealing process to increase the ductility and reduce the hardness of it. At one stage, the steel becomes yellow and temperature recorded on steel surface is 2000 °F (1093.33 °C). Find the radiative flux emitted by steel, if steel has emissivity 0.94.

$$[\text{Ans: } E = 76.158 W/m^2]$$

13P-14 A polished gold sphere at 293 K (ε=0.025, c_p=129 J/kg·K, k=316 W/m·K, ρ=19300 kg/m^3) with gray surface behaviour has diameter 50 mm. The sphere is hanging in a room where walls are at 500 K, and the air between sphere and wall has convective environment T_∞=350 K and h_∞=20 W/m^2 · K. Find the net radiative heat transfer to sphere.

$$[\text{Ans: } q_{rad}=7.368 \text{ W/m}^2]$$

REFERENCES

M. F. Modest, Radiative Heat Transfer. New York: McGraw-Hill, USA, 1993.

M. Planck, The Theory of Heat Radiation, Dover Publications, Inc., New York, USA, 1959.

H. C. Hottel and A. F. Sarofin, Radiative Transfer, McGraw-Hill Book Co., Inc., New York, USA, 1967.

A. K. Oppenheim, Radiation Analysis by the Network Method, Trans. ASME, 78(4): 725-735, 1956.

E. Schmidt and E. R. G. Eckert, Über die Richtungsverteilung der Wärmestrahlung, Forsch. Gebiete Ingenieurw. 6,175, Germany, 1935.

14 Thermal Radiation Exchange between Surfaces

In this chapter, we will discuss the radiative exchange between surfaces at different orientations. The radiative exchange between surfaces depends on how the surfaces are placed with respect to each other. This aspects is mainly due to nature of the electromagnetic waves. In this chapter, we consider surfaces that transfer radiation to each other, and we assume that gases and liquids are acting transparent to thermal radiation. Hence, the emission and absorption of thermal radiation occur without the involvement of gas or liquid present between surfaces. After finishing this chapter, one will be able to:

The learning outcomes:
- Define the solid angle.
- Compute the view factors for some configurations.
- Estimate the radiation exchange within an enclosure of black and gray surfaces.

The radiative exchange between surfaces cannot be calculated unless we discuss the concept of the view factor or radiative shape factor. We will also discuss the radiative exchange between surfaces that are shaping an enclosure. For radiative exchange with gray surfaces, we will restrict our analysis only to opaque, and diffuse surfaces.

14.1 SOLID ANGLE

When radiative emission from one surface enters into another medium the energy flux has different strength levels depending on the orientation of the surface. Likewise, the photons or the electromagnetic wave going through any point inside the medium can change with direction orientation. To overcome this issue, a direction vector in terms of a spherical polar coordinate system is usually defined. To explain this further, we consider an opaque surface dA, with point **P** on the surface, as shown in Figure 14.1. The surface can radiate in any direction with the rays penetrating through the hemisphere of the unit radius. Figure 14.1 also shows the emission direction as well as the solid angles as related to a unit hemisphere.

DOI: 10.1201/9781003428404-14

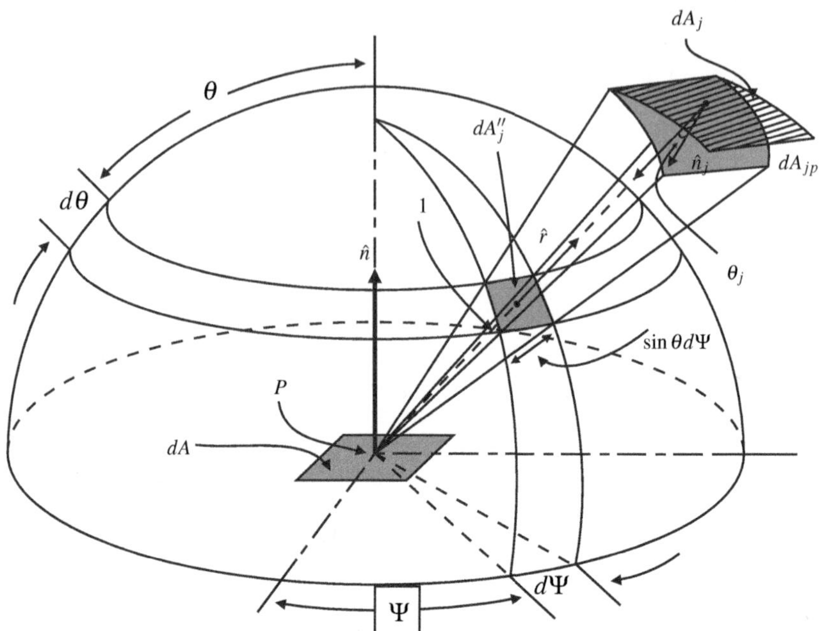

Figure 14.1   Solid angle representation in the spherical coordinate system. ψ is an azimuthal angle and θ is a polar angle.

We can express the arbitrary emission direction from the surface (shown by a unit vector $\hat{r}$) in terms of the polar angle θ, and the azimuthal angle ψ. Note that the polar angle θ is measured from the surface normal vector $\hat{n}$ and azimuthal angle ψ is measured between an arbitrary axis on the surface and the projection of $\hat{r}$ onto the surface. The total solid angle above the surface is the surface area of the hemisphere, which is equivalent to 2π. The azimuthal angle ψ is defined as an angle between an arbitrary axis on the surface and the projection of the unit direction vector onto the surface. It can be observed that for a hemispherical coordinate system, the polar (θ) varies from 0 to $\pi/2$ and the azimuthal angle (ψ) varies from 0 to 2π.

The solid angle is equal to the projected area, if the surface is projected onto the unit hemisphere above the point, and this indicates that an infinitesimal solid angle is simply an infinitesimal area on a unit sphere. The solid angle may vary between 0 and 2π, measured in dimensionless *steradians*, **sr**, equivalent to the surface area on a hemisphere. In spherical coordinates (r, θ,ψ), we have

$$d\Omega = \frac{dA_{jp}}{r^2} = \frac{\cos\theta_j dA_j}{r^2} = dA''_j$$

$$d\Omega = dA''_j$$

$$d\Omega = \sin\theta d\theta d\psi$$

Integrating in all directions, we have the total solid angle above the surface

$$\int_{\psi=0}^{2\pi} \int_{\theta=0}^{\pi/2} \sin\theta d\theta d\psi = 2\pi$$

The solid angle, with which a finite surface A_j is seen from point P, is

$$\Omega = A''_j$$

The solid angle, with which a finite surface A_j is seen from point P, i.e., the projection of A_j onto the hemisphere above P.

> **Solid Angle**
>
> A solid angle can be represented as a segment of the space inside a sphere encompassed by a conical surface with the vertex of the cone at the center of the sphere.

The emission of radiation from an area can be categorised into following four radiation quantities [see Baehr and Stephan (2011)]:

i The spectral intensity $I_\lambda(\lambda, \Psi, \theta, T)$ represents the distribution of the emitted radiation over the spectrum of wavelengths and the solid angles of the hemisphere.

ii The hemispherical spectral emissive power $E_\lambda(\lambda, T)$ indicates the dependence of of the radiated energy in the entire hemisphere on the wavelengths.

iii The total intensity $I(\Psi, \theta, T)$ represents the dependence the radiated energy on direction at all wavelengths. It thus provides the radiative energy distribution over the solid angles of the hemisphere.

iv The emissive power $E(T)$ represents the radiative flux emitted at all wavelengths and in the entire hemisphere, and hence, it is a hemispherical total quantity.

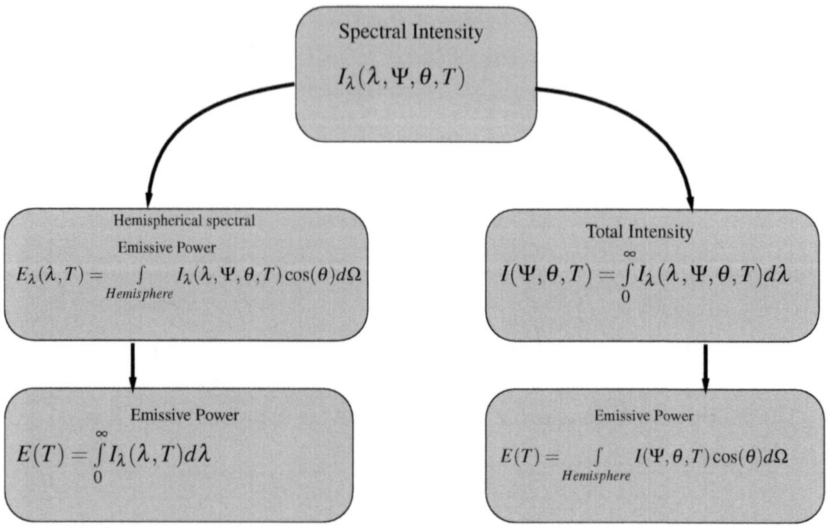

Lambert Cosine Law

Johann Heinrich Lambert (1728–1777) proposed that the directional emitted flux of a blackbody varies with the cosine of the polar angle. The net heat flux from the surface may be calculated as

$$q_{\lambda,net} = \int\limits_{4\pi} I_\lambda(\hat{r})\cos\theta d\Omega$$

where $\hat{r}$ is a single direction vector and it describes the total range of solid angles, i.e. 4π. We integrate this equation over the spectrum to obtain the total radiative heat flux at the surface

$$q.\hat{n} = \int_0^\infty \int_{4\pi} I_\lambda(\hat{r})\hat{n}.\hat{r}\ d\Omega d\lambda$$

14.2 VIEW FACTOR

Consider Figure 14.2 where radiation exchange is happening between surfaces.
In previous section, we developed the relation for radiative heat flux:

$$dq_{\lambda,net} = I_\lambda \cos\theta d\Omega$$

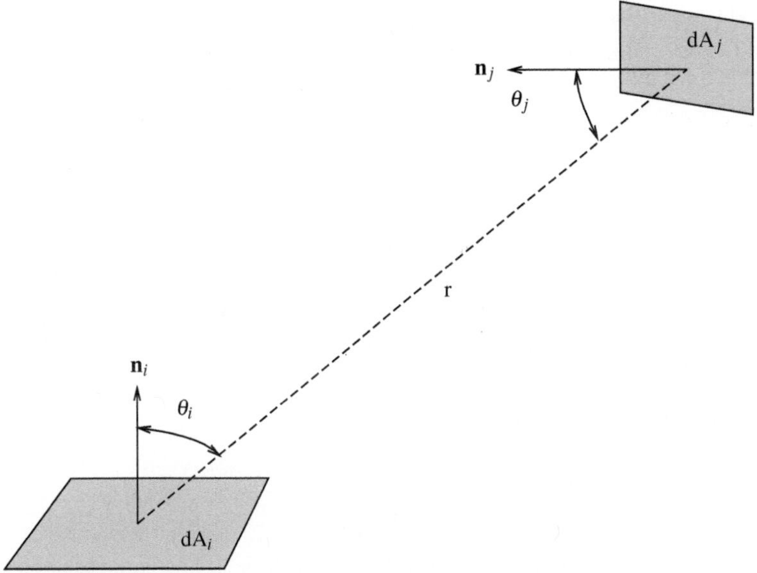

Figure 14.2 Radiation exchange between infinitesimally small surfaces.

The heat transfer associated with surface

$$dQ_{\lambda,net} = I_\lambda \cos\theta \, dA \, d\Omega \qquad (14.1)$$

The solid angle is

$$d\Omega = \frac{\cos\theta_j \, dA_j}{r^2}$$

The radiation intensity for emitted radiation I is defined as the rate at which radiation energy is emitted in the (θ, Ψ) direction per unit area normal to this direction and per unit solid angle about this direction. The rate at which radiative energy leaves a unit area of a surface in all directions is called **radiosity** (J), and for diffused surface we can write

$$J = \int_{\theta=0}^{\pi/2} \int_{\psi=0}^{2\pi} I \cos\theta \cos\Psi \, d\theta \, d\Psi = \pi \cdot I_\lambda \qquad (14.2)$$

For ith surface we may write

$$J_i = \pi \cdot I_{\lambda,i} \qquad (14.3)$$

and since a blackbody absorbs the entire radiation incident on it we can conclude that for a blackbody, radiosity J is equals to the emissive power of the blackbody (E_b) and hence for a blackbody $J=E_b$. Inserting I_λ from 14.3 into equation 14.1 and , we have

$$dQ_{\lambda,net} = I_\lambda \cos\theta dA \left(\frac{\cos\theta_j dA_j}{r^2} \right)$$

The radiation exchange between i and j surfaces is

$$Q_{i\to j} = \left(\frac{J_i}{\pi} \right) \int_{A_i} \int_{A_j} \left(\frac{\cos\theta_j \cos\theta_i}{r^2} \right) dA_j dA_i$$

We can cast this equation into fraction of the radiation that leaves the surface A_i and is intercepted by surface A_j is

$$F_{ij} = \frac{Q_{i\to j}}{A_i J_i} = \left(\frac{1}{\pi A_i} \right) \int_{A_i} \int_{A_j} \left(\frac{\cos\theta_j \cos\theta_i}{r^2} \right) dA_j dA_i$$

The fraction of the radiation that leaves the surface A_i and is intercepted by surface A_j is also called the view factor.

The view factor for radiation transfer from surface i to j is

$$F_{ij} = \left(\frac{1}{\pi A_i} \right) \int_{A_i} \int_{A_j} \left(\frac{\cos\theta_j \cos\theta_i}{r^2} \right) dA_j dA_i$$

Similarly, the view factor for radiation transfer from surface j to i is

$$F_{ji} = \left(\frac{1}{\pi A_j} \right) \int_{A_i} \int_{A_j} \left(\frac{\cos\theta_j \cos\theta_i}{r^2} \right) dA_j dA_i$$

Rearranging above equations, we can have

$$A_i F_{ij} = A_j F_{ji}$$

In the following tables, we will list the view factors for different geometric configurations. In literature, one may find view factors for more complex geometries. Here in tables only few are listed to make reader familiar with their usage.

Example 14.1

Consider a circular disk of Radius R with area A_j, facing a surface of area A_i, and positioned at a gap of ℓ from the center of A_j as shown in Figure 14.3. Assuming that surfaces are diffuse, derive an expression for the view factor F_{ij}.

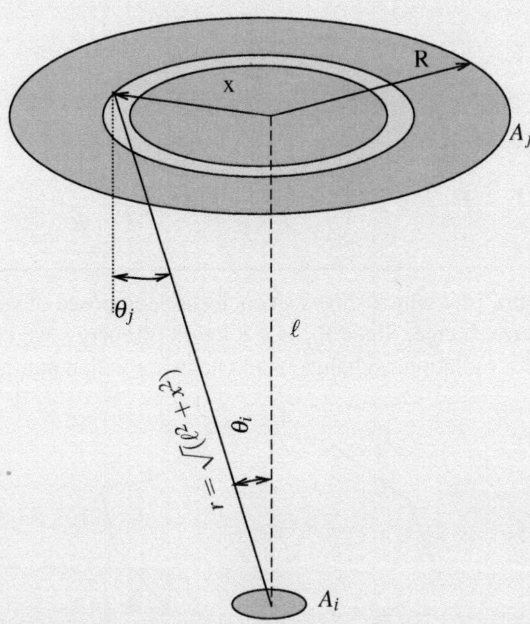

Figure 14.3   Two diffused surfaces placed parallel to each other.

Solution

We start from equation

$$F_{ij} = \left(\frac{1}{\pi A_i}\right) \int_{A_i} \int_{A_j} \left(\frac{\cos\theta_j \cos\theta_i}{r^2}\right) dA_j dA_i$$

Assuming that area A_i is not changing with ℓ we have

$$F_{ij} = \left(\frac{1}{\pi A_i}\right) \left[\int_{A_j} \left(\frac{\cos\theta_j \cos\theta_i}{r^2}\right)\right] dA_j A_i$$

and

$$F_{ij} = \int_{A_j} \left(\frac{\cos\theta_j \cos\theta_i}{\pi r^2}\right) dA_j$$

From the figure it is clear that

$$\theta_j = \theta_i = \theta$$

$$dA_j = 2\pi x\,dx$$

$$r = \sqrt{x^2 + \ell^2}$$

$$\cos\theta = \frac{\ell}{\sqrt{x^2 + \ell^2}}$$

This leads to

$$F_{ij} = 2\int_0^R \left(\frac{\ell^2}{(x^2 + \ell^2)^2}\right)x\,dx = \frac{R^2}{\ell^2 + R^2}$$

Consider Figure 14.4 which shows an enclosure composed of walls that are having radiation interexchange. Since F_{ij} is a fraction of energy, we can write the following equality for radiation exchange in an enclosure with n number of surfaces:

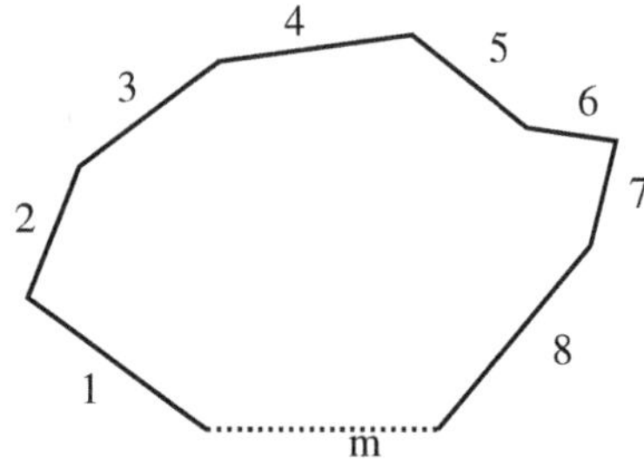

Figure 14.4   Radiation in enclosure.

$$\boxed{\sum_{n=1}^{m} F_{1n} = 1} \tag{14.4}$$

$$F_{11} + F_{12} + F_{13} + \cdots F_{1m} = 1$$

Let say there are two surfaces, i and j which are exchanging radiation as

$$\boxed{A_i F_{ij} = A_j F_{ji}} \tag{14.5}$$

The above relation is known as **Reciprocity Relation** in heat transfer literature. For two interacting surfaces, we may write:

$$A_1 F_{12} = A_2 F_{21}$$

If we divide the are into multiple areas still the relationship of view factor are applicable as for the case shown in Figure 14.5

$$F_{12}A_1 = F_{21}A_2$$

$$F_{12}A_1 = F_{1-2a}A_1 + F_{1-2b}A_1$$

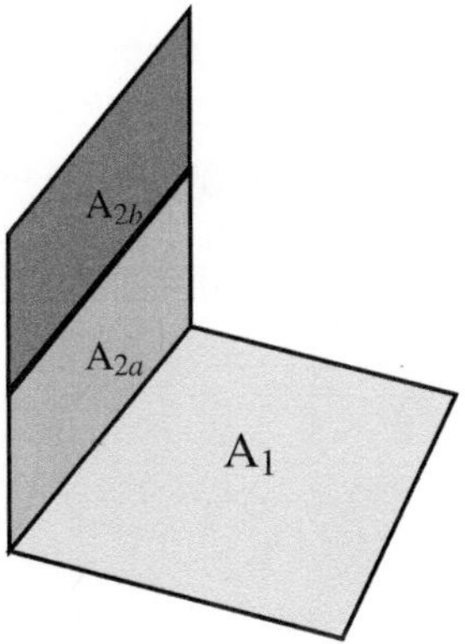

Figure 14.5   Radiation on subdivision of the receiving surface.

14.2.1 HOTTEL'S STRING RULE

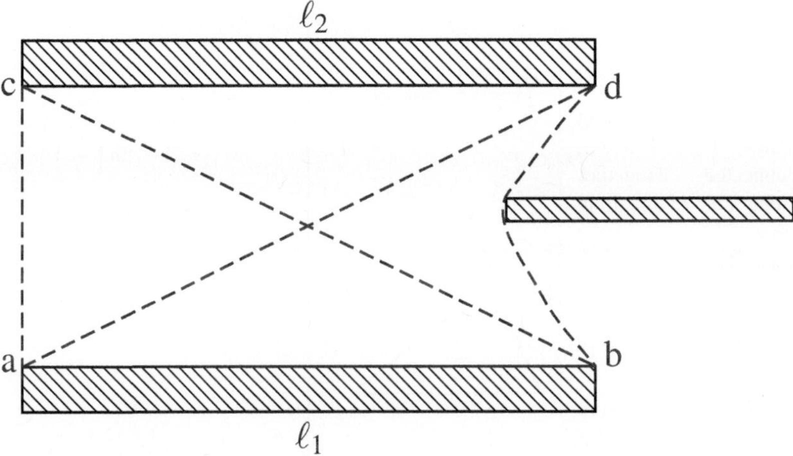

Figure 14.6   Hottel's string rule for view factor determination for a 2D geometric configurations.

Hottel and Sarofim (1967) proposed a simple relation for view factor calculation, although it has gained popularity as Hottel's method. According to his procedure, we just need to measure the string lengths $\overline{ad}$, $\overline{bc}$, $\overline{ac}$, and $\overline{bd}$:

$$F_{12} = \frac{1}{2\ell_1}\left[\overline{ad} + \overline{bc} - \overline{ac} - \overline{bd}\right]$$

The lengths in this equation are depicted in Figure 14.6 for Hottel's string rule.

The view factor, which is also referred to as radiation shape and geometric factors is tabulated in Table 14.1 to 14.3.

The radiation view factor for some commonly occuring geometrical configurations are plotted in Figure 14.7 to 14.10.

TABLE 14.1

View Factor for Some Geometric Configurations (also Called Radiation Shape and Geometric Factors)

Description	Geometric Configuration	View Factor
Case I Infinitely long plates sharing one common edge at right angle		$F_{12} = \frac{1}{2}\left[1 + \chi - \sqrt{(1+\chi^2)}\right]$ where $\chi = h/L$
Case II Two infinitely long plates of equal width, connected together along one of the long edge		$F_{12} = 1 - \sin\left(\dfrac{\theta}{2}\right)$
Case III Triangular enclosure formed by three infinitely long plates of different widths		$F_{12} = \dfrac{\ell_1 + \ell_2 - \ell_3}{2\ell_1}$
Case IV Two parallel co-axial plates		$F_{12} = \frac{1}{2}\left\{ S - \sqrt{S^2 - 4(x\cdot y)^2} \right\}$ $x = L/r_1$ $y = r_2/L$ $S = 1 + x^2(y^2 + 1)$

TABLE 14.2

View Factor for Some Geometric Configurations (also Called Radiation Shape and Geometric Factors)

Description	Geometric Configuration	View Factor
Case V Coaxial parallel discs of equal radius	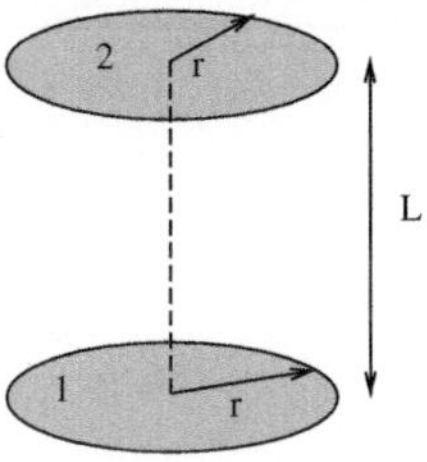	$F_{12} = 1 + \left(\dfrac{L^2}{2r^2}\right)\left[1 - \sqrt{\left(1 + \dfrac{4r^2}{L^2}\right)}\right]$
Case VI Sphere near a coaxial disc	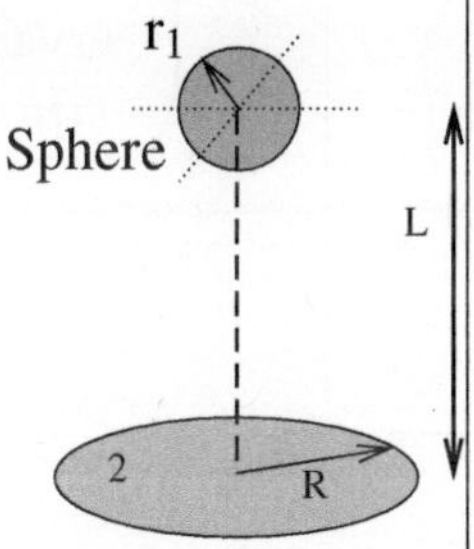	$F_{12} = \dfrac{1}{2}\left[1 - \dfrac{L}{\sqrt{R^2 + L^2}}\right]$
Case VII Long cylinder placed close to a large plane area	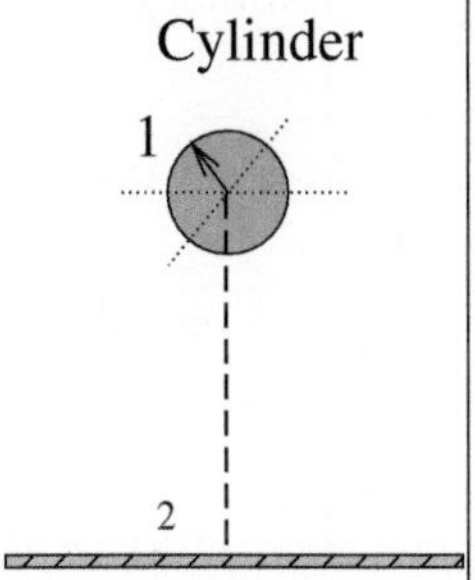	$F_{12} = 0.5$

TABLE 14.3

View Factor for Some Geometric Configurations (also Called Radiation Shape and Geometric Factors)

Description	Geometric Configuration	View Factor
Case VIII Parallel Aligned Rectangles		$F_{12} =$ $$\frac{2}{\pi x.y}\left(\begin{array}{l}\dfrac{\ln\left(\dfrac{(x^2+1)(y^2+1)}{x^2+y^2+1}\right)}{2}\\[6pt] +x\sqrt{y^2+1}\,\arctan\left(\dfrac{x}{\sqrt{y^2+1}}\right)\\[6pt] +y\sqrt{x^2+1}\,\arctan\left(\dfrac{y}{\sqrt{x^2+1}}\right)\\[6pt] -x\arctan(x)-y\arctan(y)\end{array}\right)$$
Case IX Row of infinite number of cylinders placed parallel to an infinity long plate		$F_{12} = 1 + \chi\arctan\left(\sqrt{\dfrac{1-\chi^2}{\chi^2}}\right) - \sqrt{[1-\chi^2]}$ where, $\chi = D/p$
Case X Area inside a sphere		$F_{12} = \dfrac{A_2}{4\pi R^2}$

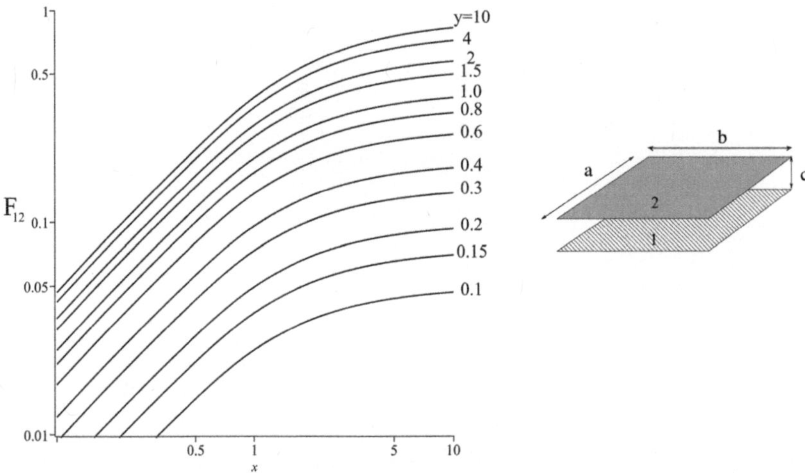

Figure 14.7 Radiation view factor for parallel rectangular plates; x=a/c and y=b/c.

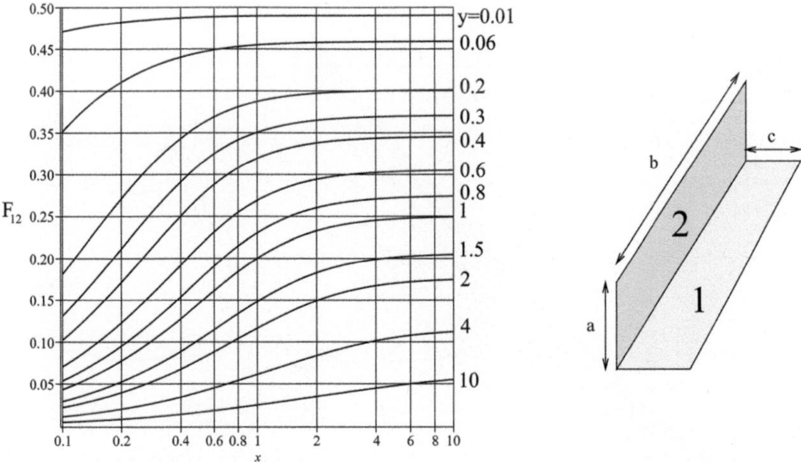

Figure 14.8 Radiation view factor for rectangular plates with common edge; x=a/b and y=c/b.

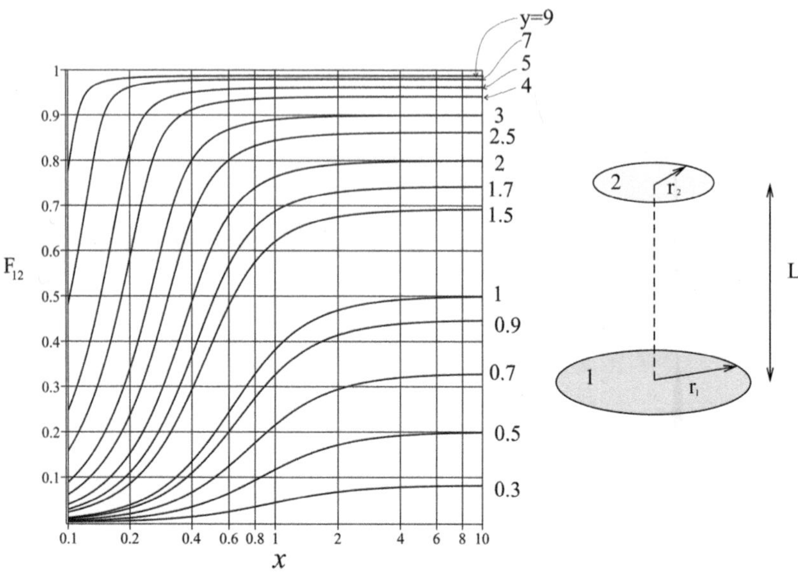

Figure 14.9   Radiation view factor for two parallel co-axial plates; $x=L/r_1$ and $y=r_2/L$.

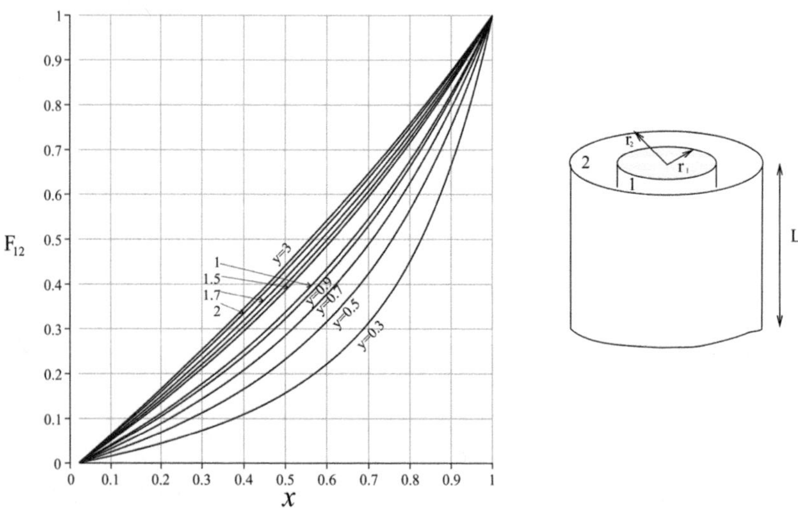

Figure 14.10   Radiation view factor for concentric cylinders; $x=r_1/r_2$ and $y=L/r_2$.

14.3 RADIATION EXCHANGE WITHIN AN ENCLOSURE OF BLACK SURFACES

Figure 14.11 shows an enclosure comprised of a number of blackbody surfaces. Assuming that the a surface k, with area A_k and temperature T_k is exchanging radiative energy with other m surfaces in an enclosure ($m=1$ to n). The corresponding areas are A_1 to A_m with temperatures T_1 to T_m. Further, the kth surface also receives radiative thermal energy from all other surfaces in an enclosure which are emitting energy as a backbody.

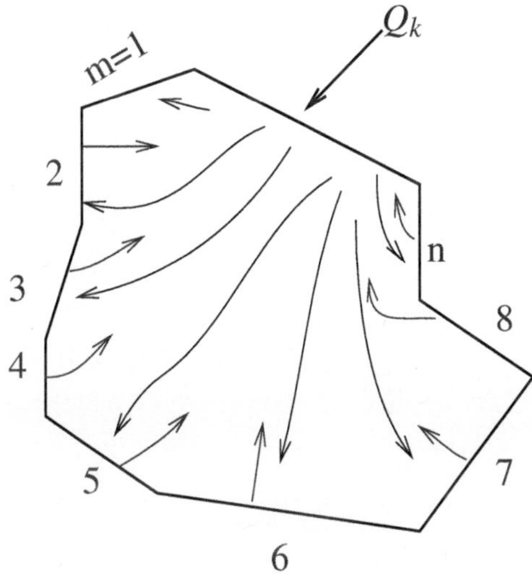

Figure 14.11   Radiative exchange in an isothermal enclosure comprised of n surfaces. Q_k is the energy received by kth surface from an external source.

Note that the Q_k indicated in Figure 14.11 is the energy received by kth surface from an external source to maintain its temperature at T_k. For an enclosure, view factors holds the relation:

$$F_{k-1} + F_{k-2} + F_{k-3} + \cdots + F_{k-n} = \sum_{m=1}^{n} F_{k-m} = 1$$

The radiative energy incident on area A_k, from the other surfaces is

$$F_{1-k}\sigma T_1^4 A_1 + F_{2-k}\sigma T_2^4 A_2 + F_{3-k}\sigma T_3^4 A_3 + \cdots + F_{k-k}\sigma T_k^4 A_k \cdots + F_{n-k}\sigma T_n^4 A_n$$

The term $F_{k-k}\sigma T_k^4 A_k$ is indicating that k can be a concave surface as well. We can sum up the above equation for the energy received from enclosing surfaces as

$$E_{from\ surfaces} = \sum_{m=1}^{n} F_{m-k} \cdot \sigma \cdot T_m^4 A_m$$

$$\left\{ \begin{array}{c} Energy\ emitted \\ by\ kth\ surface \end{array} \right\} = \sigma T_k^4 A_k$$

In case of steady state, we may write energy balance as

$$\left\{ \begin{array}{c} Energy\ from \\ External\ source\ on\ kth\ surface \end{array} \right\} + \left\{ \begin{array}{c} Energy\ coming\ from \\ n\ surfaces\ in\ enclosure \end{array} \right\}$$

$$= \left\{ \begin{array}{c} Energy\ emitted \\ by\ kth\ surface \end{array} \right\}$$

$$Q_k + \sum_{m=1}^{n} F_{m-k}\sigma T_m^4 A_m = \sigma T_k^4 A_k$$

We can write the equation again after multiplying 1 with RHS of above equation:

$$Q_k + \sum_{m=1}^{n} F_{m-k}\sigma T_m^4 A_m = \sigma T_k^4 A_k\,[1]$$

Since,

$$\sum_{m=1}^{n} F_{k-m} = 1$$

We introduce above expression in place of one:

$$Q_k + \sum_{m=1}^{n} F_{m-k}\sigma T_m^4 A_m = \sigma T_k^4 A_k \left[\sum_{m=1}^{n} F_{k-m} \right]$$

Rearranging we have

$$Q_k = \sum_{m=1}^{n} F_{k-m}\sigma T_k^4 A_k - \sum_{m=1}^{n} F_{m-k}\sigma T_m^4 A_m$$

Bringing the two terms on RHS into one summation sign, we have a form

$$\boxed{Q_k = \sigma A_k \sum_{m=1}^{n} F_{k-m}(T_k^4 - T_m^4)} \tag{14.6}$$

For an enclosure of n surfaces, we will have a number of algebraic nonlinear equations which must be solved simultaneously to obtain temperature. For an enclosure with three surfaces, we have equation for k=1, and m=2,3:

$$Q_1 = \sigma A_1 \left(F_{1-2}(T_1^4 - T_2^4) + F_{1-3}(T_1^4 - T_3^4) \right)$$

Example 14.2

Figure 14.12 shows a triangular enclosure formed with three surfaces. One side of the enclosure is insulated. The surface 1 is heated and is maintained at temperature of 1500 °R. The surface 2 is insulated and surface 3 is maintained at temperature of 600 °R. Determine the temperature of the insulated surface and the heat exchanges among the surfaces.

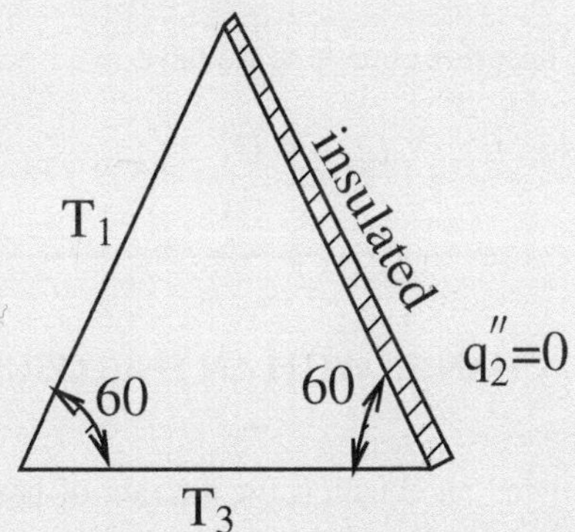

Figure 14.12   Radiative exchange in an isothermal enclosure comprised of three surfaces.

Solution

The three surfaces are involved so the generalised relation for heat transfer is

$$Q_k = \sigma A_k \left(\left(-T_1^4 + T_k^4 \right) F_{k,1} + \left(-T_2^4 + T_k^4 \right) F_{k,2} + \left(-T_3^4 + T_k^4 \right) F_{k,3} \right)$$

where, k=1,2,3. The Stefan-Boltzmann constant in USCU units is

$$\sigma = 1.71344 \times 10^{-9}$$

For the enclosure, the exchange of energy is among themselves we can write for view factors

$$F_{1,3} = F_{2,3} = F_{3,1} = 0.5$$
$$F_{1,2} = F_{2,1} = 0.5$$

Now for surface 1 energy balance is

$$\frac{Q_1''}{A_1} = 8563.259088 - 8.56720 \times 10^{-10} T_2^4$$

for insulated, surface 2 energy balance is

$$\frac{Q_2''}{A_2} = 1.713440 \times 10^{-9} T_2^4 - 4448.175912 = 0$$

Solving we have

$$T_2 = 1269.34°R$$

For surface 3 energy balance is

$$\frac{Q_3''}{A_3} = -4115.083176 - 8.56720 \times 10^{-10} T_2^4$$

Substituting T_2 into heat flux equations we have

$$\frac{Q_1''}{A_1} = -6339.17 W/m^2, \frac{Q_3''}{A_3} = 6339.17 W/m^2$$

14.4   RADIATIVE EXCHANGE WITH AN ENCLOSURE OF GRAY SURFACES

We have already formulated an equation for the rate of heat transfer from a gray surfaces (see equation 13.5). For the gray surfaces enclosure, we can repeat the above analysis. For kth gray surface, radiative heat transfer with other gray surfaces in enclosure is

$$Q_k = \left[\frac{E_{bk} - J_k}{A_k \left(\frac{1-\varepsilon_k}{\varepsilon_k} \right)} \right]$$

$$Q_k = A_k \left[J_k - \sum_{m=1}^{n} J_m F_{k-m} \right]$$

If in an enclosure there are only three surfaces which are exchanging thermal radiation, then we have heat transfer from surface 1, 2, and 3 as follows:

For surface k=1:

$$Q_1 = \left[\frac{E_{b1} - J_1}{\frac{1}{A_1} \left(\frac{1-\varepsilon_1}{\varepsilon_1} \right)} \right] \tag{14.7}$$

$$Q_1 = A_1 \left[J_1 - (J_1 F_{1-1} + J_2 F_{1-2} + J_3 F_{1-3}) \right] \tag{14.8}$$

For surface k=2:

$$Q_2 = \left[\frac{E_{b2} - J_2}{\frac{1}{A_2} \left(\frac{1-\varepsilon_2}{\varepsilon_2} \right)} \right] \tag{14.9}$$

$$Q_2 = A_2 \left[J_2 - (J_1 F_{2-1} + J_2 F_{2-2} + J_3 F_{2-3}) \right] \tag{14.10}$$

For surface k=3:

$$Q_3 = \left[\frac{E_{b3} - J_3}{\frac{1}{A_3}\left(\frac{1-\varepsilon_3}{\varepsilon_3}\right)} \right] \tag{14.11}$$

$$Q_3 = A_3 \left[J_3 - (J_1 F_{3-1} + J_2 F_{3-2} + J_3 F_{3-3}) \right] \tag{14.12}$$

The equation from to can solved simultaneously to get Q_1, Q_2, Q_3, J_1, J_2, and J_3. We can also use the thermal network to solve this problem.

14.5 HEAT EXCHANGE WITH INSULATED SURFACE AND GRAY SURFACES WITH LARGE AREAS

The (E_b-J) represents the potential difference for radiative heat transfer through the surface resistance. In case if one of the surfaces is perfectly insulated, or re-radiates all the energy incident upon it, it has zero heat flow. For such insulating surface, the potential difference across the surface resistance is zero, which gives J = E_b. For three surfaces, we can represent it in a network diagram shown in Figure 14.13.

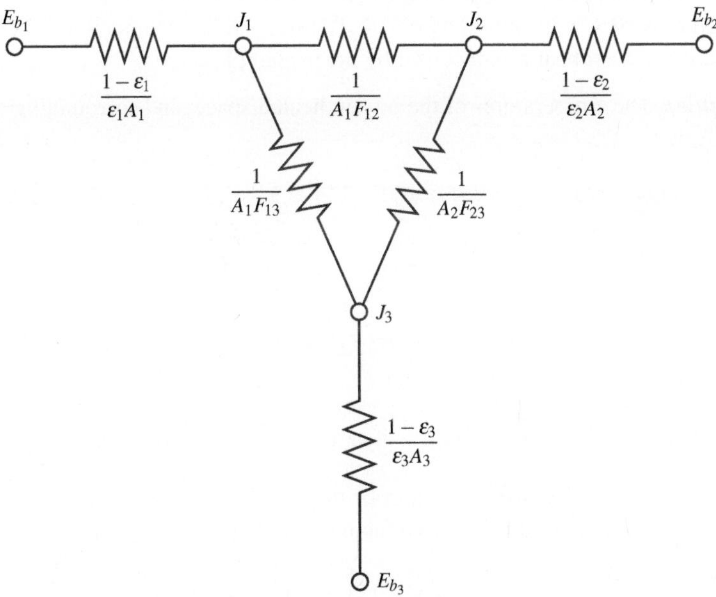

Figure 14.13 Resistive circuit for a three gray surface enclosure.

Thermal network for a gray surface enclosure with one surface insulated is shown in Figure 14.14.

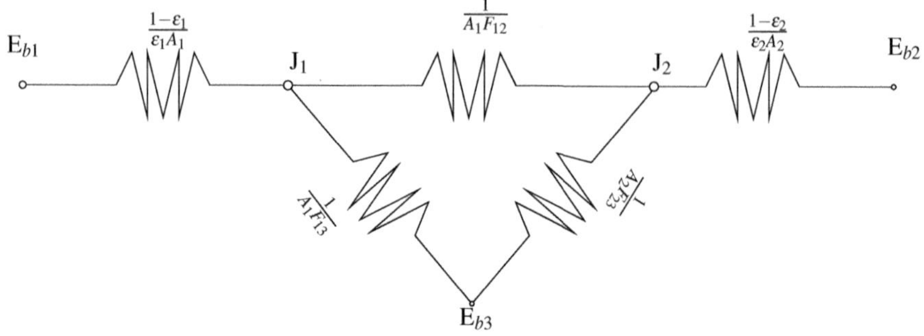

Figure 14.14   Resistive circuit for a gray surface enclosure with one surface insulated.

Example 14.3

A circular heating plate (radius = 1m) is used to heat a circular disk of radius 2 m. The gap between the plates is 4 m. The heater plate is maintained at constant temperature of 800°C with emissivity 0.92. The heated plate temperature is 80°C with emissivty of 0.4. The surrounding temperature is 25°C. Find the radiant heat flux received by the heated plate.

Solution The temperatures of the heater, heated space, and surrounding are

$$T_1 = 800 + 273 = 1073K, T_2 = 80 + 273 = 353K, T_3 = 25 + 273 = 298K$$

The blaclboday emissions are

$$E_{b1} = \sigma \cdot T_1^{4} = 75159.16 \quad E_{b2} = \sigma \cdot T_2^{4} = 880.40 \quad E_{b3} = \sigma \cdot T_3^{4} = 447.14$$

$$r_1 = 1m, \ r_2 = 2\,m, L = 4\,m, x = \frac{L}{r_1} = 4, y = \frac{r_2}{L} = 0.5$$

We can find view factor from chart or directly compute it. Here in solution we directly compute it for the two facing plates case:

$$S = 1 + x^2 \cdot \left(1 + y^2\right) = 21$$

$$F_{12} = \left(\frac{1}{2} \cdot \left(S - \sqrt{\left(S^2 - 4 \cdot (x \cdot y)^2\right)} \right) \right) = 0.1922$$

The area of heater and heated plate is

$$A_1 = \pi \cdot r_1^{2} = \pi \quad A_2 = \pi \cdot r_2^{2} = 4\pi$$

Using view factor algebra

$$\cancel{F_{11}} + F_{12} + F_{13} = 1$$
$$F_{13} = 1 - F_{12} = 0.8078$$
$$F_{21} = \left(\frac{A_1}{A_2}\right) \cdot F_{12} = 0.048$$
$$\cancel{F_{22}} + F_{21} + F_{23} = 1$$
$$F_{23} = 1 - F_{21} = 0.9519$$

Energy balance at node 1, node 2, and node 3 gives

Node 1

$$\frac{(E_{b1}-J_1)}{\frac{(1-\varepsilon_1)}{\varepsilon_1 \cdot A_1}} + \frac{(J_2-J_1)}{\left(\frac{1}{(A_1 \cdot F_{12})}\right)} + \frac{(J_3-J_1)}{\left(\frac{1}{(A_1 \cdot F_{13})}\right)} = 0$$

$$eq1 := 2.716 \times 10^6 - 39.269 J_1 + 0.6220 J_2 = 0$$

Node 2

$$\frac{(E_{b2}-J_2)}{\frac{(1-\varepsilon_2)}{\varepsilon_2 \cdot A_2}} + \frac{(J_1-J_2)}{\left(\frac{1}{(A_1 \cdot F_{12})}\right)} + \frac{(J_3-J_2)}{\left(\frac{1}{(A_2 \cdot F_{23})}\right)} = 0$$

$$eq2 := 12716.49 - 20.943 J_2 + 0.622 J_1 = 0$$

Node 3

$$\frac{(E_{b3}-J_3)}{\frac{(1-\varepsilon_3)}{\varepsilon_3 \cdot A_3}} + \frac{(J_1-J_3)}{\left(\frac{1}{(A_1 \cdot F_{13})}\right)} + \frac{(J_2-J_3)}{\left(\frac{1}{(A_2 \cdot F_{23})}\right)} = 0$$

Taking $J_3 = E_{b3}$ we have

$$eq3 := -6467.452 + 2.519 J_1 + 11.944 J_2 = 0$$

Using Maple to solve the equations 1 and 2, we get

```
solve([eq1, eq2], {J1, J2})
```

$$J_1 = 18264.87666, J_2 = 1148.990690$$

$$Q_1 = \frac{(E_{b1}-J_1)}{\frac{(1-\varepsilon_1)}{\varepsilon_1 \cdot A_1}} = 2.0554 \times 10^6$$

14.6 RADIATIVE EXCHANGE WITH SEMI-INFINITE GRAY SURFACES

Consider the case of radiative exchange from two semi-infinite flat parallel gray surfaces which are identified as surface 1 and surface 2. The areas of surfaces are A_1 and A_2.

The radiation leaving surface 1 is $\sigma T^4 A \varepsilon_1 F_{12}$, and the absorption at surface 2 is $\left(\sigma T_1^4 A \varepsilon_1 F_{12}\right) \alpha_2$. Assuming for first incident radiation $\alpha_2 = \varepsilon_2$, we have $\left(\sigma T_1^4 A \varepsilon_1 F_{12}\right) \varepsilon_2$. Also, the part of reflected radiation is $\left(\sigma T^4 A \varepsilon_1 F_{12}\right) \rho_2$. The component of reflected energy $\left(\sigma T_1^4 A \varepsilon_1 F_{12}\right) \rho_2$ shall be returned back to surface 1, where some of it will be absorbed and a second reflection of energy will happen. The reflected component shall be the second incident

$$\left(\sigma T_1^4 A \varepsilon_1 F_{12} \rho_2\right) \rho_1$$

For sake of simplification, we take $F_{12}=1$.

$$\left(\sigma T_1^4 A \varepsilon_1 \rho_2\right) \rho_1$$

This energy on reaching surface 2 shall be be absorbed and reflected again:

$$\left(\sigma T_1^4 A \varepsilon_1 \rho_2 \rho_1\right) \alpha_2 = \left(\sigma T_1^4 A \varepsilon_1 \rho_2 \rho_1\right) \varepsilon_2 = \sigma T_1^4 A \cdot \varepsilon_1 \varepsilon_2 \cdot \rho_1 \rho_2$$

We can infer that with multiple reflections, we have

$$Q_1 = \sigma T_1{}^4 A \cdot \varepsilon_1 \varepsilon_2 \left[1 + \rho_1 \rho_2 + (\rho_1 \rho_2)^2 + \cdots + (\rho_1 \rho_2)^n\right]$$

$$Q_2 = \sigma T_2{}^4 A \cdot \varepsilon_1 \varepsilon_2 \left[1 + \rho_1 \rho_2 + (\rho_1 \rho_2)^2 + \cdots + (\rho_1 \rho_2)^n\right]$$

Using geometric progression formula, we reduce the bracketed term into form

$$\left[1 + \rho_1 \rho_2 + (\rho_1 \rho_2)^2 + \cdots + (\rho_1 \rho_2)^n\right] = \frac{1}{1 - \rho_1 \rho_2} \tag{14.13}$$

For opaque surfaces $\alpha + \rho = 1$, assuming absorptivity (α) equals to emissivity (ε), we have

$$\rho = 1 - \alpha = 1 - \varepsilon$$

We introduce $\rho_1 = 1 - \varepsilon_1$, and $\rho_2 = 1 - \varepsilon_2$ into equation 14.13, and the denominator is adjusted as

$$1 - \rho_1 \rho_2 = (1 - \varepsilon_1)\varepsilon_2 + \varepsilon_1$$

This gives

$$\left[1 + \rho_1 \rho_2 + (\rho_1 \rho_2)^2 + \cdots + (\rho_1 \rho_2)^n\right] = \frac{1}{(\varepsilon_1 + \varepsilon_2) - \varepsilon_1 \varepsilon_2}$$

The net radiative exchange is

$$Q_{net} = \sigma A \cdot \varepsilon_1 \varepsilon_2 (T_1{}^4 - T_2{}^4)\left[\frac{1}{(\varepsilon_1 + \varepsilon_2) - \varepsilon_1 \varepsilon_2}\right]$$

Finally, for two long smi-infinite long plates the net radiative heat exchange will lead to equation:

$$\boxed{Q_{net} = \sigma A \cdot (T_1{}^4 - T_2{}^4)\left[\frac{1}{\left(\frac{1}{\varepsilon_1} + \frac{1}{\varepsilon_2}\right) - 1}\right]} \tag{14.14}$$

In case we have two long concentric infinite cylinders, the radiative heat exchange will lead to equation:

$$Q_{net} = \sigma A_1 \cdot (T_1{}^4 - T_2{}^4) \left\{ \dfrac{1}{\dfrac{1}{\varepsilon_1} + \left(\dfrac{r_1}{r_2}\right)\left[\dfrac{1}{\varepsilon_2} - 1\right]} \right\} \tag{14.15}$$

In case we have two long concentric infinite spheres, the radiative heat exchange will lead to equation:

$$Q_{net} = \sigma A_1 \cdot (T_1{}^4 - T_2{}^4) \left\{ \dfrac{1}{\dfrac{1}{\varepsilon_1} + \left(\dfrac{r_1}{r_2}\right)^2\left[\dfrac{1}{\varepsilon_2} - 1\right]} \right\} \tag{14.16}$$

14.7 RADIATION SHIELDS TO REDUCE GRAY BODY RADIATIVE EXCHANGE

Radiative heat transfer between surfaces can be decreased by placing highly reflective thin plates of low emissivity between the two surfaces. These thin plates are called radiation shields. It is important that lower the emissivity of the shield as this will create higher thermal resistance. We now consider heat transfer between two plates represented as 1 and 2. The emissivity of plate 1 is ε_1, and the emissivity of plate 2 is ε_2. The shield placed between plate 1 and 2 has side facing plate 1 has area A_{1S}, and side facing plate 2 has area A_{2S} (see Figure 14.15). The emissivity of radiation shield on surface side $S1$ is ε_{S1}, and the emissivity of radiation shield on surface side $S2$ is ε_{S2}. The temperatures of plates 1 and 2 are T_1, and T_2. The temperature of radiation shield on surface side $S1$ is T_{S1}, and temperature of radiation shield on surface side $S2$ is T_{S2}.

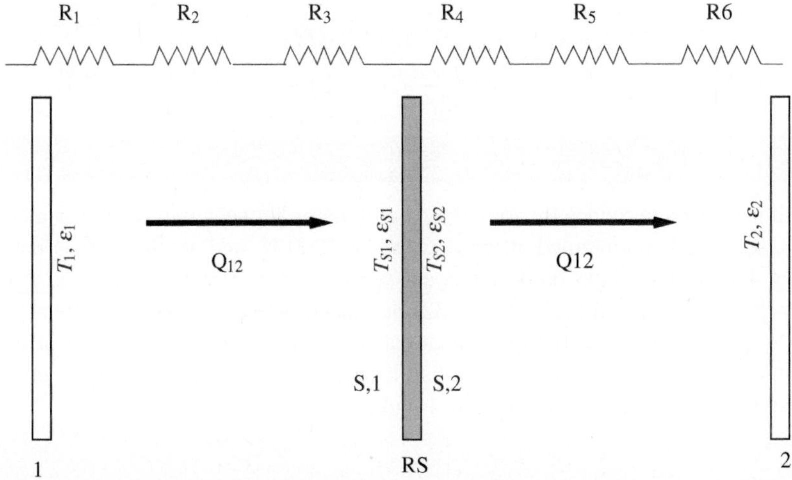

Figure 14.15   Thermal network when the radiation shield (RS) is between two parallel plates.

The thermal resistances are

$$R_1 = \frac{1-\varepsilon_1}{A_1 \cdot \varepsilon_1}, \quad R_2 = \frac{1}{A_1 F_{1-S1}}$$

$$R_3 = \frac{1-\varepsilon_{S1}}{A_{S1} \cdot \varepsilon_{S1}}, \quad R_4 = \frac{1-\varepsilon_{S2}}{A_{S2} \cdot \varepsilon_{S2}}$$

$$R_5 = \frac{1}{A_{S2} F_{S2-2}}, \quad R_6 = \frac{1-\varepsilon_2}{A_2 \cdot \varepsilon_2}$$

The total resistance is

$$R_{tot} = R_1 + R_2 + R_3 + R_4 + R_5 + R_6$$

$$R_{tot} = \frac{1}{\varepsilon_1} + \frac{1}{\varepsilon_{s1}} + \frac{1}{\varepsilon_{s2}} + \frac{1}{\varepsilon_2} - 2$$

We take

$$F_{1-S1} = F_{S2-2} = 1$$

and

$$A_1 = A_2 = A_{S1} = A_{S2} = A$$

The heat transfer from plate 1 to 2 is

$$Q = \frac{\sigma A(T_1^4 - T_2^4)}{\left(\frac{1}{\varepsilon_1} + \frac{1}{\varepsilon_2} - 1\right) + \left(\frac{1}{\varepsilon_{s1}} + \frac{1}{\varepsilon_{s2}} - 1\right)}$$

In certain applications, one shield is not enough. If we insert n number of shields between plate 1 and 2, we have

$$Q_{12,n} = \frac{\sigma A(T_1^4 - T_2^4)}{\left(\frac{1}{\varepsilon_1} + \frac{1}{\varepsilon_2} - 1\right) + \left(\frac{1}{\varepsilon_{s1}} + \frac{1}{\varepsilon_{s2}} - 1\right) + \cdots + \left(\frac{1}{\varepsilon_{n1}} + \frac{1}{\varepsilon_{n2}} - 1\right)}$$

Example 14.4

It is desired to reduce the radiative heat transfer between two parallel plates. One plate is maintained at temperature of 1200 K, and another one is kept at 400 K. The plates emissivities are $\varepsilon_1 = 0.3$ and $\varepsilon_1 = 0.9$. It is proposed that a plate of polished brass ($\varepsilon = 0.032$) be placed between these plate 1 and 2. What percentage reduction in heat transfer will be achieved, if area of all plates is 2 m^2? Also, how much shall be heat transfer if two radiative shields of brass are placed between plate 1 and 2.

Solution We first estimate the heat transfer without radiative shield.

$$Q_{12,no-shield} = \frac{\sigma \cdot \left(T_1{}^4 - T_2{}^4\right) \cdot A}{\frac{1}{\varepsilon_1} + \frac{1}{\varepsilon_2} - 1} = 67.425kW$$

We assume that polished brass plate has same emissivity on both sides of plate

$$Q_{12,1-shield} = \frac{\sigma A\left(T_1^4 - T_2^4\right)}{\left(\frac{1}{\varepsilon_1} + \frac{1}{\varepsilon_2} - 1\right) + \left(\frac{1}{\varepsilon_{s1}} + \frac{1}{\varepsilon_{s2}} - 1\right)} = 3.576kW$$

The radiative shield will bring 94.69 % reduction in heat transfer from plate 1 to 2. Putting another radiative shield of brass with bring forth the 97.27 % reduction in heat transfer:

$$Q_{12,n} = \frac{\sigma A\left(T_2^4 - T_1^4\right)}{\left(\frac{1}{\varepsilon_1} + \frac{1}{\varepsilon_2} - 1\right) + \left(\frac{1}{\varepsilon_{s1}} + \frac{1}{\varepsilon_{s2}} - 1\right) + \left(\frac{1}{\varepsilon_{s1}} + \frac{1}{\varepsilon_{s2}} - 1\right)} = 1.836kW$$

PROBLEMS

14P-1 Two circular plates are placed parallel to each other at distance of L=5 m a part. The lower plate is indicated as 1 in the figure and the top plate is indicated as 1. The plates are maintained at T_1=700 K and T_2=1000 K. The emissivities of the plates are ε_1=0.3 and ε_1=0.45, respectively. The plates are in a place where surrounding temperature is 300 K. Find the heat transfer.

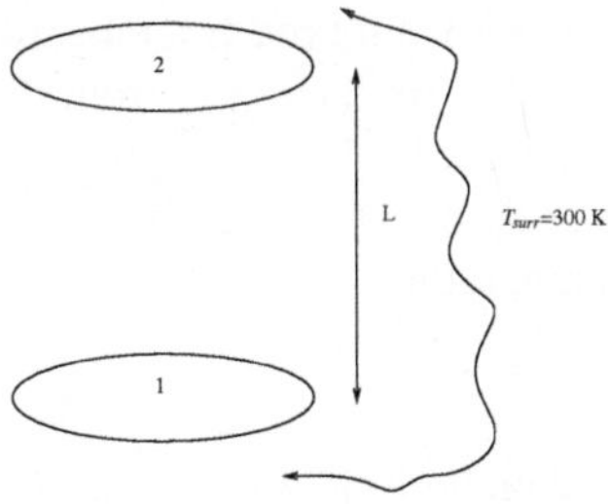

14P-2 Find the shape factor for the wall and the floor in the following figure.

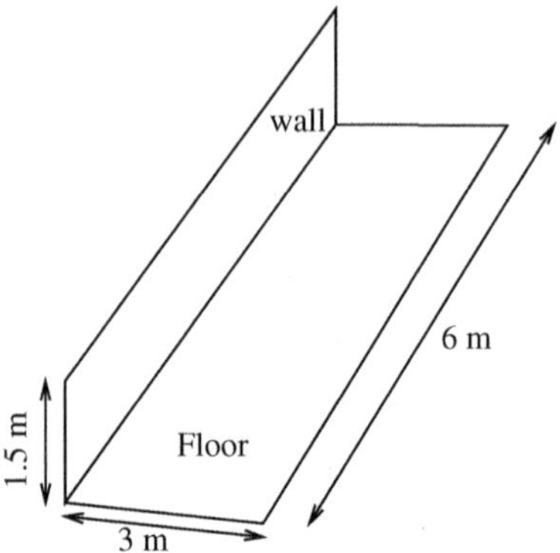

14P-3 Two parallel but offset plates are facing each other (see Figure 14.16). The length of the plate is 8 cm and the minimum distance between the plates is 1 cm.

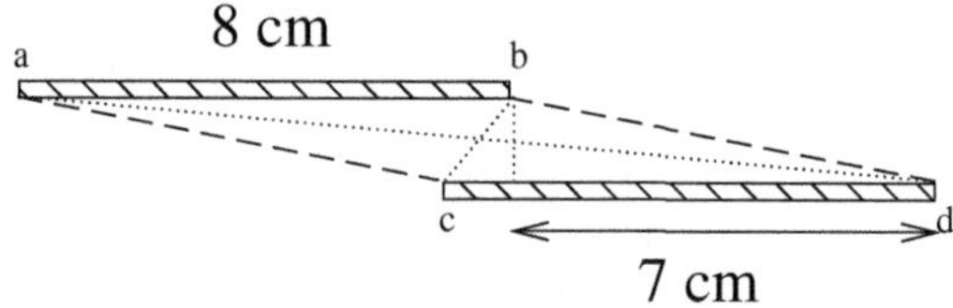

Figure 14.16   Plates with off-set.

[Ans: $F_{12} = 0.1440$]

14P-4 An array of infrared lamps are used to cure the painted surface in an industrial application. The plate is placed in a rectangular cross-section chamber where it is expected that the plate surface would reach temperature higher than 150°C. The emissivity of the plate is 0.7. The irradiation G from the chamber walls and lamps are 450 and 5000 W/m², respectively. The absorptivity of chamber walls and lamps are 0.6 and 0.5, respectively. Ignore any convective and conductive heat transfer and estimate the plate surface temperature.

[Ans T=241 °C]

14P-5 A sample is treated on both sides with a laser beam in an evacuated chamber. The walls of the chamber are isothermal having temperature 300 K. If the sample surface temperature is 2000 K, find the the laser beam flux. The absorptivity associated with the laser beam and chamber walls are $\alpha = 0.8$ and 0.4. The sample surface emissivity is 0.2. The irradiance from surrounding walls is 200 W/m². Ignore the sample thickness.

[Ans $G = 0.113 MW/m^2$]

14P-6 A circular metallic bar (ε=0.5) of 50 mm diameter and 300 mm length is treated in a furnace. The air is circulated and natural convection heat transfer coefficient is 30 W/m^2·K. The furnace walls are at 1200 K and air temperature is 500 K. Find the cylinder surface temperature if cylinder surface is diffuse and gray.

[Ans $T = 1096K$]

14P-7 Assuming that cylinder has the spectrally selective behaviour:

$$\alpha_\lambda = \begin{cases} 0.23 & for\ 0\,\mu m < \lambda < 3\mu m \\ 0.85 & for\ 3\,\mu m < \lambda < \infty \mu m \end{cases}$$

and using data of problem 14P6, find the temperature of cylinder.

14P-8 A hot parallel upper plate at 800 K is used to heat up the liquid flowing through an array of long, circular, thick-walled tubes. The insulated lower plate is at 300 K, and the space between the plates has convective environment conditions T_∞=472 K, h_∞=232 W/m^2·K. The plate and tube surfaces can be approximated as blackbodies. Take $p = 25$ cm and $d = 8$ cm. Ignore internal tube convection, tube wall conduction, and estimate the tube surface temperature.

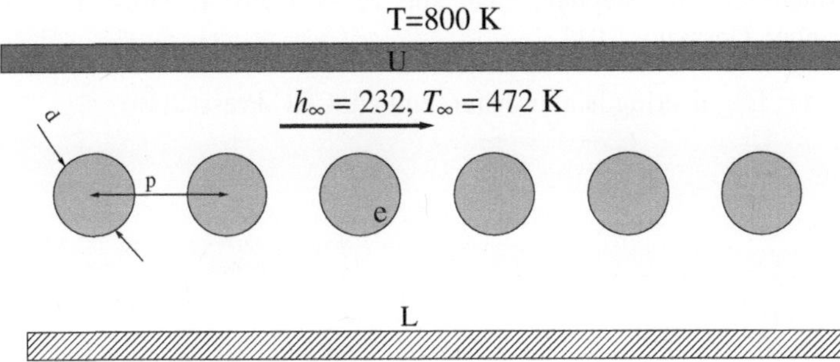

[Ans: $T_e = 1402K$]

14P-9 An evacucated space is formed by assembly of two concentric long metallic pipes. A cryogenic fluid is flowing in the inner pipe. The inner pipe is made up of aluminium (ε=0.04), and outer pipe tin (ε=0.07). The inner and outer pipe wall temperatures are 60 K and 400 K. Estimate the heat transferred to the flowing cryogenic fluid, if length of pipes are 100 m. The pipe radii are 30 and 60 cm, respectively. Ignore convective and conductive heat transfer. How much shall be the heat transfer if both pipes are of aluminium?

[Ans: $Q_{net} = 8.781kW$]

14P-10 Liquified nitrogen at -260 °F is stored in a spherical steel vessel of 2 ft diameter. The tank is covered with another spherical layer of steel having 2.5 ft diameter thus forming an evacuated space between two concentric steel spheres. The emissivity of steel is ε=0.520. The temperature of outer spherical casing is 40 °F. Estimate the radiative heat transferred to the liquified nitrogen. How much is heat transfer if aluminium (ε=0.04) spheres are used instead of steel.

[Ans: $Q_{net} = 608.297Btu/h,\ Q_{net} = 32.75Btu/h$]

14P-11 In a space application where only mode of heat transfer encountered is the radiative heat transfer, is to reduce heat transfer between two surfaces as much as possible. One surface is maintained at temperature of 1000 K, and another one is kept at 600 K. The surface emissivities are $\varepsilon_1=0.7$ and $\varepsilon_1=0.6$. It is planned to insert several sheets of polished aluminium ($\varepsilon=0.056$) be placed between these surfaces 1 and 2. How many aluminium sheets are enough to achieve minimal heat transfer?

REFERENCES

H. C. Hottel, Radiant heat transfer. In: McAdams W. H. (ed.) Heat Transmission, 3rd edn. McGraw-Hill Book Co, New York, USA, 1954.

H. C. Hottel and A. E. Sarofim, Radiative Transfer, McGraw-Hill, New York, USA, 1967.

J. T. Kiehl and K. E. Trenberth, Earth's Annual Global Mean Energy Budget. Bulletin of the American Meteorological Society, 78(2), 197–208, 1997.

H. D. Baehr and K. Stephan, Heat and Mass Transfer, Third, revised edition, Springer, Germany, 2011.

W. S. Janna, Engineering heat transfer, 2nd ed, CRC Press, 2000.

15 Heat Exchangers

A heat exchanger is a device that is used to transfer thermal energy from one fluid to to another fluid. To design a highly efficiency and the low cost heat exchanger, the knowledge of the heat transfer processes is necessary. There are variety of heat exchangers. Heat exchangers are unusually classified according to the type of construction and flow arrangements. Heat exchanger performance also deteriorate with time as deposits can happen, and this will change the overall heat transfer coefficient. There are two popular methods of heat exchanger design, which will be covered in this chapter. After finishing this chapter, one will be able to:

- Compute the overall heat transfer coefficient using fouling resistance.
- Design heat exchanger using Log mean temperature difference.
- Design heat exchanger using Number of Transfer Units.

Heat exchangers are devices that facilitate the transfer of heat from the hot to the cold fluid. They are designed in many sizes and shapes due to space limitations and technical requirements. They handle different fluids ranging from air to organic fluids and oil to high-temperature exhaust gases, and now added to this list are the nanofluids. A heat exchanger can be described by the *heat duty* they are handling. Heat Duty is the heat required to be removed or added from the fluid to create the desirable change in temperature at the inlet and outlet.

Note that heat exchangers are not mixing chambers, as they do not allow the internal mixing of two fluids. In terms of modes of heat transfer, heat exchangers are mostly designed because of conduction-convection analysis and the use of the overall heat transfer coefficient in a design process is a must. With persistent use and low maintenance, the dirt and chemicals might deposit on the heat transfer surfaces, resulting in an off-performing device. This deposition is called fouling of the heat transfers, and they are considered as a resistance to the heat transfer process.

We start our discussion with the types of heat exchangers, followed by the discussion on fouling process, and finally we exhibit and discuss two most commonly used design approaches.

15.1 TYPES AND DESIGN ASPECTS OF HEAT EXCHANGERS

A multitude of heat exchangers is used in industry for heat transfer applications depending on the flow configuration, the heat transfer surface, the heat transfer fluids, and the materials of heat exchanger. See Figures 15.1 and 15.2 for variety of heat exchangers used in industrial applications.

DOI: 10.1201/9781003428404-15

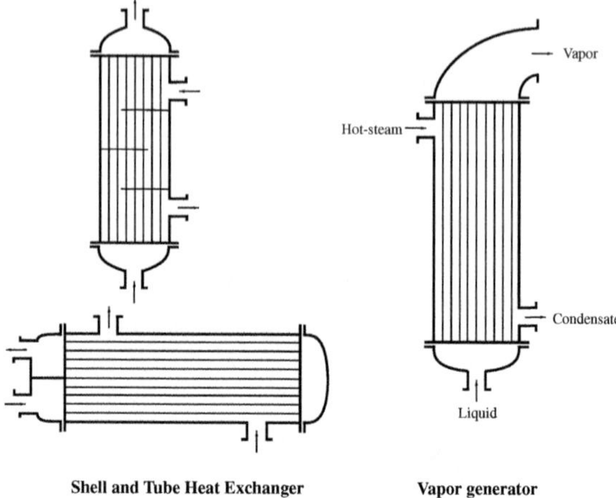

Figure 15.1　Schematic presentation of shell and tube heat exchanger.

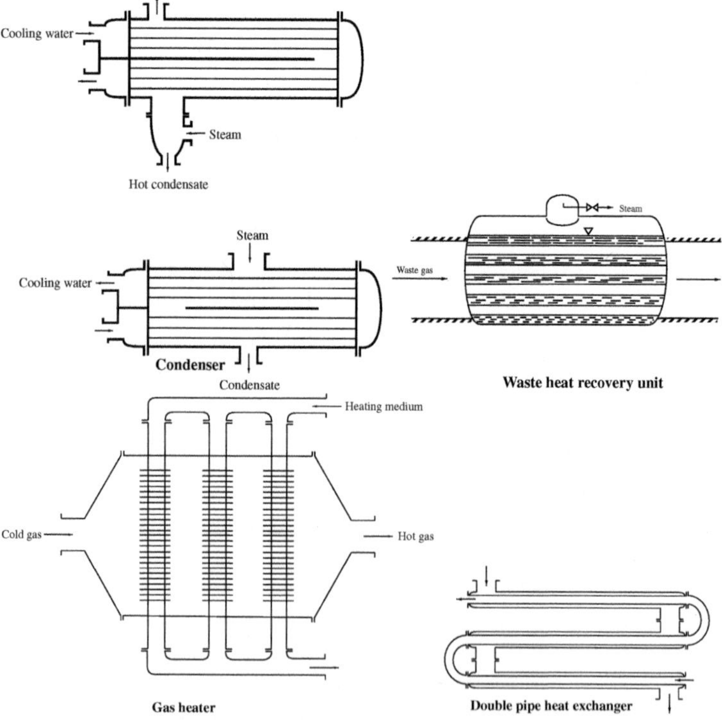

Figure 15.2　Schematics diagram of a some more heat exchangers.

Heat transfers are classified on the basis of direction of flow of heat transfer fluids. In parallel flow, both the hot and cold fluids enter the heat exchanger from the same side and move parallel to each other in same direction. In counter flow, the heat transfer fluids enter the heat exchanger from opposite sides and flow in opposite directions.

Sometimes in heat transfer literature we call heat exchanger with titles like Regenerator, Recuperator, Re-heater, evaporator, condenser, etc.

Regenerator: A regenerator is a type of heat exchanger that uses a porous medium to transfer heat between two fluid streams. It is often used in applications where the temperature difference between the two fluids is relatively small, and where a large amount of thermal energy needs to be transferred over a relatively small surface area.

Recuperator: A recuperator is a type of heat exchanger used to recover waste heat from exhaust gases or other sources and transfer it to a separate fluid stream. It is commonly used in industrial processes, power generation, and HVAC systems to improve energy efficiency and reduce operating costs. They are commonly used in applications such as gas turbines, furnaces, boilers, and other high-temperature processes.

Reheaters are used to heat up steam further in large steam power plants such as coal-fired and nuclear power plants. They are also used in some industrial processes where high-temperature steam is required for various applications. Reheating process helps in increasing efficiency and power output for the same amount of fuel input.

An *agitated film evaporator* is suitable for the evaporation of heat-sensitive materials, viscous materials, and materials with fouling tendencies. In an agitated film evaporator, the liquid to be evaporated is spread onto the walls of a heated cylindrical vessel or jacket. As the liquid flows down the walls of the vessel, it forms a thin film. The heating surface is typically agitated to maintain a uniform film thickness and prevent fouling.

Condenser is a heat exchanger that is used in a power plant or a refrigeration system. It is designed to condense a substance from its gaseous state into its liquid state by removing heat.

Shell-and-tube heat exchanger (STHE) consists of a bank of tubes that are mounted inside a cylindrical shell. One fluid flows through the tubes while the other fluid flows around the outside of the tubes in the shell. The baffle-plates are added to give flow an intricate path for flow which helps in maximising the contact and heat transfer. The tubes in shell may have many passes.

Multiple passes: In a multiple pass heat exchanger, the heat transfer fluids flow through the exchanger multiple times before exiting. This helps in increasing the total amount of time they spend in contact with each other and, hence improve the overall heat transfer efficiency. The number of passes is determined by the design and can range from 2 to over 10.

Also, we need to provide allowance in physical design for differential expansion and contraction of the heat exchanger components.

Floating heads are a type of mechanical design feature used in heat exchangers to allow for differential expansion and contraction of the heat exchanger components. In design, the tube sheet is not fixed in place, but rather it is allowed to move within a housing or shell. The tubes are typically sealed to the tube sheet, and the tube sheet is allowed to move freely within the housing or casing. This design allows for differential expansion and contraction of the tubes and tube sheet without causing stress or damage. The floating head design is commonly used in high-pressure and high-temperature applications, such as in chemical processing, oil refining, and power generation. Such an arrangement mitigate the the risk of equipment failure and leaks.

The STHEs require a large area for installation and thus they are not suitable for application where we have limitations in area, like in trains, helicopters, , airplane, vehicles and offshore platforms. To address these limitations, a solution is to design a compact heat exchanger.

Compact heat exchangers (CHE) are heat exchangers who have heat transfer area-to-volume ratio higher than 300 m^2/m^3. Thus, they are designed to maximise the surface area available for heat transfer while minimising the volume of the heat exchanger. Compact heat exchangers are typically designed with finned surfaces, small diameter tubes, with pressure drop as low as possible. A STHE requires a large area for installation and a great robust supporter for holding it; thus it is not suitable for application where the area is limited like vehicles and offshore platforms. Compact heat exchangers are considered to resolve such problems.

Some common examples of compact heat exchangers are: Car radiators 100 m^2/m^3, Cheetah lung 4000 m^2/m^3, African elephant lung 3000 m^2/m^3, Gas turbine heat exchangers 6000 m^2/m^3, Human lung 12,000 m^2/m^3 (12 square meters per liter), Regenerator of a Stirling engine 15,000 m^2/m^3.

Recently, there is growing interest in *gyroid* structure based heat exchanger which can be printed using 3D-printers. A *Schoen gyroid* or gyroid, is a highly complex and interconnected three-dimensional lattice structure, which is named after the American mathematician and computer scientist, Alan Schoen, who first described the structure in the 1970s. Gyroid consists of minimal iso-surfaces containing no straight lines. The gyroid structures are known for their high surface area to volume ratio, and that bring them into the category of compact heat exchangers. Gyroid heat exchangers are typically made from lightweight materials such as metal or plastic. They can be manufactured using a variety of techniques like 3D printing, casting, and milling, and other additive manufacturing techniques. A gyroid structures can have a high surface area to volume ratio of around 20,000 m^2/m^3.

15.2   FOULING RESISTANCE

The heat transfer fluids contain amounts of undissolved or suspended material or they can allow the growth of microorganisms or macro organisms in the heat exchangers. This leads to the formation of a stable deposition on the heat transfer surfaces (see Figure 15.3). This eventually reduced the effectiveness of heat exchanger. The narrowing of passages would cause the flow pressure loss and the increase in inner wall roughness. It is estimated that 3 mm of scaling inside a pipe would lead to 25% increase in energy costs. The process of deposit formation on the heating surface is called fouling and the thermal resistance caused by the deposit is referred to as *fouling resistance*. The term named as fouling factor often used in the heat transfer literature is incorrectly named, since this is an additional, additive Thermal resistance and is by no means a factor.

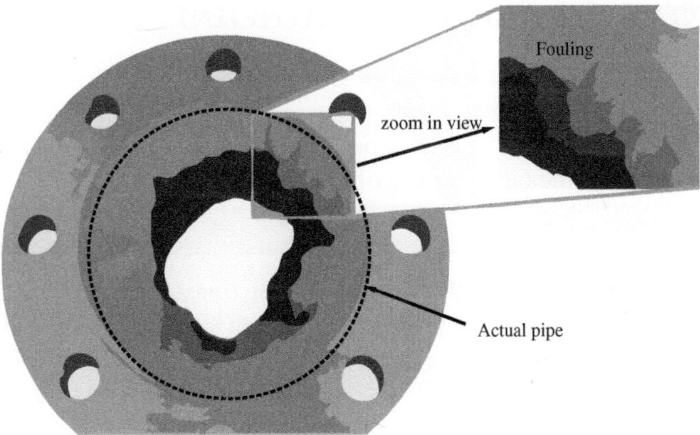

Figure 15.3   Schematic presentation of fouling in a pipe with zoom-in view.

These fouling resistances are generally not known a-priori and can be roughly estimated using the Tubular Exchanger Manufacturers Association (TEMA) recommendations. Some of the values of fouling resistances are tabulated in Tables 15.1 and 15.2.

TABLE 15.1

Fouling Resistance (R_f) for Some Viscous Fluids (m^2 · K/kW)

Heating oil (No. 2)	0.35
Heating oil (No. 6)	0.88
Transformer oil	0.18
Lubricating oil	0.18
Hydraulic oil	0.18
Quench oil	0.7
Pitch	0.8
Tar	0.9
Vegetable oil	0.53

Care should be exercised in assuming fouling resistance as it is noted by heat exchanger designers that heat exchangers are 50–500% oversized, largely due to erroneous fouling resistance assumptions during the design process. These oversized heat exchanges are not only turn out to be expensive, but they are also often prone to increased deposit formation due to suboptimal flow conditions. Note that TEMA tables do not account for the the time-dependent deterioration of the fouling resistances.

15.3 OVERALL HEAT TRANSFER COEFFICIENT

$$\frac{1}{U_{fouled}} - \frac{1}{U_{clean}} = R_{fi} + R_{fo}\left(\frac{r_i}{r_o}\right)$$

where, R_{fi} is the fouling resistance on the inner surface of the tube and R_{fo} is the fouling resistance on the outer surface of the tube.

Example 15.1

A new heat exchanger with convective heat transfer coefficient of 1933 W/m^2·K is feed with the sea water at 70 °C with speed of 2 m/s.

Solution From table 15.2, we have R_f=0.18 m^2 ·K/kW.

$$\frac{1}{h_{dirty}} = \frac{1}{h_{clean}} + R_f$$

$$\frac{1}{h_{dirty}} = \frac{1}{1933} + \frac{0.18}{1000}$$

$$h_{clean} = 1434.040 W/m^2 \cdot K$$

TABLE 15.2

Fouling Resistance (R_f) for Water Flow in Shell and Tube Heat Exchangers (m^2 · K/kW)

Water speeds	Under 50 °C		Above 50 °C	
	< 1 m/s	> 1 m/s	< 1 m/s	> 1 m/s
Sea water	0.09	0.09	0.18	0.18
Brackish water	0.35	0.18	0.53	0.35
Cooling circuit:				
Processed water	0.18	0.18	0.35	0.35
Untreated water	0.53	0.53	0.9	0.75
City water	0.18	0.18	0.35	0.35
Clean river water	0.35	0.18	0.53	0.35
Normal river water	0.53	0.35	0.7	0.53
Sewer water	1.41	0.9	1.75	1.41
Muddy water	0.53	0.35	0.75	0.53
Engine coolant	0.18	0.18	0.18	0.18
Condensate water	0.09	0.09	0.09	0.09
Boiler feed water	0.18	0.09	0.18	0.18
Boiler blowdown water	0.35	0.35	0.35	0.35

15.4   DESIGN OF HEAT EXCHANGER

The two most frequently employed methods to solve heat exchanger design problems are the LMTD method and the effectiveness-number of the transfer unit (NTU) method. If inlet and outlet temperatures are known and the overall heat transfer coefficient can be estimated, then the LMTD method can be used to estimate the heat duty and the surface area required for the exchanger. The LMTD method is an easy method for the sizing of heat exchangers.

On the other hand, if only the inlet temperatures/ flow rates are known, then the LMTD cannot be used as outlet conditions are unclear. To execute the LMTD method in such cases, we need to do a trial and error method and this results in a lengthy iterations procedure. The NTU method offers here a way out for such problems where the outlet temperatures are unknown at the beginning of the design process.

15.4.1   LOG MEAN TEMPERATURE DIFFERENCE METHOD

We recall here the LMTD equation 8.26 developed in section 8.6.

$$LMTD = \frac{[(T_w - T_{b2}) - (T_w - T_{b1})]}{\ln\left(\frac{T_w - T_{b2}}{T_w - T_{b1}}\right)} \tag{15.1}$$

Figure 15.4 shows the Log mean temperature difference for the case of flow through pipe at constant wall temperature conditions.

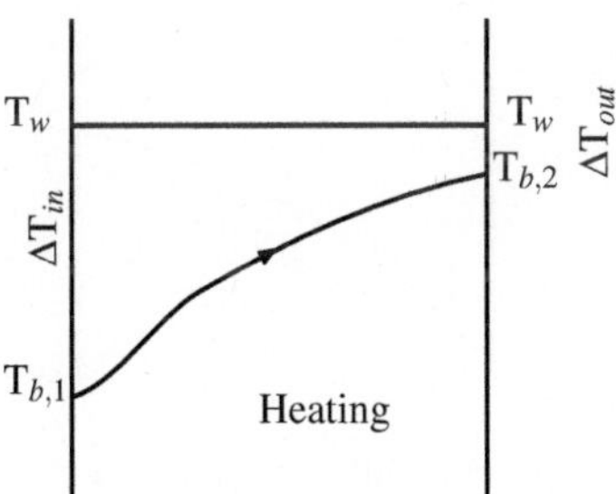

Figure 15.4   Log mean temperature difference for the case of flow through pipe at constant wall temperature condition.

We notice that equation is representing the temperature differences at inlet and outlet. We can define the temperature differences as $\Delta T_{in} = (T_w - T_{b,1})$, and $\Delta T_{out} = (T_w - T_{b,2})$, respectively.

$$LMTD = \frac{[\Delta T_{out} - \Delta T_{in}]}{\ln\left(\frac{\Delta T_{out}}{\Delta T_{in}}\right)} \tag{15.2}$$

For the case of heat exchanger, where two fluids are exchanging heat, the temperature change can be graphically represented as in Figure 15.6.

Figure 15.5 shows the Log mean temperature difference for the case of counter-flow heat exchanger.

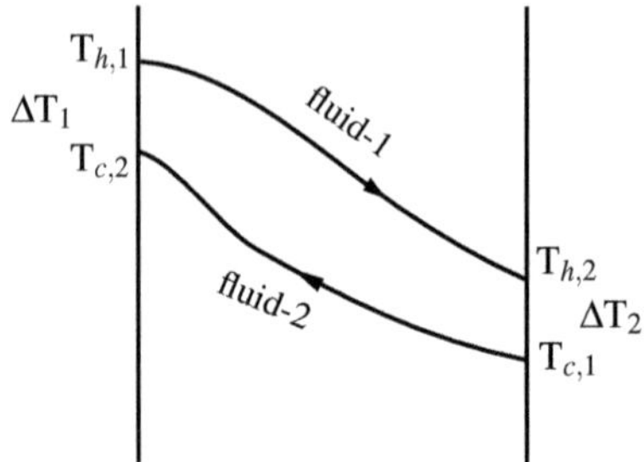

Figure 15.5   Log mean temperature difference for the case of counter flow heat exchanger.

This shows that for the heat exchanger we can cast a similar equation in the form

$$LMTD = \frac{(\Delta T)_1 - (\Delta T)_2}{\ln\left(\frac{(\Delta T)_1}{(\Delta T)_2}\right)} \tag{15.3}$$

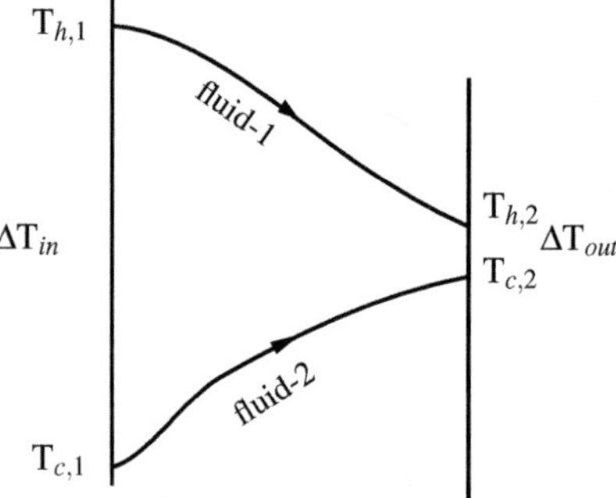

Figure 15.6   Log mean temperature difference for the case of parallel flow heat exchanger.

Note that the counter flow exchangers are preferred over parallel exchangers as the LMTD for the counter-flow eat exchanger is larger than the LMTD for the parallel flow exchanger for the same heat duty.

Example 15.2

Consider the parallel flow heat exchanger as shown in Figure 15.7. One fluid is entering at 20°C and going out at 50°C, whereas the another fluid is entering at 140°C and going out at 75°C. Find the log mean temperature difference.

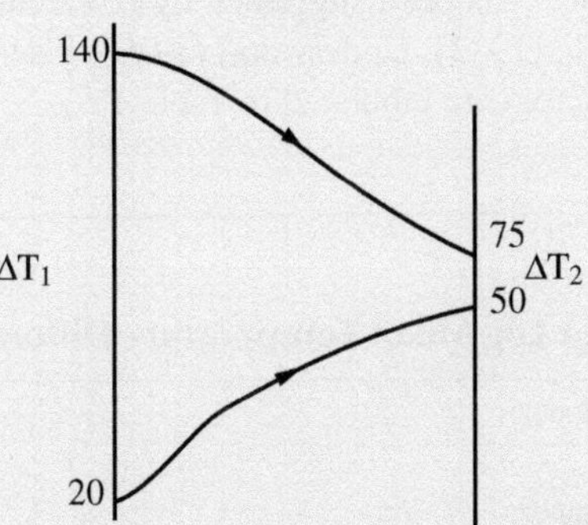

Figure 15.7   The parallel flow heat exchanger.

Solution

We calculate the temperature difference:

$$(\Delta T)_1 = 140 - 20 = 120°C$$

$$(\Delta T)_2 = 75 - 50 = 25°C$$

$$LMTD = \frac{(\Delta T)_1 - (\Delta T)_2}{\ln\left(\frac{(\Delta T)_1}{(\Delta T)_2}\right)}$$

$$LMTD = \frac{120 - 25}{ln(120/25)}$$

$$LMTD = 60.5629°C$$

The log mean temperature difference has two basic assumptions:

(i) the fluid specific heats do not vary with temperature, and

(ii) the convective heat transfer coefficients are constant for both heated and cold fluids inside the heat exchanger.

For counter-flow double-pipe heat exchanger, the heat transfer is calculated as

$$Q = U \cdot A \cdot (\cdot LMTD_{cfdp})$$

where, subscript cfdp with LMTD indicates the LMTD that is calculated for counter-flow double-pipe heat exchanger. For the a heat exchanger other than the double-pipe heat exchangers, the heat transfer is calculated by using a correction factor (F). The correction is used account for the deviation of the mean temperature difference from the LMTD value in complex configurations. The heat transfer can be calculated as

$$Q = U \cdot A \cdot (F \cdot LMTD_{cfdp})$$

The correction factor F has been suggested by researchers as early as 1933, like those suggested by Nagle (1933), Underwood (1934), Fischer (1938), and Bowman et al. (1940). These F factors are tabulated in Table 15.3.

TABLE 15.3

Correction Factor F for Log Mean Temperature Difference (LMTD) Method

Parallel flow double-pipe	$F = 1$
Counterflow double-pipe	$F = 1$
One or both fluids condensing or evaporating	$F = 1$
	In the below relations, $$P = \frac{t_0 - t_i}{T_i - t_i} \quad R = \frac{T_i - T_0}{t_0 - t_i} \ (t = \text{tube}, \quad T = \text{shell})$$
One-shell-pass and multiple of two-tube-passes	$$F = \frac{\sqrt{1+R^2}}{1-R} \frac{\ln\left[(1-PR)/(1-P)\right]}{\ln\left[\dfrac{2-P\left(1+R-\sqrt{1+R^2}\right)}{2-P\left(1+R+\sqrt{1+R^2}\right)}\right]}$$
	$$F = \frac{P}{1-P} \frac{\sqrt{2}}{\ln\left[\dfrac{2/P-2+\sqrt{2}}{2/P-2-\sqrt{2}}\right]} \quad \text{for } R = 1$$
Two-shell-passes and multiple of four-tube-passes	$$F = \frac{\sqrt{1+R^2}}{2(1-R)} \frac{\ln\left[(1-PR)/(1-P)\right]}{\ln\left[\dfrac{2-P\left(1+R-\sqrt{1+R^2}\right)+2\sqrt{(1-P)(1-PR)}}{2-P\left(1+R+\sqrt{1+R^2}\right)+2\sqrt{(1-P)(1-PR)}}\right]}$$
	$$F = \frac{\sqrt{2}}{2} \frac{P/(1-P)}{\ln\left[\dfrac{2-P(2-\sqrt{2})+2\sqrt{(1-P)^2}}{2-P(2+\sqrt{2})+2\sqrt{(1-P)^2}}\right]} \quad \text{for } R = 1$$
Crossflow with one fluid mixed one fluid unmixed	$$F = -\frac{1}{1-R} \frac{\ln\left[(1-PR)/(1-P)\right]}{\ln\left[1+(1/R)\ln(1-PR)\right]}$$
	$$F = -\frac{P/(1-PR)}{\ln\left[1+\ln(1-P)\right]} \quad \text{for } R = 1$$

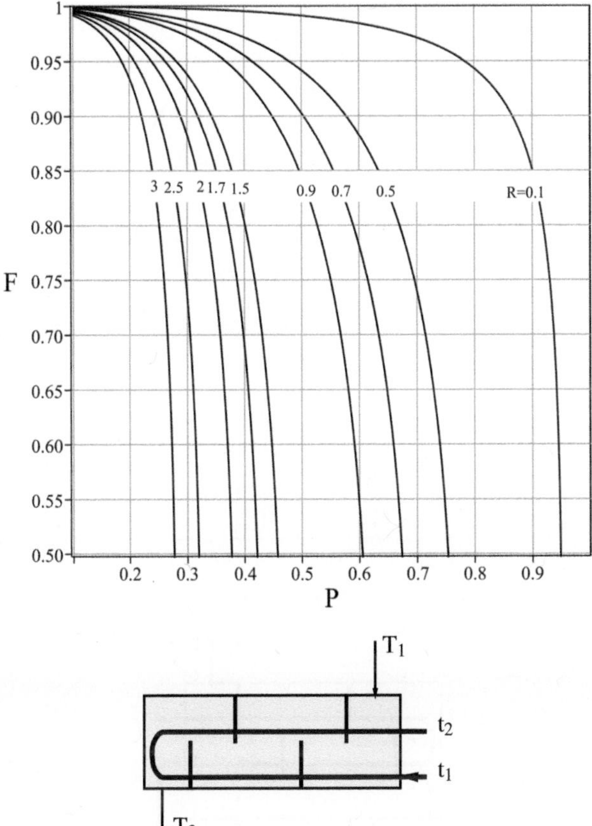

Figure 15.8   Log mean temperature difference (LMTD) correction factor F to be used for one shell pass and one or more tube passes.

$$P = \frac{t_2 - t_1}{T_1 - t_1}, \ R = \frac{T_1 - T_2}{t_2 - t_1}$$

where P is defined as the cold-side effectiveness and function R can be defined as capacity rate ratio:

$$R = C_{cold}/C_{hot}$$

The correction factor F for the Log Mean Temperature Difference (LMTD) is illustrated in Figure 15.8. This factor is applicable for systems with a single shell pass and one or multiple tube passes.

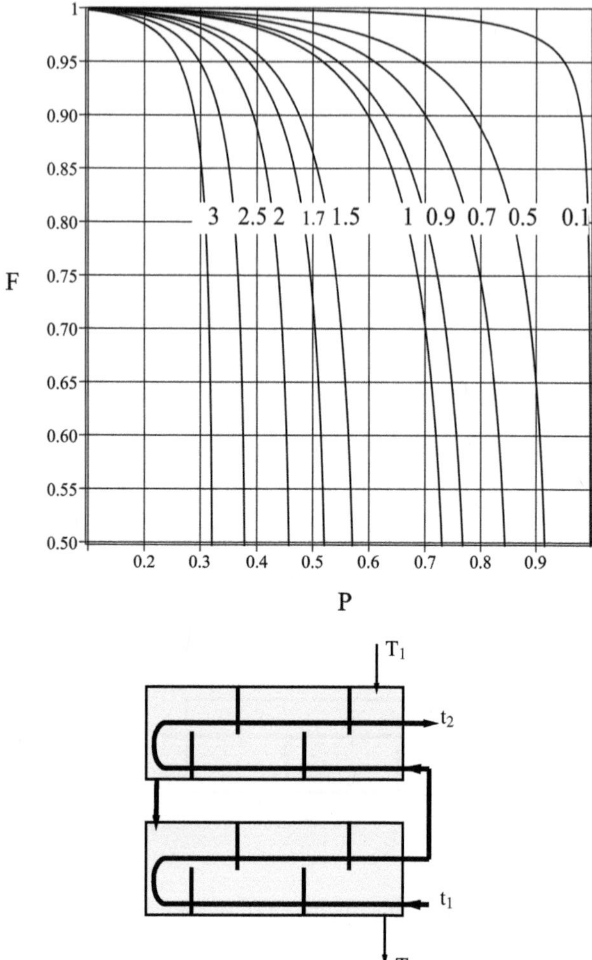

Figure 15.9 Log mean temperature difference (LMTD) correction factor F to be used for two shell pass and one or more tube passes.

Figure 15.9 shows the plot of the Correction Factor for the Log Mean Temperature Difference (LMTD) in Heat Exchangers with Two Shell Passes and One or Multiple Tube Passes. This factor is used to refine the LMTD value for specific geometric and flow conditions.

LMTD approach for double pipe heat exchanger
Hot water at 80°C is used in a double-pipe counter flow heat exchanger to heat up the cold water, which is entering the heat exchanger at 10°C. The mass flow rate of the cold water is 2.3 kg/s and that of hot water is 1.5 kg/s. The overall heat transfer coefficient (U) for this heat exchanger is estimated to be 450 W/m²·C. If the surface area of the heat exchanger is 9 m², find the heat duty of the exchanger in kW and the exit temperatures of hot and cold water. Take c_p= 4187 J/kg·K.

Solution

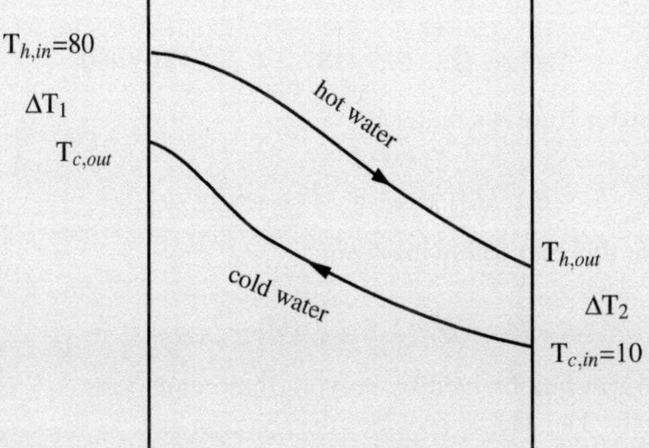

Figure 15.10   The counter flow heat exchanger with temperatures unknown at outlets.

We calculate the temperature differences (see Figure 15.10):

$$\Delta T_1 = 80 - Tc, out$$

$$\Delta T_2 = T_{h,out} - 10$$

The LMTD is

$$LMTD = \frac{90 - T_{c,out} - T_{h,out}}{\ln\left(\frac{80 - T_{c,out}}{T_{h,out} - 10}\right)}$$

$$Q = U \cdot A_w \cdot LMTD$$

We formulate the equation:

$$eq1 := Q - U \cdot A_w \cdot LMTD = 0$$

$$eq1 := Q - 450 \cdot 9 \cdot \frac{90 - T_{c,out} - T_{h,out}}{\ln\left(\frac{80 - T_{c,out}}{T_{h,out} - 10}\right)}$$

The heat transfer from cold water is

$$Q = \dot{m}_{cold} \cdot cp \cdot (T_{c,out} - T_{c,in})$$

We formulate this into following form

$$eq2 := Q - 2.3 \cdot 4187 \cdot (T_{c,out} - 10) = 0$$

The heat transfer from hot water is

$$Q = \dot{m}_{hot} \cdot c_p \cdot (T_{h,in} - T_{h,out})$$

We formulate this into following form

$$eq3 := Q - 1.5 \cdot c_p \cdot (80 - T_{h,out}) = 0$$

The solution can be obtained by solving three equations eq1, eq2, and eq3 simultaneously. The Maple provides the solution as

```
solve({eq1, eq2, eq3}, [Q, T_c_out, T_h_out])
```

This gives the solution:

$$Q = 184.46kW, T_{c,out} = 29.154°C, T_{h,out} = 50.62°C$$

Example 15.4

LMTD Method: when all temperatures are known

A one shell and two tube passes, heat exchanger with the water inside shell is used to heat up oil from 30°C to 85°C. The water temperatures in shell decreases from 120°C to 80°C. If the overall heat-transfer coefficient inside heat exchanger as 350 W/m^2·°C and heat transferred as 15.65 MJ/min, find the area required.

Solution

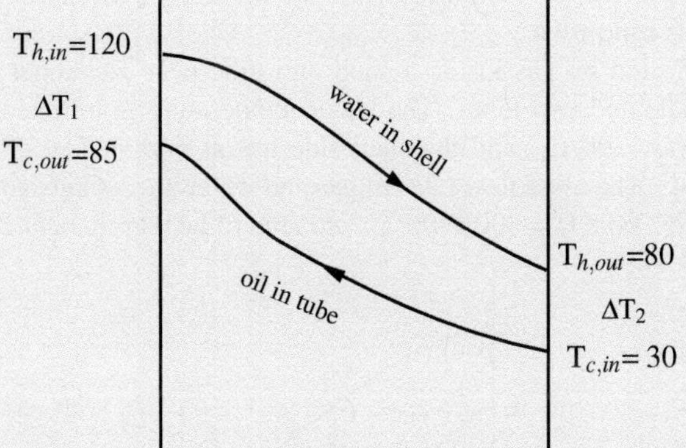

We calculate the log mean temperature as (see figure above)

$$\Delta T_1 = 120 - 85 = 35, \quad \Delta T_2 = 80 - 30 = 50$$

$$LMTD = \left(\frac{\Delta T_1 - \Delta T_2}{\ln\left(\frac{\Delta T_1}{\Delta T_2}\right)} \right) = 42.055$$

Labeling the shell side temperatures as T and tube side temperature as t, we calculate P and R factors for one shell and two tube passes:

$$P = ((t_2 - t_1)/(T_1 - t_1)) = 0.4444$$

$$R = ((T_1 - T_2)/(t_2 - t_1)) = 1.375$$

Referring to Figure 15.8, we have F = 0.73. Alternatively, we can use equation:

$$F = \frac{\sqrt{1+R^2}}{1-R} \cdot \frac{\ln\left(\frac{1-P\cdot R}{1-P}\right)}{\ln\left(\frac{2-P\cdot\left(1+R-\sqrt{1+R^2}\right)}{2-P\cdot\left(1+R+\sqrt{1+R^2}\right)}\right)} = 0.7358$$

The heat transfer is 15.65 MJ/min = 260.833 kW:

$$Q = F \cdot U \cdot A \cdot LMTD$$

$$260833.3333W = 0.7358 \cdot 350 \cdot A \cdot 42.055$$

This gives the area, A as

$$A = 24.082m^2$$

Example 15.5

LMTD approach for shell and tube heat exchanger when one outlet temperature is unknown

Water is heated by gas inside a shell and tube heat exchanger comprising of one shells and two tubes. The water temperature in tube side at inlet is $t_1 = 15°C$, $t_2 = 90°C$, and the shell side gas at inlet is $T_1 = 250°C$. $c_{p,g} = 1000$ J/kg·C. The mass flow rates of gas and water are 2.4 kg/s and 1.55 kg/s, respectively. Take U = 400 W/m^{2}C, and area of heat exchanger as A=7m^2.

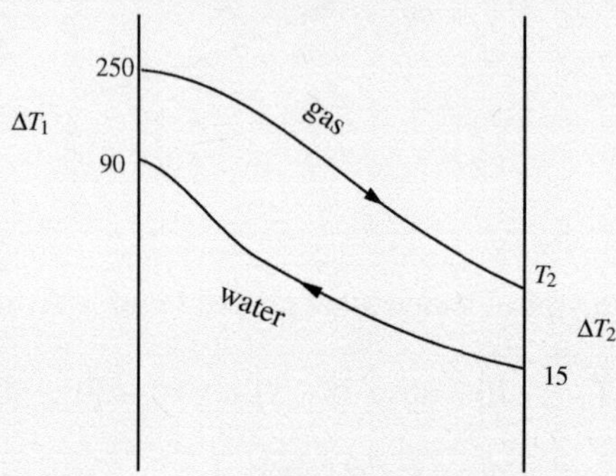

Solution

$$t_1 = 15°C, t_2 = 90°C, T_1 = 250°C$$

Referring to above figure we have

$$\Delta T_1 = T_1 - t_2, \Delta T_2 = T_2 - t_1$$

$$LMTD = \left(\frac{\Delta T_1 - \Delta T_2}{\ln\left(\frac{\Delta T_1}{\Delta T_2}\right)} \right)$$

$$LMTD = \frac{175.0 - T_2}{\ln\left(\frac{160.0}{T_2 - 15.0}\right)}$$

$$Q = F \cdot U \cdot A \cdot LMTD$$

For first iteration we set F=1, this gives

$$Q = \frac{2800\,(175.0 - T_2)}{\ln\left(\frac{160.0}{T_2 - 15.0}\right)}$$

The bulk energy transfer is

$$Q = c_{pg} \cdot \dot{m}_g \cdot (T_1 - T_2) = 1000 \cdot 2.4 \cdot (250 - T_2)$$

Solving the above two equations simultaneously for Q and T_2 we arrive at two solutions:

$$Q = 343.70 kW, T_2 = 106.789°C$$

and

$$Q = 180 kW, T_2 = 175°C$$

For T_2 = 106.789:

$$P = ((t_2 - t_1)/(T_1 - t_1)) = 0.3191$$

$$R = ((T_1 - T_2)/(t_2 - t_1)) = 1.909$$

Using equation:

$$F = \frac{\sqrt{1+R^2}}{1-R} \cdot \frac{\ln\left(\frac{1-P \cdot R}{1-P}\right)}{\ln\left(\frac{2-P \cdot \left(1+R-\sqrt{1+R^2}\right)}{2-P \cdot \left(1+R+\sqrt{1+R^2}\right)}\right)} = 0.864$$

For T_2 = 175:

$$P = ((t_2 - t_1)/(T_1 - t_1)) = 0.3191$$

$$R = ((T_1 - T_2)/(t_2 - t_1)) = 1$$

Using $T_2 = 175$, we have $\Delta T_1 = \Delta T_2 = 60$ and LMTD is undefined. Hence, LMTD approach is only applicable if we consider

$$Q = 343.70 kW, T_2 = 106.789°C$$

as solution. We now repeat the calculations, and we use the heat transfer equation as

$$Q = F \cdot U \cdot A \cdot LMTD$$

with $T_2 = 106.789°C$ and F computed as 0.864, we have the new solution of the set of equations:

$$T_2 = 118.924°C, \text{ and } F = 0.892$$

We repeat the procedure which gives

$$T_2 = 116.25°C, \text{ and } F = 0.887$$

We repeat the procedure which gives the solution to this problem:

$$T_2 = 116.8°C, Q = 319.677 kW \text{ and } F = 0.887$$

15.4.2 THE NUMBER OF TRANSFER UNITS (NTU) METHOD

The parameter P which is required for the estimation of logarithmic mean temperature difference correction factor (F) requires that both the cold and hot fluid stream temperatures are specified. However, when the cold side outlet temperature is unknown, a trial-and-error procedure is required to estimate P (as exhibited in previous examples). This trial and error can be cumbersome in some cases. In 1955, Kays and London proposed a method for designing of heat exchanger, which is now known as the effectivenessNTU or simply NTU method. This new method is based on the following quantities:

i. **Effectiveness:** The dimensionless quantity effectiveness of the heat transfer is used.

$$\varepsilon = \frac{Q_{actual}}{Q_{\max}} = \frac{Actual\ heat\ transfer\ rate}{Maximum\ heat\ transfer\ rate}$$

ii. **Capacity rate ratio:** The heat capacity (C) is the product of mass flow rate and specific heat of the fluid. The dimensionless capacity ratio is used in NTU approach, which is defined as

$$C^* = C_{ratio} = C_{\min}/C_{\max}$$

where, the capacity can take values $0 \leq C^* \leq 1$.

The parallel-flow heat exchangers have the lowest effectiveness values, whereas the counter-flow heat exchangers have the highest effectiveness values. The second in tier are the cross-flow heat exchangers with both fluids unmixed. The effectiveness shall be a maximum for C=0 and a minimum for C=1 cases. In case of phase change like that in boiler or condenser, $C^* \rightarrow 0$, which corresponds to $C_{max} \rightarrow \infty$.

iii. **Number of transfer units:** is defined as

$$NTU = \frac{1}{C_{\min}} \int_{Aw} U \cdot dA_w \equiv \frac{U \cdot A_w}{C_{\min}}$$

Although a heat exchanger with a very high effectiveness value may look desirable from a heat transfer point of view, a heat exchanger with values of NTU greater than 3 would not result into a substantial improvement in heat transfer. Therefore, NUT should practically be taken less than and equal to 3. For cases when NUT<0.3, the effectiveness of a heat exchanger is independent of the capacity ratio.

The effectiveness relation in terms of NTU is reported in Table 15.4.

Note that ε_1 is The effectiveness of each of the two shell passes, i.e. ε_1 is the effectiveness of a one-shell-pass exchanger.

TABLE 15.4

Effectiveness ε for (ε-NTU) Method

$NTU = (U \cdot A)/C_{\min}$	$C_{ratio} = C_{\min}/C_{\max}$
Double-pipe parallel flow	$\varepsilon = \dfrac{1 - \exp\left[-NTU(1 + C_{\text{ratio}})\right]}{1 + C_{\text{ratio}}}$
Double-pipe counter-flow	$\varepsilon = \dfrac{1 - \exp\left[-NTU(1 - C_{\text{ratio}})\right]}{1 - C_{\text{ratio}}\exp\left[-NTU(1 - C_{\text{ratio}})\right]}$ $\varepsilon = \dfrac{NTU}{NTU + 1}$ for $C_{\text{ratio}} = 1$
One-shell-pass and multiple of two-tube-passes	$\varepsilon = \dfrac{2}{1 + C_{\text{ratio}} + \sqrt{1 + C_{\text{ratio}}^2}\,(1 + e^{-\Lambda})/(1 - e^{-\Lambda})}$ where $\Lambda = NTU\sqrt{1 + C_{\text{ratio}}^2}$
Two-shell-passes and multiple of four-tube-passes	$\varepsilon = \dfrac{[(1 - \varepsilon_1 C_{\text{ratio}})/(1 - \varepsilon_1)]^2 - 1}{[(1 - \varepsilon_1 C_{\text{ratio}})/(1 - \varepsilon_1)]^2 - C_{\text{ratio}}}$ $\varepsilon = \dfrac{2\varepsilon_1}{1 + \varepsilon_1}$ for $C_{\text{ratio}} = 1$ ε_1 is The effectiveness of a one-shell-pass exchanger (given above).
Crossflow with both fluids unmixed	$\varepsilon = 1 - \exp\left\{\dfrac{NTU^{0.22}}{C_{\text{ratio}}}\left[\exp\left(-C_{\text{ratio}}NTU^{0.78}\right) - 1\right]\right\}$
Crossflow with One fluid mixed one fluid unmixed	For $C_{\max}$ mixed and $C_{\min}$ unmixed $\varepsilon = \dfrac{1}{C_{\text{ratio}}}\left(1 - \exp\left\{-C_{\text{ratio}}\left[1 - \exp(-NTU)\right]\right\}\right)$ For $C_{\min}$ mixed and $C_{\max}$ unmixed $\varepsilon = 1 - \exp\left\{-\dfrac{1}{C_{\text{ratio}}}\left[1 - \exp\left(-C_{\text{ratio}}NTU\right)\right]\right\}$
One or both fluids condensing or evaporating ($C_{\text{ratio}} = 0$)	$\varepsilon = 1 - \exp(-NTU)$

The actual heat transfer in the heat exchanger is

$$Q_{actual} = c_{p,cold}\left[T_{c,out} - T_{c,in}\right] = c_{p,hot}\left[T_{h,in} - T_{h,out}\right]$$

The heat capacity associated with cold and hot fluids is

$$C_{cold} = \dot{m}_{cold} \cdot c_{p,cold}$$

$$C_{hot} = \dot{m}_{hot} \cdot c_{p,hot}$$

The maximum possible temperature difference is $\Delta T_{max} = T_{h,in} - T_{c,in}$, hence the maximum heat transfer is

$$Q_{max} = C_{min}\left[T_{h,in} - T_{c,in}\right]$$

where $C_{min} = min\left[C_{cold}, C_{hot}\right]$ i.e. minimum of either of these two heat capacity values. Note that in terms of shell and tube heat exchanger nomenclature, if

$$C_h > C_c \text{ then } (T_1 - T_2) < (t_2 - t_1)$$

and

$$C_c > C_h \text{ then } (t_2 - t_1) < (T_1 - T_2)$$

then the fluid that may experience the maximum temperature change, $(T_1 - t_1)$, and hence, the maximum possible heat transfer can be expressed as

$$Q_{max} = C_c(T_1 - t_1) \quad (C_c < C_h)$$

$$Q_{max} = C_h(T_1 - t_1) \quad (C_h < C_c)$$

We will now exhibit few cases for the solution methodology to be used for using NTU method.

Example 15.6

NTU approach for shell and tube heat exchanger when one outlet temperature is unknown

Water is heated by gas inside a shell and tube heat exchanger comprising of one shells and two tubes. The water temperature in tube side at inlet is $t_1 = 15°C$, $t_2 = 90°C$, and the shell side gas at inlet is $T_1 = 250°C$. $c_{p,g} = 1000$ J/kg·C. The mass flow rates of gas and water are 2.4 kg/s and 1.55 kg/s, respectively. Take $U = 400$ W/m^{2}C, and area of heat exchanger as A=7m^2.

Solution

$$C_{water} = c_{p,w} \cdot \dot{m}_{water} = 6510 \, W/K$$

$$C_{gas} = c_{p,g} \cdot \dot{m}_{gas} = 2400 W/K$$

$$C_{gas} < C_{water}$$

Hence we set:

$$C_{min} = C_{gas} = 2400 W/K$$

We also compute the capacity ratio:

$$C_{ratio} = C_{min}/C_{max} = 0.3686$$

The number of transfer of units is

$$NTU = \frac{U \cdot A}{C_{min}} = \frac{400 \cdot 7}{2400} = 1.1666$$

We estimate the effectiveness using equation for one-shell-pass and multiple of two-tube-passes

$$\varepsilon = \left(\frac{2}{1 + C_{ratio} + \frac{\chi \cdot (1 + \exp(-\Lambda))}{(1 - \exp(-\Lambda))}} \right)$$

where,

$$\chi = \sqrt{1 + C_{ratio}^2} = 1.0657; \Lambda = \frac{NTU}{\chi} = 1.0946$$

This gives

$$\varepsilon = 0.5703511718$$

Since by definition

$$\varepsilon = \frac{Q}{Q_{max}} = \left(\frac{C_{gas} \cdot (T_1 - T_2)}{C_{min} \cdot (T_1 - t_1)} \right)$$

$$\varepsilon = 1.0638 - 0.00425 T_2$$

Solving for T_2, we arrive at the values

$$T_2 = 115.96^\circ C, Q = 321.678 kW$$

This solution is quite close to the solution that was obtained via LMTD example ($T_2 = 116.8^\circ C$, Q = 319.677 kW), but we observe the benefit of NTU procedure that no further iterations are required.

Example 15.7

NTU Method when one fluid is going through phase change

Consider the case of a shell-and-tube heat exchanger in which steam is condensing over tubes carrying water. We assume that both the steam and the water are doing only one pass through this device. The water flow rate is 5 kg/s and it is heated up from 20°C to 65°C. The steam is condensing at 110°C. If the overall heat transfer coefficient of th is heat exchanger is 1232 W/m^{2}K, find the sizing for this heat exchanger? Take $c_{p,w}$=4200 J/kg·K.

Solution

It is given that

$$t_1 = 20°C, t_2 = 65°C, \text{ and } T_1 = 110°C$$

The non-condensing fluid is water and its capacity will be the minimum:

$$C_{min} = \dot{m}_w \cdot c_{p,w} = 5 \cdot 4200 = 21000 \; W/K$$

The number of transfer units is computed as

$$NTU = \frac{U \cdot A}{C_{min}} = \frac{22 \cdot A}{375}$$

For a condenser, $C_{min} = C_{max} = 0$, we use the relation:

$$\varepsilon = (1 - \exp(-NTU)) = 1 - exp(-0.05866 \cdot A)$$

Using now the definition of effectiveness we have

$$\varepsilon = \left(\frac{\dot{m}_w \cdot c_{p,w} \cdot (t_2 - t_1)}{C_{min} \cdot (T_1 - t_1)} \right) = 0.5$$

Solving for A, we arrive at result

$$A = 11.8150 \; m^2$$

The effectiveness for various heat exchanger situations is plotted along NTU in Figure 15.11–15.16.

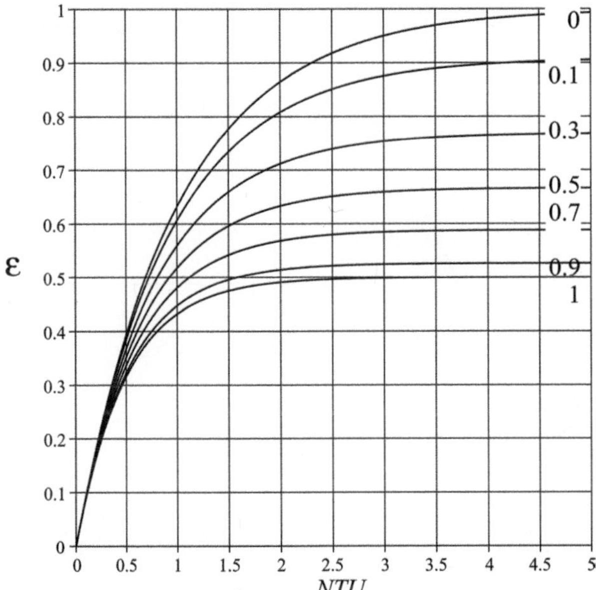

Figure 15.11   Effectiveness ε for parallel flow.

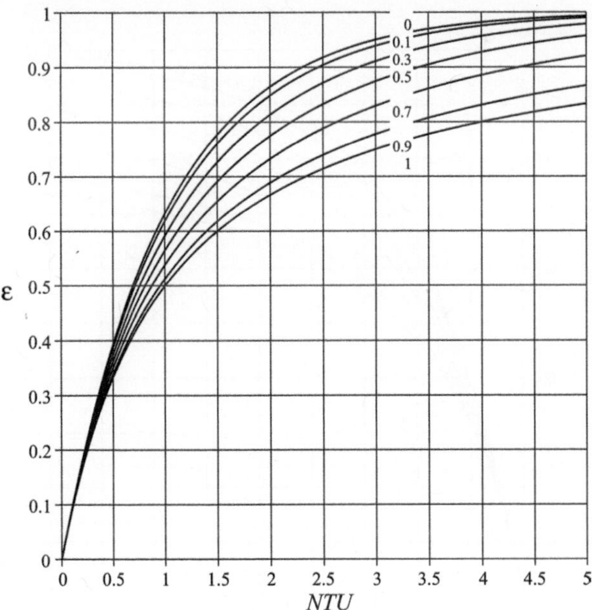

Figure 15.12   Effectiveness ε for double pipe counter flow.

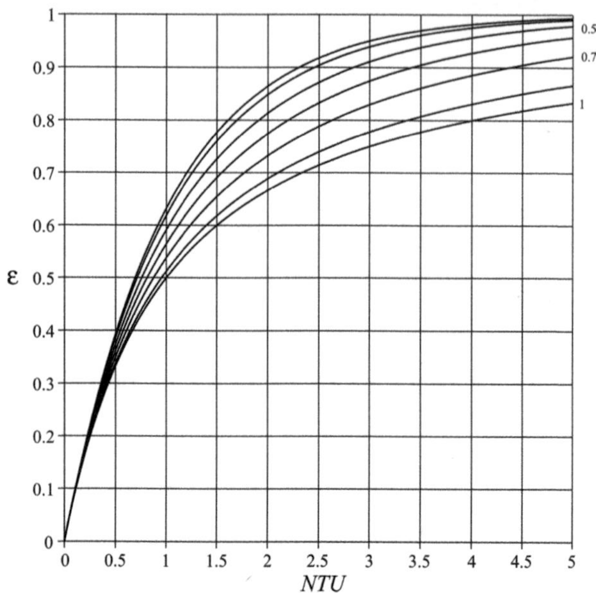

Figure 15.13　Effectiveness ε for one-shell-pass and multiple of two-tube-passes $C_{ratio}= 0$, 0.1, 0.3, 0.5, 0.7, 0.9, 1.

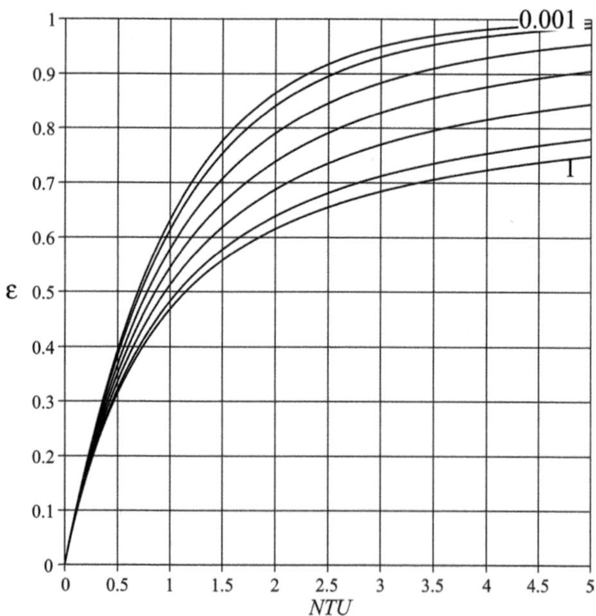

Figure 15.14　Effectiveness ε for Cross-flow with both fluids unmixed $C_{ratio}= 0.001$, 0.1, 0.3, 0.5, 0.7, 0.9, 1.

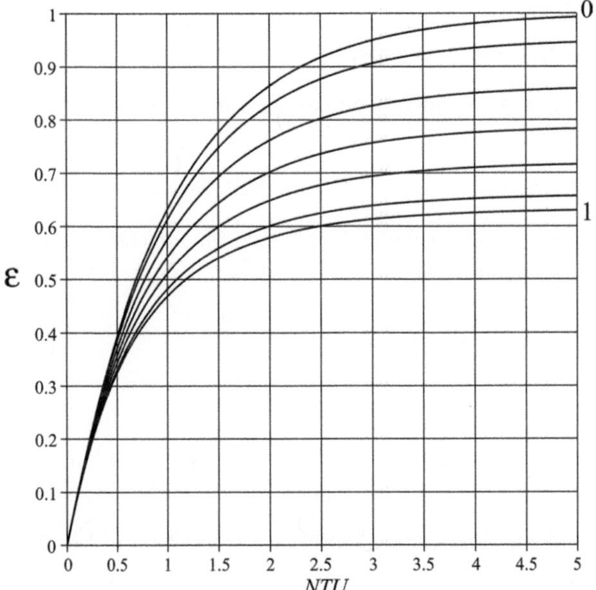

Figure 15.15 Effectiveness ε for Cross-flow with C_{max} fluid mixed and C_{min} fluid unmixed; C_{ratio} = 0, 0.1, 0.3, 0.5, 0.7, 0.9, 1.

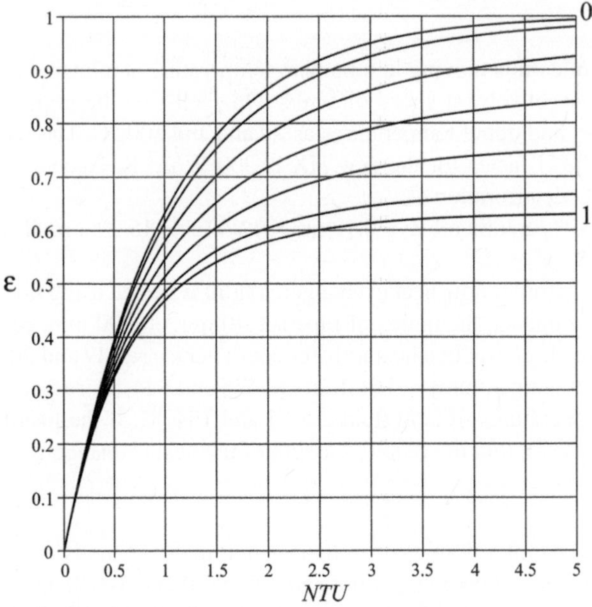

Figure 15.16 Effectiveness ε for Cross-flow with C_{min} fluid mixed and C_{max} fluid unmixed; C_{ratio} = 0, 0.1, 0.3, 0.5, 0.7, 0.9, 1.

REFERENCES

W. Kays and A.L. London, Compact Heat Exchangers, McGraw-Hill, 1955.

W. M. Kays and A. L. London, Compact heat exchangers, 2nd edn. McGraw-Hill, New York, 1964.

D. Q. Kern, Process heat transfer. McGraw-Hill, New York, 1950.

TEMA: Standards of the Tubular Exchanger Manufacturers Association, 6. Ed,. Tubular Exchanger Manufacturers Association, New York, 1978.

VDI-Wärmeatlas, 12th ed., VDI, Springer, 2019.

W. M. Nagle, Mean Temperature Difference in Multipass Heat Exchangers, Ind. Eng. Chem., 25, 604–609, 1933.

A. J. V. Underwood, The Calculation of Mean Temperature Difference in Multipass Heat Exchangers, J. Inst. Pet. Technol., 20, 145–158, 1934.

F. K. Fischer, Mean Temperature Difference Correction in Multipass Exchangers, Ind. Eng. Chem., 30(4), 377–383, 1938.

R. A., Bowman, A. C. Mueller, and W. M. Nagle, Mean Temperature Difference in Design, Trans. ASME, 62, 283–294, 1940.

PROBLEMS

15P-1 Hydraulic oil is moving in tube with temperature at inlet and outlet as $120\,°C$ and $10\,°C$, respectively, at speed of 5 m/s. On shell-side the normal river water is used with inlet and outlet temperatures as 20 and and $100\,°C$. The river water speed is 1.2 m/s. The U factor for the new HX is 300 W/m^2·K. Account for the fouling conditions and deterioration U.

Table 15.1 Hydraulic oil $R_f = 0.18$ m^2·K/kW Table 15.2 Normal river water $R_f = 0.53$ m^2·K/kW

15P-2 In a parallel flow heat exchanger, a fluid is heated using hot exhaust gases. The inside and outside diameters of pipe are 70 mm and 90 mm, respectively. The inside and outside convective heat transfer coefficients are 149 and 205 W/m^{2}K. The inlet and outlet temperatures of hot fluid are 476 and $324\,°C$, respectively. The inlet and outlet temperatures of cold fluid are 55 and $134\,°C$. If the heat transferred per hour is 30,000 kJ/h, find the required length of the heat exchanger.

$$[\text{Ans: } L = 1.37m]$$

15P-3 In a parallel flow heat exchanger hot and cold fluids are entering at 200 and $20\,°C$. The mass flow rates of hot and cold fluid are 20,000 and 2000 kg/h. The specific heats of hot and cold fluids are 2000 and 500 J/kg·K. If the overall heat transfer coefficient is 300 W/m^{2}K, find the exit temperature of hot fluid. The area of heat exchanger is 20 m^2.

$$[\text{Ans: } T_{h,o} = 195.6\,°C]$$

15P-4 In a double pipe heat exchanger cold and hot water enters parallel with 40,000 and 44,000 kg/h. The hot and cold water temperatures at inlet are 93 and 25 °C. The outlet hot temperature is 62 °C. The overall heat transfer coefficient U=2280 W/m^{2}K. Find the heat transfer area and the heat exchanger effectiveness. Take c_p=4181 J/kg.K for both hot and cold streams.

$$[\text{Ans: } A = 33.67m^2, \varepsilon = 0.455]$$

15P-5 In an heat exchanger hot oil and cold water streams having specific heats 2090 and 4180 J/kg·K, respectively. The oil inlet and outlet temperatures are 85 and 35 °C, whereas the water inlet temperature is 30 °C. The mass flow rate for both oil and water is 0.277 kg/s. If the overall U factor is 260 W/m^{2}K, find the area of this heat exchanger.

$$[\text{Ans: } A = 2.7m^2]$$

15P-6 In a counter-flow heat exchanger the dry saturated steam ($c_{p,s}$=2710 J/kg·K) enters at 350 °C with mass flow rate of 745 kg/min, and leaves at 200 °C. The tube diameter is 43 mm and they are 2 m long. The gas ($c_{p,g}$=1002 J/kg·K) is entering at 800 °C with mass flow rate of 1200 kg/min. Estimate the number of tubes required if convective heat transfer on steam and gas sides are 650 and 232 W/m^2·K.

$$[\text{Ans: } N_{tubes} = 169]$$

15P-7 A counter flow heat exchanger has 3200 tubes and each having diameter of 35 mm. The air is being cold having mass flow rate of 13 kg/s. The inlet and outlet temperatures for air are 530 and 136 °C. If the resistance on the cooling fluid side negligible, and T_{co}=20 °C and T_{co}=80 °C. Estimate the tube length required for this heat duty. Properties of air at the mean temperature are ρ= 1.23904 kg/m^3, c_p= 1027 J/kg·K, ν=0.0000197703 m^2/s, k=0.02615 W/m·K, Pr=0.697.

$$[\text{Ans: } L = 3.719m]$$

15P-8 A shell-and-tube heat exchanger with one shell and 40 tubes/8 passes is carrying water on shell side and unused engine oil is being cooled in the tubes. The mass flow rate of cooling water is 3 kg/s as its temperature changes from 10 to 90 °C. The oil inlet and outlet temperatures are 150 and 90 °C. If the over all heat transfer coefficient U=300 W/m^2·K, estimate the oil flow rate and length of tubes to accomplish this heat duty. The diameter of tube is 30 mm. Take $c_{p,water}$=4154, and $c_{p,oil}$=2309 J/kg·K, and use NTU method.

$$[\text{Ans: } L = 15.8m]$$

15P-9 A feed-water heater designed as a shell-and-tube type heat exchanger is supplying hot water to the boiler. Two hundred tubes ($d = 25mm, L = 3m$) with two-passes have total water mass flow rate of 15 kg/s flowing through them. The inlet water temperature is 296 K, and it is required to find the water outlet temperature. Steam is condensing over the tubes with the temperature and h values of 373.15 K and 20,000 W/m^2·K. Also, find the outlet temperature distribution as function of total water mass flow rate.

$$[\text{Ans: } T_{c,o} = 350K, Tc, o = 373.15 - 77.15exp(17.859/\dot{m}_w)]$$

15P-10 Consider a single-pass cross-flow heat exchanger with both fluids unmixed. The device is used to warm up water from 30 to 80 °C. The hot exhaust gases inlet temperature is 430 °C. The mass flow rate of water is 2.5 kg/s. The specific heats of water and gases are $c_{p,w}$=4200 J/kg.K, and $c_{p,g}$=1300 J/kg·K. The area of heat exchanger is 35 m^2 and the U factor is 152.57 W/m^2·K. Find the mass flow rate of exhaust gases.

[Ans: $\dot{m} = 1.05 kg/s$]

15P-11 A nanofluid comprised of φ=5% Titanium oxide-water solution is used on the shell side of a one pass shell and tube heat exchanger. In the 50 tubes/8 passes the oil is entering at $T_{h,i}$=400 K. The nanofluid side inlet and outlet temperatures are $T_{c,i}$=333 K and $T_{c,o}$=353 K, respectively. The mass flow rates of nanofluid and oil are 2.5 kg/s and 4 kg/s, respectively. If both the fluids are unmixed, find the area of device and the outlet temperature of the oil. The specific heat of oil is c_p=2308 J/kg·K and $U = 346$ W/m^{2}K. For nanofluid properties use the relations given in chapter one.

[Ans: $T_{c,o} = 378.3K$]

15P-12 A one shell and tube heat exchanger has 30 tubes with water inside tubes and oil on the shell side. The tubes have inner and outer diameter of 22 and 24 mm. The water enters at 86 C and leaves at 28 °C. The oil enters at 11 °C and leaves at 30 °C. The length per pass is 10 m. The water particles properties are μ_w= 487E-6 Pa.s, $c_{p,w}$= 4183 J/kg·K, k_w= 0.653 W/m·K. Find the shell side convective heat transfer coefficient.

[Ans: h_o= 109.87 W/m^2·K]

Properties Dataset

A: THERMAL PROPERTIES OF SELECTED METALLIC ELEMENTS

TABLE A1
Thermal Properties of Selected Metallic Elements at 293 K (20°C) or 528°R (65°F)

Element	Specific grav-ity	Specific heat, c_p		Thermal conductivity, k		Diffusivity, α		Melting temperature	
		J/(k·K)	BTU/(lbm·°R)	W/(m·K)	BTU/(hr·ft·°R)	m²/s × 10⁶	ft²/s × 10³	K	°R
Aluminium	2.702	896	0.214	236	136	97.5	1.05	933	1,680
Beryllium	1.850	1750	0.418	205	118	63.3	0.681	1550	2,790
Chromium	7.160	440	0.105	91.4	52.8	29.0	0.312	2118	3,812
Copper	8.933	383	0.0915	399	231	116.6	1.26	1356	2,441
Gold	19.300	129	0.0308	316	183	126.9	1.37	1336	2,405
Iron	7.870	452	0.108	31.1	18.0	22.8	0.245	1810	3,258
Lead	11.340	129	0.0308	35.3	20.4	24.1	0.259	601	1,082
Magnesium	1.740	1017	0.243	156	90.1	88.2	0.949	923	1,661
Manganese	7.290	486	0.116	7.78	4.50	2.2	0.0236	1517	2,731
Molybdenum	10.240	251	0.0600	138	79.7	53.7	0.578	2883	5,189
Nickel	8.900	446	0.107	91	52.6	22.9	0.246	1726	3,107
Platinum	21.450	133	0.0318	71.4	41.2	25.0	0.269	2042	3,676
Potassium	0.860	741	0.177	103	59.6	161.6	1.74	337	607
Silicon	2.330	703	0.168	153	88.4	93.4	1.01	1685	3,033
Silver	10.500	234	0.0559	427	247	173.8	1.87	1234	2,221
Tin	5.750	227	0.0542	67.0	38.7	51.3	0.552	505	909
Titanium	4.500	611	0.146	22.0	12.7	8.0	0.0861	1953	3,515
Tungsten	19.300	134	0.0320	179	103	69.2	0.745	3653	6,575
Uranium	19.070	113	0.0270	27.4	15.8	12.7	0.137	1407	2,533

Continued on next page

Element	Specific gravity	Specific heat, c_p		Thermal conductivity, k		Diffusivity, α		Melting temperature	
		$J/(k \cdot K)$	$BTU/(lbm \cdot {}^\circ R)$	$W/(m \cdot K)$	$BTU/(hr \cdot ft \cdot {}^\circ R)$	$m^2/s \times 10^6$	$ft^2/s \times 10^3$	K	${}^\circ R$
Vanadium	6.100	502	0.120	31.4	18.1	10.3	0.111	2192	3,946
Zinc	7.140	385	0.0920	121	69.9	44.0	0.474	693	1,247

Source: Data from several sources.

Notes: Density = ρ = Specific gravity $\times$ 62.4 lbm/ft^3 = Specific gravity $\times$ 1000 kg/m^3

Diffusivity = α; for aluminium, α m^2/s $\times$ 10^6 = 97.5; So $\alpha = 97.5 \times 10^{-6}$ m^2/s

Also, $\alpha = k/\rho c_p$

TABLE A2
Thermal Properties of Selected Alloys at 293 K (20°C) or 528°R (65°F)

Base metal	% Composition	Specific gravity	Specific heat, c_p		Thermal conductivity, k		Diffusivity, α	
			$\dfrac{\text{J}}{(\text{k} \cdot \text{K})}$	$\dfrac{\text{BTU}}{(\text{lbm} \cdot °\text{R})}$	$\dfrac{\text{W}}{(\text{m} \cdot \text{K})}$	$\dfrac{\text{BTU}}{(\text{hr} \cdot \text{ft} \cdot °\text{R})}$	$\text{m}^2/\text{s} \times 10^5$	$\text{ft}^2/\text{s} \times 10^4$
Aluminium								
Duralumin	94-96 Al, 3-5 Cu. tr Mg	2.787	833	0.199	164	94.7	6.676	7.187
Silumin	87 Al, 13 Si 7	2.659	871	0.208	164	94.7	7.099	7.642
Copper								
Al-bronze	92 Cu. 5 Al	8.666	410	0.09/9	83	47.9	2.330	2.508
Bronze	75 Cu, 25 Sn	8.666	343	0.0819	26	15.0	0.859	0.925
Red brass	85 Cu, 9 Sn, 6 Zn	8.714	385	0.0920	61	35.2	1.804	1.942
Brass	70 Cu, 30 Zn	8.522	385	0.0920	111	64.1	3.412	3.673
German silver	62 Cu, 15 Ni, 22 Zn	8.618	394	0.0941	24.9	14.4	0.733	0.789
Constantan	60 Cu, 40 Ni	8.922	410	0.0979	22.7	13.1	0.612	0.659
Iron								
Cast iron	4 c	7.272	420	0.100	52	30.0	1.702	1.832
Wrought iron	0.5 CH	7.849	460	0.1 10	59	34.1	1.626	1.750
Steel								
Carbon steel	1 C	7.801	473	0.113	43	24.8	1.172	1.262
	1.5 C	7.753	486	0.116	36	20.8	0.970	1.04
Chrome steel	1 Cr	7.865	460	0.110	61	35.2	1.665	1.792
	5 Cr	7.833	460	0.110	40	23.1	1.110	1.195
	10 Cr	7.785	460	0.110	31	17.9	0.867	0.933
Chrome	15 Cr, 10 Ni	7.865	460	0.110	19	11.0	0.526	0.577
nickel steel	20 Cr, 15 Ni	7.833	460	0.110	15.1	8.72	0.415	0.447
Nickel steel	10 Ni	7.945	460	0.110	26	15.0	0.720	0.775
	20 Ni	7.993	460	0.110	19	11.0	0.526	0.566

Continued on next page

Base metal	% Composition	Specific gravity	Specific heat, c_p		Thermal conductivity, k		Diffusivity, α	
			$\dfrac{\text{J}}{\text{(k}\cdot\text{K)}}$	$\dfrac{\text{BTU}}{\text{(lbm}\cdot{}^\circ\text{R)}}$	$\dfrac{\text{W}}{\text{(m}\cdot\text{K)}}$	$\dfrac{\text{BTU}}{\text{(hr}\cdot\text{ft}\cdot{}^\circ\text{R)}}$	$\text{m}^2/\text{s}\times 10^5$	$\text{ft}^2/\text{s}\times 10^4$
	40 Ni	8.169	460	0.110	10	5.78	0.279	0.300
	60 Ni	8.378	460	0.110	19	11.0	0.493	0.531
Nickel-	80 Ni, 15 C	8.522	460	0.110	17	9.82	0.444	0.478
chrome steel	40 Ni, 15 C	8.073	460	0.110	11.6	6.70	0.305	0.328
Manganese	1 Mn	7.865	460	0.110	50	28.9	1.388	1.494
steel	5 Mn	7.849	460	0.110	22	12.7	0.637	0.686
Silicon	1 Si	7.769	460	0.110	42	24.3	1.164	1.164
steel	5 Si	7.417	460	0.110	19	11.0	0.555	0.597
Stainless	Type 304	7.817	461	0.110	14.4	8.32	0.387	0.417
steel	Type 347	7.817	461	0.110	14.3	8.26	0.387	0.417
Tungsten	1 W	7.913	448	0.107	66	38.1	1.858	2.000
steel	5 W	8.073	435	0.104	54	31.2	1.525	1.642

Notes: Density $= \rho =$ Specific gravity $\times$ 62.4 lbm/ft^3 $=$ Specific gravity $\times$ 1000 kg/m^3

Diffusivity $= \alpha$; fix Duralumin, $\alpha \times 10^5 = 6.676$ m^2/s; So $\alpha = 6.676 \times 10^{-5}$ m^2/s

TABLE A3
Thermal Properties of Selected Building Materials and Insulations at 293 K (20°C) or 528°R (65°F)

Material	Specific gravity	Specific heat, c_p		Thermal conductivity, k		Diffusivity, α	
		$\dfrac{J}{(k \cdot K)}$	$\dfrac{BTU}{(lbm \cdot °R)}$	$\dfrac{W}{(m \cdot K)}$	$\dfrac{BTU}{(hr \cdot ft \cdot °R)}$	$m^2/s \times 10^5$	$ft^2/s \times 10^6$
Asbestos	0.383	816	0.195	0.113	0.0653	0.036	3.88
Asphalt	2.120			0.698	0.403		
Bakelite	1.270			0.233	0.135		
Brick							
Carborundum							
(50% SiC)	2.200			5.82	3.36		
Common	1.800	840	0.201	0.38-0.52	0.22-0.30	0.028-0.034	3.0-3.66
Magnesite							
(50% MgO)	2.000			2.68	1.55		
Masonry	1.700	837	0.200	0.658	0.38	0.046	5.0
Silica							
(95% SiO_2)	1.90			1.07	0.618		
Cardboard				0.14-0.35	0.08-0.2		
Cement (hard)				1.047	0.605		
Clay (48.7% moist)	1.545	880	0.210	1.26	0.728	0.101	10.9
Coal (anthracite)	1.370	1260	0.301	0.238	0.137	0.013-0.015	1.4-1.6
Concrete (dry)	0.500	837	0.200	0.128	0.074	0.049	5.3
Cork board	0.150	1880	0.449	0.042	0.0243	0.015-0.044	1.6-4.7

Continued on next page

Material	Specific gravity	Specific heat, c_p		Thermal conductivity, k		Diffusivity, α	
		$\dfrac{\text{J}}{(\text{k}\cdot\text{K})}$	$\dfrac{\text{BTU}}{(\text{lbm}\cdot{}^\circ\text{R})}$	$\dfrac{\text{W}}{(\text{m}\cdot\text{K})}$	$\dfrac{\text{BTU}}{(\text{hr}\cdot\text{ft}\cdot{}^\circ\text{R})}$	$\text{m}^2/\text{s}\times 10^5$	$\text{ft}^2/\text{s}\times 10^6$
Cork (expanded)	0.120			0.036	0.0208		
Earth (diatomaceous)	0.466	879	0.210	0.126	0.072	0.031	3.3
Earth (clay with 28% moist)	1.500			1.51	0.872		
Earth (sandy with % moist)	1.500			1.05	0.607		
Glass fiber	0.220			0.035	0.02		
Glass (window pane)	2.800	800	0.191	0.81	0.47	0.034	3.66
Glass (wool)	0.200	670	0.160	0.040	0.023	0.028	3.0
Granite	2.750			3.0	1.73		
Ice at 0°	0.913	1830	0.437	2.22	1.28	0.124	13.3
Kapok	0.025			0.035	0.02		
Linoleum	0.535			0.081	0.047		
Mica	2.900			0.523	0.302		
Pine bark	0.342			0.080	0.046		
Plaster	1.800			0.814	0.47		
Plexiglas	1.180			0.195	0.113		
Plywood	0.590			0.109	0.063		
Polystyrene	1.050			0.157	0.0907		
Rubber							
Buna	1.250			0.465	0.269		
Ebonite	1.150	2009	0.480	0.163	0.0942	0.0062	0.67
Spongy	0.224			0.055	0.0318		

Continued on next page

Material	Specific gravity	Specific heat, c_p			Thermal conductivity, k		Diffusivity, α	
		$\dfrac{\text{J}}{(\text{k}\cdot\text{K})}$	$\dfrac{\text{BTU}}{(\text{lbm}\cdot{}^\circ\text{R})}$		$\dfrac{\text{W}}{(\text{m}\cdot\text{K})}$	$\dfrac{\text{BTU}}{(\text{hr}\cdot\text{ft}\cdot{}^\circ\text{R})}$	$\text{m}^2/\text{s}\times10^5$	$\text{ft}^2/\text{s}\times10^6$
Sand								
Dry				0.582	0.336			
Moist	1.640				1.13	0.653		
Sawdust	0.215				0.071	0.041		
Wood								
Fir,pine,and								
spruce	0.444	2720	0.650		0.15	0.087	0.012 4	1.33
Oak	0.705	2390	0.571		0.19	0.11	0.01 13	1.22
Celotex	0.400				0.055	0.03 18		
Fiber sheets	0.200				0.047	0.0172		
Wool	0.200				0.038	0.0220		

Notes: Density $= \rho =$ Specific gravity $\times$ 62.4 $\text{lbm}/\text{ft}^3 =$ Specific gravity $\times$ 1000 kg/m^3

Diffusivity $= \alpha$; for asbestos, $\alpha \times 10^5 = 0.036\ \text{m}^2/\text{s}$; So $\alpha = 0.036 \times 10^{-5}\text{m}^2/\text{s}$

Also, $\alpha = k/\rho c_p$

TABLE A4
Properties of Saturated Liquids: Ammonia NH_3[*]

Temp, T		Specific gravity	Specific heat, c_p		Kinematic viscosity, ν		Thermal Conductivity, k		Thermal diffusivity, α		Prandtl number, Pr	β	
°C	°F		$\dfrac{J}{kg \cdot K}$	$\dfrac{BTU}{lbm \cdot °R}$	$\dfrac{m^2/s}{\times 10^6}$	$ft^2/s \times 10^5$	$\dfrac{W}{m \cdot K}$	$\dfrac{BTU}{hr \cdot ft \cdot °R}$	$m^2/s \times 10^7$	$ft^2/hr \times 10^3$		1/K	1/°R
-50	-58	0.703	4463	1.066	0.435	0.468	0.547	0.316	1.742	6.75	2.60		
-40	-40	0.691	4467	1.067	0.406	0.437	0.547	0.316	1.775	6.88	2.28		
-30	-22	0.679	4476	1.069	0.387	0.417	0.549	0.317	1.801	6.98	2.28		
-20	-4	0.666	4509	1.077	0.381	0.410	0.547	0.316	1.819	7.05	2.09		
-10	14	0.653	4564	1.090	0.378	0.407	0.543	0.314	1.825	7.07	2.07		
0	32	0.640	4635	1.107	0.373	0.402	0.540	0.312	1.819	7.05	2.05		
10	50	0.626	4714	1.126	0.368	0.396	0.531	0.307	1.801	6.90	2.04		
20	68	0.611	4798	1.146	0.359	0.386	0.521	0.301	1.775	6.88	2.02	2.45×10^{-3}	1.36×10^{-3}
30	86	0.595	4890	1.168	0.349	0.376	0.507	0.293	1.742	6.75	2.01		
40	104	0.580	4999	1.194	0.340	0.366	0.493	0.285	1.701	6.59	2.00		
50	122	0.564	5116	1.222	0.330	0.355	0.476	0.275	1.654	6.41	1.99		

[*]**Source:** Data taken from *Analysis of Heat and Mass Transfer* by E. R. G. Eckert and R. M. Drake, Jr. Taylor & Francis Inc. 1972. Used with permission.

TABLE A5

Properties of Saturated Liquids: Carbon Dioxide CO_2[*]

Temp, T			Specific heat, c_p		Kinematic viscosity, ν		Thermal Conductivity, k		Thermal diffusivity, α			β	
°C	°F	Specific gravity	$\dfrac{J}{kg \cdot K}$	$\dfrac{BTU}{lbm \cdot °R}$	$\dfrac{m^2/s}{\times 10^6}$	$\dfrac{ft^2/s}{\times 10^5}$	$\dfrac{W}{m \cdot K}$	$\dfrac{BTU}{hr \cdot ft \cdot °R}$	$m^2/s \times 10^7$	$ft^2/hr \times 10^3$	Prandtl number, Pr	1/K	1/°R
-50	-58	1.156	1840	0.44	0.119	0.128	0.0855	0.0494	0.4021	1.558	2.96		
-40	-40	1.117	1880	0.45	0.118	0.127	0.1011	0.0584	0.4810	1.864	2.46		
-30	-22	1.076	1970	0.47	0.117	0.126	0.1116	0.0654	0.5272	2.043	2.22		
-20	-4	1.032	2050	0.49	0.115	0.124	0.1151	0.0665	0.5445	2.110	2.12		
-10	14	0.983	2180	0.52	0.113	0.122	0.1099	0.0635	0.5133	1.989	2.20		
0	32	0.926	2470	0.59	0.108	0.117	0.1045	0.0604	0.4578	1.774	2.38		
10	50	0.860	3140	0.75	0.101	0.109	0.0971	0.0561	0.3608	1.398	2.80		
20	68	0.772	5000	1.2	0.091	0.098	0.0872	0.0504	0.2019	0.860	4.10	14.00×10^{-3}	3.67×10^{-3}
30	86	0597	36400	8.7	0.080	0.086	0.0703	0.0406	0.0279	0.108	28.7		

TABLE A6

Properties of Saturated Liquids: Dichlorodifluoromethane (Freon-1 2) CCl_2F_2[*]

Temp, T		Specific gravity	Specific heat, c_p		Kinematic viscosity, v		Thermal Conductivity, k		Thermal diffusivity, α		Prandtl number, Pr	β	
°C	°F		$\dfrac{J}{kg \cdot K}$	$\dfrac{BTU}{lbm \cdot °R}$	$m^2/s \times 10^6$	$ft^2/s \times 10^5$	$\dfrac{W}{m \cdot K}$	$\dfrac{BTU}{hr \cdot ft \cdot °R}$	$\dfrac{m^2/s}{\times 10^7}$	$ft^2/hr \times 10^3$		1/K	1/°R
-50	-58	1.546	875.0	0.2090	0.310	0.334	0.067	0.039	0.501	1.94	6.2	2.63×10^{-3}	1.4×10^{-4}
-40	-40	1.518	884.7	0.2113	0.279	0.300	0.069	0.040	0.514	1.99	5.4		
-30	-22	1.489	895.6	0.2139	0.253	0.272	0.069	0.040	0.526	2.04	4.8		
-20	-4	1.460	907.3	0.2167	0.235	0.253	0.071	0.041	0.539	2.09	4.4		
-10	14	1.429	920.3	0.2198	0.221	0.238	0.073	0.042	0.550	2.13	4.0		
0	32	1.397	934.5	0.2232	0.214	0.230	0.073	0.042	0.557	216	3.8		
10	50	1.364	949.6	0.2268	0.203	0219	0.073	0.042	0.560	2.17	3.6		
20	68	1.330	965.9	0.2307	0.198	0.213	0.073	0.042	0.560	2.17	3.5		
30	86	1.295	983.5	0.2349	0.194	0.209	0.071	0.041	0.560	2.17	3.5		
40	104	1.257	1001.9	0.2393	0.191	0.206	0.069	0.040	0.555	2.15	3.5		
50	122	1.215	1021.6	0.2440	0.190	0.204	0.067	0.039	0.545	2.11	35		

TABLE A7
Properties of Saturated Liquids: Engine oil (Unused)*

Temp, T		Specific heat, c_p			Kinematic viscosity, ν		Thermal Conductivity, k		Thermal diffusivity, α			β	
°C	°F	Specific gravity	$\dfrac{\text{J}}{\text{kg}\cdot\text{K}}$	$\dfrac{\text{BTU}}{\text{lbm}\cdot°\text{R}}$	m^2/s	ft^2/s	$\dfrac{\text{W}}{\text{m}\cdot\text{K}}$	$\dfrac{\text{BTU}}{\text{hr}\cdot\text{ft}\cdot°\text{R}}$	$\text{m}^2/\text{s}\times10^7$	$\text{ft}^2/\text{hr}\times10^3$	Prandtl number, Pr	1/K	1/°R
0	32	0.899	1796	0.429	0.00428	0.0461	0.147	0.085	0.911	3.53	47100		
20	68	0.888	1880	0.449	0.00090	0.0097	0.145	0.084	0.872	3.38	10400	0.70×10^{-3}	0.39×10^{-3}
40	104	0.876	1964	0.469	0.00024	0.0026	0.144	0.083	0.834	3.23	2870		
60	140	0.864	2047	0.489	0.839E-4	0.903E-3	0.140	0.081	0.800	3.10	1050		
80	176	0.852	2131	0.509	0.375E-4	0.404E-3	0.138	0.080	0.769	2.98	490		
100	212	0.840	2219	0.530	0.203E-4	0.219E-3	0.137	0.079	0.738	2.86	276		
120	248	0.828	2307	0.551	0.124E-4	0.133E-3	0.135	0.078	0.710	2.75	175		
140	284	0.816	2395	0.572	0.080E-4	0.086E-3	0.133	0.077	0.686	2.66	116		
160	320	0.805	2483	0.593	0.056E-4	0.060E-3	0.132	0.076	0.663	2.57	84		

TABLE A8
Properties of Saturated Liquids: Ethylene Glycol $C_2H_4(OH_2)$ *

Temp, T			Specific heat, c_p		Kinematic viscosity, ν		Thermal Conductivity, k		Thermal diffusivity, α			β	
°C	°F	Specific gravity	$\dfrac{J}{kg \cdot K}$	$\dfrac{BTU}{lbm \cdot °R}$	$m^2/s \times 10^6$	$ft^2/s \times 10^5$	$\dfrac{W}{m \cdot K}$	$\dfrac{BTU}{hr \cdot ft \cdot °R}$	$m^2/s \times 10^7$	$ft^2/hr \times 10^3$	Prandtl number, Pr	1/K	1/°R
0	32	1.130	2294	0.548	57.53	61.92	0.242	0.140	0.934	3.62	615		
20	68	1.116	2382	0.569	19.18	20.64	0.249	0.144	0.939	3.64	204	0.65×10^{-3}	0.36×10^{-3}
40	104	1.101	2474	0.591	8.69	9.35	0.256	0.148	0.939	3.64	93		
60	140	1.087	2562	0.612	4.75	5.11	0.260	0.150	0.932	3.61	51		
80	176	1.077	2650	0.633	2.98	3.21	0.261	0.151	0.921	3.57	32.4		
100	212	1.058	2742	0.655	2.03	2.18	0.263	0.152	0.908	3.52	22.4		

TABLE A9

Properties of Saturated Liquids: Eutectic Calcium Chloride Solution (29.9% CaCl₂)*

Temp, T			Specific heat, c_p		Kinematic viscosity, ν		Thermal Conductivity, k		Thermal diffusivity, α			β	
°C	°F	Specific gravity	$\dfrac{\text{J}}{\text{kg}\cdot\text{K}}$	$\dfrac{\text{BTU}}{\text{lbm}\cdot°\text{R}}$	m²/s×10⁶	ft²/s×10⁵	$\dfrac{\text{W}}{\text{m}\cdot\text{K}}$	$\dfrac{\text{BTU}}{\text{hr}\cdot\text{ft}\cdot°\text{R}}$	m²/s×10⁷	ft²/hr×10³	Prandtl number, Pr	1/K	1/°R
-50	-58	1.319	2608	0.623	36.35	39.13	0.402	0.232	1.166	4.52	312		
-40	-40	1.314	2635.6	0.6295	24.97	26.88	0.415	0.240	1.200	4.65	208		
-30	-22	1.310	2661.1	0.6356	17.18	18.49	0.429	0.248	1.234	4.78	139		
-20	-4	1.305	2688	0.642	11.04	11.88	0.445	0.257	1.267	4.91	87.1		
0	32	1.290	2738	0.654	4.39	4.73	0.472	0.273	1.332	5.16	33.0		
10	50	1.291	2763	0.660	3.35	3.61	0.485	0.280	1.363	5.28	24.6		
20	68	1.286	2788	0.666	2.72	2.93	0.498	0.288	1.394	5.40	19.6		
20	86	1.281	2814	0.672	2.27	2.44	0.511	0.295	1.419	5.50	16.0		
40	104	1.277	2839	0.678	1.92	2.07	0.523	0.302	1.445	5.60	13.3		
50	122	1.272	2868	0.685	1.6	1.78	0.535	0.309	1.468	5.69	11.3		

TABLE A10

Properties of Saturated Liquids: Glycerin $C_3H_5(OH)_3$[*]

Temp, T			Specific heat, c_p		Kinematic viscosity, ν		Thermal Conductivity, k		Thermal diffusivity, α			β	
°C	°F	Specific gravity	$\dfrac{J}{kg \cdot K}$	$\dfrac{BTU}{lbm \cdot °R}$	m^2/s	ft^2/s	$\dfrac{W}{m \cdot K}$	$\dfrac{BTU}{hr \cdot ft \cdot °R}$	$m^2/s \times 10^7$	$ft^2/hr \times 10^3$	Prandtl number, Pr	1/K	1/°R
0	32	1.276	2261	0.540	0.00831	0.0895	0.282	0.163	0.983	3.81	84.7×10^3		
10	50	1.270	2319	0.554	0.00300	0.0323	0.284	0.164	0.965	3.74	31.0		
20	68	1.264	2386	0.570	0.00118	0.0127	0.286	0.165	0.947	3.67	12.5	0.50×10^{-3}	0.28×10^{-3}
30	86	1.258	2445	0.584	0.00050	0.0054	0.286	0.165	0.929	3.60	5.38		
40	104	1.252	2512	0.600	0.00022	0.0024	0.286	0.165	0.914	3.54	2.45		
50	122	1.244	2583	0.617	0.00015	0.0016	0.287	0.166	0.893	3.46	1.63		

TABLE A11
Properties of Saturated Liquids: Mercury Hg[*]

Temp, T			Specific heat, c_p		Kinematic viscosity, ν		Thermal Conductivity, k		Thermal diffusivity, α			β	
°C	°F	Specific gravity	$\dfrac{\text{J}}{\text{kg}\cdot\text{K}}$	$\dfrac{\text{BTU}}{\text{lbm}\cdot\text{°R}}$	$\text{m}^2/\text{s}\times10^6$	$\text{ft}^2/\text{s}\times10^5$	$\dfrac{\text{W}}{\text{m}\cdot\text{K}}$	$\dfrac{\text{BTU}}{\text{hr}\cdot\text{ft}\cdot\text{°R}}$	$\text{m}^2/\text{s}\times10^7$	$\text{ft}^2/\text{hr}\times10^3$	Prandtl number, Pr	1/K	1/°R
0	32	13.628	140.3	0.0335	0.124	0.133	8.20	4.74	42.99	166.6	0.0288		
20	68	13.579	139.4	0.0333	0.114	0.123	8.69	5.02	46.06	178.5	0.0249	1.82×10^{-4}	1.01×10^{-4}
50	122	13.505	138.6	0.0331	0.104	0.112	9.40	5.43	50.22	194.6	0.0207		
100	212	13.384	137.3	0.0328	0.0928	0.0999	10.51	6.07	57.16	221.5	0.0162		
150	302	13.264	136.5	0.0326	0.0853	0.0918	11.49	6.64	63.54	246.2	0.0134		
200	392	13.144	135.0	0.0325	0.0802	0.0863	12.34	7.13	69.08	267.7	0.0116		
250	482	13.025	135.7	0.0324	0.0765	0.0823	13.07	7.55	74.06	287.0	0.0103		
315.5	600	12.847	134.0	0.032	0.0673	0.0724	14.02	8.10	81.5	316	0.0083		

TABLE A12
Properties of Saturated Liquids: Methyl Chloride Ch_3Cl^*

Temp, T			Specific heat, c_p		Kinematic viscosity, ν		Thermal Conductivity, k		Thermal diffusivity, α			β	
°C	°F	Specific gravity	$\dfrac{J}{kg \cdot K}$	$\dfrac{BTU}{lbm \cdot °R}$	$m^2/s \times 10^6$	$ft^2/s \times 10^5$	$\dfrac{W}{m \cdot K}$	$\dfrac{BTU}{hr \cdot ft \cdot °R}$	$m^2/s \times 10^7$	$ft^2/hr \times 10^3$	Prandtl number, Pr	1/K	1/°R
-50	-58	1.052	1475.9	0.3525	0.320	0.344	0.215	0.124	1.388	5.38	2.31		
-40	-40	1.033	1482.6	0.3541	0.318	0.342	0.209	0.121	1.368	5.30	1.32		
-30	-22	1.016	1492.2	0.3564	0.314	0.338	0.202	0.117	1.337	5.18	2.35		
-20	-4	0.999	1504.3	0.3593	0.309	0.333	0.196	0.113	1.301	5.04	2.38		
-10	14	0.981	1519.4	0.3629	0.306	0.329	0.187	0.108	1.257	4.87	2.43		
0	32	0.962	1537.8	0.3673	0.302	0.325	0.178	0.103	1.213	4.70	2.49		
10	50	0.942	1560.0	0.3726	0.297	0.320	0.171	0.099	1.166	4.52	2.55		
20	68	0.923	1586.0	0.3788	0.293	0.315	0.163	0.094	1.112	4.31	2.63		
30	86	0.903	1616.1	0.3860	0.288	0.310	0.154	0.089	1.058	4.10	2.72		
40	104	0.883	1650.4	0.3942	0.281	0.303	0.144	0.083	0.989	3.86	2.83		
50	122	0.861	1689.0	0.4034	0.274	0.295	0.133	0.077	0.921	3.57	2.97		

TABLE A13

Properties of Saturated Liquids: Sulfur Dioxide SO_2[*]

Temp, T		Specific gravity	Specific heat, c_p		Kinematic viscosity, ν		Thermal Conductivity, k		Thermal diffusivity, α		Prandtl number, Pr	β	
°C	°F		$\frac{J}{kg \cdot K}$	$\frac{BTU}{lbm \cdot °R}$	$m^2/s \times 10^6$	$ft^2/s \times 10^5$	$\frac{W}{m \cdot K}$	$\frac{BTU}{hr \cdot ft \cdot °R}$	$m^2/s \times 10^7$	$ft^2/hr \times 10^3$		1/K	1/°R
-50	-58	1.560	1359.5	0.3247	0.484	0.521	0.242	0.140	1.141	4.42	4.24		
-40	-40	1.536	1360.7	0.3250	0.424	0.456	0.235	0.136	1.130	4.38	3.74		
-30	-22	1.520	1361.6	0.3252	0.371	0.399	0.230	0.133	1.117	4.33	3.31		
-20	-4	1.488	1362.4	0.3254	0.324	0.349	0.225	0.130	1.107	4.29	2.93		
-10	14	1.463	1362.8	0.3255	0.288	0.310	0.218	0.126	1.097	4.25	2.62		
0	32	1.438	1363.6	0.3257	0.257	0.277	0.211	0.122	1.081	4.19	2.38		
10	50	1.412	1364.5	0.3259	0.232	0.250	0.204	0.118	1.066	4.13	2.18		
20	68	1.386	1365.3	0.3261	0.210	0.226	0.199	0.115	1.050	4.07	2.00	1.94×10^{-3}	1.08×10^{-3}
30	86	1.359	1366.2	0.3263	0.190	0.204	0.192	0.111	1.035	4.01	1.83		
40	104	1.329	1367.4	0.3266	0.173	0.186	0.185	0.107	1.019	3.95	1.70		
50	122	1.299	1368.3	0.3268	0.162	0.174	0.177	0.102	0.999	3.87	1.61		

TABLE A14
Properties of Saturated Liquids: Water H_2O^*

Temp, T			Specific heat, c_p		Kinematic viscosity, ν		Thermal Conductivity, k		Thermal diffusivity, α			β	
°C	°F	Specific gravity	$\dfrac{J}{kg \cdot K}$	$\dfrac{BTU}{lbm \cdot °R}$	$m^2/s \times 10^6$	$ft^2/s \times 10^5$	$\dfrac{W}{m \cdot K}$	$\dfrac{BTU}{hr \cdot ft \cdot °R}$	$m^2/s \times 10^7$	$ft^2/hr \times 10^3$	Prandtl number, Pr	1/K	1/°R
0	32	1.002	4217	1.0074	1.788	1.925	0.552	0.319	1.308	5.07	13.6		
20	68	1.000	4181	0.9988	1.006	1.083	0.597	0.345	1.430	5.54	7.02	0.18×10^{-3}	0.10×10^{-3}
40	104	0.994	4178	0.9980	0.658	0.708	0.628	0.363	1.512	5.86	4.34		
60	140	0.985	4184	0.9994	0.478	0.514	0.651	0.376	1.554	6.02	3.02		
80	176	0.974	4150	1.0023	0.364	0.392	0.668	0.386	1.636	6.34	2.22		
100	212	0.960	4216	1.0070	0.294	0.316	0.680	0.393	1.680	6.51	1.74		
120	248	0.945	4250	1.015	0.247	0.266	0.685	0.394	1.708	6.62	1.446		
140	284	0.928	4283	1.023	0.214	0.230	0.684	0.395	1.724	6.68	1.241		
160	320	0.909	4342	1.037	0.190	0.204	0.680	0.393	1.729	6.70	1.099		
180	356	0.889	4417	1.055	0.173	0.186	0.675	0.390	1.724	6.68	1.004		
200	392	0.866	4505	1.076	0.160	0.172	0.665	0.384	1.706	6.61	0.937		
220	428	0.842	4610	1.101	0.150	0.161	0.652	0.377	1.680	6.51	0.891		
240	464	0.815	4756	1.136	0.143	0.154	0.635	0.367	1.639	6.35	0.871		
260	500	0.785	4949	1.182	0.137	0.148	0.611	0.353	1.577	6.11	0.874		
280	537	0.752	5208	1.244	0.135	0.145	0.580	0.335	1.481	5.74	0.910		
300	572	0.714	5728	1.368	0.135	0.145	0.540	0.312	1.324	5.13	1.019		

TABLE A15

Properties of Gases at Atmospheric Pressure (101.3 = 14.7 Air [Gas Constant = 286.8 J/kg·K) = 53.3 ft·lbf/lbm·°R; $\gamma = c_p/c_v = 1.4$][*]

Temp, T		Density, ρ		Specific heat, c_p		Kinematic viscosity, ν		Thermal Conductivity, k		Thermal diffusivity, α		Prandtl number, Pr
K	°R	kg/m³	lbm/ft³	$\frac{\text{J}}{\text{kg}\cdot\text{K}}$	$\frac{\text{BTU}}{\text{lbm}\cdot°\text{R}}$	m²/s×10⁶	ft²/s×10⁵	$\frac{\text{W}}{\text{m}\cdot\text{K}}$	$\frac{\text{BTU}}{\text{hr}\cdot\text{ft}\cdot°\text{R}}$	m²/s×10⁴	ft²/hr	
100	180	3.601	0.225	1026.6	0.245	1.923	2.070	0.009246	0.005342	0.02501	0.0969	0.786
150	270	2.368	0.148	1009.9	0.241	4.343	4.674	0.013735	0.007936	0.05745	0.223	0.758
200	360	1.768	0.110	1006.1	0.240	7.490	8.062	0.01809	0.01045	0.10165	0.394	0.739
250	450	1.413	0.0882	1005.3	0.240	9.49	10.2	0.02227	0.01287	0.13161	0.510	0.722
300	540	1.177	0.0735	1005.7	0.240	15.68	16.88	0.02624	0.01516	0.22160	0.859	0.708
350	630	0.998	0.0623	1009.0	0.241	20.76	22.35	0.03003	0.01735	0.2983	1.156	0.697
400	720	0.883	0.0551	1014.0	0.242	25.90	27.88	0.03365	0.01944	0.3760	1.457	0.689
450	810	0.783	0.0489	1020.7	0.244	28.86	31.06	0.03707	0.02142	0.4222	1.636	0.683
500	900	0.705	0.0440	1029.5	0.245	37.90	40.80	0.04038	0.02333	0.5564	2.156	0.680
550	990	0.642	0.0401	1039.2	0.248	44.34	47.73	0.04360	0.02519	0.6532	2.531	0.680
600	1080	0.589	0.0367	1055.1	0.252	51.34	55.26	0.04659	0.02692	0.7512	2.911	0.680
650	1170	0.543	0.0339	1063.5	0.254	58.51	62.98	0.04953	0.02862	0.8578	3.324	0.682
700	1260	0.503	0.0314	1075.2	0.257	66.25	71.31	0.05230	0.03022	0.9672	3.748	0.684
750	1350	0.471	0.0294	1085.6	0.259	73.91	79.56	0.05509	0.03183	1.0774	4.175	0.686
800	1440	0.441	0.0275	1097.8	0.262	82.29	88.58	0.05779	0.03339	1.1951	4.631	0.689
850	1530	0.415	0.0259	1109.5	0.265	90.75	97.68	0.06028	0.03483	1.3097	5.075	0.692
900	1620	0.393	0.0245	1121.2	0.268	99.3	107	0.06279	0.03628	1.4271	5.530	0.695
950	1710	0.372	0.0232	1132.1	0.270	108.2	116.5	0.06525	0.03770	1.5510	6.010	0.699
1000	1800	0.352	0.0220	1141.7	0.273	117.8	126.8	0.06752	0.03901	1.6779	6.502	0.702
1100	1980	0.320	0.0120	1160	0.277	138.6	149.2	0.0732	0.0423	1.969	7.630	0.704
1200	2160	0.295	0.0184	1179	0.282	159.1	171.3	0.0782	0.0452	2.251	8.723	0.707
1300	2340	0.271	0.0169	1197	0.286	182.1	196.0	0.0837	0.0484	2.583	10.01	0.705
1400	2520	0.252	0.0157	1214	0.290	205.5	221.3	0.0891	0.0515	2.920	11.32	0.705
1500	2700	0.236	0.0147	1230	0.294	229.1	246.6	0.0946	0.0547	3.262	12.64	0.705
1600	2880	0.221	0.0138	1248	0.298	254.5	273.9	0.100	0.0578	3.609	13.98	0.705
1700	3060	0.208	0.0130	1267	0.303	280.5	301.9	0.105	0.0607	3.977	15.41	0.705
1800	3240	0.197	0.0123	1287	0.307	308.1	331.6	0.111	0.0641	4.379	16.97	0.704

TABLE A16

Properties of Gases at Atmospheric Pressure (101.3 = 14.7 Air [Gas Constant = 286.8 J/kg·K) = 53.3 ft·lbf/lbm·°R; $\gamma = c_p/c_v = 1.4$][*]

| Temp, T | | Density, ρ | | Specific heat, c_p | | Kinematic viscosity, ν | | Thermal Conductivity, k | | Thermal diffusivity, α | | Prandtl number, Pr |
K	°R	kg/m^3	lbm/ft^3	$\dfrac{J}{kg \cdot K}$	$\dfrac{BTU}{lbm \cdot °R}$	m^2/s×10^6	ft^2/s×10^5	$\dfrac{W}{m \cdot K}$	$\dfrac{BTU}{hr \cdot ft \cdot °R}$	m^2/s×10^4	ft^2/hr	
1900	3420	0.186	0.0115	1309	0.313	338.5	364.4	0.117	0.0676	4.811	18.64	0.704
2000	3600	0.176	0.0110	1338	0.320	369.0	397.2	0.124	0.0716	5.260	20.38	0.702
2100	3780	0.168	0.0105	1372	0.328	399.6	430.1	0.131	0.0757	5.715	22.15	0.700
2200	3960	0.160	0.0100	1419	0.339	432.6	465.6	0.139	0.0803	6.120	23.72	0.707
2300	4140	0.154	0.00955	1482	0.354	464.0	499.4	0.149	0.0861	6.540	25.34	0.710
2400	4320	0.146	0.00905	1574	0.376	504.0	542.5	0.161	0.0930	7.020	27.20	0.718
2500	4500	0.139	0.00868	1688	0.403	543.5	585.0	0.175	0.101	7.441	28.83	0.730

[*]Source: *Engineering Heat Transfer*, William S. Janna, Taylor and Francis, 2000

TABLE A17

Properties of Gases at Atmospheric Pressure (101.3 KPa = 14.7 psia): Carbon Dioxide
[Gas Constant = 188.9 J/(Kg·K) = 35.11 ft·lbf/lbm·°R; $y = c_p/c_v = 1.30$]*

Temp, T		Density, ρ		Specific heat, c_p		Kinematic viscosity, v		Thermal Conductivity, k		Thermal diffusivity, α		Prandtl number, Pr
K	°R	kg/m^3	lbm/ft^3	$\dfrac{\text{J}}{\text{kg}\cdot\text{K}}$	$\dfrac{\text{BTU}}{\text{lbm}\cdot°\text{R}}$	m^2/s$\times10^6$	ft^2/s$\times10^5$	$\dfrac{\text{W}}{\text{m}\cdot\text{K}}$	$\dfrac{\text{BTU}}{\text{hr}\cdot\text{ft}\cdot°\text{R}}$	m^2/s$\times10^4$	ft^2/hr	
220	396	2.4733	0.1544	783	0.187	4.490	4.833	0.010805	0.006243	0.05920	0.2294	0.818
250	450	2.1657	0.1352	804	0.192	5.813	6.257	0.012884	0.007444	0.07401	0.2868	0.793
300	540	1.7973	0.1122	871	0.208	8.321	8.957	0.016572	0.009575	0.10588	0.4103	0.770
350	630	1.5362	0.0959	900	0.215	11.19	12.05	0.02047	0.01183	0.14808	0.5738	0.755
400	720	1.3424	0.0838	942	0.225	14.39	15.49	0.02461	0.01422	0.19463	0.7542	0.738
450	810	1.1918	0.0744	980	0.234	17.90	19.27	0.02897	0.01674	0.24813	0.9615	0.721
500	900	1.0732	0.0670	1013	0.242	21.67	23.33	0.03352	0.01937	0.3084	1.195	0.702
550	990	0.9739	0.0608	1047	0.250	25.74	27.71	0.03821	0.02208	0.3750	1.453	0.685
600	1080	0.8938	0.0558	1076	0.257	30.02	32.31	0.04311	0.02491	0.4483	1.737	0.668

TABLE A18

Properties of Gases at Atmospheric Pressure (101.3 KPa = 14.7 psia): Helium
[Gas Constant = 2 077 J/(Kg·K) = 386 ft·lbf/lbm·°R; $y = c_p/c_v = 1.66$]*

Temp, T		Density, ρ		Specific heat, c_p		Kinematic viscosity, v		Thermal Conductivity, k		Thermal diffusivity, α		Prandtl number, Pr
K	°R	kg/m^3	lbm/ft^3	$\frac{J}{kg \cdot K}$	$\frac{BTU}{lbm \cdot °R}$	m^2/s×10^6	ft^2/s×10^5	$\frac{W}{m \cdot K}$	$\frac{BTU}{hr \cdot ft \cdot °R}$	m^2/s×10^4	ft^2/hr	
33	60	1.4657	0.0915	5200	1.242	3.42	3.68	0.0353	0.0204	0.04625	0.1792	0.74
144	260	0.3380	0.0211	5200	1.242	37.11	39.95	0.0928	0.0536	0.5275	2.044	0.70
200	360	0.2435	0.0152	5200	1.242	64.38	69.30	0.1177	0.0680	0.9288	3.599	0.69
255	460	0.1906	0.0119	5200	1.242	95.50	102.8	0.1357	0.0784	1.3675	5.299	0.70
366	660	0.13280	0.00829	5200	1.242	173.6	186.9	0.1691	0.0977	2.449	9.490	0.71
477	860	0.10204	0.00637	5200	1.242	269.3	289.9	0.197	0.114	3.716	14.40	0.72
589	1060	0.08282	0.00517	5200	1.242	375.8	404.5	0.225	0.130	5.215	20.21	0.72
700	1260	0.07032	0.00439	5200	1.242	494.2	531.9	0.251	0.145	6.661	25.81	0.72
800	1440	0.06023	0.00376	5200	1.242	634.1	682.5	0.275	0.159	8.774	34.00	0.72
900	1620	0.05286	0.00330	5200	1.242	781.3	841.0	0.298	0.172	10.834	41.98	0.72

TABLE A19

Properties of Gases at Atmospheric Pressure (101.3 KPa = 14.7 psia): Hydrogen
[Gas Constant = 4 126 J/(Kg·K) = 767 ft·lbf/lbm·°R; $\gamma = c_p/c_v = 1.405$]*

Temp, T		Density, ρ		Specific heat, c_p		Kinematic viscosity, ν		Thermal Conductivity, k		Thermal diffusivity, α		Prandtl number, Pr
K	°R	kg/m³	lbm/ft³	$\frac{J}{kg \cdot K}$	$\frac{BTU}{lbm \cdot °R}$	$\frac{m^2/s}{\times 10^6}$	ft²/s×10⁵	$\frac{W}{m \cdot K}$	$\frac{BTU}{hr \cdot ft \cdot °R}$	m²/s×10⁴	ft²/hr	
50	90	0.50955	0.03181	10501	2.508	4.880	5.253	0.0362	0.0209	0.0676	0.262	0.721
100	180	0.24572	0.01534	11229	2.682	17.14	18.45	0.0665	0.0384	0.2408	0.933	0.712
150	270	0.16371	0.01002	12602	3.010	34.18	36.79	0.0981	0.0567	0.475	1.84	0.718
200	360	0.12270	0.00766	13540	3.234	55.53	59.77	0.1282	0.0741	0.772	2.99	0.719
250	450	0.09819	0.00613	14059	3.358	80.64	86.80	0.1561	0.0902	1.130	4.38	0.713
300	540	0.08185	0.00511	14314	3.419	109.5	117.9	0.182	0.105	1.554	6.02	0.706
350	630	0.07016	0.00438	14436	3.448	141.9	152.7	0.206	0.119	2.031	7.87	0.697
400	720	0.06135	0.00383	14491	3.461	177.1	190.6	0.228	0.132	2.568	9.95	0.690
450	810	0.05462	0.00341	14499	3.463	215.6	230.1	0.251	0.145	3.164	12.26	0.682
500	900	0.04918	0.00307	14507	3.465	257.0	276.6	0.272	0.157	3.817	14.79	0.675
550	990	0.04469	0.00279	14232	3.471	301.6	324.6	0.292	0.169	4.516	17.50	0.668
600	1080	0.04085	0.00255	14537	3.472	349.7	376.4	0.315	0.182	5.306	20.56	0.664
700	1260	0.03492	0.00218	14574	3.481	455.1	489.9	0.351	0.203	6.903	26.75	0.659
800	1440	0.03060	0.00191	14675	3.505	569	612	0.384	0.222	8.563	33.18	0.664
900	1620	0.02723	0.00170	14801	3.540	690	743	0.412	0.238	10.217	39.59	0.676
1000	1800	0.02451	0.00153	14968	3.575	822	885	0.440	0.254	11.997	46.49	0.686
1100	1980	0.02227	0.00139	15165	3.622	965	1039	0.464	0.268	13.726	53.19	0.703
1200	2160	0.02050	0.00128	15366	3.670	1107	1192	0.488	0.282	15.484	60.00	0.715
1300	2340	0.01890	0.00118	15575	3.720	1273	1370	0.512	0.28	17.394	67.40	0.733
1333	2400	0.01842	0.00115	15638	3.735	1328	1429	0.519	0.300	18.013	69.80	0.736

TABLE A20

Properties of Gases at Atmospheric Pressure (101.3 KPa = 14.7 psia): Nitrogen
[Gas Constant = 296.8 J/(Kg·K) = 55.16 ft·lbf/lbm·°R; $\gamma = c_p/c_v = 1.40$]*

Temp, T		Density, ρ		Specific heat, c_p		Kinematic viscosity, ν		Thermal Conductivity, k		Thermal diffusivity, α		Prandtl number, Pr
K	°R	kg/m^3	lbm/ft^3	$\dfrac{J}{kg\cdot K}$	$\dfrac{BTU}{lbm\cdot °R}$	$m^2/s\times10^6$	$ft^2/s\times10^5$	$\dfrac{W}{m\cdot K}$	$\dfrac{BTU}{hr\cdot ft\cdot °R}$	$m^2/s\times10^4$	ft^2/hr	
100	180	3.4808	0.2173	1072.2	0.2561	1.97	2.122	0.009450	0.005460	0.025319	0.09811	0.786
200	360	1.7108	0.1068	1042.9	0.2491	7.568	8.146	0.01824	0.01054	0.10224	0.3962	0.747
300	540	1.1421	0.0713	1040.8	0.2486	15.63	16.82	0.02620	0.01514	0.22044	0.8542	0.713
400	720	0.8538	0.0533	1045.9	0.2498	25.74	27.71	0.03335	0.01927	0.3734	1.447	0.691
500	900	0.6824	0.0426	1055.5	0.2521	37.66	40.54	0.03984	0.02302	0.5530	2.143	0.684
600	1080	0.5687	0.0355	1075.6	0.2569	51.19	55.10	0.04580	0.02646	0.7486	2.901	0.686
700	1260	0.4934	0.0308	1096.9	0.2620	65.13	70.10	0.05123	0.02960	0.9466	3.668	0.691
800	1440	0.4277	0.0267	1122.5	0.2681	81.46	87.68	0.05609	0.03241	1.1685	4.528	0.700
900	1620	0.3796	0.0237	1146.4	0.2738	91.06	98.02	0.06070	0.03507	1.3946	5.404	0.711
1000	1800	0.3412	0.0213	1167.7	0.2789	117.2	126.2	0.06475	0.03741	1.6250	6.297	0.724
1100	1980	0.3108	0.0194	1185.7	0.2832	136.0	146.4	0.06850	0.03958	1.8591	7.204	0.736
1200	2160	0.2851	0.0178	1203.7	0.2875	156.1	168.0	0.07184	0.04151	2.0932	8.111	0.748

TABLE A21

Properties of Gases at Atmospheric Pressure (101.3 KPa = 14.7 psia): Oxygen
[Gas Constant = 260 J/(Kg·K) = 48.3 ft·lbf/lbm·°R; $\gamma = c_p/c_v = 1.40$][*]

Temp, T		Density, ρ		Specific heat, c_p		Kinematic viscosity, ν		Thermal Conductivity, k		Thermal diffusivity, α		Prandtl number, Pr
K	°R	kg/m^3	lbm/ft^3	$\dfrac{\text{J}}{\text{kg}\cdot\text{K}}$	$\dfrac{\text{BTU}}{\text{lbm}\cdot°\text{R}}$	m^2/s×10^6	ft^2/s×10^5	$\dfrac{\text{W}}{\text{m}\cdot\text{K}}$	$\dfrac{\text{BTU}}{\text{hr}\cdot\text{ft}\cdot°\text{R}}$	m^2/s×10^4	ft^2/hr	
100	180	3.9118	0.2492	947.9	0.2264	1.946	2.095	0.00903	0.00522	0.023876	0.09252	0.815
150	270	2.6190	0.1635	917.8	0.2192	4.387	4.722	0.01367	0.00790	0.05688	0.2204	0.773
200	360	1.9559	0.1221	913.1	0.2181	7.593	8.173	0.01824	0.01054	0.10214	0.3958	0.745
250	450	1.5618	0.0975	915.7	0.2187	11.45	12.32	0.02259	0.01305	0.15794	0.6120	0.725
300	540	1.3007	0.0812	920.3	0.2198	15.86	17.07	0.02676	0.01546	0.2235	0.8662	0.709
350	630	1.1133	0.0695	929.1	0.2219	20.80	22.39	0.03070	0.01774	0.2300	1.150	0.702
400	720	0.9755	0.0609	942.0	0.2250	26.18	28.18	0.03461	0.02000	0.3768	1.460	0.695
450	810	0.8682	0.0542	956.7	0.2285	31.99	34.43	0.03828	0.02212	0.4609	1.786	0.694
500	900	0.7801	0.0487	972.2	0.2322	38.34	41.27	0.04173	0.02411	0.5502	2.132	0.697
550	990	0.7096	0.0443	988.1	0.2360	45.05	48.49	0.04517	0.02610	0.6441	2.400	0.700
600	1080	0.6504	0.0406	1004.4	0.2399	55.15	56.13	0.04882	0.02792	0.7399	2.867	0.704
1200	2160	0.2851	0.0178	1203.7	0.2875	156.1	168.0	0.07184	0.04151	2.0932	8.111	0.748

TABLE A22

Properties of Gases at Atmospheric Pressure (101.3 KPa = 14.7 psia): Water Vapour or Steam
[Gas Constant = 461.5 J/(Kg·K) = 85.78 ft·lbf/lbm·°R; $\gamma = c_p/c_v = 1.33$]*

| Temp, T | | Density, ρ | | Specific heat, c_p | | Kinematic viscosity, ν | | Thermal Conductivity, k | | Thermal diffusivity, α | | Prandtl number, Pr |
K	°R	kg/m³	lbm/ft³	$\frac{\text{J}}{\text{kg}\cdot\text{K}}$	$\frac{\text{BTU}}{\text{lbm}\cdot°\text{R}}$	m²/s×10⁵	ft²/s×10⁴	$\frac{\text{W}}{\text{m}\cdot\text{K}}$	$\frac{\text{BTU}}{\text{hr}\cdot\text{ft}\cdot°\text{R}}$	m²/s×10⁴	ft²/hr	
380	684	0.5863	0.0366	2060	0.492	2.16	2.33	0.0246	0.0142	0.2036	0.789	1.060
400	720	0.5542	0.0346	2014	0.481	2.42	2.61	0.0261	0.0151	0.2338	0.906	1.040
450	810	0.4902	0.0306	1980	0.473	3.11	3.35	0.0299	0.0173	0.307	1.19	1.00
500	900	0.4405	0.0275	1985	0.474	3.86	4.16	0.0339	0.0196	0.387	1.50	0.996
550	990	0.4005	0.0250	1997	0.477	4.70	5.06	0.0379	0.0219	0.475	1.84	0.991
600	1080	0.3652	0.0228	2026	0.484	5.66	6.09	0.0422	0.0244	0.573	2.22	0.986
650	1170	0.3380	0.021	2056	0.491	6.64	7.15	0.0464	0.02608	0.666	2.58	0.995
700	1260	0.3140	0.0196	2085	0.498	7.72	8.31	0.0505	0.0292	0.772	2.99	1.000
750	1350	0.2931	0.0183	2119	0.506	8.88	9.56	0.0549	0.0317	0.883	3.42	1.005
800	1440	0.2739	0.0171	2152	0.514	10.20	10.98	0.0592	0.0342	1.001	3.88	1.000
850	1530	0.2579	0.0161	2186	0.522	11.52	12.40	0.0637	0.0368	1.130	4.38	1.019

EMISSIVITY TABLES

TABLE A23
Normal of Various Metals

Material	Surface condition	Temperature		Normal emissivity. ε
		K	°R	
Aluminium	Polished plate	296	533	0.040
		498	896	0.039
	Rolled and polished	443	797	0.039
	Rough plate	298	536	0.070
Brass	Oxidised	611	1,100	0.22
	Polished	292	526	0.05
		573	1,030	0.032
	Tarnished	329	592	0.202
Chromium	Polished	423	761	0.058
Copper	Black oxidised	293	527	0.780
	Tarnished lightly	293	527	0.037
	Polished	293	527	0.030
Gold	Not polished	293	527	0.47
	Polished	293	527	0.025
Iron	Smooth oxidised	398	716	0.78
	Ground bright	293	527	0.24
	Polished	698	1,256	0.144
Lead	Gray oxidised	293	527	0.28
	Polished	403	725	0.056
Nickel	Oxidized	373	671	0.41
	Polished	373	671	0.045
Silver	Polished	293	527	0.025
Steel	Rough oxidised	313	563	0.94
	Ground	1213	2,183	0.520
Tin	Bright	293	527	0.070
Tungsten	Filament	3300	5,940	0.39
Zinc	Tarnished	293	527	0.25
	Polished	503	905	0.045

Source: *Engineering Heat Transfer*, William S.Janna, Taylor & Francis, 2000

TABLE A24

Normal or Total (Hemispherical) of Various Nonmetallic Solids

Surface	Temperature		Emissivity	
	K	°R	Normal	Total
Aluminium oxide	600	1080	0.69	
	1000	1800	0.55	
Asphalt pavement	300	540		0.85-0.93
Building materials				
Asbestos sheet	300	540		0.93-0.96
Red brick	300	540		0.93-0.96
Gypsum or plasterboard	300	540		0.90-0.92
Wood	300	540		0.82-0.92
Concrete	300	540		0.88-0.93
Cloth	300	540		0.75-0.90
Glass, window	300	540		0.90-0.95
Ice	273	492		0.95-0.98
Paint				
Black	300	540		0.98
White	300	540		0.90-0.92
Paper, white	300	540		0.92-0.97
Skin	300	540		0.95
Snow	273	492		0.82-0.90
Soil	300	540		0.93-0.96
Teflon	300	540		0.85
	500	900		0.92
Water	300	540		0.96

Source: *Engineering Heat Transfer*, William S.Janna, Taylor & Francis, 2000

TABLE A25

Thermal Conductivities of Some Insulation (W/m·K)

Vacuum insulation panels	0.003-0.011
Vacuum glazing	0.003-0.008
Silica blankets	0.010-0.023
Silica aerogels	0.012-0.02
Clay	0.015-0.05
Phenolic foams	0.018-0.025
Gypsum foam	0.02
Silica panels	0.023-0.056
Polyurethane	0.020-0.029
Expanded polystyrene	0.029-0.055
Glass	0.031-0.043
Mineral wool	0.033-0.05
Melamine foam	0.035
Sheep wool	0.036
Loose-fill cellulose	0.039-0.042
Foam glass	0.039-0.045
Cork	0.04-0.05
Cellulose	0.04-0.05

Interpolation Using the Jupyter Notebook

The book comes with a Jupyter notebook for the interpolation of the property data. Since liquid and gas phases have different behaviour, two notebooks are provided along with dataset. For example *DataSet-SO2.xlsx* is for gas phase and *DataSet-SO2L.xlsx* is for liquid phase. Also, a separate Jupyter notebook is provided for engine oil as its properties need a different set of polynomials.

The property data can be provided via the excel file in 13 columns as follows:

$$K \; R \; density1 \; density2 \; cp1 \; cp2 \; kin1 \; kin2 \; k1 \; k2 \; alpha1 \, alpha2 \; Pr$$

where, K is temperature in Kelvin,
R is temperature in Rankine,
cp is specific heat,
kin is kinematic viscosity,
k is thermal conductivity,
alpha is thermal diffusivity,
Pr is Prandtl number.

The 1 indicates the quantities in SI units and 2 indicates the quantities in US Customary units as represented in the appendix previously (see Figure 17 below).

Properties of Gases at Atmospheric Pressure (101.3 KPa = 14.7 psia): Nitrogen [Gas constant = 296.8 J/(Kg·K) = 55.16 ft·lbf/lbm·°R; $\gamma = c_p/c_v = 1.40$].[*]

Temp, T		Density, ρ		Specific heat, c_p		Kinematic viscosity, v		Thermal Conductivity, k		Thermal diffusivity, α		Prandtl number, Pr
K	°R	kg/m³	lbm/ft³	$\frac{J}{kg \cdot K}$	$\frac{BTU}{lbm \cdot °R}$	m²/s×10⁶ ft²/s×10⁵		$\frac{W}{m \cdot K}$	$\frac{BTU}{hr \cdot ft \cdot °R}$	m²/s×10⁴	ft²/hr	
100	180	3.4808	0.2173	1072.2	0.2561	1.97	2.122	0.009450	0.005460	0.025319	0.09811	0.786
200	360	1.7108	0.1068	1042.9	0.2491	7.568	8.146	0.01824	0.01054	0.10224	0.3962	0.747

Figure 17 Property data.

In **Excel** it is inserted as shown below in figure 18:

B	C	D	E	F	G	H	I	J	K	L	M	N
K	R	density1	density2	cp1	cp2	kin1	kin2	k1	k2	alpha1	alpha2	Pr
100	180	3.4808	0.2173	1072.2	0.2561	1.97	2.122	0.00945	0.00546	0.025319	0.09811	0.786
200	360	1.7108	0.1068	1042.9	0.2491	7.568	8.146	0.01824	0.01054	0.10224	0.3962	0.747
300	540	1.1421	0.0713	1040.8	0.2486	15.63	16.82	0.0262	0.01514	0.22044	0.8542	0.713
400	720	0.8538	0.0533	1045.9	0.2498	25.74	27.71	0.03335	0.01927	0.3734	1.447	0.691
500	900	0.6824	0.0426	1055.5	0.2521	37.66	40.54	0.03984	0.02302	0.553	2.143	0.684

Figure 18 Excel data.

Note that the multiplier with certain properties is not inserted in the Excel file.

The Excel file must be used with installed Python version of 3 or higher. Also install Anaconda to run notebook:

```
https://www.anaconda.com/download/
```

The Jupyter notebook can be opened with the Anaconda Prompt. Go to the folder where **Prodata Curve Fitting** file is located and run on Anaconda prompt:

```
C:\....\Prodata Curve Fitting> Jupyter notebook
```

Note that ellipsis above in the syntax are indicating the directory path which must be correctly prescribed by the user. This will open the notebook in browser, see figure 19:

Figure 19   Jupyter Notebook in browser.

Click on **Prodata Curve Fitting.ipynb** which will open the file.

The code is performing curve fitting for a set of functions using data provided in a *Pandas dataframe*. The goal is to find the parameters of the functions that best fit the data. The code is using the *SciPy library* for curve fitting and the *Matplotlib* library for visualisation.

The code defines 6 functions for the quantities to be fitted: density1, cp1, kin1, k1, alpha1, and Pr. Each function takes a temperature T and some parameters as input, and returns the value of the corresponding quantity at that temperature.

The code then uses the `curve_fit()` function from SciPy to fit each of the 6 functions to the data in the *dataframe*. For each function, the parameters that provide the best fit to the data are returned in `popt_x`, and the covariance matrix of the parameters is returned in `pcov_x`, where x is the name of the function being fitted.

Finally, the code uses Matplotlib to visualise the data and the fits for each of the 6 quantities. For each quantity, a scatter plot of the data is shown along with the fitted curve. The x-axis shows the temperature in Kelvin, and the y-axis shows the value of the quantity being plotted. The title of each plot shows the name of the quantity being plotted.

Index